Engineering Materials

This series provides topical information on innovative, structural and functional materials and composites with applications in optical, electrical, mechanical, civil, aeronautical, medical, bio- and nano-engineering. The individual volumes are complete, comprehensive monographs covering the structure, properties, manufacturing process and applications of these materials. This multidisciplinary series is devoted to professionals, students and all those interested in the latest developments in the Materials Science field.

More information about this series at http://www.springer.com/series/4288

Inamuddin · Rajender Boddula ·
Abdullah M. Asiri
Editors

Self-standing Substrates

Materials and Applications

Springer

Editors
Inamuddin
Department of Chemistry
Faculty of Science
King Abdulaziz University
Jeddah, Saudi Arabia

Department of Chemistry
Faculty of Engineering and Technology
Aligarh Muslim University
Aligarh, India

Abdullah M. Asiri
Department of Chemistry
Faculty of Science
King Abdulaziz University
Jeddah, Saudi Arabia

Rajender Boddula
CAS Key Laboratory of Nanosystem
and Hierarchical Fabrication
National Center for Nanoscience
and Techology
Beijing, China

ISSN 1612-1317 ISSN 1868-1212 (electronic)
Engineering Materials
ISBN 978-3-030-29521-9 ISBN 978-3-030-29522-6 (eBook)
https://doi.org/10.1007/978-3-030-29522-6

This Springer imprint is published by the registered company Springer Nature Switzerland AG
The registered company address is: Gewerbestrasse 11, 6330 Cham, Switzerland

Preface

The applications of nanomaterials for applied science involve the tedious film casting or coating procedure with the help of binders and additional conductive agents, which results in the uncontrolled nanostructures, restricted active surface area, a lot of dead volumes, and unwanted interfaces. These ambiguities adversely affect the conductivity and mass transport of products. Self-supported materials directly growing on conductive self-standing substrates have attracted enormous enthusiasm over the previous decades since they show excellent optical, electrical nanostructure and flexible properties. These properties make the self-supported materials applicable in numerous modern research areas of flexible electronic, medical devices, self-supported electrodes, etc.

Self-standing Substrates: Materials and Applications aims to provide an understanding of various types of self-supported materials used for numerous technological applications. This book provides in-depth overview of the developments of fabrication techniques used for self-supported materials growing on flexible substrates, rigid substrates, and freestanding films composed of semiconductors, inorganic, polymers, and carbon hybrids. The book covers technical aspects, concepts, and versatile applications of self-supported materials and provides in-depth knowledge to the readers belonging to the fields of advanced science, material science, flexible electronics, and surface chemistry, industry, academic, and renewable energy sector. This book is an archival reference guide to students, professors, scientists, and R&D industrial specialists working in the field of advanced science, nanodevices, flexible electronics, energy conversion, and storage.

Jeddah, Saudi Arabia/Aligarh, India — Inamuddin
Beijing, China — Rajender Boddula
Jeddah, Saudi Arabia — Abdullah M. Asiri

Contents

Self-standing Nanoarchitectures

Katarzyna Siuzdak, Łukasz Haryński, Jakub Wawrzyniak, Piotr Kupracz and Katarzyna Grochowska

Abstract Despite there are structures invisible for the human eye, they mastered the world of advanced electronic devices, sensors, novel cosmetics or drugs. When the dimensions of the materials go down to the nanometres scale, their properties change dramatically comparing to the observable objects. Because of their tiny size, they gained the name of nanomaterials but simultaneously their importance has significantly grown up. Nanomaterials exhibit superb features such as a distinctive catalytic activity, hydrophobicity, photoconversion activity and biological affinity. Following that, even a small amount of nanomaterials is sufficient to provide unusual properties to the final products such as coatings, active layers in solar cells, clothes, electrodes and electrolytes used for energy storage devices. Owing to the rapid development in the synthesis methods and characterization techniques, especially those used for morphology inspection, we can investigate them in details on the molecular scale and describe the mechanism that stays behind improved antimicrobiological activity, hydrophobicity, capacitance or catalytic properties. Despite the number of usable elements is limited, the diversity of morphologies, namely rods, particles, tubes, planes and the possibility of heterostructures formation, provides researchers the wide room for maneuverer. Sometimes, only small change in the material geometry, structure or a little amount of introduced dopant atoms is enough to obtain completely new nanomaterial that has not been known so far. Therefore, we should not be surprised how fast surrounding environment is changing and our everyday life is supported by the novelties from the nano world. The aim of this chapter is to present the diversity of nanomaterials taking into account their dimensions, shape and composition. Herein, particles, tubes, wires, pores, walls, exhibiting at least one dimension within the nanoscale will be evoked. Moreover, the nanostructures that morphology reminds well known objects from nature are discussed. The description of some interesting examples is supported by the extraordinary SEM images illustrating the beauty unavailable for naked eye.

K. Siuzdak (✉) · Ł. Haryński · J. Wawrzyniak · P. Kupracz · K. Grochowska
Centre of Plasma and Laser Engineering, The Szewalski Institute of Fluid-Flow Machinery, Polish Academy of Sciences, Fiszera 14, 80-231 Gdańsk, Poland
e-mail: ksiuzdak@imp.gda.pl

Inamuddin et al. (eds.), *Self-standing Substrates*,
Engineering Materials, https://doi.org/10.1007/978-3-030-29522-6_1

1 Introduction

In the present world, more and more electronic devices that plays important role in our life owns its popularity to achievements in nanotechnology: new functional materials, advanced fabrication and characterization techniques. As we need long lasting batteries in our cell phones, highly sensitive diagnostic tests or clothes highly resistant for dirt or moisture, we have to appreciate researchers' efforts to control dimensions, shape and properties of tiny structures. Taking into account such small dimensions, the prefix "nano" originating from the Greek nomenclature and meaning dwarf has been reserved for those objects. Considering nanostructures, we should be aware that at least one of their dimension does not exceed 100 nm and because of that, their morphology cannot by simply verified by visual observation. To realize how small are these structures, the dimensions of selected objects are given in Fig. 1. This diagram

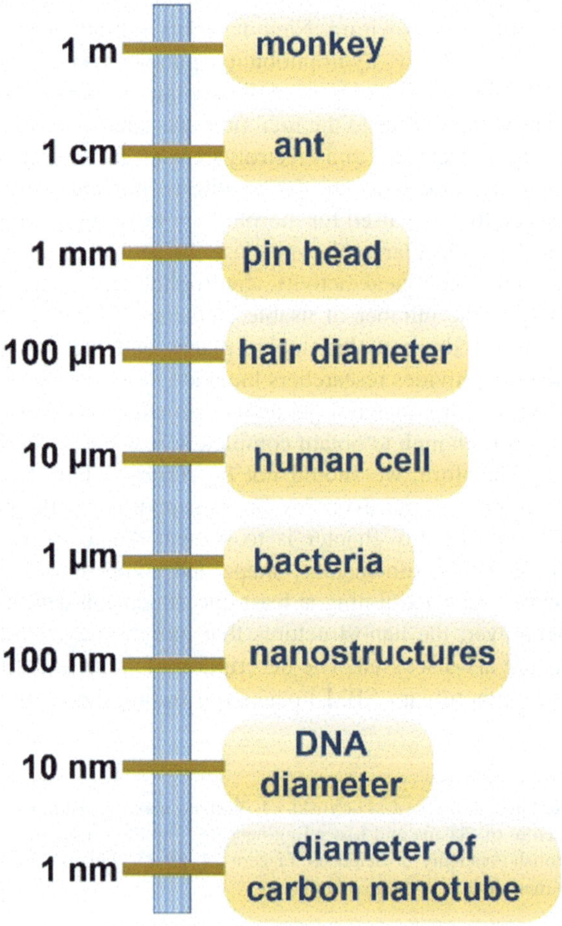

Fig. 1 Dimensions scale with the examples of objects from the 1 m to the 1 nm

illustrates in dimensional scale how nanomaterials are far from those we can just catch in our hands.

Apart from the size, comparing to typical bulk materials, like the silver plate or plastic spoon, nanomaterials exhibit unique physical and chemical properties. Those size-dependent features, like higher conductivity, degradation of organic pollution, ability to transport small drug molecules or anti-corrosion resistance, justify the importance of intense research in this field. Those nanostructures can be formed directly on the stable substrate, suspended freely in some liquid or may form the powder. Following the way how the nanomaterial is stored, one may distinguish free-standing or substrate bound nanomaterials. In the case of free-standing nanostructures, usually their larger amount looks like a powder and in such form it is stored in the closed container and offered to the customer. However, since the powder is composed of tiny particles, there is a fear of dusting that could be harmful to human health. Because of safety issues, some nanomaterials are stored as a suspension in a liquid medium, usually water or alcohol. The safest form of nanomaterials storage and further transport is their immobilization or in the best case, application of such synthesis method allowing its formation already on the stable substrate. Nevertheless, today optimism towards unique features of nanomaterials and their application in many fields should be balanced by their safe usage, storage and finally recycling.

1.1 Dimensions of Nanomaterials

Among variety of materials for which at least one dimension does not exceed 100 nm, one may distinguish four classes: zero (0-D), one (1-D), two (2-D) and three (3-D)—dimensional structures [97]. Such classification was proposed by Pokropivny and Skotokhod [76] and Table 1 goes along with his scheme.

Table 1 Classification of nanomaterials depending on the dimensions

0-D	1-D	2-D	3-D
Nanoparticles, quantum dots [22]	Nanotubes, nanorods, nanowires [68]	Nanofilms, nanolayers, nanocoatings [4]	Bundles of nanotubes, nanowires, multi-layered materials [99]
	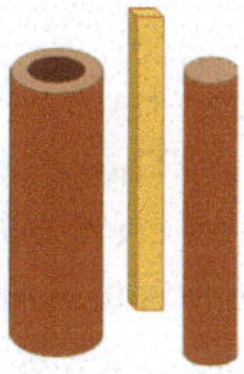		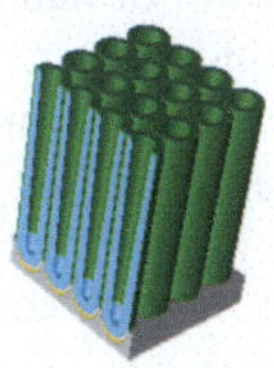

However, irrespective of the dimensions, among those structures, we can distinguish amorphous or crystalline, single crystalline or polycrystalline, chemical pure or doped in bulk or only on the surface, standalone materials or embedded within other medium. Nanostructures can be built of metallic or semiconducting units as well as polymeric molecules.

Zero-dimensional nanostructures exhibit all three dimensions in nanometres regime and have a diameter less than 100 nm. Here, we can find nanoclusters possessing irregular structure and dimensions within 1–10 nm range. As nanocrystals, monocrystalline nanostructures are defined with size from 1 to 30 nm. Commonly recognized quantum dots are semiconducting nanocrystals. Recently, apart from the quantum dots, hollow spheres, nanolenses or onion like structures are associated to 0-D group.

For **one-dimensional nanomaterial**, one dimension is outside the nanoscale range. Following that, one can find nanotubes, nanorods and nanowires, nanobelts or even nanoribbons. Nanowires exhibit elongated structures and are obtained from various metallic, semiconducting or oxide materials. They are especially applied in electronics, new generation of photovoltaic devices and sensing platforms. On the contrary, tubular structures possess a hollow interior. The most known tubes are of course carbon structures: single-, double- and multi-walled carbon nanotubes. However, recently the focus is put onto the semiconducting tubular materials, like titania or alumina nanotubes that could be easily formed during anodization.

Two-dimensional nanomaterials exhibit plate-like shape and therefore very thin layers or coatings are there included as well as some disc-like structures. However, their thickness still does not exceed 100 nm. Within 2-D nanostructures one can find branched structures, nanoplates, nanosheets or nanowalls. The most common example of 2-D is graphene due to the unique properties allowing application in different areas, from electronic devices to medicine as the crucial component of sensing electrode material. Additionally, owing to the sophisticated deposition techniques, like magnetron sputtering, chemical vapour deposition, laser ablation, precisely controlled nanofilm can be formed using targets or chemical compounds as precursors.

Three-dimensional nanomaterials are materials that are not exactly fitting nanoscale at any dimension. Nevertheless, they still have high surface area and provide many adsorption sites for all involved molecules. Within this group bulk powders, dispersion of nanoparticles or some multilayers, polycrystals and bundles of nanotubes or nanowires, nanoballs formed out of dendric structures, nanopillars and nanoflowers are included.

Apart from the dimension factor, nanomaterials can be formed by one or more components leading to the formation of composite. Those counterparts exhibit different properties but can act together in synergy to create features that are not achieved by each single part. The matrix of nanocomposite can be similar to other single materials composed of metal oxides, polymers, metals, ceramics or could state as a mixture of them, e.g. polymer with embedded metal nanoparticles or semiconductor modified with metal nanoparticles. Their dimensions could fit nanoscale range but could also be above them, whereas the reinforcing phase fits the nanoscale range. There, of course, exists several classifications of nanocomposites and following that we can find composites: (a) reinforced with discrete particles, (b) with interpenetrating skeletons of two or more structural components, (c) reinforced with discrete fibres or whiskers, (d) reinforced with continuous fibres or (e) with layered structures [43].

1.2 Difference Between Nanostructures and Bulk Structures

As was mentioned at the beginning of this chapter, nanomaterial and the bulk one formed by the same element exhibit totally different features that are caused by increasing relative surface area and quantum effects. As the size decreases, many more atoms forming the particular particle are in direct contact with an outer environment and less and less stay in its core. Taking into account sphere shape, when the particle has only 3 nm in diameter, half of its atoms are on the surface and a half form the core. Therefore, regarding chemical reactions that take place at the materials surface, much higher reactivity could be expected for those tiny structures than for larger parts, assuming that the volume remains the same. Along with the enlarged real surface area, quantum effects control materials properties. They influence optical, magnetic, electrical properties, ability to adsorb certain molecules onto the surface or to inhibit the growth of the bacteria or yeast, known as an interaction with biological creatures. For example, in macroscale, bulk gold is malleable and exhibits beautiful gloss typical for metals. However, when it comes to nanoscale, few nanometres small gold particles could form pink, red or purple colloids depending on the shape and tiny dimensions and because of the size the term malleability loses its importance.

1.3 Fabrication Methods

Some of nanomaterials can be found in the natural environment, but in the centre of interest are those that could be designed and fabricated firstly in the laboratory and then easily scaled up to the commercial scale. Fabrication of self-standing nanostructures requires utilization of methods that enable controlled connection of single molecules, ions in bigger structures or fragmentation of solid material into much smaller species. According to those different attempts, two main approaches were proposed: bottom-up and top-down. As indicated in Fig. 2, bottom-up route concerns processes where small components of atomic or molecular dimensions are put together owing to the natural physical principle or an externally applied force, to obtain complex system. In the case of top-down route, as a substrate, some large piece of material is used and the application of appropriate tools leads to its destruction into smaller components. Within those two approaches one can find various methods that were listed in Tables 2 and 3.

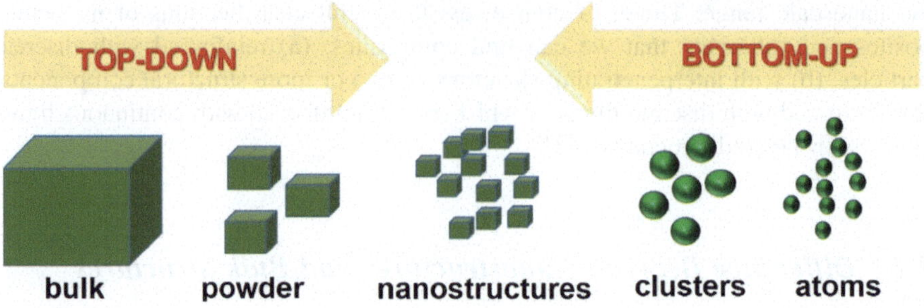

Fig. 2 The general scheme depicting the idea of bottom-up and top-down approaches for nanostructures synthesis

Table 2 Examples of top-down fabrication methods

Litography methods	Methods requiring inexpensive energy consumption and expensive facilities; are based on making printed electric circuits and computer broads
Anodization	Electrochemical oxidation of metal substrate performed usually in 2-electrode arrangement when both electrodes are immersed in the etching electrolyte. Electrodes are placed opposite to each other. Optimization of process parameters enables formation of porous layer composed of highly ordered aligned or separated nanotubes
Mechanosynthetic method	The cheapest way to material fragmentation. Most common technique is ball milling based on the mechanical attrition when kinetic energy is transferred to bigger substrate undergoing destruction
Thermal method	In general it concerns application of thermal stress to thin metallic films that upon elevated temperatures tend to form nanoislands/nanoparticles on the substrate. The thermal treatment could be realized in an electric oven or by the interaction with the laser beam
Laser ablation	Laser light interacts with the solid target and kicks of ions that could then collide with each other and be deposited on the substrate or just imprisoned in medium when the target is immersed in liquid environment

Other classification concerns the type of synthesis used for material fabrication, namely physical and chemical approaches could be distinguished. The most frequently used fabrication routes are assigned to each brand and listed in Table 4. Unfortunately, because of high cost of precursors or equipment and the complex preparation route, not all of mentioned techniques can be applied on the commercial scale. Nevertheless, upon intensive research and optimization of fabrication methods, more and more approaches used in the laboratory are moving to the large scale industry. Owing it to the fact that the control of the scaling-up process has been mastered and wide access to the knowledge is assured, the fabrication of nanomaterials comes out from the laboratories and as a result we can find nanostructures in cosmetics, drugs, bandages or household chemicals.

Table 3 Examples of bottom-up fabrication methods

Chemical vapour deposition (CVD)	Occurs in gas phase where on the substrate surface the decomposition of precursors takes place and the desired nanostructure is formed This method is usually applied for fabrication of carbon nanotubes or carbon like diamond
Atomic layer deposition (ALD)	States as an industrial process that allows to coat uniformly material of any topography. In each deposition cycle, a monolayer or few are formed. Unfortunately, used precursors are of high cost
Electrochemical deposition	Requires conductive substrate that can be polarized in a certain mode. Usually: potentiostatic or potentiodynamic (pulsed mode, cyclic voltammetry) techniques are applied to deposit the material onto the working electrode acting here as a substrate. The conductive electrolyte with easily dissolved material precursor is required. With the help of electrochemical deposition, the layers of conducting polymers, semiconductors or thin metallic films are formed
Sol-gel	Regarded as a soft chemical process, usually applied for fabrication of metal oxides or ceramic materials. Depending on established conditions, spherical or tube shape, thin coatings, membranes could be obtained

Table 4 The classification of nanostructures fabrication methods based on their chemical or physical character

Physical techniques	Chemical techniques
Sputtering techniques, lithography process, hot and cold plasma, evaporation technique, spray pyrolysis, pulsed laser ablation, sonochemical reduction	Electrochemical deposition, lyotropic liquid crystal templates, hydrothermal and solvothermal techniques, sol-gel technique, chemical vapour deposition, laser pyrolysis

1.4 How to Look at Nanostructures?

Since nanomaterials are characterized by dimensions at least one thousand times smaller than the diameter of human hair, our eyes or simple optical microscopes are insufficient to observe unique features of their morphology. Thus, special equipment has to be used to get the information about the topography and to obtain information concerning detailed dimensions and shape of nanostructure. For this purpose, advanced microscopes are used, i.e. scanning electron microscope (SEM), transmission electron microscope (TEM) and atomic force microscope (AFM). The enormous progress also in the characterization techniques, leads to the miniaturization of those pieces of laboratory equipment and in the case of AFM and SEM, only a simple office desk is needed for their installation. Despite much lower size of modern microscopes comparing to older ones, the operation principle is preserved and below some short description is given.

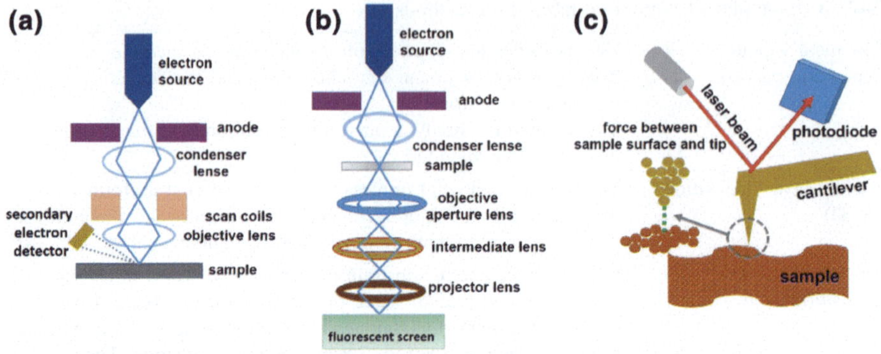

Fig. 3 The schemes depicting the operation principle of SEM (**a**), TEM (**b**) and AFM (**c**) microscopies utilized frequently to illustrate the morphology of nanostructures

Scanning electron microscopy uses the focused flux of high-energy electrons that interacts with the surface and as a result generates variety of signals at the surface of the solid specimen (see Fig. 3a). Those signals are attributed among others to the secondary electrons, backscattered electrons and characteristic X-rays. When electrons meet the surface of the material, they penetrate the sample down to few μm, depending on the optimized parameters such as accelerating voltage and sample density. Different types of generated electrons provide information about the sample including its morphology, chemical composition and even crystalline structure that is regarded as a basic knowledge on the investigated material. Usually, data is collected from the specified area of the sample and converted into two-dimensional image that looks like a black-white photo. As currently advanced optical microscope enables 1500 magnification of the sample, SEM microscope allows even for 500,000 magnification and spatial resolution lower than 1 nm can be achieved. Owing to the additional guns and particular energy recorded in selected place of the investigated samples, information concerning elemental composition using energy dispersive spectroscopy (EDS), crystalline structure or crystal orientation may be obtained. First SEM was constructed by a German technician Max Knoll in 1935, whilst the microscope was offered as a commercial product in 1965. Since time when SEM had been introduced to the laboratories, its dimensions have decreased significantly and some models require only office desk. Moreover, due to the simplification of the software the special operator is not required and almost everyone after short training can operate SEM.

Despite SEM technique allows to obtain very detailed images of the surface of nanostructures, transmission electron microscopy enables to look even deeper. In general, the construction of both microscopes is similar. However, in the case of TEM, transmitted electrons are analysed and thus the investigated sample has to be semi-transparent to the focused electron flux (see Fig. 3b). For this purpose, nanostructures are deposited onto the special copper grid that is placed in the way of the electron beam. On the contrary to SEM, TEM inspection provides information concerning internal composition covering crystalline or magnetic domains. Moreover, dislocations, tiny precipitates, grain boundaries or even defects could be illustrated. Nevertheless, transmission electron microscopy still requires a lot of space and a highly trained operator.

Other frequently used technique for morphology investigation is atomic force microscopy. This approach was initially used to visualise surface with atomic scale resolution. The operation principle of AFM is based on recording the modulation of the topographical signal originating from the interaction between materials surface and a very sensitive position detector. The role of this detector is played by the tiny cantilever looking as a needle with an extremely sharp tip at its end. The tip being of tens of nanometres in its dimension scans horizontally the surface of the material. During those scans, the position of cantilever is measured by the system composed of laser beam that is collimated onto the arm, as indicated in Fig. 3c, and reflected light is captured at diode detector. AFM imaging also provides information about the materials' topography in nm-rage, however in the investigated area, the difference between some spots cannot be too high as that lowers image quality. AFM was constructed in 1986 by the team of G. Binning, C. F. Quate and C. Gerner. Their works were forced by the SEM limitations, since isolator cannot be observed under electron microscope. Now, owing to the great improvements concerning AFM technique, sample investigation could be performed even on living cells in their natural liquid environment. As was in the case of SEM, AFM microscopes have become popular and are used as a typical equipment for students laboratory lessons whereas in industry allow for control of optical materials, semiconductors or data storages.

Summarizing, the material imaging inseparably accompanies the fabrication of nanostructures to confirm their size, expected morphology that affects the physicochemical features and in consequence future application areas. Fortunately, some new technical solutions, additional detectors or upgraded software allow to gain information also about crystal phase, magnetic character or elemental composition.

2 Nanoparticles

As has been already defined, nanoparticles (NPs) are a tiny pieces of matter which are usually between 1 and 100 nanometres in size, thus they are sometimes referred to as zero-dimensional (0-D) materials. Nanoparticles are often best described by their material, size and morphology (see Fig. 4), but different classification criteria are sometimes used in literature depending on the synthesis methods or their functionalization.

Extraordinary properties of the nanoparticles origin from their high surface-to-volume ratio. Because of high reactivity, nanoparticles could bind to each other in their immediate vicinity. This particular feature could be regarded either as an advantage or disadvantage, depending on their particular application. In order to prevent agglomeration, their surface is often modified with the surface

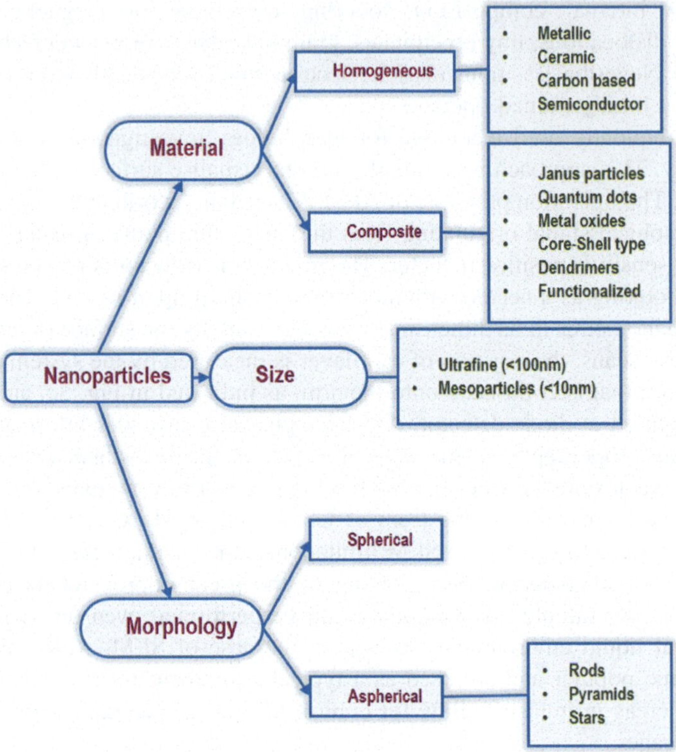

Fig. 4 General classification of nanoparticles

active agents, resulting in attenuation of chemical activity, whereas their physical properties are preserved. Similarly, nanoparticles can be functionalized using certain functional groups immobilized onto their surface [26].

2.1 Noble Metal Nanoparticles

Among others, metal nanoparticles, especially noble metal NPs, are the most broadly applied ones. They can be synthesized by variety of techniques and their size and shape can be easily tailored by choosing an adequate synthesis method. Owing to the free electrons present in metal nanoparticles, most of them exhibit clear absorbance bands within the visible waverange. Thanks to their uniform surface chemistry, they are also regarded as great candidates for functionalization, extending the field of potential application even further.

Concerning noble NPs, gold nanoparticles (see Fig. 5a) are extensively investigated. They can be synthesized via bottom-up and top-down approaches, but in most

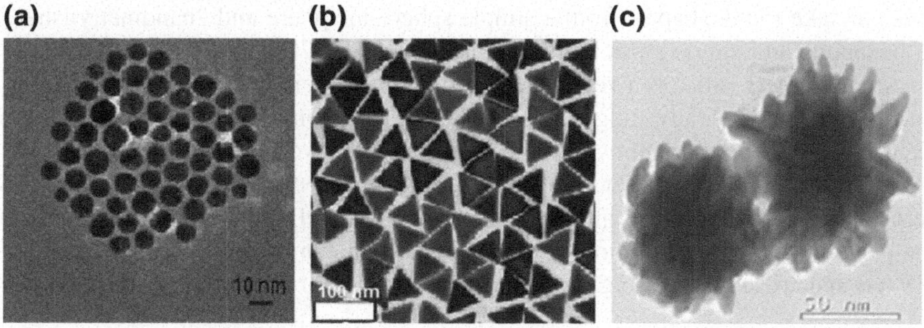

Fig. 5 Examples of gold nanoparticles synthesized through wet-chemistry methods. **a** spherical NPs (Reproduced from Liu et al. [59]), **b** triangular NPs (Reproduced from Bhattari et al. [9], **c** virus-shaped NPs (Reproduced from Jiji and Gopchandran [41])

cases, the reduction of chloroauric acid ($HAuCl_4$) is utilized for their fabrication. In a simple reaction we can obtain gold NPs exhibiting a perfect spherical shape. However, rapid agglomeration without the addition of surface capping agents, e.g. cetrimonium bromide (CTAB), may result in inhibition of their beneficial properties, namely surface enhanced Raman spectroscopy (SERS) emission [7].

In order to develop more distinctive surface morphologies, additional weak reducing agent along with metal ions, particularly Ag^+, might be added, which can selectively passivate surface of the nanoparticle. Following that procedure, at varying ions and nanoparticles ratios, one can produce different NPs shapes like triangles, bipyramids or stars (see Fig. 5b, c) affecting their physicochemical features [41].

Very similarly to gold NPs, platinum nanoparticles can be synthesized. By adding sodium citrate and citric acid to boiling, aqueous solution containing chloroplatinic acid hexahydrate, we can initiate precipitation of Pt nanostructures and finally NPs of about 5 nm in diameter are formed [10]. Those tiny particles can then be used as seeds for further particle growth, opening up further possibilities to develop their size and form.

2.2 *Core-Shell Type Nanoparticles*

Apart from the monometallic NPs, particles could be built of different elements. Within this group, one can distinguish the core-shell type NPs. According to the naming convention, one material stays in the particle's core, and a different one, uniformly shields the core. Although both organic and inorganic materials can be used, in practical applications often inorganic nanoparticles are being applied because of their superior stability compared to organic compounds. The core-shell

NPs can take many shapes, from a simple sphere-in-sphere and "nanomatryushka" to hexagons and single particle in an empty core [61, 78].

Among others, silica is a very common shell material in a core-shell arrangement that provides many advantages. For example, its chemical inertness allows the occurrence of redox reactions through its protective layer without interfering with them, which is beneficial for catalysis or sensing purposes. However, it can also serve as a core material and alongside with metallic shell may be used to modulate the position of surface plasmon resonance band that can be registered by means of UV-vis spectroscopy, see Fig. 6 [52].

2.3 Dendrimers

Another interesting NPs examples are dendrimers whose name comes from Greek word "dendron" meaning "tree". Those nanoparticles are recognized as nanosized macromolecules, with characteristic tree-like dendrons homogeneously cascading from its centre. The core of a dendrimer contains a small molecule or linear polymer that acts as an ovule for consecutive generations of dendrons bounded to the centre. Generally, dendrimers are synthesized using one of two approaches: divergent or convergent one (see Fig. 7).

Divergent method of dendrimer synthesis concerns the growth outwards from the dendrimer core, which reacts with monomer molecules forming further off-shoots. Usage of this approach for production of more branched dendrimers requires even hundreds of steps. Therefore, as some statistical errors may to happen during the synthesis, finally non-perfect molecule could be obtained, e.g. the branches distribution is not homogenous. The question about the molecular shape of the dendrimer also arises, depending on the environmental conditions, as it is impossible to know if the dendrons are pointing outwards from their centre or if they are bending inside.

Convergent approach, however, relies on the synthesis of the outer molecular network first and subsequent reactions towards the core. This method allows higher degree of control over the synthesis because of significantly lower number of simultaneous reactions. The resulting dendrimer created by convergent approach can often be defect-free, although production of higher-generation dendrimers can be troublesome [1, 12].

Multitude of molecules can be used to create dendrimers, providing ways to tailor their properties with a specific task in mind. Because of the versatility of dendrimers, they are widely used in chemistry, pharmacy, bioengineering, and physics.

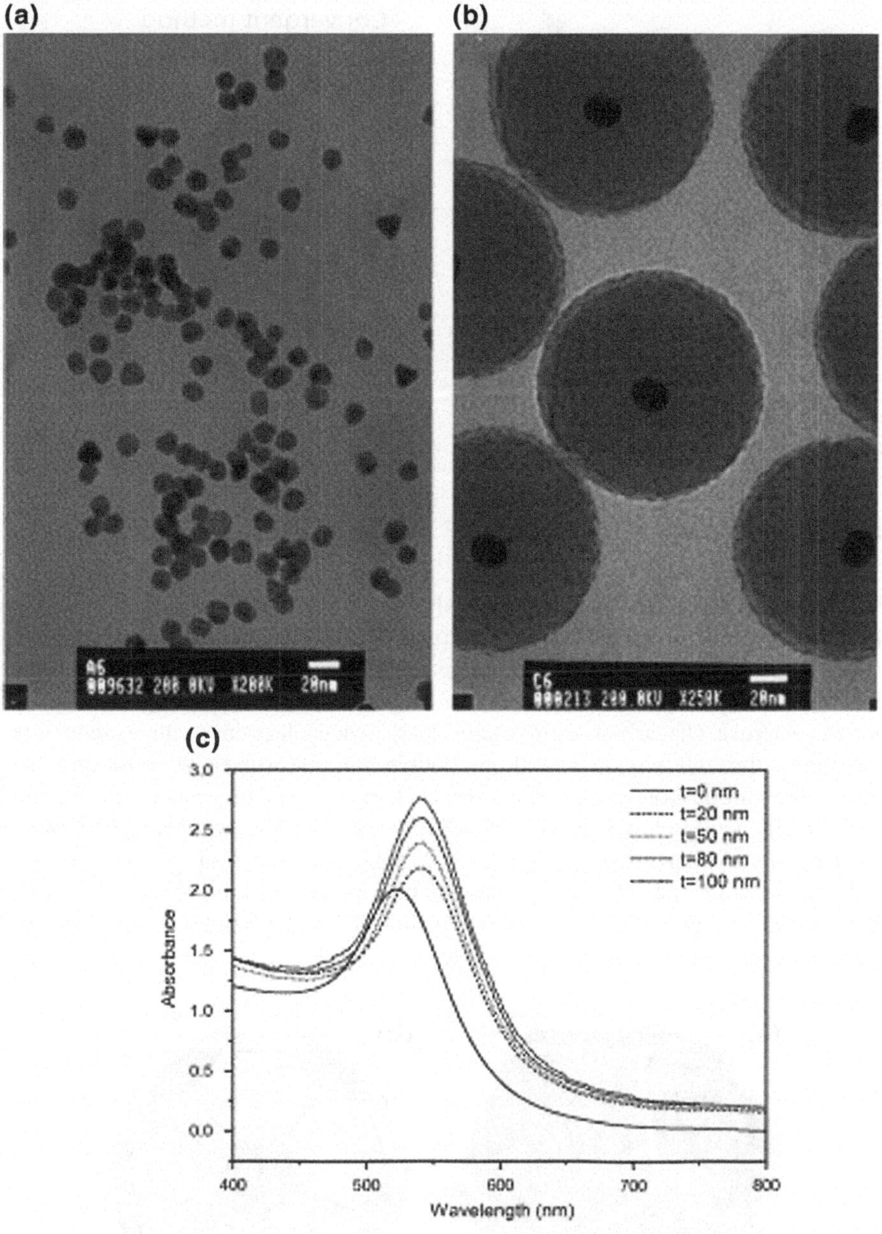

Fig. 6 Absorbance spectra of gold nanoparticles (**a**) is heavily dependent on the thickness of outer silica shell (**b**), therefore spectroscopical measurements (**c**) can be a useful tool for determining shell parameters. Reproduced from Lee et al. [52]

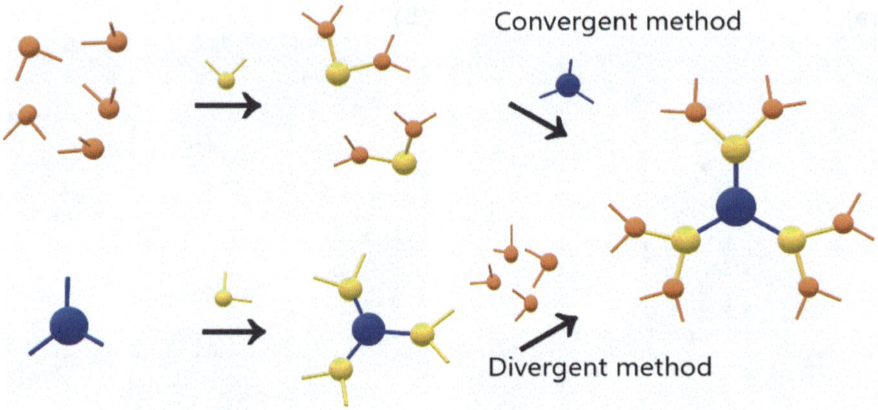

Fig. 7 Schematic representation of convergent and divergent methods of dendrimer synthesis

2.4 Fullerenes

Fullerenes are one of the many extensively studied allotropes of carbon. Although the most common and stable fullerene consists of 60 carbon atoms forming truncated icosahedron, their number can range from 20 up to a couple of thousands (see Fig. 8). The stability of C60 fullerene is associated with the strong covalent sp^2 bonds, where each carbon atom shares its valence electrons with exactly three neighbours through two single and one double bond. However, the position where the double carbon bonds appear, can be exchanged without structure destruction. Additionally, C60 fullerene is the smallest fullerene in which two pentagonal planes do not share an edge, as they can have a destabilizing effect on the structure [66].

In nature, fullerenes of varying sizes can be found in soot. It is also the source of their commercial production. One of the methods of synthesizing fullerenes is based

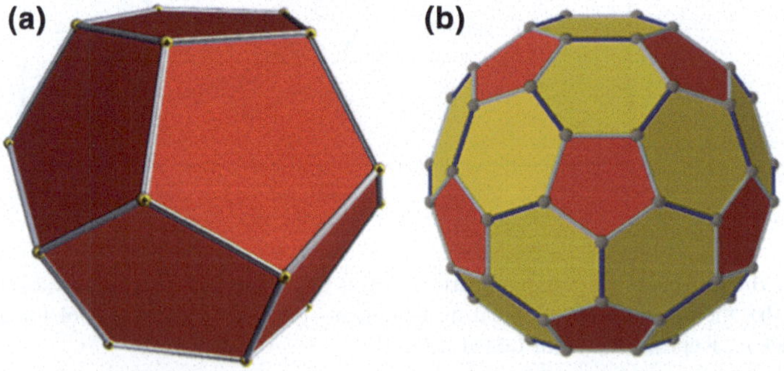

Fig. 8 Visual representation of the structure of C20 (**a**) and C60 fullerenes (**b**) Red colour indicates pentagonal planes, while yellow shows hexagonal planes. (created using Stella Software)

on electric arc discharge between two graphite electrodes in the inert atmosphere. As the product, a fullerene-rich soot is produced and desirable fullerenes are extracted using appropriate organic solvents.

Applications for the fullerenes are wide. They can be used as MRI contrast agents, in drug delivery and for the inhibition of antibiotic-resistant pathogens proliferation [95]. Particularly, in the medical application field, fullerenes are used as a photosensitizers for tumour removal process.

3 Nanowires

The nanowires (NWs) manifest promising properties as being the 1-D quantum systems nanoscopic in diameter and microscopic in length. Their unique properties originate from phonon and electron size confinement in the direction perpendicular to the nanowire length. However, due to the different materials properties, observed effects of the size confinement differ enormously. For instance, metals exhibit a significant decrement of electrical conductivity when the nanowire diameter approaches the mean free path of electrons in the bulk material [84, 88], semimetals—a reduction of the U(I) dependency, semiconductors and insulators—an increment of the bandgap, and superconductors—an increment of the critical temperature.

The current section is divided into several parts, each describing another type of nanowire material. The first part is devoted to nanowires made of metals. Next, unique properties of nanowires made of oxides and semiconductors are discussed. In the fourth part, the variety of organic (including molecular) nanowires and their characteristics are shown. Finally, we conclude with the discussion of the superconducting nanowires.

3.1 Metallic Nanowires

One of the commonly used transparent electrode materials is indium tin oxide (ITO). Despite its good optoelectronic properties, scientists try to find its substitute. Such an approach results from the fact that ITO is expected to become increasingly expensive and its brittleness is not compatible with the strong demands for flexible electronics [84]. One of the promising candidates is a film made of conducting nanowires so thin that their light transparency exceeds 90%. Herein, a few examples of metallic nanowires are presented.

3.2 Noble Metals

Noble metal nanowires exhibit precious set of properties as high conductivity, environmental inertness, and flexibility [84]. Additionally, owing to the nanoscale diameter, metal nanowires are almost invisible for a human eye. Thus, the most perspective applications of noble metals are transparent, flexible electrodes, which should exhibit as high as possible light transparency. Yu et al. [107] have shown that elongation of NWs length increases the strength of light back-scattering. As they showed, for the application as transparent electrodes of solar cells, silver NWs should be shorter than 20 μm. However, the decrement of NWs length may reduce the conductivity of the electrode due to the weakening junction between nanowires. The issue solution was proposed by Kim et al. [47] who applied intense pulsed illumination for increasing silver nanowires temperature, and in a consequence, for welding of individual Ag NWs without damaging the PET substrate.

Noble metal nanomaterials are also known to be efficient catalysts for electrochemical reactions in the field of energy conversion and for chemical reactions in the chemical industry. However, due to their high price and decreasing resources, the elaboration of new highly effective modified noble metal based materials including noble-transition metal alloys and transition metal wires covered by the noble metal is required [28].

For instance, Dai et al. [19] obtained $PtCu_3$ [48] synthesized ruthenium nanowires as thin as 44 nm utilizing a template-based method under ambient conditions. However in 2016, Zhao et al. [116] obtained even thinner Ru and Ru/Pt nanowires (with a Ru/Pt relation from 50:50 to 73:27) of an extremely low diameter of 1.5 nm via an aqueous approach. Using the element mapping analysis, the authors revealed that the Ru and Pt were uniformly distributed in the nanowire networks forming a solid-solution alloy. The obtained material was tested for the electro-catalytic activity towards the methanol oxidation. Results have shown that the RuPt nanowires displayed promising composition dependent properties.

3.3 Non-noble Metallic Nanowires

Due to the high price of noble metals, transition metal NWs are examined as the cheaper substitution. Kim et al. [45] demonstrated a synthesis route to fabricate 60 μm-long nickel nanowires, which displayed to be a good candidate for use in transparent conducting electrodes. A transparent electrode made of Ni NWs showed transparency as high as an ITO electrode for the same effective resistivity. The authors even suggest that obtained electrodes are more suitable as a bottom OLED layer than ITO or Ag nanowires.

Another material considered as a promising candidate for transparent conducting electrodes is copper as nanowires. Until now, Cu NWs with a diameter of 60–80 nm and length over 50 μm were successfully synthesized by a modified self-catalytic method [109]. Although their transparency is not as high as that of the noble metal, it is still relevant to use in semi-transparent solar cells.

Another interesting group of materials is semimetals characterised by a narrow bandgap. Bismuth is a semimetal frequently serving as a typical model for investigations on semimetals. Due to its band structure, small electron concentration and electron scattering, the effective electron mass is very small and the Fermi wavelength (electron wavelength at Fermi level) equals 40 nm. As a consequence, free electron path length becomes extremely large and reaches even up to 1 mm at $T = 4.2$ K. Additionally, in semimetal nanowires, as in other metals, decreasing the size causes other effects like a cut off the phonon spectra, and the drop of the thermal conductance.

3.4 Oxide Nanowires

As well as metallic, metal oxides attract great attention among nanowires. They are characterized by lower conductivity but exhibit other useful properties like tunable bandgap, wavelength dependent transparency and catalytic properties. For instance, ZnO NWs of ~ 5 μm in length and ~ 120 nm in diameter on a p-type GaN (Fig. 9) find an application as a nanowire nanolasers [37]. The wires play a double role, both as an optical resonator when the wire length is divisible on the integer number of half-waves and an active medium. A spontaneous emission arises when current power overcomes a lasing barrier, which under irradiation of the laser beam of 355 nm wavelength is 180 mW/cm^2. Such a ZnO nanowire assisted laser possesses a relatively small lasing threshold owing to bulk ZnO parameters and operates in a single-mode regime due to the low length of wires.

Another unique property of NWs made of semiconductors and insulators is ultra-low thermal conductivity. Because of the high surface-to-volume ratio of nanowire, the surface phonons modes start to play an important role in phonon scattering. Their interaction with the longitudinal and transverse phonons as the boundary scattering within the nanostructures, leads to a decrease in the phonon lifetime. Thus, the phonon transports and the thermal properties of the nanowires significantly decrease compared to that of the bulk material.

Spinel nanowires have found an application in the field of the electrochemistry cells and catalysts. Thanks to a high specific surface and a high defect concentration, spinel nanowires are considered being an effective substitute for platinum catalysts. For example, Liu et al. [62] have found that hydrothermally synthesized $NiCo_2O_4$ nanowires after transformation to the $NiCoO_2$ became a very active catalyst due to a high concentration of oxygen vacancies.

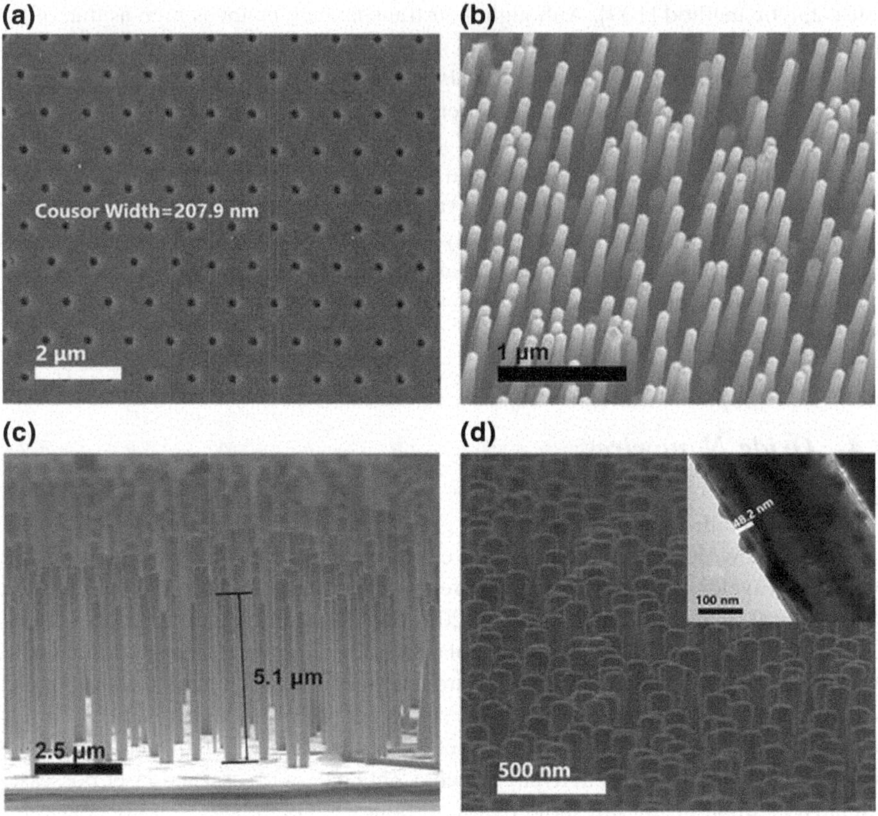

Fig. 9 SEM characterization of the structure used in laser based nanodevice: **a** SEM image of a regular hexagonal array of nanoholes on the substrate, **b** top view SEM image of n-type bare ZnO NWs (the diameter is of 120 ± 10 nm), **c** side view SEM image of n-type vertically aligned ZnO NWs (the length is about 5.1 µm) array on the p-type GaN substrate, **d** top view SEM image of n-type vertically aligned ZnO NWs, the inset is the TEM image of a single NW coated by a hybrid SiO_2/Ag film with a thickness of 50 nm. Reproduced from Huang et al. [37]

3.5 Non-oxide Semiconducting Nanowires

Bendable electronics have attracted widespread attention because of their potential application in many smart electronic devices and flexible displays. Therefore, high attention is paid to the research on bendable electronic elements, like diodes, transistors, capacitors, etc. Because of their elasticity, 1-D and 2-D nanostructures are intensively studied. One of the study achievements are bendable NAND logic gates with silicon nanowire channels capable of operating with a supply voltage as low as 0.8 V [108]. The gate was based on a Si nanowires with a 40 nm diameter prepared via low-pressure chemical vapour deposition, ion implantation doping, and finally deposition onto a plastic substrate.

III–V semiconductors are considered as a beneficial material for optoelectronic devices such as multi-junction solar cells due to their broadly tunable bandgaps. Thus, III–V nanowires are even more interesting, since they add new benefits for nanophotonics devices like increased optical absorption compared with the bulk material caused by light-trapping effects. Moreover, NWs are expected to reduce the materials consumption of these relatively expensive semiconductor materials in the future. For example, a comparison of planar and nanophotonic geometries of InPb [18] NWs with the same material quality, indicates that the external radiative efficiency of the nanostructured layer is 14-times enhanced compared to those found for the planar coating. The higher efficiency arises from the enhanced out-coupling of photons promoting radiative recombination and the lower active material volume, which reduces bulk recombination.

Another common approach used for the preparation of III–V semiconductors is the metal-seeded vapour-liquid-solid (VLS) technique. Berg et al. [8] synthesized InGaP nanowires as low as 30 nm in diameter (see Fig. 10). They have shown that the uniform growth was achieved in a temperature range of 675–750 °C, while increasing of the synthesis temperature leads to the increasing of a nanowire deformation degree and decreasing of the nanowires length simultaneously.

The regularly dispersed nanowires can also be obtained by the self-assembled formation processes. One example is the self-assembled formation of GaN nanostructures obtained by the plasma-assisted molecular beam epitaxy on the TiN substrate [98]. Using this approach, van Treeck et al. [98] were able to obtain nanowires of the diameter less than 50 nm and of a homogeneous length of more than 1 μm. It is worth noting that the separation distance between nanowires is caused by the diffusion-induced repulsion of neighbouring NWs, without using a templated substrate.

It is known that the composition defects could substantially alter the properties of semiconductors. Therefore, incorporation of foreign non-equilibrium atoms in the host semiconductor lattice may be used for new or added functionalities in the existing semiconductor system [106], namely strain engineering, controlled defect

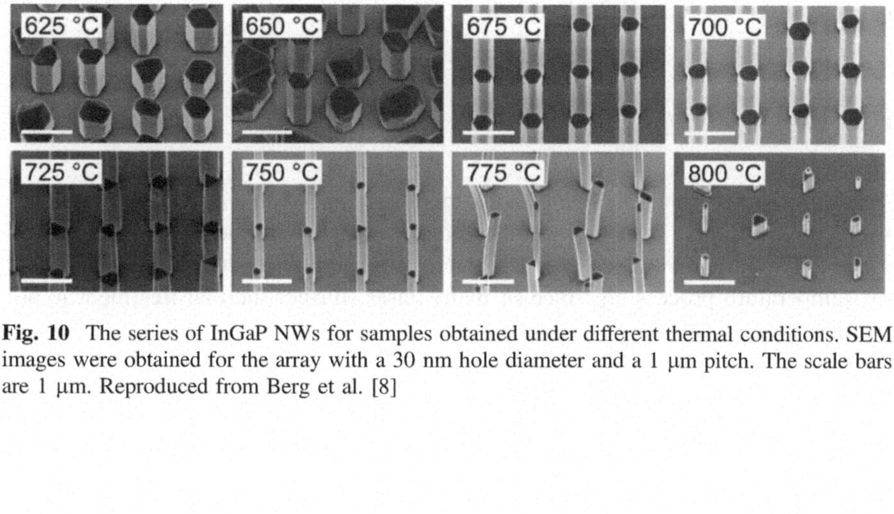

Fig. 10 The series of InGaP NWs for samples obtained under different thermal conditions. SEM images were obtained for the array with a 30 nm hole diameter and a 1 μm pitch. The scale bars are 1 μm. Reproduced from Berg et al. [8]

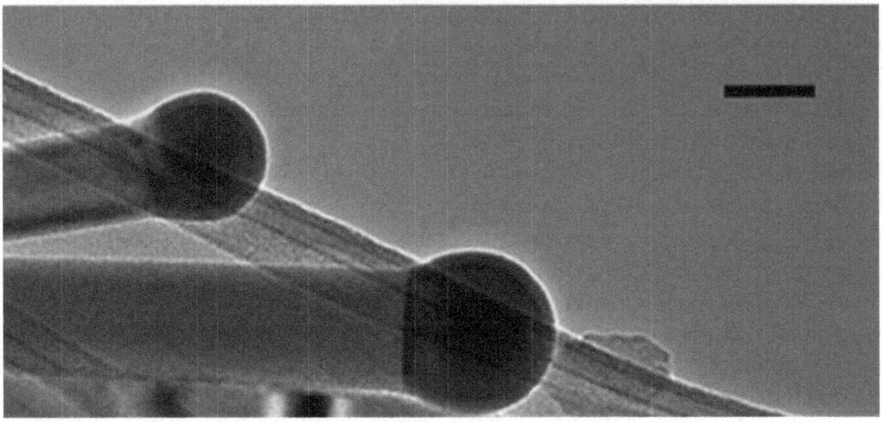

Fig. 11 TEM image confirming the participation of VLS growth mechanism with dark-contrast spherical seed at the tip of the nanowire with AuAg catalysts from precursor solution containing 15 at% Sn. Scale bar is 100 nm. Reprinted with permission from. Reproduced from Biswas et al. [11]

formation, band structure modulation, etc. For example, a non-equilibrium incorporation of Sn in Ge induces the small energy band variation, which makes them a direct bandgap semiconductor [106]. The non-equilibrium Sn-Ge alloy was obtained by a "solute trapping" process. At the certain solidification rate, during the vapour-liquid deposition growth of nanowire (Fig. 11) a local chemical equilibrium at the liquid–solid interface cannot be settled due to a large interface velocity. Therefore, Sn atoms can be trapped on the unfavourable energy sites of the nanowire crystal lattice leading to the formation of metastable solids at the nanowire growth front.

Some optical properties of III–V semiconductor nanowires may also be improved by their surface treatment. For example, a deposition of parylene-C, a highly resistive transparent polymer, onto the GaAs followed by an annealing, decreases the system reflectivity below 1%, what strongly improves its absorbance and photoluminescence intensity [29].

3.6 *Organic Nanowires*

Despite inorganic materials being much more popular when considering a quantity of scientific reports, such elongated structures are also formed by organic compounds. Organic nanowires have inherent advantages related to low-cost and low-temperature processing, since in many cases, further thermal treatment is not required.

The high success of carbon nanotubes led to the initiation of research devoted to other organic elongated forms of nanomaterials. For example, Lentz et al. [56] developed a method which allows to obtain amorphous carbon nanowires. The semiconducting nanowires with an average diameter of 150 nm were synthesized by a pyrolysis of polymerized polyfurfuryl alcohol inside the pores of an anodized alumina template. The thinner NWs were obtained from the paralyzed bacterial cellulose. Nitrogen- and phosphorous-doped 50 nm thick carbon nanowires obtained by Hu et al. [35] showed promising properties for supercapacitors like capacitance density of 258 F/g at a current density of 1 A/g and exhibited an excellent cycling stability over 30,000 cycles.

Pyrolysis, as a synthesis method of organic NWs, limits control of the morphology, especially, the fabrication of molecular nanowires with a diameter below 100 nm is very hard to control. Therefore, the application of organic nanowires in optoelectronic devices is a big challenge. However, in the work of Tan et al. [90] the first demonstration of the highly ordered organic single-crystalline nanowires can be found. The nanowires are fully composed of self-assembled small organic molecules, which next are successfully used as a Fabry-Pérot optical resonance cavity for 720 nm NIR nanolasers.

In most cases, nanowires are obtained via chemical, electrochemical or physical methods. However, in recent years, a new branch of nanowires fabrication methods has been elaborated. Among them, biological synthesis is potentially attractive because it allows the obtaining of nanowires from a fully renewable resources. Moreover, the final product may be obtained without the need for toxic solvents or harsh chemical processes [90]. An attention is paid to *Geobacter sulfurreducens* [81] as one of the electrically conductive pili, due to its high aspect ratio (3 nm 15 μm). Tan et al. [90] have shown that manipulation within the genotype of *G. sulfurreducens* allows to produce conductive nanowires of a diameter below 2 nm and over 3 μm in length. It is worth noting that the authors measured the final conductivity of the nanowires only one order of magnitude lower than observed for carbon nanotubes [93].

3.7 Superconducting Nanowires

The last materials discussed in the current subsection are superconductors. These specific group exhibits its unique properties below the critical temperature, therefore a discussion about its high-temperature characteristic is neglected.

Similarly to other properties, critical temperature of superconductivity is size-dependent as well [70]. For example, transport measurements of individual cylindrical single-crystal Bi NWs, 20 and 32 nm in diameter, have shown that, in contrast to non-superconducting bulk or Bi nanoribbons, cylindrical Bi nanostructures exhibit superconductivity at the temperature as high as 1.3 K [96].

One of the most interesting phenomenon observed in the superconducting nanowires is a manifestation of Majorana fermions [71]. They are specific particles identical to their own antiparticle, hypothesized by Ettore Majorana in 1937. In 2008 Fu and Kane [25] have shown that these specific particles may be observed on the junction of superconductor and metal, what was finally proven a few years later.

4 Nanowalls

The nanowalls, often named nanosheets, manifest promising properties being the 2-D quantum systems nanoscopic in thickness but microscopic in length and width. A great interest in two-dimensional material arose after the exploration of graphene. Nowadays, other classes of 2-D nanomaterials are known, like single- or few-layered transition metal dichalcogenides, metal oxides, layered double hydroxides, hexagonal boron nitride (h-BN), graphitic carbon nitride (g-C_3N_4), metal carbides and nitrides, and monoelemental compounds [16]. Their specific area is theoretically not as large as nanowires. However, this feature, sometimes regarded as a disadvantage, is diminished by much more facile synthesis compared to 1-D structures [118]. The outstanding properties of selected 2-D nanomaterials are described in four sections, each devoted to one family of nanowalls/nanosheets: semiconductors, oxides, carbon-based materials and composites.

4.1 Semiconducting Nanowalls

2-D materials are found to be chemically inert, have high mechanical strength and thermal conductivity in a parallel to the plane direction, while their parameters may urgently differ in the perpendicular direction. The most well-known 2-D and widely applied nanomaterial is graphene. However, other compounds, including hexagonal boron nitride (h-BN) can be obtained as a single- or few-walled nanomaterial. Besides typical for 2-D materials properties, its multi-layered structure exhibits deep UV luminescence, and, what is more interesting, vertically aligned nanowalls of h-BN were found to be cytotoxic against *E. Coli* [67, 111]. Although several methods of h-BN preparation have been elaborated, the most successful way to obtain nanowalls is a CVD method [67]. The growth mechanism of the <15 nm thick nanowalls is shown in Fig. 12. At the beginning, the interfacial layer is

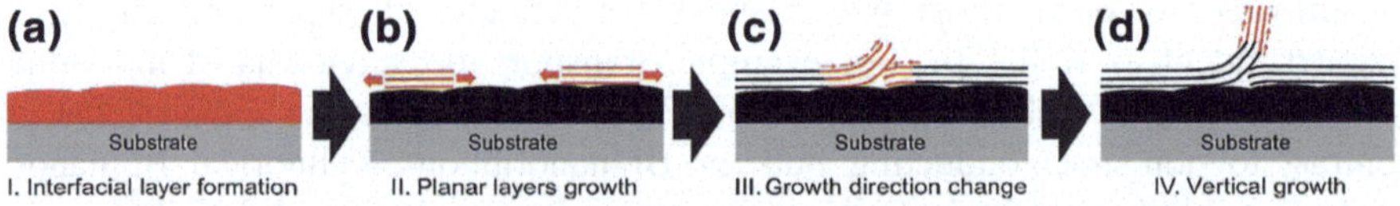

Fig. 12 Schematic representation of the h-BN nanowall growth mechanism. Reprinted with permission from. Reproduced from Merenkov et al. [67]

formed. Then the nucleation of several plane nanocrystallites takes place, which grows in the plane until neighbouring layers contact each other. Finally, one of the layers turns upward and the vertical growth appears.

It is known that the effectiveness of catalysts depends on their specific surface area, therefore, 2-D semiconducting nanomaterials are expected to be highly effective. For example, CuS nanowalls were used for the thermal decomposition of ammonium perchlorate [104]. The presence of CuS nanowalls decreases the temperature of the ammonium perchlorate decomposition for over 150 °C, from 312 to 145 °C. It is worth to note, that the synthesis of CuS nanowalls was performed not on the solid substrate, but at the liquid-liquid interface of chloroform and water, which leads to the formation of a 2 μm layer composed of weakly joined CuS nanowalls.

The bandgap of nanomaterials undergoes a widening with decreasing size due to quantum confinement effects. This statement was a background for the study performed by Ichikawa et al. [38] on the 2-D Si quantum well structure. Using anisotropic wet etching, the authors were able to fabricate high-quality Si nanowalls as thin as 3 nm with vertical flat sidewalls at the Si/SiO_2 interface. However, they were unable to record the variation of the optical bandgap because of the light dispersion on the regular nanowalls structure.

4.2 Oxide Nanowalls

Nanowalls were also formed by oxide materials which find applications as effective gas sensors, highly capacitive electrodes, optical modulators, an active transistor channels material, and others. For example, ZnO is a direct-bandgap n-type semiconductor material ($E_g = 3.37$ eV) which may find application as a hazardous gas sensor. However, its functionality is limited by high working temperature, low electrical conductivity, and poor thermal stability. Kim et al. [44] tested Nb doping of ZnO nanowalls, what increases the specific surface area of the sensor in comparison to that of pristine ZnO nanowall by ca. 10%, improves its conductivity and decreases possible operating temperature.

The interconnected nanowall structure can over excellent charge transfer path and provide large contact specific areas of electrolyte/electrode interfaces. Therefore, highly capacitive electrodes made of transition metal oxides nanowalls, like CoO, NiO or MnO_2, are considered being good candidates for electrodes in supercapacitors. Tang et al. [92] examined properties of the CoO porous nanowall structure obtained by a combination of solvothermal growing and annealing. It was shown that the morphology of CoO nanowalls can be tuned by adjusting the annealing temperature. In particular, nano-sized pores below 50 nm are formed on CoO nanowalls during thermal treatment. The unique porous nanowall structure leads to a large specific surface area (65 m^2g^{-1}), which increases the electrolyte/electrode interface. Following that, enhanced electrochemical performance of CoO was obtained.

Electrochromism is another interesting property of transition metal oxides. Zhang et al. [114] synthesized Co_3O_4 nanowalls via hydrothermal method. The product exhibited highly porous nature and large surface area. Due to that, fast and large optical modulations are realized in the Co_3O_4 nanowalls.

Transition metal oxide nanowalls are considered finding applications in nanoelectronic devices. However, due to strong electrons correlation in metal oxides, a rich variety of electronic phases with unique properties including superconductivity, colossal magnetoresistance or metal-insulator transition, is observed. Therefore, control of that electric phases, especially in a field-effect transistor seems to be very promising. Hattori et al. [33] fabricated an electric double layer transistor with a (La, Pr, Ca)MnO_3 channel and investigated the gating effect. The nanowall was fabricated by combining UV-nanoimprint lithography, pulsed laser deposition, etching and annealing processes. The reported transistor is the first based on the control of the insulator-metal transition.

4.3 Carbon-Based Nanowalls

A distinctive family of nanomaterials is a transition metal oxide nanowalls deposited onto the carbon nanowalls. Due to the high conductivity of carbon-based materials and high electrical capacity of transition metal structures, they found an application as electrodes in the Li-ion batteries or supercapacitors [20]. Zhu et al. [119] reported on ultrathin Co oxides nanowall arrays deposited on reduced graphene oxide (rGO) sheets. In that nanohybrid system, the Co oxides exhibited a porous structure and the grain size of the Co oxide was tailored by varying the annealing temperature.

Carbon nanowalls attract attention also due to their specific electrical properties, including a resistive switching phenomenon. Russo et al. [82] have reported for the first time the resistive switching behaviour of a device made of carbon nanowalls deposited on a fluorine-doped tin oxide substrate. The carbon nanowalls were synthesized by electrophoretic deposition from the solution of polyynes. The obtained memory resistive switching device showed an operation voltage of 2 V and long retention time (over 10^4 s). Those parameters suggest that carbon nanowalls may be used in the future as a basis of commercially available memory devices.

5 Nanopores

Ordered self-standing porous materials offer a wide range of possible applications, since they can be used as a ready product for certain processes, e.g. drug transport, catalysis, investigations of adsorption or separation processes and a substrate for further synthesis of materials that morphology can project the substrate shape.

Following that, nanopores can act as a template for nanotubes or nanowires production as well as a patterning mask. Furthermore, elaboration of novel manufacturing techniques that allow to control geometric parameters of internal pore structure can develop other materials based on the template assisted synthesis.

5.1 Porous Anodic Aluminium Oxides

Aluminium, as typical oxidizing materials, exposed to air spontaneously forms a thin barrier oxide layer. The very thin barrier oxide film, called passive layer can also be obtained via electrochemical oxidation in much more controlled way. Anodizing in the electrolytes of pH = 5–7, such as borate, oxalate or citrate with the respect to the other conditions forms compact passive layer on aluminium plate. However, usage of acidic solutions such as selenic, sulfuric or oxalic partially dissolves aluminium oxide whereas simultaneous electrochemical oxidization under the optimized conditions enables formation of porous anodic aluminium oxide (AAO) layer [54].

Self-ordered porous AAO with a uniform pore size and distance can be achieved via two-step anodization, so-called mild anodization (MA). The first anodization typically takes over 24 h. In the next step photochemical etching is carried out that concerns oxide layer removal by the immersion of as anodized alumina in sulfuric acid and chromium oxide mixture at the elevated temperature. The process results in the wells-like morphology formation onto the remaining aluminium substrate. Owing to such procedure, nanopores grown during the second anodization are uniform and parallel to each other. Moreover, pores diameter, distance between them and thickness of barrier layer can be controlled by an applied voltage and electrolyte composition. MA allows to reach the growth rate within 2–10 μm/h range. On the contrary, hard anodization (HA) performed at the voltage exceeding breakdown potential allows for much faster growth rates up to 50–100 μm/h. Such a high voltage generates very high current flow, 1 or 2 orders of magnitude higher comparing to MA. As a consequence, large amount of heat causes spatial damages in morphology. Since the geometric parameters of the material are out of control and the morphology destructions are present, HA has not found applications in nanotechnology [53].

Sequential combination of MA and HA by applying alternately low (mild anodization voltage) and high (hard anodization voltage) potentials allowed to obtain modulated internal pore diameters within AAO. This concept, so called pulsed anodization (PA) is regarded as a beginning for the further structural modifications in AAOs by cyclic anodization (CA). The technique requires applying oscillating current signals with different parameters such as period or amplitude to electrodes. It allows formation of different pore geometry [51]. According to PA approach, tailor-made AAO pore structure can be prepared as shown in Fig. 13a whereas polymer nanopillars obtained inside such template is shown in Fig. 13b.

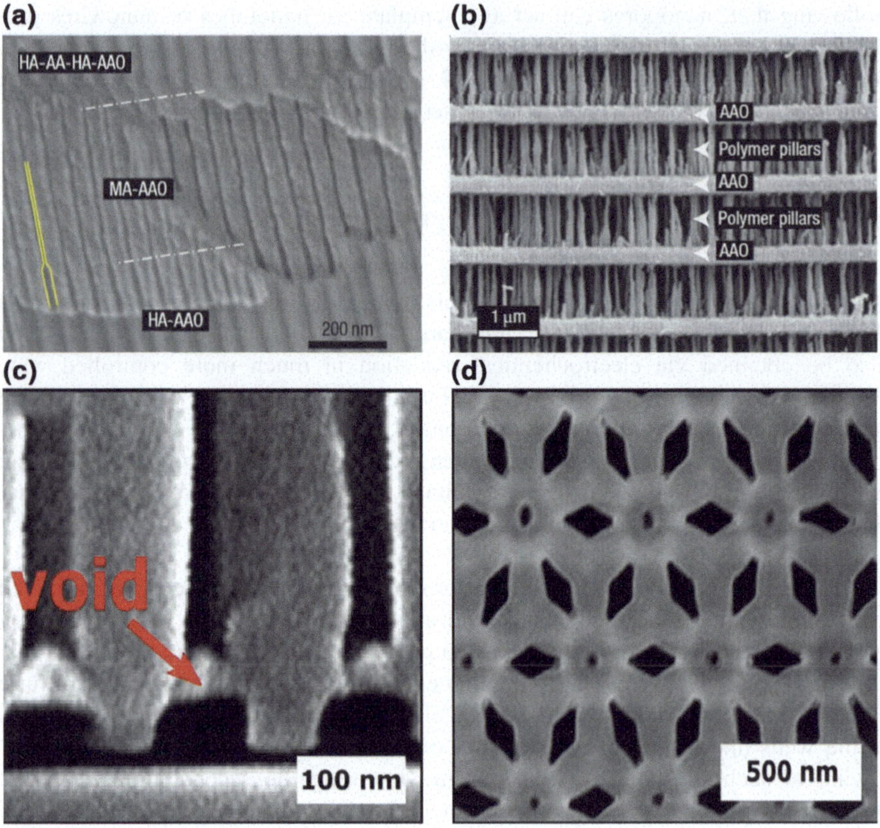

Fig. 13 SEM image showing **a** the cross section view of as-prepared AAO produced by H_2SO_4 pulsed anodization, **b** polymer nanopillars/AAO composite microstructure, with the MA/AAO slabs supported by PS nanopillars (Reproduced from Lee et al. [51]); **c** cross-sectional SEM image of the edge profile of pore bottoms that becomes angular with void growth by further anodization (Reproduced from Seo et al. [85]); **d** FESEM image of diamond pattern obtained after etching of alumina template nanoimprinted pattern in aluminium. Reproduced from Smith et al. [87]

Apart from anodization of aluminum foil itself, ordered structures can be obtained by anodization of sputtered Al films on different substrates such as insulators, semiconductors, non-valve metals, valve metals or transparent indium thin oxide. The most widely investigated was anodization of Al thin film deposited onto the Si substrate resulting in interfacial voids formation at the bottom of pores (see Fig. 13c). This phenomenon is related to the stress accumulation and its minimizing. It is worth noting that for longer anodization times SiO_2 nanodots are formed underneath the voids due to the Si oxidation by electrolyte migrating through the small hollow spheres located below elongated pores.

Long-range arrangement with a high degree of uniformity that is required for the most applications can be achieved by anodizing nanoimprinted pattern on aluminum substrate. The shape and position of that pattern define very precisely further nanopore arrangement. However, still obtaining almost ideally ordered porous AAO depends on appropriate adjustment of the anodization parameters. According to that procedure, AAO diamond pattern template was fabricated by Seo et al. [85], see Fig. 13d.

5.2 Porous Anodic Titanium Oxide

As was already discussed, anodization of aluminum enables AAO formation. Whereas, anodization of other valve transition metal, namely titanium may provide two different morphologies: nanotubes or nanopores depending on the anodization parameters such as: anodization time, temperature, applied voltage or fluoride ions concentration (see Fig. 14a) [89]. Pore formation mechanism is directly related to the chemical etching of titanium during anodization. Suitably low titanium etching rate causes growth of TiO_2 on the sides of the nanotubes entrances [103]. As the consequence, pores form interconnected nanotube architecture. Schematic diagram depicting pore and tube growth is given in Fig. 14b.

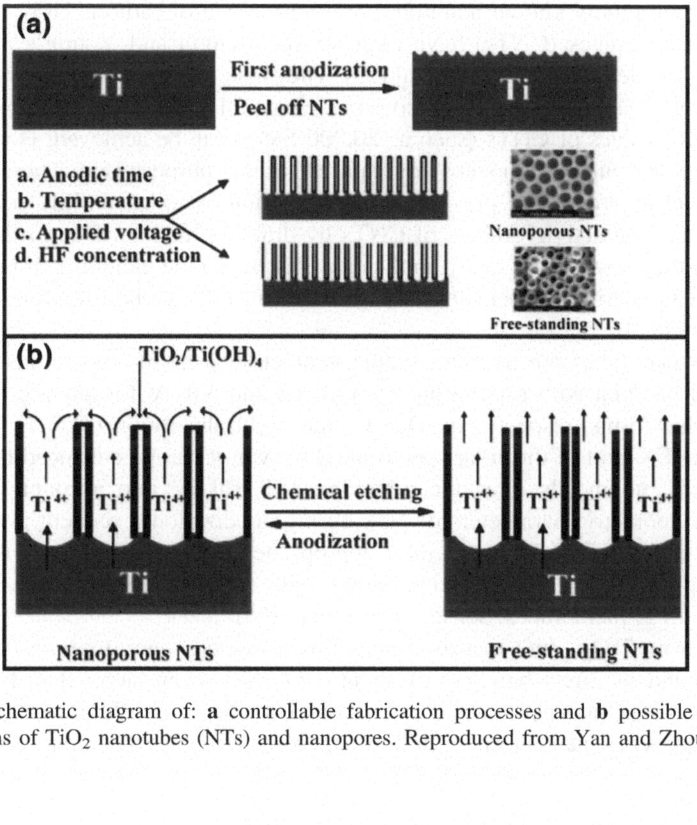

Fig. 14 Schematic diagram of: **a** controllable fabrication processes and **b** possible formation mechanisms of TiO_2 nanotubes (NTs) and nanopores. Reproduced from Yan and Zhou [103]

Titania structures i.e. nanotubes and nanopores obtained via anodization are in the amorphous phase. Therefore, calcination in a furnace is usually performed to ensure the crystalline phase. It is worth mentioning that the whole process covering the heating up of samples and their cooling down to room temperature can take over 10 h. Since there is a competition between pore and tube formation during anodization and as-anodized titania requires phase conversion, the number of reports on TiO_2 nanopores synthesized via electrochemical oxidation is not that much as for AAO.

6 Nanotubes

As has been already underlined, nanostructures with an increasing specific-surface area are of high importance for catalysis, sensing, or light harvesting. Nanotubes (NTs) because of hollow-like structure and the high aspect ratio (length-to-diameter) have an available surface on the inner and outer wall that makes surface area higher comparing to nanorod where only outer sphere could form interface. Moreover, the 1-D tubular geometry improves electron transport for longer distances and facilitates light absorption [83]. Due to unique properties, different NTs have found various applications, especially in solar cells and sensors. The most commonly known nanotubes are those made of carbon, TiO_2, and ZnO.

Carbon nanotubes (CNTs) have high tensile strength and Young's module at relatively low density of this material [75]. They also exhibit extraordinary thermal conductivity [31] and unique anisotropic electrical properties. Extremely high conductivity values of CNTs (such as 20,000 S/cm) can be achieved. However, in this case, many purification steps are required for this purpose [55]. Therefore, it is important to regard all the pre- and post-processing treatments while considering the reported conductivity values of CNTs by different authors. Due to the unique properties they have found many applications for: medicine including drug delivery or tissue engineering, biosensors [79], gas sensing [32], carbon-reinforced matrix composites [72].

Titania nanotubes are as other titania geometric forms n-type semiconducting functional material with a large bandgap of 3.2 and 3.0 eV for anatase and rutile polymorphic forms, respectively. Due to that, light absorption of TiO_2 is limited only to the UV part of the solar spectrum. However, it can be enhanced by doping with metals, nonmetals or semiconductors [14]. Taking into account the facile control of geometric parameters of NTs during anodization, excellent photocorrosion resistance of TiO_2NTs and their appropriate band edge positions they have found applications in dye-sensitized solar cells (DSSC), pollutants degradation, water splitting, membranes, sensors and others [53].

Apart from TiO_2, ZnO is also regarded as a low cost, non-toxic, n-type semiconductor but its direct bandgap of about 3.37 eV is even larger than for titania. Moreover, it has higher electronic mobility and synthesis of different morphologies such as nanowires, nanosheets or hollow spheres is possible in comparison to TiO_2.

Since ZnO has suitable band structure for both water reduction and oxidation, ZnO NTs are investigated as a photoanode material for hydrogen production via photoelectrochemical water splitting process [64], in solar cells [86] or for gas sensing [17]. As it can be seen ZnO NTs and TiO_2 NTs have similar properties and applications. However, the huge advantages of the electrochemical anodization over ZnO NTs manufacturing methods make this material only to be considered as a replacement for TiO_2 NTs.

6.1 Carbon Nanotubes

CNTs were synthesized for the very first time by Sumio Iijima in 1991 [39]. They can be considered a "rolled" graphene sheets while graphene is a single layer of carbon atoms arranged in a hexagonal network. Depending on the number of rolled layers CNTs can be single-(SWCNTs), double-(DWCNTs), or multi-walled (MWCNTs). Figure 15 shows SW- and MWCNTs structures whereas their comparison is shown in Table 5. Taking into account the direction of how graphene sheet is rolled different CNTs arrangements can be formed such as: zig-zag, chiral or armchair.

Among SWCNTs, one can find three various structures (see Fig. 16) obtained by rolling graphene sheet in different directions in an accordance with the values of a chiral a_1 and a_2 vector pair. Knowing that the spatial arrangement impacts the material properties, a diversity of carbon tubular forms offers a wide range of applications.

In the literature there are three main approaches based on carbon precursor decomposition allowing fabrication of CNTs: arc discharge, laser ablation and chemical vapour deposition. Classification of CNTs manufacturing methods with more detailed distinction of CVD approach is shown in Fig. 17.

Concerning arc discharge, the method involves electrical breakdown by DC current of a gas to generate plasma. It results in temperature increase at the graphite

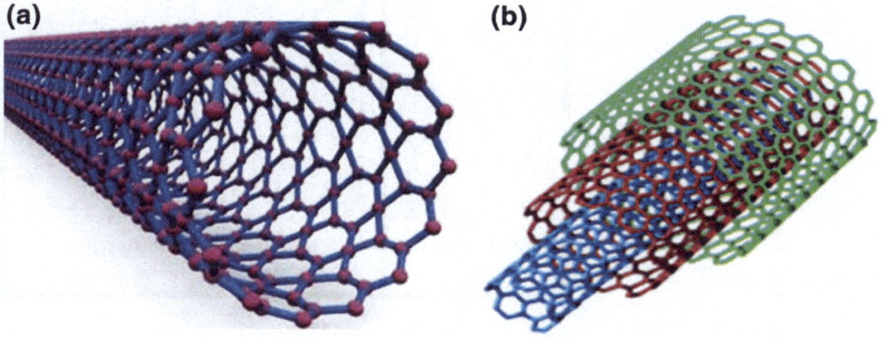

Fig. 15 **a** SWCNT, **b** MWCNT. Reproduced from Wang et al. [101]

Table 5 Differences between SWCNTs and MWCNTs

SWCNTs	MWCNTs
Single layer of graphene	Multiple layers of graphene
Catalyst is required for synthesis	Can be produced without catalyst
Bulk synthesis is difficult as it requires proper control over growth and atmospheric conditions	Bulk synthesis is easier
Purity is poor	Purity is high
High risk of defects during functionalization	Low risk of defects and if occurred, can be improved
Can be twisted easily due to a single later of graphene	Cannot be twisted easily due to multiple layers

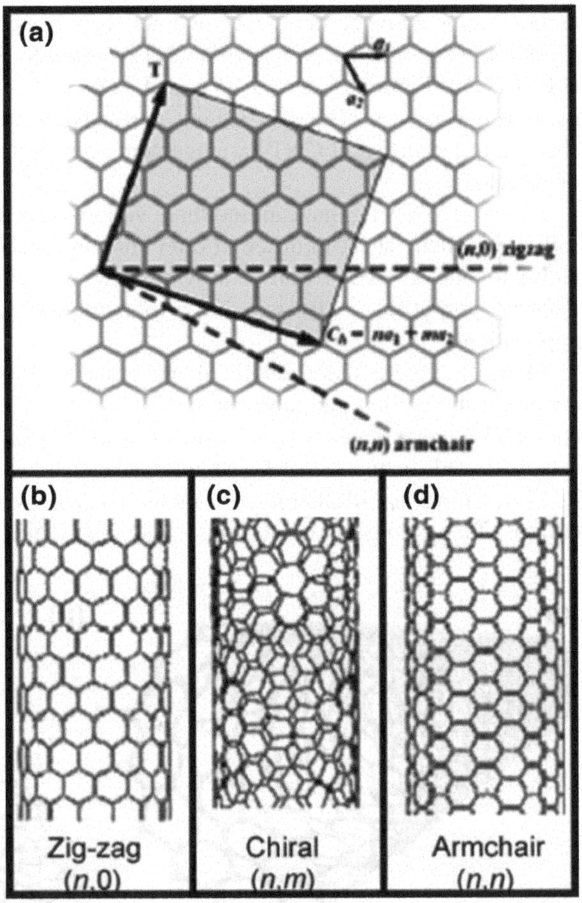

Fig. 16 Schematic diagram showing different CNTs structures. Reproduced from Zhang and Zhao [115]

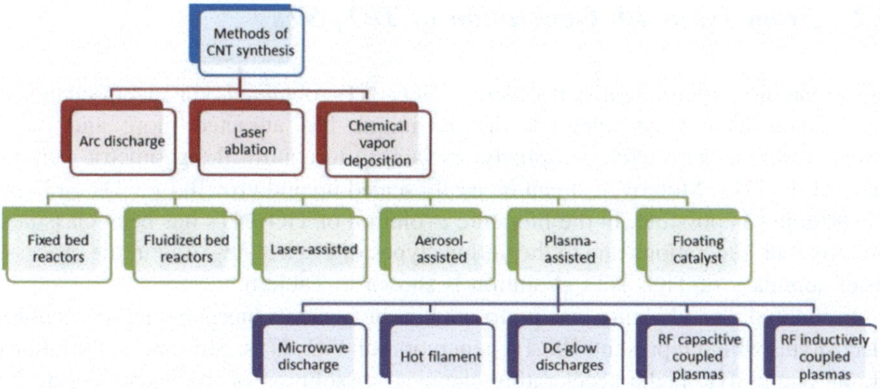

Fig. 17 Classification of CNTs manufacturing methods. Reproduced from Kumar et al. [49]

anode where the carbon precursor is evaporated and deposited onto the graphite cathode. The process takes place in a vacuum chamber with a controlled supply of inert gas, argon, or helium. Both SWCNTs and MWCNTs can be produced via an arc discharge approach. However, in the case of SWCNTs anode filled with catalysts such as Fe, Co, Ni is essential. Catalyst facilitates the decomposition of carbon from the precursor at a lower temperature. Nevertheless, it makes the production process more expensive [49, 69, 120].

The laser ablation technique is a quite similar approach to arc discharge. It requires high energy laser radiation to ablate graphite from the target under the inert gas atmosphere which is further deposited on the substrate. As was in the case of the arc discharge, SWCNTs and MWCNTs can be produced. However, this method allows to fabricate smaller quantities of CNTs compared to the arc discharge [49, 69, 120].

Nowadays, the most widely used manufacturing technique for different CNTs is CVD. This method involves the pyrolysis of carbon precursors such as acetylene, ethylene, propylene, methane, polymers, or carbon monoxide. Vapors generated in this process are transported using the inert gas to the chamber containing the catalyst. It is worth mentioning that catalysts act here as nucleation sites for CNTs growth. CVD enables large-scale production due to the high yield rate and relatively simple equipment compared to machines needed for arc discharge and laser ablation methods. It also enables control of the growth process and ensures good alignment. The diameter of the NTs can be adjusted simply by the temperature knowing that the CNT diameter is directly proportional to the applied temperature. Temperature also determines whether SW- or MWCNTs are formed. Higher temperature (ca. 900 °C) favors SWCNTs formation, whereas lower temperature (630 °C) supports the synthesis of MWCNTs [49].

6.2 *From 1st to 4th Generation of TiO_2 NTs*

Since the first reports on self-ordered TiO_2 NTs obtained via electrochemical anodization have been released, this technique has attracted more and more attention due to its relative simplicity and ability to control the geometric parameters of the NTs. Moreover, it can be easily scaled up and growth of TiO_2 on large Ti substrates is possible. In the literature evolution of TiO_2 NTs has been classified into the four generations due to the various types of electrolytes used in the process. Brief summary of TiO_2 NTs evolution is shown in Table 6.

Initially, the anodization was performed in the aqueous inorganic HF-containing electrolytes which represents the 1st generation of TiO_2 NTs. Since acidic solutions are an aggressive media accelerating chemical dissolution of the barrier layer, the length of 1st generation NTs reaches only a maximum of ca. 500 nm. Moreover, NTs exhibit a rippled structure due to the presence of fluoride ions originating from HF dissolved in a water bath. The 2nd generation is based on the usage of buffered solutions with the addition of fluoride ions in the form of an inorganic salt such as Na_2SO_4/NaF or $(NH_4)_2SO_4/NH_4F$. Reduction of acidity by excluding HF from the solution allowed to obtain NTs even up to 5 μm length. However, the morphology of 2nd gen. NTs is still rippled due to aqueous solutions. Irregularities in the 1st and 2nd generations of the NTs might be related to oxygen bubbling in water solutions, which occurs on the anode surface during anodization. The presence of water is a key factor in ribs formation. Contrary to previous 1st and 2nd approaches, 3rd generation of TiO_2 NTs is based on organic electrolytes with the addition of F^- ions and low water content. In this type, both smooth or rippled structures can be formed depending on the water content. Below 0.5 wt% of the water threshold the NTs are smooth and no ripples are formed. The 3rd generation of TiO_2 NTs allows to manufacturing up to 1000 μm-long NTs. Morphology can be controlled by potential, water content and time in ethylene glycol solution. It is worth noting that depending on organic solvent type it is possible to space out NTs. Ozkan et al. [74] obtained spaced (SP) NTs in diethylene glycol and dimethyl sulfoxide solutions. Such a change in the morphology affects significantly optical, electrochemical and photoelectrochemical properties of the NTs. Close-packed (CP) NTs provide higher surface area and may exhibit higher electrochemical activity. However, in the case of photoelectrochemical performance, SP NTs are more efficient due to the unique morphology in which spaces between NTs act as traps for incident light. Moreover, SP NTs show faster charge transport compared to CP ones. Therefore, due to the enhanced light harvesting by free-standing NTs they might be a promising material for photoelectrochemical water splitting or solar panels.

Concerning the 4th generation of TiO_2 NTs, the classification is not completely clear. In the literature some researchers classify NTs obtained in fluoride-free electrolytes to the 4th generation [36, 100]. This unique approach is called rapid breakdown anodization (RBA) and it is performed in perchloric acid ($HClO_4$). RBA allows to obtain several dozen micrometers long NTs within few minutes. However, RBA provides irregularities in the NTs morphology and high aspect ratio

Table 6 Brief summary of TiO_2 NTs evolution

TiO_2 NTs generation	Electrolyte	Short description	SEM	Reproduced from
1st	Aqueous HF solution	Limited length to 0.5 μm Ribbed structure	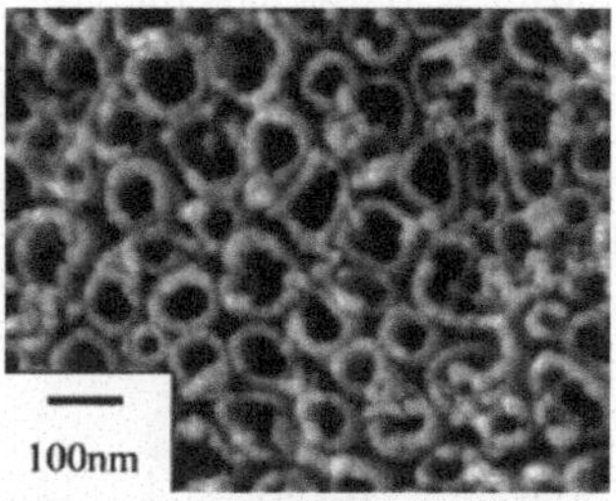	Gong et al. [27]
2nd	Buffered aqueous solutions with F^- ions	Increased length compared to 1st gen. up to 5 μm Ribbed structure	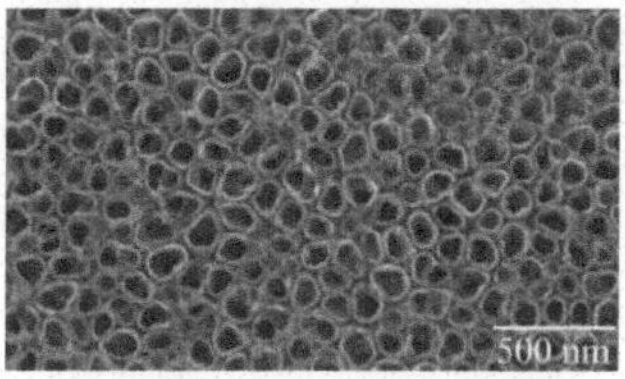	Taveira et al. [94]
3rd	Ethylene glycol (EG) solution	Increased length compared to 1st and 2nd generation up to 1000 μm Uniform and smooth layer of TiO_2 NTs	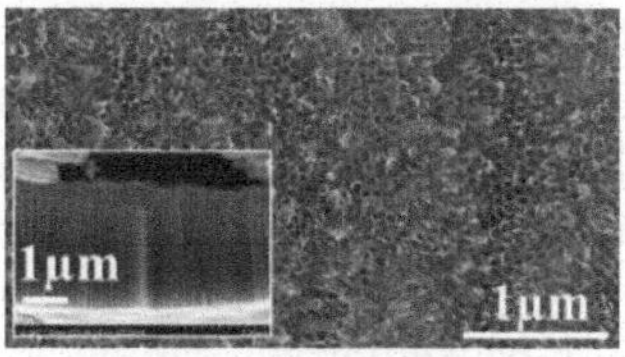	Ozkan et al. [74]
	Diethylene glycol (DEG)	Spaced NTs Not so uniform Ribs present at the bottom	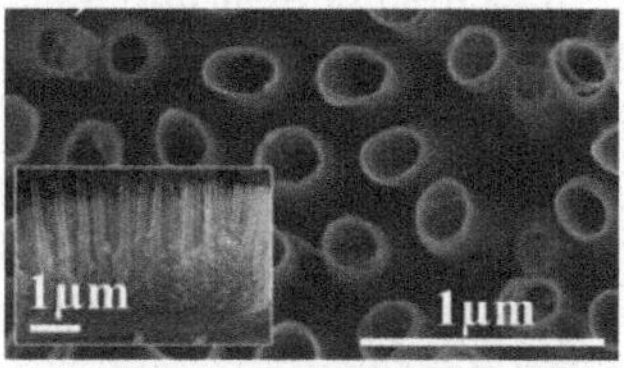	
	Dimethyl sulfoxide (DMSO)	Spaced NTs Over-etched walls (compared to DEG) Not so uniform Ribs at the bottom	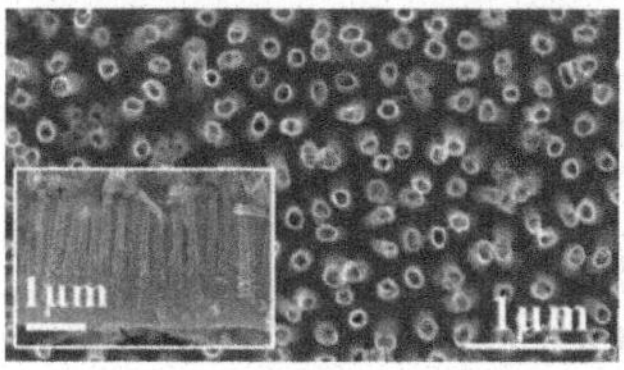	

(continued)

Table 6 (continued)

TiO_2 NTs generation	Electrolyte	Short description	SEM	Reproduced from
4th	$HClO_4$ (fluoride-free solution)	Non uniform structure High aspect ratio achieved only within local areas		Hahm et al. [30]
	EG with optimized anodization parameters	Highly ordered hexagonal structure	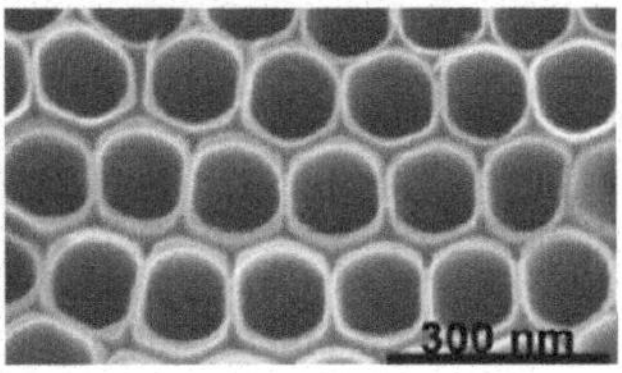	

is achieved only within bundles [30]. Therefore, RBA is rather used to manufacture TiO_2 NTs powders [3, 5]. On the other hand, there are reports classifying 4th generation of NTs to fluoride-containing organic electrolytes (similar to 3rd generation) but with a strict consideration of anodization parameters such as fluoride concentration, anodizing potential and time [80]. In this case highly ordered hexagonally close-packed arrays can be achieved by a multi-step approach, where anodization steps and etching of as-formed titania are repeated.

Summarizing, electrochemical anodization is a relatively simple technique that allows to synthesize a wide range of TiO_2 NTs structures with various morphologies. The development of the four TiO_2 NTs generations has allowed to manufacture short, long, ribbed, smooth, spaced or close-packed NTs. Such a facile control over the NTs morphology allows to elaborate almost ideal geometry for particular applications. Due to that, TiO_2 NTs have been one of the most studied materials in the last decade.

6.3 *ZnO Nanotubes*

As has been already mentioned, ZnO NTs do not have one leading manufacturing method as in the case of TiO_2 NTs. There are several approaches for ZnO NTs formation that can be divided into three groups: physical, chemical (hydrothermal) and electrochemical methods. A brief summary of ZnO NTs manufacturing

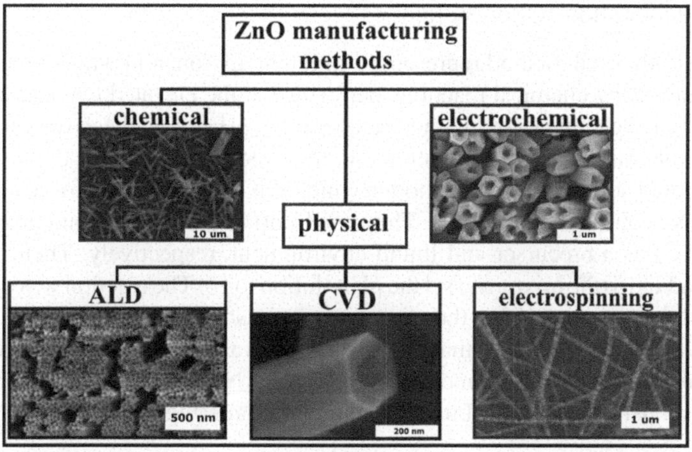

Fig. 18 The classification of ZnO NTs synthesis methods. Reproduced from Samadipakchin et al. [83], Long et al. [64], Choi and Chang [17], Zhang et al. [113, 110]

methods with a SEM micrographs of morphology is shown schematically in a diagram given in Fig. 18.

Physical methods are based on deposition of the ZnO from its precursor onto the template using a thin layer coating techniques such as atomic layer deposition (ALD), magnetron sputtering (MS) or CVD. Within the first approach core-shell structures are manufactured and the following template removal ZnO NTs remain. Zhang et al. [113] synthesized ZnO NTs via ALD on AAO templates. After the material deposition, AAO template was removed by etching in a 0.1 mol/L NaOH water solution [113]. Choi and Chang [17] obtained porous and nonporous ZnO NTs by deposition of Zn via radio frequency magnetron sputtering on the polyvinyl acetate (PVA) core nanowires obtained by the electrospinning method. The PVA nanowire core was subsequently removed by calcination in the furnace at 550 °C for an hour at the reduced pressure. Basing on another approach, Zhang et al. [110] synthesized Ag-ZnO cored NTs on Si wafers decorated with Au particles via CVD inside a quartz tube furnace. As source materials, they used the mixtures of Ag and ZnO powders. Once the temperature in the furnace reached 800 °C, Ag vaporizes and is deposited onto the Au particles. In this method, Ag forms a wired structure on the Au particles. Next, the temperature in the furnace was set to 1300 °C. Zn vaporized and following deposition on Ag-core led to Ag-ZnO NTs structures.

ZnO NTs growth can be also supported by the electrochemical deposition. Long et al. [64] proposed a simple three-step strategy to obtain ZnO/$ZnFe_2O_4$ NTs onto FTO glass. In this procedure, ZnO NTs are achieved in the following steps:

(1) Synthesis of ZnO nanorods through electrochemical deposition in $Zn(NO_3)_2$-containing electrolyte at the elevated temperature for 1.5 h,
(2) Etching of the inner part of ZnO nanorods by soaking ZnO nanorods in 0.18M KOH warm solution for an hour, washing with ultrapure water, drying

(3) Annealing at 400 °C for 30 min.

The hydrothermal method represents a different approach in which tubular structure is achieved by chemical reactions performed at the elevated temperature. Here, concentration of chemical compounds, temperature, pH and synthesis duration play an important role on nanotubes formation and their morphology. Samadipakchin et al. [83] elaborated solution-based method in which ZnO formation occurred in the series of reactions of $Zn(NO_3)_2$ and NH_3. They used $Zn(NO_3)_2 \times 6H_2O$ and polyethylene glycol (PEG) as a precursor and liquid environment, respectively. Then, NH_3 was gradually added to the mixture to set the pH solution to 10. The reaction was carried out at elevated temperature. After the reaction was finished, the residue reactants were cleaned by distilled water and ethanol. The product was finally dried in an oven at 60 ° C under air atmosphere. The morphology of ZnO NTs obtained via hydrothermal method is a bush-like and exhibits a low degree of ordering.

7 Mimetic Nanomaterials

The word mimetic comes from Greek mīmēsis, which means "to imitate", therefore mimetic materials ought to imitate nature in some way, shape or form. In nanotechnology, the term is used when material, molecule or particle looks like an object or performs a function similar to one present in natural world.

Materials visually mimicking nature are usually not the goal of the research itself. Scientists tend to pursue improved properties rather than visual appeal of new structures, nonetheless, developed nanostructures sometimes yield shapes that look oddly familiar. Thanks to their unique combination of morphology and properties, these kinds of structures have been gathering interest in both the press and scientists throughout the years.

Herein we describe selected examples of mimetic nanomaterials, their synthesis, properties and possible applications.

7.1 Nanotrees

Indium tin oxide is a transparent, conductive material, commonly used in touchscreens, light-emitting and photovoltaic devices. Chemical vapour deposition method, often used for production of ITO coatings, allows only the formation of a thin layer on the surface. However, it is widely accepted, that in order to achieve improved photovoltaic devices, high surface area is necessary. To reach this goal, nanotrees were formed using modified CVD method, called vapour-liquid-solid (VLS) combined with glancing angle physical vapour deposition (GLAD). While VLS allows for rapid adsorption of deposited material, the angle of deposition can be directed using GLAD. The combination of both techniques allows for growth of tree-like branching nanostructures. Nanotrees produced with

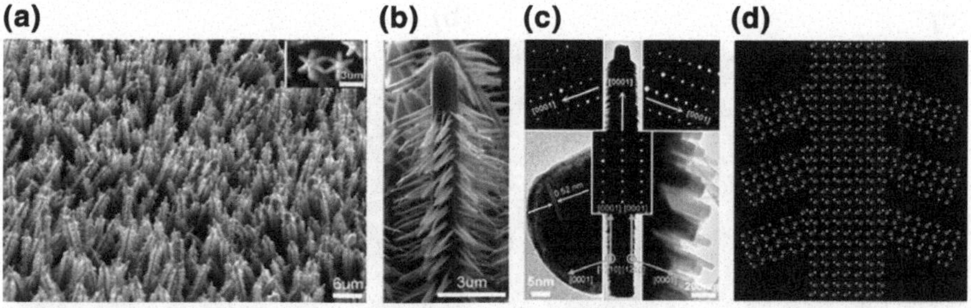

Fig. 19 **a** SEM image of typical branched ZnO nanotrees, **b** an enlargement of a side-view single nanotree, **c** TEM image of a single ZnO tree (middle) recorded along the [1210] selected-area electron diffraction patterns (top left, top right and centre) taken respectively from the branches and the trunk and lattice image (bottom left) and high-magnification TEM image (bottom right) of the branches. The green arrows indicate the major growth directions of the trunk and the branch, which share the [1210] direction, **d** a simple structural model of a branched ZnO nanotree projected along the [1210] direction. Relaxation at the interface between the trunk and the branch to achieve minimum interface energy is ignored here. Reproduced from Zhao et al. [117]

VLD-GLAD technique can be successfully applied in photovoltaic devices. Developed surface area improves photoactivity by an increasing number of excitons generated within the material, compared to devices based on flat ITO. Additionally, they can be used as a protein or gas sensing devices or UV light generators [6]. On the contrary to much demanding CVD, simple hydrothermal oxidation has been proposed for tree-like ZnO arrays formation [117]. Branched nanotrees shown in Fig. 19 were grown onto the cheap Zn supports and detailed TEM analysis confirmed wurtzite crystals growing mainly in [0001] direction. Usage ethylenediamine as a co-solvent allows for the morphology control and complexing of ZnO unique structures. What is of high importance, the proposed method is one-step, low-cost and environmentally benign to produce large scale and densely packed arrays exhibiting pure phase whereas their morphology could be easily tuned. Owing to the applied synthesis conditions, at a relatively low temperature much more complex nano-/microstructures might be obtained (see Fig. 20). The formation of developed ZnO structures is reached by modification of three main parts of nanotree: top, branches and stem. For example, the stem structure can be controlled with single, coalescent and splitting growth behaviour. Such expanded nanostructures can find application in mechanical, chemical, optical and electronic fields.

7.2 *Nanoflowers*

Nanoflowers are the most diverse family of mimetic nanoparticles, owing it to the multitude of easy ways to synthesize them and many possible applications of this kind of materials. They are usually synthesized by reducing noble metal salts and their entrapment with a surfactant in order to prevent their agglomeration and degradation.

Fig. 20 SEM images show **a** the top-view of a complex ZnO array grown at 100 °C with a concentration of 3.75 mol L^{-1} and **b–e** several unique individual ZnO crystalline trees. Reproduced from Zhao et al. [117]

Adding dopamine hydrochloride to chloroauric acid results in the rapid precipitation of gold nanoparticles. In this process dopamine acts both as reductant and shaping agent, so its amount determines shape and size of the nanoflowers (Fig. 21). These structures are particularly efficient as a platform for surface-enhanced Raman spectroscopy (SERS) and thanks to their low c-toxicity they can be used in cellular research and biomedical applications in particular [34].

Another way to create flowery, organic nanogarden is to incorporate rolling-circle replication (RCR) method to create DNA-based nanoflowers, see Fig. 22 [34]. RCR, as an isothermal enzymatic reaction, can rapidly synthesize multiple DNA or RNA molecules which are used as a building blocks for the flowers synthesis through liquid crystallization.

Flowers-like nanomaterials obtained in such way exhibit exceptional stability in biological and denaturing conditions, therefore could be recognized as promising candidates for research in biotechnology. Additionally, adsorption of molecular

Fig. 21 **a** SEM image of nanoflowers synthesized by reducing $HAuCl_4$ with dopamine. **b** Inset presents morphology of a single flower. Reproduced from Yi et al. [105]

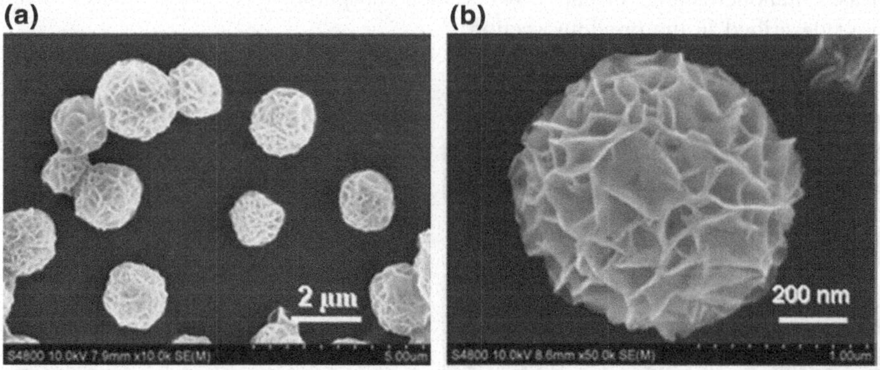

Fig. 22 DNA-based nanoflowers created with RCR method. Reproduced from Hu et al. [34]

dyes onto the surface of nanoflowers, allows them to be used as a traceable fluorescent drug carriers.

There are also reported examples of organic-inorganic flower structures (see Fig. 23). Streptavidin-horseradish peroxidase-$Cu_3(PO_4)_2$ nanoflowers have been synthesized and combined with immunosorbent assay and afterwards used for protein detection [60].

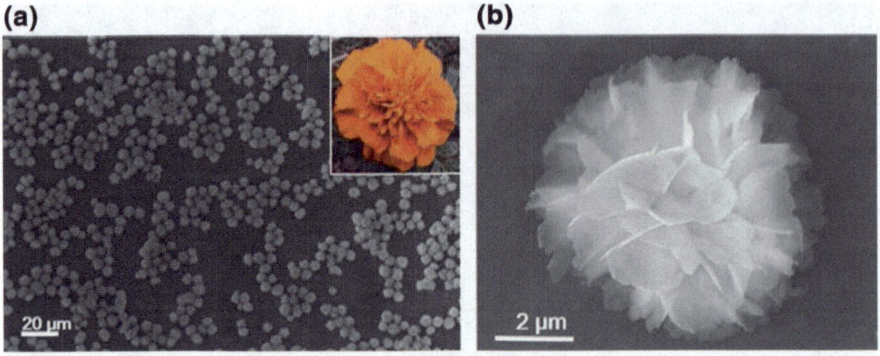

Fig. 23 Hybrid protein-inorganic nanoflowers synthesized using one-pot protocol. Reproduced from Liu et al. [60]

7.3 Nanourchins

This particular type of nanoparticles is characterized by long, thin spikes coming out of the central part of nanoparticle. Although they are sometimes called nanoviruses, nanohedgehogs or nanoflowers, their morphology is clearly different from those described in the previous section.

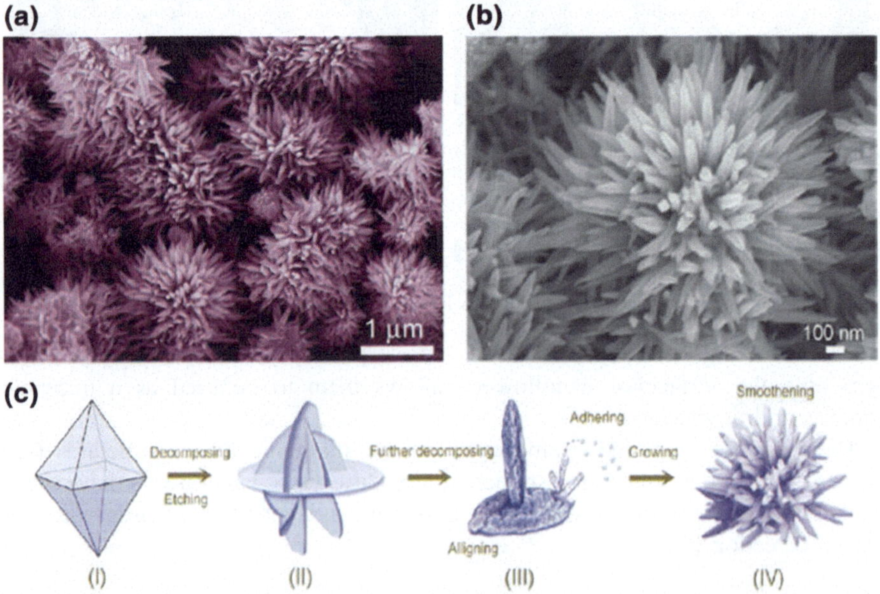

Fig. 24 **a**, **b** SEM images of urchin-like nanostructures and **c** schematic representation of the stages of their synthesis. Reproduced from Dou et al. [21]

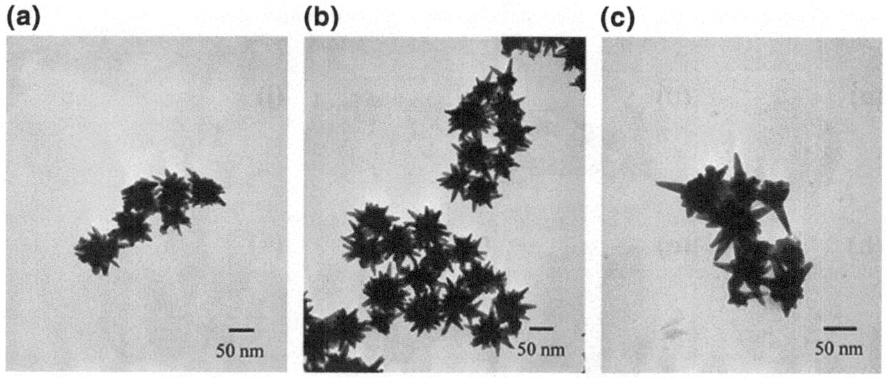

Fig. 25 AuAg nanourchin particles synthesized using different molar ratios of $HAuCl_4$: $AgNO_{3-3}$, 10, 15 for (**a**), (**b**) and (**c**) consequently. Reproduced from Cheng et al. [15]

One example of the nanourchin type of structures would be coral-like Zn doped SnO_2 nanoparticles (Fig. 24a, b) synthesized using hydrothermal method and further verified as a possible candidate for dye sensitized solar cells. As the synthesis was performed in an elevated temperature and pressure, at first $ZnSn(OH)_6$ bipyramids are formed. As the synthesis proceeds, bipyramids are then decomposed through OH^- etching and reformed as a crystalline wires springing from the etched plate left from the initial shape, see Fig. 24c [21].

Slightly different, AuAg bimetallic nanourchins could be easily synthesized from the solution of gold tetrachloride, silver nitrate and ascorbic acid (Fig. 25). In this case, ascorbic acid acts here as a reducing agent for the metal ions. When gold particles precipitate, silver adsorbs on their surface selectively modifying their further growth. As a result, nanoparticles with clearly defined pikes emerging from gold centre are obtained [15]. These kind of nanoparticles can be used as SERS active substrates as well as in biomedical applications.

8 Heterostructures

Different nanostructures, namely nanotubes, nanowires or nanoparticles were described so far. However, one may notice that their description was limited mostly to the structures composed of one specific chemical compound/single element. In here, the focus is put onto heterostructures that could be prepared from materials characterized with different properties than their counterparts. It should be here clarified also that by heterostructure we do not understand only "heterogeneous semiconductor structure built from two or more semiconductors, in such a way that the transition region or interface between the different materials plays an essential role in any device action" as was said by Herbert Kroemer during Nobel Lecture given on December 8th, 2000, but every structure composed of at least two various

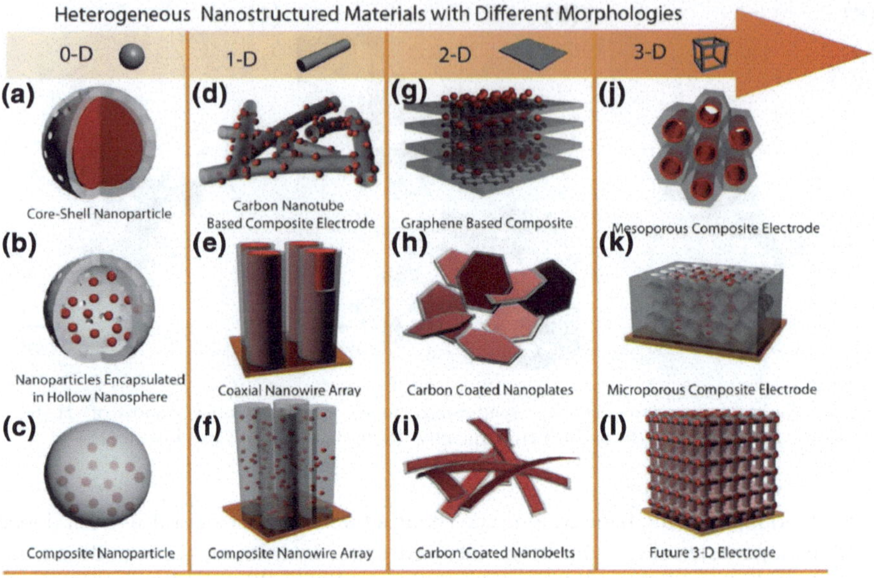

Fig. 26 Exemplary hybrid materials based on different structure configuration. Reproduced from Liu et al. [63]

materials of different physical/chemical nature. Hetero-, composite and hybrid structure will be used as synonyms.

The obvious examples of composite materials are various kinds of nanoparticles, nanowires or nanowalls. Due to the structural features they can be divided into following categories: (i) simple hybrids (e.g. quantum dots, bimetallic nanoparticles), (ii) core-shell structures and (iii) functionalized complex systems (e.g. nanoflowers). As almost all groups of such structures have been already discussed, only quantum dots will be described in this part. Nevertheless, by hybrid nanostructure one can also understand different classes of nanomaterials combined together: e.g. zero-dimensional and three-dimensional. The classification of exemplary heterostructure materials is shown in Fig. 26. As only in 2018 more than 166,000 scientific papers were published in the field of nanotechnology (https://statnano.com/news/65056), one should take into account that only few selected examples will be referred in here.

8.1 *Quantum Nanodots*

It is well known and accepted that quantum nanodots (QD) can be considered as the representative example of heterostructures. They are typically not only composed of two different materials but also QD can be regarded as semiconductors. Such

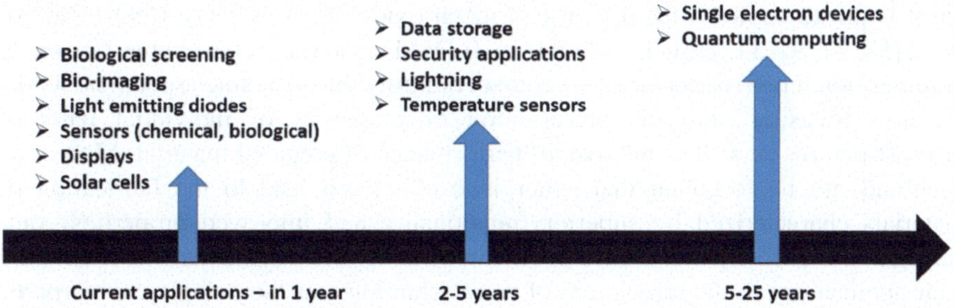

Fig. 27 Possible application of QD

nanocrystals were first discovered in the early 1980s by Alexey Ekimov and Louis Brus in glass matrix and colloidal solutions, respectively [13, 23]. Generally, QD can be divided into six groups that corresponds to the groups of periodic table of elements: (i) IV (e.g. GeSn), (ii) IV–VI (e.g. PbS, PbTe), (iii) III–V (e.g. InAs, InSb), (iv) II–VI (e.g. HgCdTe, HgSe), (v) I–VI (e.g. Ag_2S, Ag_2Se) and (vi) I–III–VI (e.g. $CuInS_2$, $CuInSe_2$) [65]. There are two usually used approaches for fabrication of QD: (i) solution synthesis, in which control over nucleation and growth of the nanoparticle can be assured and (ii) nanoscale pattering (mostly lithography based technologies) and epitaxial growth onto the particular substrate [73]. Nevertheless, it should be remembered that in the wet chemical route the hot solvents are used and therefore such a method is limited by the boiling temperature of the solution. While in the case of the latter approach, the system composed of QD and supporting platform can be hardly described as self-standing. However, for application purposes QD can be applied both in colloidal solution and in immobilized form. Due to their unique properties such as: (i) well defined composition, (ii) large bandgap tenability, (iii) bright emission response, (iv) photostability and (v) high selectivity towards targeted material, QD found applications in wide range of fields. It should be also underlined that Quantum Dots Market was valued to be 316 million USD in 2013 and is expected to reach over 5040 million USD in 2020 (https://www.alliedmarketresearch.com/quantum-dots-market). The possible applications of QD are presented in Fig. 27.

8.2 Dimensional Hybrid Nanostructures

As mentioned before zero-, one-, two- and three-dimensional structures can be combined with each other forming much more complex structures. Such combination allows for overcoming limitations of single materials. From strict economical point of view, the biggest limitation of materials is their price. For example taking into account only noble metal NPs, like Au, Ag, or Pt one can agree that they are quite expensive especially in terms of their practical applications. Therefore, the effort is put to combine them with e.g. metal oxides such as TiO_2, CeO_2, ZnO or WO_3 [58, 77, 91, 112]. Such material, i.e. 0-D/0-D material, is the simplest example of dimensional heterostructures. Of course, omitting the economic aspect, the works are also focusing onto the preservation of properties of individual parts of heterostructures as well as the overall performance of prepared material. Moreover, it should not be forgotten that synergistic effect can lead to the fabrication of materials characterized by superior functionality and improved properties comparing to the counterparts. Therefore, it can be understood that heterostructures find their applications in the same areas of nanotechnology, as their single counterparts, e.g. in photovoltaic devices, batteries, supercapacitors, sensors, surface enhanced Raman scattering platforms or fuel cell.

One of the most represented groups seems to be 0-D/2-D structures. The obvious examples of such materials are: nanoparticles located at the thin film. The most common shape of nanoparticles is spherical or nearly spherical as it is the easiest one to obtain as free energy of such system is minimal. Nevertheless, more visually interesting are nanoparticles of different shapes. For instance, Lewis et al. [57] explored triangle cobalt nanoparticles over a copper support (Fig. 28).

Even more attractive example is reported by Lai et al. [50]. The researchers studied star-shaped Au/Ag NPs deposited onto silver film as a SERS substrate (Fig. 29) proving that such structures possess improved sensing properties.

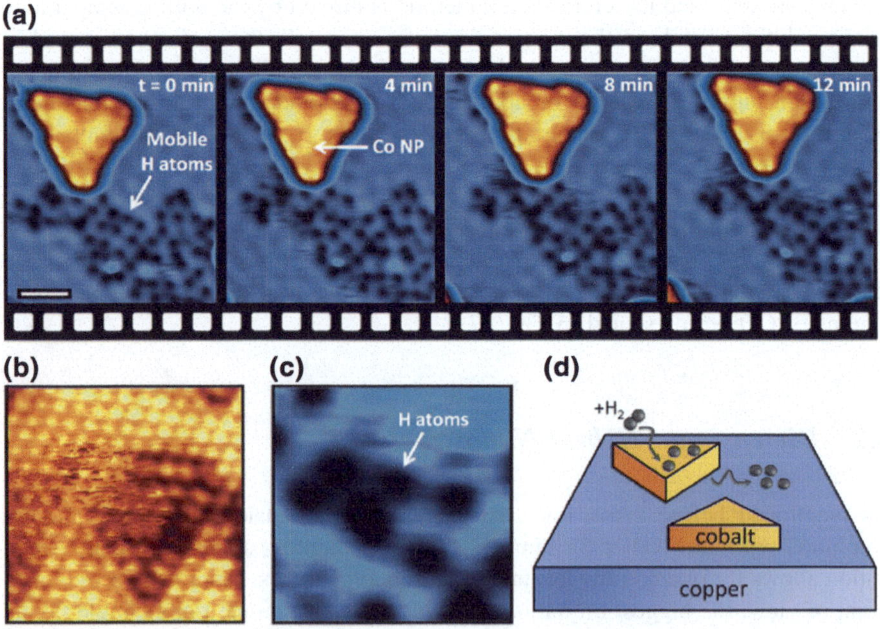

Fig. 28 Heterostructure composed of Co nanotriangles over Cu coating. Reproduced from Lewis et al. [57]

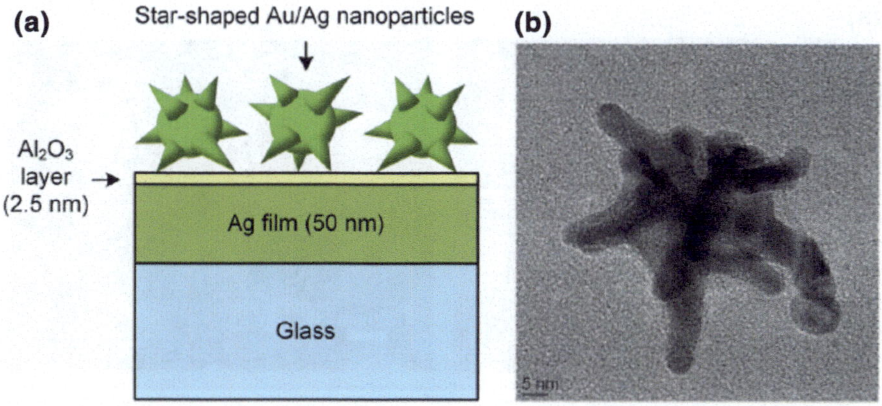

Fig. 29 **a** Schematic of star-shaped nanoparticle based SERS substrate. **b** TEM image of a single Au/Ag nanoparticle. Reproduced from Lai et al. [50]

On the other hand, less obvious configuration is the opposite one, i.e. 2-D material located onto 0-D one. Ji et al. [40] worked onto silica microcapsules composed of flake-shells. Under hydrothermal synthesis conditions, silica spherical nanoparticles slowly dissolve and those dissolved silicate species form flakes surrounding the maintained spheres—see Fig. 30. Nevertheless, it should be noted that prolonged fabrication process leads to the creation of completely hollow spheres covered with flakes that can be used as capsules for drug delivery.

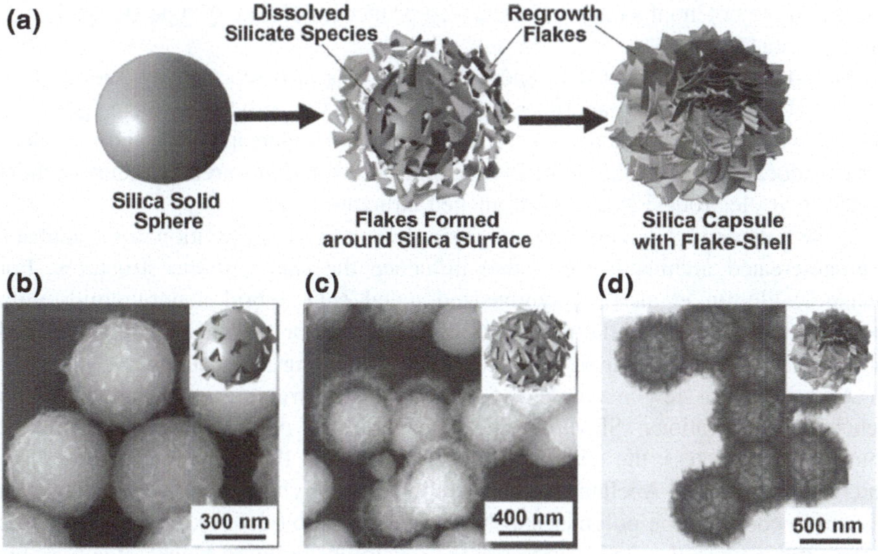

Fig. 30 **a** Scheme of the formation route of flake-shell silica structures. **b–d** SEM images of silica spheres overgrown by nanosheets. Reproduced from Ji et al. [40]

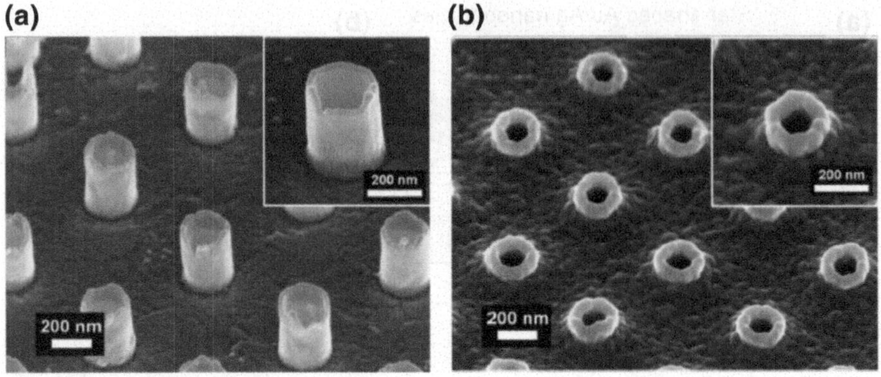

Fig. 31 SEM images of: **a** 3-D hybrid nanostructure on ITO and **b** the same nanostructure coated with polymeric material. Reproduced from Kim et al. [44]

Another rich group of hybrid materials contains 3-D structures. For example Kim et al. [46] reported on fabrication of 3-D/2-D hybrid nanostructures composed of Au particles and cylindrical patterns onto indium tin oxide substrate (Fig. 31). It is stated that surface-structuring processes do not impact onto initial properties of the support and after covering with polymeric material obtained structures could be potentially applied in photovoltaics devices.

Other interesting example of 3-D heterostructure was given by Ke et al. [42]. The authors prepared TiO_2@$Ni(OH)_2$ core-shell nanostructures onto carbon fiber paper (Fig. 32) for supercapacitors applications. The prototype device was fabricated and the potential strategy to enhanced properties of battery type electrode was also introduced.

Material that can be used in energy storage was also reported by Xiong et al. [102]. Hybrid carbon nanotubes/gold nanoparticles structures were prepared (Fig. 33) and studied as active materials for efficient supercapacitors and templates for pseudocapacitors. It should be in here underlined that, inspired by nature authors called their electrodes: leaves-on branched structure.

Quite interestingly, nature is not a single source of inspiration for scientists. Human-created architecture can also influence the shape of the structures. For instance, Flomin et al. [24] synthesized metal-ZnO hybrid nanopyramids with possibility to attach metallic nanoparticle to the base or the ZnO tip (Fig. 34) and obtained material was shown to be used as a promising nanocatalyst.

Above, just a few examples of various heterostructures were given. With schematic illustrations, SEM or TEM images, also possible applications were quoted. Nevertheless, the authors would like to finish this chapter with one more interesting material. Abellan et al. [2] reported on hybrid material consisting of cadmium coordination polymer and NiAl-layered double hydroxide (Fig. 35). It is stated in this work that "with appropriate selection of building blocks, it is possible to engineer the assembly of different nanomaterials while keeping their initial functionalities intact." This sentence is an essence of motivation behind all studies

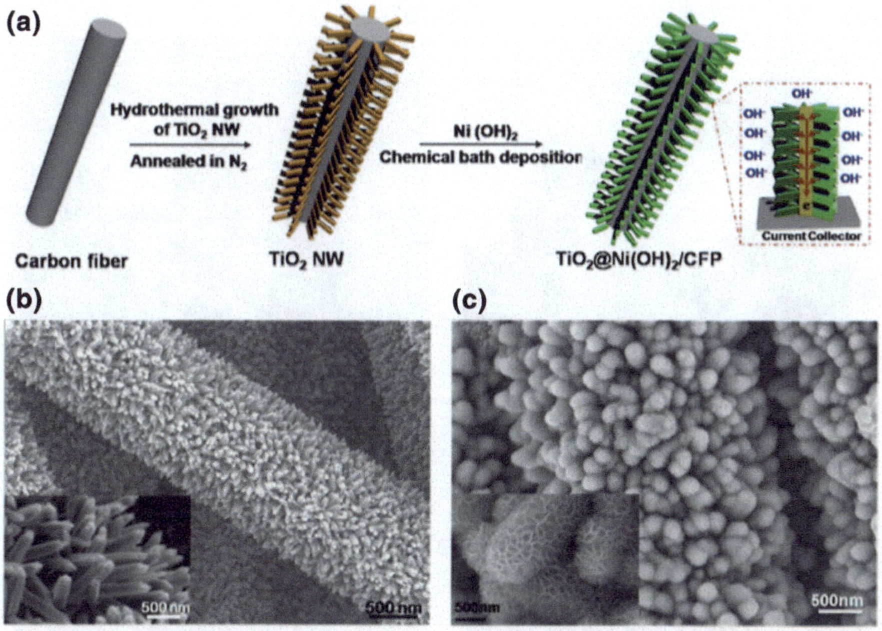

Fig. 32 **a** Scheme of fabrication process of $TiO_2@Ni(OH)_2$ core-shell nanowire arrays onto carbon fiber paper (CFP). SEM images of: **b** TiO_2 nanowire array and **c** 3-D hybrid material. Insets: magnified images. Reproduced from Ke et al. [42]

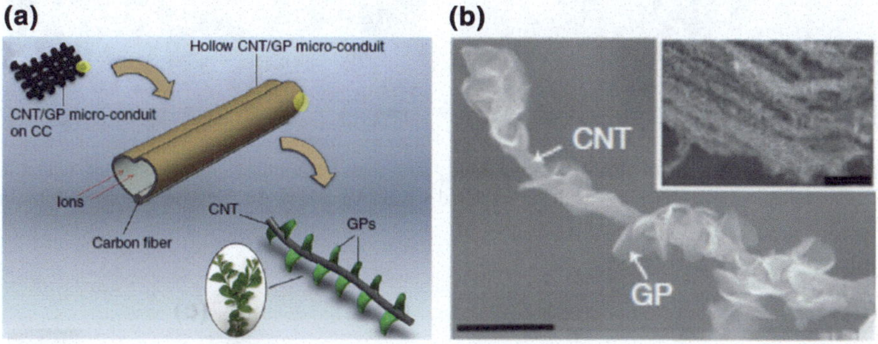

Fig. 33 **a** Schematic illustration of 3-D heteromaterial. **b** SEM image of carbon nanotubes covered with gold nanoparticles (GPs). Reproduced from Xiong et al. [102]

on hybrid nanostructures. It can be said without any doubt that this field of science has not reached its limitation. Therefore, it is expected more works to be published in the future showing materials that will outperform the present ones.

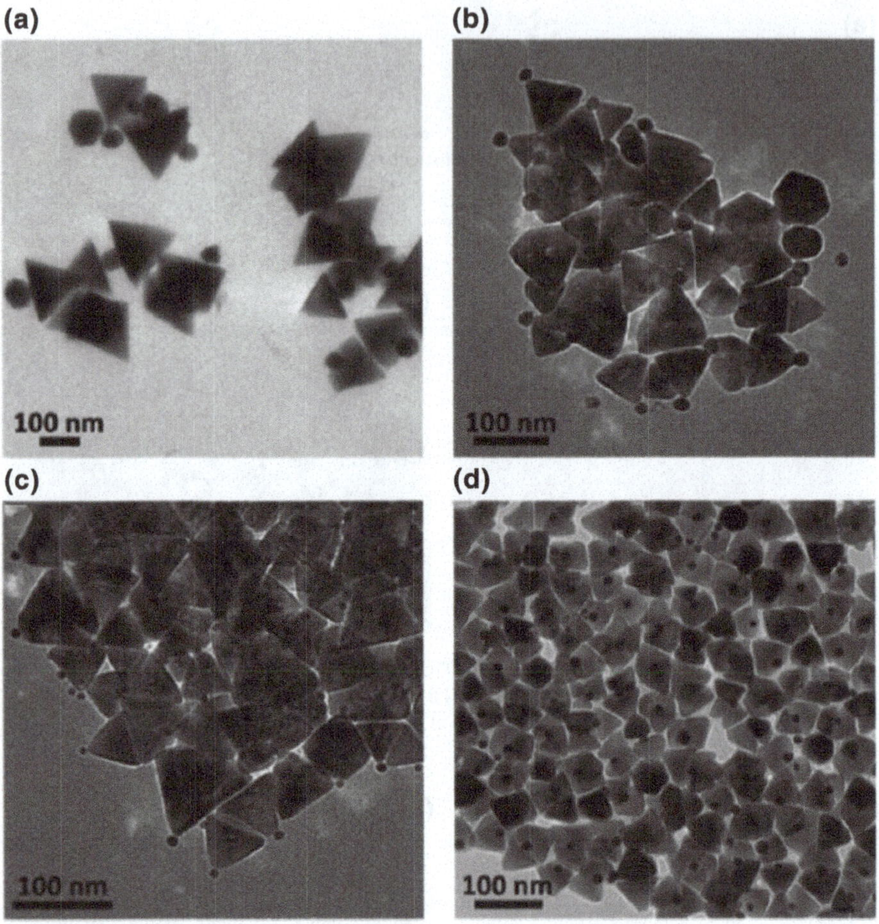

Fig. 34 TEM images of ZnO nanopyramid structures combined with different metallic nanoparticles: **a** base-attached Cu, **b** tip-attached Cu, **c** Ag and **d** Au. Reproduced from Flomin et al. [24]

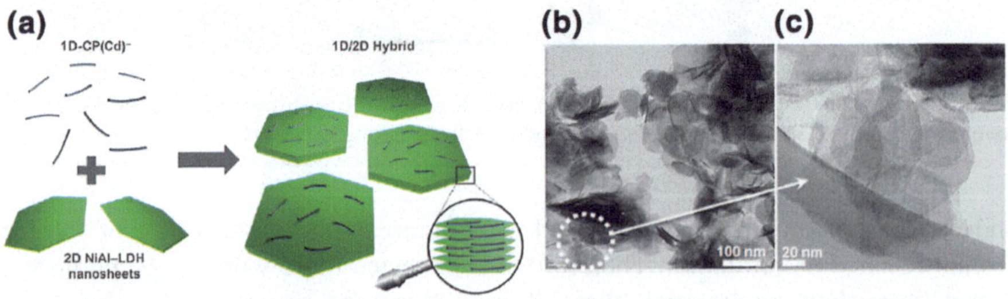

Fig. 35 **a** Illustration of 1-D/2-D hybrid material preparation and its TEM image (**b**), **c** Magnification of (**b**). Reproduced from Abellan et al. [2]

References

1. Abbasi, E., Aval, S.F., Akbarzadeh, A., Milani, M., Nasrabadi, H.T., Joo, S.W., Hanifehpour, Y., Nejati-Koshki, K., Pashaei-Asl, R.: Dendrimers: synthesis, applications, and properties. Nanoscale Res. Lett. **9**, 247–257 (2014). https://doi.org/10.1186/1556-276X-9-247
2. Abellan, G., Amo-Ochao, P., Fierro, J.L.G., Ribera, A., Coronado, E., Zamora, F.: Self-assembly of 1D/2D hybrid nanostructures consisting of a Cd(II) coordination polymer and NiAl-layered double hydroxides. Polymers **8**, 5, 13 (2016). https://doi.org/10.3390/polym8010005
3. Ali, S., Hannula, S.: Titania nanotube powders obtained by rapid breakdown anodization in perchloric acid electrolytes. J. Solid State Chem. **249**, 189–198 (2017). https://doi.org/10.1016/j.jssc.2017.03.007
4. Ariga, K., Watanabe, S., Mori, T., Takeya, J.: Soft 3D nanoarchitecures. NPG Asia Mater. **10**, 90–106 (2018). https://doi.org/10.1038/s41427-018-0022-
5. Arunchandran, C., Ramya, S., George, R.P., Kamachi Mudali, U.: Corrosion inhibitor storage and release property of TiO_2 nanotube powder synthesized by rapid breakdown anodization method. Mater. Res. Bull. **48**, 635–639 (2013). https://doi.org/10.1016/j.materresbull.2012.11.034
6. Beaudry, A.L., Tucker, R.T., LaForge, J.M., Taschuk, M.T., Brett, M.J.: Indium tin oxide nanowhisker morphology control by vapour-liquid-solid glancing angle deposition. Nanotechnology **23**, 105608 (2012). https://doi.org/10.1088/0957-4484/23/10/105608
7. Benz, F., Chikkaraddy, R., Salmon, A., Ohadi, H., de Nijs, B., Mertens, J., Carnegie, C., Bowman, R.W., Baumberg, J.J.: SERS of individual nanoparticles on a mirror: size does matter, but so does shape. J. Phys. Chem. Lett. **7**, 2264–2269 (2016). https://pubs.acs.org/doi/10.1021/acs.jpclett.6b00986
8. Berg, A., Caroff, P., Shahid, N., Lockrey, M.N., Yuan, X., Borgström, M.T., Tan, H.H., Jagadish, C.: Growth and optical properties of In_xGa_{1-x} P nanowires synthesized by selective-area epitaxy. Nano Res. **10**, 672–682 (2017). https://doi.org/10.1007/s12274-016-1325-1
9. Bhattarai, S.R., Derry, P.J., Aziz, K., Singh, P.K., Khoo, A.M., Chadha, A.S., Liopo, A., Zubarevc, E.R., Krishnan, S.: Gold nanotriangles: scale up and X-ray radiosensitization effects in mice. Nanoscale **9**, 5085–5093 (2017). https://doi.org/10.1039/c6nr08172j
10. Bigall, N.C., Härtling, T., Klose, M., Simon, P., Eng, L.M., Eychmüller, A.: Monodisperse platinum nanospheres with adjustable diameters from 10 to 100 nm: synthesis and distinct optical properties. Nano Lett. **8**, 4588–4592 (2008). https://pubs.acs.org/doi/10.1021/nl802901t
11. Biswas, S., Doherty, J., Saladukha, D., Ramasse, Q., Majumdar, D., Upmanyu, M., Singha, A., Ochalski, T., Morris, M.A., Holmes, J.D.: Non-equilibrium induction of tin in germanium: towards direct bandgap Ge1−xSnx nanowires. Nat. Commun. **7**, 11405 (2016). https://doi.org/10.1038/ncomms11405
12. Bosman, A.W., Janssen, H.M., Meijer, E.W.: About dendrimers: structure, physical properties, and applications. Chem. Rev. **99**, 1665–1688 (1999). https://pubs.acs.org/doi/10.1021/cr970069y
13. Brus, L.E.: Electron-electron and electron-hole interactions is small semiconductor crystallites: the size dependence of the lowest excited electronic state. J. Chem. Phys. **80**, 4403 (1984). https://doi.org/10.1063/1.447218
14. Chen, N., Hou, J., Lu, K.: Formation mechanism of TiO_2 nanotubes and their applications in photoelectrochemical water splitting and supercapacitors. Langmuir **29**, 5911–5919 (2013). https://doi.org/10.1021/la400586r

15. Cheng, L.-C., Huang, J.-H., Chen, H.M., Lai, T.-C., Yang, K.-Y., Liu, R.-S., Hsiao, M., Chen, C.-H., He, L.-J., Tsai, D.P.: Seedless, silver-induced synthesis of star-shaped gold/silver bimetallic nanoparticles as high efficiency photothermal therapy reagent. J. Mater. Chem. **22**, 2244–2253 (2012). https://doi.org/10.1039/c1jm13937a
16. Chia, X., Pumera, M.: Characteristics and performance of two-dimensional materials for electrocatalysis. Nat. Catal. **1**, 909–921 (2018). https://doi.org/10.1038/s41929-018-0181-7
17. Choi, K., Chang, S.: Effect of structure morphologies on hydrogen gas sensing by ZnO nanotubes. Mater. Lett. **230**, 48–52 (2018). https://doi.org/10.1016/j.matlet.2018.07.031
18. Cui, Y., van Dam, D., Mann, S.A., van Hoof, N.J.J., van Veldhoven, P.J., Garnett, E.C., Bakkers, E.P.A.M., Haverkort, J.E.M.: Boosting solar cell photovoltage via nanophotonic engineering. Nano Lett. **16**, 6467–6471 (2016). https://doi.org/10.1021/acs.nanolett.6b02971
19. Dai, L., Mo, S., Qin, Q., Zhao, X., Zheng, N.: Carbon monoxide-assisted synthesis of ultrathin PtCu 3 alloy wavy nanowires and their enhanced electrocatalysis. Small **12**, 1572–1577 (2016). https://doi.org/10.1002/smll.201502741
20. Dai, E., Xu, J., Qiu, J., Liu, S., Chen, P., Liu, Y.: Co@Carbon and Co_3O_4@Carbon nanocomposites derived from a single MOF for supercapacitors. Sci. Rep. **7**, 12588 (2017). https://doi.org/10.1038/s41598-017-12733-5
21. Dou, X., Sabba, D., Mathews, N., Wong, L.H., Lam, Y.M., Mhaisalkar, S.: Hydrothermal synthesis of high electron mobility Zn-doped SnO_2 nanoflowers as photoanode material for efficient dye-sensitized solar cells. Chem. Mater. **23**, 3938–3945 (2011). https://pubs.acs.org/doi/abs/10.1021/cm201366z
22. Ealias, A.M., Saravanakumar, M.P.: A review on the classification, characterisation, synthesis of nanoparticles and their application. IOP Conf. Ser. Mater. Sci. Eng. **263**, 032019 (2017). https://doi.org/10.1088/1757-899X/263/3/032019
23. Ekimov, A.I., Efros, Al.L., Onushchenko, A.A.: Quantum size effect in semiconductor microcrystals. Solid State Commun. **56**, 921–924 (1985). https://doi.org/10.1016/S0038-1098(85)80025-9
24. Flomin, K., Plante, I.J.-L., Moshofsky, B., Diab, M., Mokari, T.: Selective growth of metal particles on ZnO nanopyramids via a one-pot synthesis. Nanoscale **6**, 1335–1339 (2014). https://doi.org/10.1039/C3NR05661A
25. Fu, L., Kane, C.L.: Superconducting proximity effect and Majorana Fermions at the surface of a topological insulator. Phys. Rev. Lett. **100**, 096407 (2008). https://doi.org/10.1103/PhysRevLett.100.096407
26. Gangadoo, S., Stanley, D., Hughes, R.J., Moore, R.J., Chapman, J.: The synthesis and characterisation of highly stable and reproducible selenium nanoparticles. Inorg. Nano-Metal Chem. **47**, 1568–1576 (2017). https://doi.org/10.1080/24701556.2017.1357611
27. Gong, D., Grimes, C.A., Varghese, O.K., Hu, W., Singh, R.S., Chen, Z., Dickey, E.C.: Titanium oxide nanotube arrays prepared by anodic oxidation. J. Mater. Res. **16**, 3331–3334 (2001). https://doi.org/10.1557/JMR.2001.0457
28. Greeley, J., Stephens, I.E.L., Bondarenko, A.S., Johansson, T.P., Hansen, H.A., Jaramillo, T. F., Rossmeisl, J., Chorkendorff, I., Nørskov, J.K.: Alloys of platinum and early transition metals as oxygen reduction electrocatalysts. Nat. Chem. **1**, 552–556 (2009). https://doi.org/10.1038/nchem.367
29. Haggren, T., Shah, A., Autere, A., Kakko, J.-P., Dhaka, V., Kim, M., Huhtio, T., Sun, Z., Lipsanen, H.: Nanowire encapsulation with polymer for electrical isolation and enhanced optical properties. Nano Res. **10**, 2657–2666 (2017). https://doi.org/10.1007/s12274-017-1468-8
30. Hahm, R., Macak, J.M., Schmuki, P.: Rapid anodic growth of TiO_2 and WO_3 nanotubes in fluoride free electrolytes. Electrochem. Commun. **9**, 947–952 (2007). https://doi.org/10.1016/j.elecom.2006.11.037
31. Han, Z., Fina, A.: Thermal conductivity of carbon nanotubes and their polymer nanocomposites: a review. Prog. Polym. Sci. **36**, 914–944 (2011). https://doi.org/10.1016/j.progpolymsci.2010.11.004

32. Han, T., Nag, A., Mukhopadhyay, S.C., Xu, Y.: Carbon nanotubes and its gas-sensing applications: a review. Sens. Actuators, A **291**, 107–143 (2019). https://doi.org/10.1016/j.sna.2019.03.053
33. Hattori, A.N., Nakazawa, H., Nakamura, T., Tanaka, H.: Fabrication of the electric double layer transistor with (La, Pr, Ca)MnO_3 nanowall wire channel. Mod. Phys. Lett. B **32**, 1840058 (2018). https://doi.org/10.1142/S0217984918400584
34. Hu, R., Zhang, X., Zhao, Z., Zhu, G., Chen, T., Fu, T., Tan, W.: DNA nanoflowers for multiplexed cellular imaging and traceable targeted drug delivery. Angew. Chem. Int. Ed. **53**, 5821–5826 (2014). https://doi.org/10.1002/anie.201400323
35. Hu, Z., Li, S., Cheng, P., Yu, W., Li, R., Shao, X., Lin, W., Yuan, D.: N, P-co-doped carbon nanowires prepared from bacterial cellulose for supercapacitor. J. Mater. Sci. **51**, 2627–2633 (2016). https://doi.org/10.1007/s10853-015-9576-x
36. Huang, J., Zhang, K., Lai, Y.: Fabrication, modification, and emerging applications of TiO_2 nanotube arrays by electrochemical synthesis: a review. Int. J. Photoenergy **2013**, 1–19 (2013). https://doi.org/10.1155/2013/761971
37. Huang, X., Zhang, P., Lin, E., Wang, P., Mei, M., Huang, Q., Jiao, J., Zhao, Q.: Fabrication and optically pumped lasing of plasmonic nanolaser with regular ZnO/GaN nanoheterojunction array. Appl. Phys. A **123**, 605 (2017). https://doi.org/10.1007/s00339-017-1211-z
38. Ichikawa, Y., Yoshiba, S., Hirai, M., Konagai, M.: Fabrication of two-dimensional quantum well structure consisting of vertical Si nano-wall arrays. J. Electrochem. Soc. **164**, H293–H298 (2017). https://doi.org/10.1149/2.0251706jes
39. Iijima, S.: Helical microtubules of graphitic carbon. Nature **354**, 56–58 (1991). https://doi.org/10.1038/354056a0
40. Ji, Q., Guo, C., Yu, X., Ochs, C.J., Hill, J.P., Caruso, F., Nakazawa, H., Ariga, K.: Flake shell capsules: adjustable inorganic structures. Small **8**, 2345–2349 (2012). https://doi.org/10.1002/smll.201200317
41. Jiji, S.G., Gopchandran, K.G.: Virus shaped gold nanoparticles with tunable near infrared plasmon as SERS substrates. Mater. Res. Express **2**, 075005 (2015)
42. Ke, Q., Zheng, M., Liu, H., Guan, C., Mao, L., Wang, J.: 3D TiO_2@$Ni(OH)_2$ core-shell arrays with tunable nanostructure for hybrid supercapacitor application. Nat. Sci. Rep. **5**, 13940 (2015). https://doi.org/10.1038/srep13940
43. Khan, I., Saeed, K., Khan, I.: Nanoparticles: properties, applications and toxicities. Arab. J. Chem. (2017). https://doi.org/10.1016/j.arabjc.2017.05.011
44. Kim, S.-H., Shim, G.-I., Choi, S.-Y.: Fabrication of Nb-doped ZnO nanowall structure by RF magnetron sputter for enhanced gas-sensing properties. J. Alloy. Compd. **698**, 77–86 (2017). https://doi.org/10.1016/j.jallcom.2016.11.377
45. Kim, J., da Silva, W.J., bin Mohd Yusoff, A.R., Jang, J.: Organic devices based on nickel nanowires transparent electrode. Sci. Rep. **6**, 19813 (2016). https://doi.org/10.1038/srep19813
46. Kim, J.W., Jeon, H.-J., Lee, C.-L., Ahn, C.W.: Fabrication of three-dimensional hybrid nanostructure-embedded ITO and its application as a transparent electrode for high-efficiency solution processable organic photovoltaic devices. Nanoscale **9**, 3033–3039 (2017). https://doi.org/10.1039/c6nr06552
47. Kim, S.-W., Kim, K.-S., Park, M., Nah, W., Kim, D.U., Lee, C.-R., Jung, S.-B., Kim, J.-W.: 1.4 μm-thick transparent radio frequency transmission lines based on instant fusion of polyethylene terephthalate through surface of Ag nanowires. Electron. Mater. Lett. **14**, 599–609 (2018). https://doi.org/10.1007/s13391-018-0069-3
48. Koenigsmann, C., Semple, D.B., Sutter, E., Tobierre, S.E., Wong, S.S.: Ambient synthesis of high-quality ruthenium nanowires and the morphology-dependent electrocatalytic performance of platinum-decorated ruthenium nanowires and nanoparticles in the methanol oxidation reaction. ACS Appl. Mater. Interfaces **5**, 5518–5530 (2013). https://doi.org/10.1021/am4007462

49. Kumar, S., Rani, R., Dilbaghi, N., Tankeshwar, T., Kim, K.: Carbon nanotubes: a novel material for multifaceted applications in human healthcare. Chem. Soc. Rev. **46**, 158–196 (2017). https://doi.org/10.1039/C6CS00517A
50. Lai, C.H., Wang, G.-A., Ling, T.-K., Wang, T.-J., Chiu, P.-K., Chau, Y.-F.C., Huang, C.-C., Chiang, H.-P.: Near infrared surface-enhanced Raman scattering based on star-shaped gold/silver nanoparticles and hyperbolic metamaterial. Sci. Rep. **7**, 5446 (2017). https://doi.org/10.1038/s41598-017-05939-0
51. Lee, W., Schwirn, K., Steinhart, M., Pippel, E., Scholz, R., Gosels, U.: Structural engineering of nanoporous anodic aluminium oxide by pulse anodization of aluminium. Nat. Nanotechnol. **3**, 234–239 (2008). https://doi.org/10.1038/nnano.2008.54
52. Lee, H.B., Yoo, Y.M., Han, Y.-H.: Characteristic optical properties and synthesis of gold-silica core-shell colloids. Scripta Mater. **55**, 1127–1129 (2006). https://doi.org/10.1016/j.scriptamat.2006.08.044
53. Lee, K., Mazare, A., Schmuki, P.: One-dimensional titanium dioxide nanomaterials: nanotubes. Chem. Rev. **114**, 9385–9454 (2014). https://doi.org/10.1021/cr500061m
54. Lee, W., Park, S.: Porous Anodic aluminum oxide: anodization and templated synthesis of functional nanostructures. Chem. Rev. **114**, 7487–7556 (2014). https://doi.org/10.1021/cr500002z
55. Lekawa-Raus, A., Patmore, J., Kurzepa, Ł., Bulmer, J., Kozioł, K.: Electrical properties of carbon nanotube based fibers and their future use in electrical wiring. Adv. Func. Mater. **24**, 3661–3682 (2014). https://doi.org/10.1002/adfm.201303716
56. Lentz, C.M., Samuel, B.A., Foley, H.C., Haque, M.A.: Synthesis and characterization of glassy carbon nanowires. J. Nanomater. **2011**, 1–8 (2011). https://doi.org/10.1155/2011/129298
57. Lewis, E.A., Marcinkowski, M.D., Murphy, C.J., Liriano, M.L., Sykes, E.C.H.: Hydrogen dissociation, spillover, and desorption from Cu-supported Co nanoparticles. J. Phys. Chem. Lett. **5**, 3380–3385 (2014). https://doi.org/10.1021/jz5016789
58. Lin, W.-C., Yang, W.-D., Huang, I.-L., Wu, T.-S., Chung, Z.-J.: Hydrogen production from methanol/water photocatalytic decomposition using $Pt/TiO_{2-x}N_x$ catalyst. Energy Fuels **23**, 2192–2196 (2009). https://doi.org/10.1021/ef801091p
59. Liu, X., Atwater, M., Wang, J., Huo, Q.: Extinction coefficient of gold nanoparticles with different sizes and different capping ligands. Colloids Surf. B **58**, 3–7 (2007). https://doi.org/10.1016/j.scriptamat.2006.08.044
60. Liu, Y., Chen, J., Du, M., Wang, X., Ji, X., He, Z.: The preparation of du-al-functional hybrid nanoflower and its application in the ultrasensitive detection of disease-related biomarker. Biosens. Bioelectron. **92**, 68–73 (2017). https://doi.org/10.1016/j.bios.2017.02.004
61. Liu, J., Qiao, S.Z., Chen, J.S., Lou, X.W., Xing, X., Lu, G.Q.: Yolk/shell nanoparticles: new platforms for nanoreactors, drug delivery and lithium-ion batteries. Chem. Commun. **47**, 12578 (2011). https://doi.org/10.1039/C1CC13658E
62. Liu, D., Zhang, C., Yu, Y., Shi, Y., Yu, Y., Niu, Z., Zhang, B.: Hydrogen evolution activity enhancement by tuning the oxygen vacancies in self-supported mesoporous spinel oxide nanowire arrays. Nano Research **11**, 603–613 (2018). https://doi.org/10.1007/s12274-017-1670-8
63. Liu, R., Duay, J., Lee, S.B.: Heterogeneous nanostructured electrode materials for electrochemical energy storage. Chem. Commun. **47**, 1384–1404 (2011). https://doi.org/10.1039/C0CC03158E
64. Long, J., Wang, W., Fu, S., Liu, L.: Hierarchical architectures of wrinkle-like $ZnFe_2O_4$ nanosheet-enwrapped ZnO nanotube arrays for remarkably photoelectrochemical water splitting to produce hydrogen. J. Colloid Interface Sci. **536**, 408–413 (2019). https://doi.org/10.1016/j.jcis.2018.10.074
65. Lu, H., Carroll, G.M., Neale, N.R., Beard, M.C.: Infrared quantum dots: progress, challenges and opportunities. ACS Nano **13**, 939–953 (2019). https://doi.org/10.1021/acsnano.8b09815

66. Manolopoulos, D.E., Fowler, P.W.: Structural proposals for endohedral metal-fullerene complexes. Chem. Phys. Lett. **187**, 1–7 (1991). https://doi.org/10.1016/0009-2614(91)90475-O
67. Merenkov, I.S., Myshenkov, M.S., Zhukov, Y.M., Sato, Y., Frolova, T.S., Danilov, D.V., Kasatkin, I.A., Medvedev, O.S., Pushkarev, R.V., Sinitsyna, O.I., Terauchi, M., Zvereva, I. A., Kosinova, M.L., Ostrikov, K.: Orientation-controlled, low-temperature plasma growth and applications of h-BN nanosheets. Nano Res. **12**, 91–99 (2019). https://doi.org/10.1007/s12274-018-2185-7
68. Mikkelsen, A., Lundgren, E.: Surface science of free standing semiconductor nanowires. Surf. Sci. **607**, 97–2015 (2013). https://doi.org/10.1016/j.susc.2012.08.002
69. Mohamed, A.: Synthesis, characterization and applications carbon nanofibers. In: Carbon-Based Nanofillers and Their Rubber Nanocomposites, pp. 243–257 (2019). https://doi.org/10.1016/B978-0-12-813248-7.00008-0
70. Mooij, J.E., Schön, G., Shnirman, A., Fuse, T., Harmans, C.J.P.M., Rotzinger, H., Verbruggen, A.H.: Superconductor–insulator transition in nanowires and nanowire arrays. New J. Phys. **17**, 033006 (2015). https://doi.org/10.1088/1367-2630/17/3/033006
71. Mourik, V., Zuo, K., Frolov, S.M., Plissard, S.R., Bakkers, E.P.A.M., Kouwenhoven, L.P.: Signatures of Majorana Fermions in hybrid superconductor-semiconductor nanowire devices. Science **336**, 1003–1007 (2012). https://doi.org/10.1126/science.1222360
72. N. Ra. Je. Hynes, R. Sankaranarayanan, M. Kathiresan, P. Senthamaraikannan, Anish Khan, Abdullah Mohamed Asiri, Imran Khan, Synthesis, properties, and characterization of carbon nanotube-reinforced metal matrix composites, 27-Nanocarbon and its Composites (2019) 805–830, https://doi.org/10.1016/B978-0-08-102509-3.00027-4
73. Nozik, A.J., Beard, M.C., Luther, J.M., Law, M., Ellingson, R.J., Johnson, J.C.: Semiconductor quantum dots and quantum dots arrays and applications of multiple excitation generation to third-generation photovoltaic solar cells. Chem. Rev. **110**, 6873–6890 (2010). https://doi.org/10.1021/cr900289f
74. Ozkan, S., Nguyen, N.T., Mazare, A., Schmuki, P.: Optimized spacing between TiO_2 nanotubes for enhanced light harvesting and charge transfer. Chem. Electro. Chem. **5**, 1–9 (2018). https://doi.org/10.1002/celc.201801136
75. Paradise, M., Goswami, T.: Carbon nanotubes—production and industrial applications. Mater. Des. **28**, 1477–1489 (2007). https://doi.org/10.1016/j.matdes.2006.03.008
76. Pokropivny, V.V., Skorokhod, V.V.: Classification of nanostructures by dimensionality and concept of surface forms engineering in nanomaterial science. Mater. Sci. Eng. C Biomim. Mater. Sens. Syst. **27**, 990–993 (2007). https://doi.org/10.1016/j.msec.2006.09.023
77. Primo, A., Marino, T., Corma, A., Molinari, R., García, H.: Efficient visible-light photocatalytic water splitting by minute amounts of gold supported on nanoparticulate CeO_2 obtained by a biopolymer templating method. J. Am. Chem. Soc. **133**, 6930–6933 (2011). https://doi.org/10.1021/ja2011498
78. Radloff, C., Halas, N.J.: Plasmonic properties of concentric nanoshells. Nano Lett. **4**, 1323–1327 (2004). https://pubs.acs.org/doi/10.1021/nl049597x
79. Raphey, V.R., Henna, T.K., Nivitha, K.P., Mufeedha, P., Sabu, C., Pramod, K.: Advanced biomedical applications of carbon nanotube. Mater. Sci. and Eng. C **100**, 616–630 (2019). https://doi.org/10.1016/j.msec.2019.03.043
80. Regonini, D., Bowen, C.R., Jaroenworaluck, A., Stevens, R.: A review of growth mechanism, structure and crystallinity of anodized TiO_2 nanotubes. Mater. Sci. Eng. R Rep. **74**, 377–406 (2013). https://doi.org/10.1016/j.mser.2013.10.001
81. Reguera, G., McCarthy, K.D., Mehta, T., Nicoll, J.S., Tuominen, M.T., Lovley, D.R.: Extracellular electron transfer via microbial nanowires. Nature **435**, 1098 (2005). https://doi.org/10.1038/nature03661
82. Russo, P., Xiao, M., Zhou, N.Y.: Carbon nanowalls: a new material for resistive switching memory devices. Carbon **120**, 54–62 (2017). https://doi.org/10.1016/j.carbon.2017.05.004

83. Samadipakchin, P., Mortaheb, H.R., Zolfaghari, A.: ZnO nanotubes: preparation and photocatalytic performance evaluation. J. Photochem. Photobiol. A Chem. **337**, 91–99 (2017). https://doi.org/10.1016/j.jphotochem.2017.01.018
84. Sannicolo, T., Lagrange, M., Cabos, A., Celle, C., Simonato, J.-P., Bellet, D.: Metallic nanowire-based transparent electrodes for next generation flexible devices: a review. Small **12**, 6052–6075 (2016). https://doi.org/10.1002/smll.201602581
85. Seo, H., Jung, Y., Jee, S., Yang, J.M., Lee, J.: Compositionally bilayered feature of interfacial voids in a porous anodic alumina template directly formed on Si. Scripta Mater. **57**, 968–971 (2007). https://doi.org/10.1016/j.scriptamat.2007.06.067
86. Shen, W., Zhang, J., Wang, S., Du, H., Tang, Y.: Improve the performance of the quantum dot sensitized ZnO nanotube solar cells with inserting ZnS-MnS composites layers. J. Alloy. Compd. **787**, 751–758 (2019). https://doi.org/10.1016/j.jallcom.2019.02.108
87. Smith, J.T., Hang, Q., Franklin, A.D., Janes, D.B., Sands, T.D.: Highly ordered diamond and hybrid triangle-diamond patterns in porous anodic alumina thin films. Appl. Phys. Lett. **93**, 1–3 (2008). https://doi.org/10.1063/1.2957991
88. Sofiah, A.G.N., Samykano, M., Kadirgama, K., Mohan, R.V., Lah, N.A.C.: Metallic nanowires: mechanical properties—theory and experiment. Appl. Mater. Today **11**, 320–337 (2018). https://doi.org/10.1016/j.apmt.2018.03.004
89. Sulka, G.D., Kapusta-Kołodziej, J., Brzózka, A., Jaskuła, M.: Fabrication of nanoporous TiO2 by electrochemical anodization. Electrochem. Acta **55**, 4359–4367 (2010). https://doi.org/10.1016/j.electacta.2009.12.053
90. Tan, Y., Adhikari, R.Y., Malvankar, N.S., Pi, S., Ward, J.E., Woodard, T.L., Nevin, K.P., Xia, Q., Tuominen, M.T., Lovley, D.R.: Synthetic biological protein nanowires with high conductivity. Small **12**, 4481–4485 (2016). https://doi.org/10.1002/smll.201601112
91. Tanaka, A., Hashimoto, K., Kominami, H.: Visible-light-induced hydrogen and oxygen formation over Pt/Au/WO_3 photocatalyst utilizing two types of photoabsorption due to surface plasmon resonance and band-gap excitation. J. Am. Chem. Soc. **136**, 586–589 (2014). https://doi.org/10.1021/ja410230u
92. Tang, N., Wang, W., You, H., Zhai, Z., Hilario, J., Zeng, L., Zhang, L.: Morphology tuning of porous CoO nanowall towards enhanced electrochemical performance as supercapacitors electrodes. Catal. Today **330**, 240–245 (2019). https://doi.org/10.1016/j.cattod.2018.03.024
93. Tans, S.J., Devoret, M.H., Dai, H., Thess, A., Smalley, R.E., Geerligs, L.J., Dekker, C.: Individual single-wall carbon nanotubes as quantum wires. Nature **386**, 474 (1997). https://doi.org/10.1038/386474a0
94. Taveira, L.V., Macak, J.M., Tsuchiya, H., Dick, L.F.P., Schmuki, P.: Initiation and growth of self-organized TiO_2 nanotubes anodically formed in NH_4F /$(NH_4)_2SO_4$ electrolytes. J. Electrochem. Soc. **152**, B405–B410 (2005). https://doi.org/10.1149/1.2008980
95. Tegos, G.P., Demidova, T.N., Arcila-Lopez, D., Lee, H., Wharton, T., Gali, H., Hamblin, M. R.: Cationic fullerenes are effective and selective antimicrobal photosensitizers. Chem. Biol. **12**, 1127–1135 (2005)
96. Tian, M., Wang, J., Ning, W., Mallouk, T.E., Chan, M.H.W.: Surface superconductivity in thin cylindrical Bi nanowire. Nano Lett. **15**, 1487–1492 (2015). https://doi.org/10.1021/nl503398d
97. Tiwari, J.N., Tiwari, R.N., Kim, K.S.: Zero-dimensional, one-dimensional, two-dimensional and three-dimensional nanostructured materials for advanced energy devices. Prog. Mater Sci. **67**, 724–803 (2012). https://doi.org/10.1016/j.pmatsci.2011.08.003
98. van Treeck, D., Calabrese, G., Goertz, J.J.W., Kaganer, V.M., Brandt, O., Fernández-Garrido, S., Geelhaar, L.: Self-assembled formation of long, thin, and uncoalesced GaN nanowires on crystalline TiN films. Nano Res. **11**, 565–576 (2018). https://doi.org/10.1007/s12274-017-1717-x
99. van Gough, D., Juhl, A.T., Braun, P.V.: Programming structure into 3D nanomaterials. Mater. Today **12**, 28–35 (2009). https://doi.org/10.1016/S1369-7021(09)70178-6

100. Wang, Q., Huang, J., Li, H., Zhao, A.Z., Wang, Y., Zhang, K., Sun, H., Lai, Y.: Recent advances on smart TiO2 nanotube platforms for sustainable drug delivery applications. Int. J. Nanomed. **12**, 151–165 (2017). https://doi.org/10.2147/IJN.S117498
101. Wang, R., Tao, J., Yu, B., Dai, L.: Characterization of multiwalled carbon nanotube-polymethyl methacrylate composite resins as denture base materials. J. Prosthet. Dent. **111**, 318–326 (2014). https://doi.org/10.1016/j.prosdent.2013.07.017A
102. Xiong, G., He, P., Lyu, Z., Chen, T., Huang, B., Chen, L., Fisher, T.S.: Bioinspired leaves-on-branchlet hybrid carbon nanostructure for supercapacitors. Nat. Commun. **9**, 790 (2018). https://doi.org/10.1038/s41467-018-03112-3
103. Yan, J., Zhou, F.: TiO_2 nanotubes: Structure optimization for solar cells. J. Mater. Chem. **21**, 9406–9418 (2011). https://doi.org/10.1039/C1JM10274E
104. Yao, K., Zhao, C., Sun, N., Lu, W., Zhang, Y., Wang, H., Wang, J.: Freestanding CuS nanowalls: ionic liquid-assisted synthesis and prominent catalytic performance for the decomposition of ammonium perchlorate. Cryst. Eng. Commun. **19**, 5048–5057 (2017). https://doi.org/10.1039/C7CE01119A
105. Yi, S., Sun, L., Lenaghan, S.C., Wang, Y., Chong, X., Zhang, Z., Zhang, M.: One-step synthesis of dendritic gold nanoflowers with high surface-enhanced Raman scattering (SERS) properties. RSC Adv. **3**, 10139–10144 (2013). https://doi.org/10.1039/C3RA40716K
106. Yin, W.-J., Gong, X.-G., Wei, S.-H.: Origin of the unusually large band-gap bowing and the breakdown of the band-edge distribution rule in the SnxGe1-x alloys. Phys. Rev. B **78**, 161203 (2008). https://doi.org/10.1103/PhysRevB.78.161203
107. Yu, X., Yu, X., Zhang, J., Chen, L., Long, Y., Zhang, D.: Optical properties of conductive silver-nanowire films with different nanowire lengths. Nano Res. **10**, 3706–3714 (2017). https://doi.org/10.1007/s12274-017-1583-6
108. Yun, J., Lee, M., Jeon, Y., Kim, M., Kim, Y., Lim, D., Kim, S.: Nanowatt power operation of silicon nanowire NAND logic gates on bendable substrates. Nano Res. **9**, 3656–3662 (2016). https://doi.org/10.1007/s12274-016-1235-2
109. Zhai, H., Li, Y., Chen, L., Wang, X., Shi, L., Wang, R., Sun, J.: Semi-transparent polymer solar cells with all-copper nanowire electrodes. Nano Res. **11**, 1956–1966 (2018). https://doi.org/10.1007/s12274-017-1812-z
110. Zhang, K., Cai, Y., Yao, C., Wen, X., Han, Y., Yin, H., Sun, W., Li, Q.: Structures, growth mechanism and optical properties of Ag-ZnO cored nanotubes. Mater. Lett. **234**, 191–195 (2019). https://doi.org/10.1016/j.matlet.2018.09.100
111. Zhang, K., Feng, Y., Wang, F., Yang, Z., Wang, J.: Two dimensional hexagonal boron nitride (2D-hBN): synthesis, properties and applications. J. Mater. Chem. C **5**, 11992–12022 (2017). https://doi.org/10.1039/C7TC04300G
112. Zhang, X., Liu, Y., Kang, Z.: 3D branched ZnO nanowire arrays decorated with plasmonic Au nanoparticles for high-performance photoelectrochemical water splitting. ACS Appl. Mater. Interfaces **6**, 4480–4489 (2014). https://doi.org/10.1021/am500234v
113. Zhang, Y., Liu, M., Ren, W., Ye, Z.: Well-ordered ZnO nanotube arrays and networks grown by atomic layer deposition. Appl. Surf. Sci. **340**, 120–125 (2015). https://doi.org/10.1016/j.apsusc.2015.02.176
114. Zhang, C., Zheng, C., Zhou, S., Shen, Y., Zuo, C.: Enhanced electrochromic porous cobalt oxides nanowall electrodes: a new way for fast modulation of yellow-brown light. Mater. Res. Bull. **89**, 204–209 (2017). https://doi.org/10.1016/j.materresbull.2017.01.037
115. Zhang, L.L., Zhao, X.S.: Carbon-based materials as supercapacitor electrodes. Chem. Soc. Rev. **38**, 2520–2531 (2009). https://doi.org/10.1039/B813846J
116. Zhao, W., Du, N., Xiao, C., Wu, H., Zhang, H., Yang, D.: Large-scale synthesis of Ag–Si core–shell nanowall arrays as high-performance anode materials of Li-ion batteries. J. Mater. Chem. A. **2**, 13949–13954 (2014). https://doi.org/10.1039/C4TA03238A
117. Zhao, F., Zheng, J.G., Yang, X., Li, X., Wang, J., Zhao, F., Wong, K.S., Liang, C., Wu, M.: Complex ZnO nanotree arrays with tunable top, stem and branch structures. Nanoscale **2**, 1674–1683 (2010). https://doi.org/10.1039/C0NR00076K

118. Zhou, M., (David) Lou, X.W., Xie, Y.: Two-dimensional nanosheets for photoelectrochemical water splitting: Possibilities and opportunities. Nano Today **8**, 598–618 (2013). https://doi.org/10.1016/j.nantod.2013.12.002
119. Zhu, J., Sharma, Y.K., Zeng, Z., Zhang, X., Srinivasan, M., Mhaisalkar, S., Zhang, H., Hng, H.H., Yan, Q.: Cobalt oxide nanowall arrays on reduced graphene oxide sheets with controlled phase, grain size, and porosity for Li-ion battery electrodes. J. Phys. Chem. C **115**, 8400–8406 (2011). https://doi.org/10.1021/jp2002113
120. Zulhairum, A.K., Abdullah, M.S., Ismail, A.F., Goh, P.S.: Graphene and CNT technology. In: Current Trends and Future Developments on (Bio-)Membranes, pp. 3–26 (2019). https://doi.org/10.1016/B978-0-12-813551-8.00001-2
121. http://www.software3d.com/Stella.php, Distributed under GNU General Public License, Created using Stella Software; https://statnano.com/news/65056. Accessed 07 May 2019
122. https://www.alliedmarketresearch.com/quantum-dots-market. Accessed 07 May 2019

Application of Self-supported Materials for Photo and Photoelectrocatalysis

Mohd Nur Ikhmal Salehmin, Mohamad Azuwa Mohamed, Rosmahani Mohd Shah, Rozan Mohamad Yunus and Zul Adlan Mohd Hir

Abstract Self-supported materials or often called as self-standing, free-standing, and self-assembly materials is an emerging class of materials that have been applied in various application. Owing to a unique structured or sometimes known as hierarchical structures have improved the specific properties of the materials. In the case of catalysis application, the self-supported catalyst has significantly enhanced the specific chemical reaction due to the ability to provide more reaction active site. In this chapter, the application of self-supported materials for photo and photoelectrolysis application is highlighted and discussed. The various promising fabrication self-supported photocatalyst and photoelectrode is introduced and discussed. On the of that, the various potential application of self-supported materials in the field of photocatalysis technology such as water and wastewater treatment, hydrogen production, and carbon dioxide reduction are discussed in detail.

Keywords Self-supported · Self-standing · Free-standing · Self-assembly · Photocatalysis · Photoelectrocatalysis

M. N. I. Salehmin · R. Mohamad Yunus
Fuel Cell Institute, Universiti Kebangsaan Malaysia, 43600 UKM Bangi, Selangor, Malaysia

M. A. Mohamed (✉)
Centre for Advanced Materials and Renewable Resources (CAMARR), Faculty of Science and Technology, Universiti Kebangsaan Malaysia, 43600 UKM Bangi, Selangor, Malaysia
e-mail: mazuwa@ukm.edu.my

R. Mohd Shah
Research Center for Sustainable Process Technology (CESPRO), Faculty of Engineering and Built Environment, Universiti Kebangsaan Malaysia, 43600 UKM Bangi, Selangor, Malaysia

Z. A. Mohd Hir
Faculty of Applied Sciences, Universiti Teknologi MARA, 26400 Bandar Tun Abdul Razak Jengka, Pahang, Malaysia

Inamuddin et al. (eds.), *Self-standing Substrates*,
Engineering Materials, https://doi.org/10.1007/978-3-030-29522-6_2

1 Introduction

The growing concern on the efficiency of photocatalytic materials for a different type of applications has made the scientists indulge more into research and bringing up some room for improvement. Most reported work tackles the fabrication and modification of the photocatalytic materials with a combination of two or several kinds of materials for enhancement. The strategies are to produce superior catalytic photo-efficiency, outstanding chemical, and thermal stability, environmentally friendly, cost-effective and presents a compatible band gap energy for a long lifetime of electron-hole pairs. Among these features, the spacing between two band orbitals (conduction and valence bands) is a very crucial characteristic relative to applicability of the materials in both photocatalysis and photoelectrocatalysis. Recently, the fabrication of the photocatalytic materials through organized nanostructures self-supported materials for photocatalytic improvement have attracted reasonable consideration as they revolutionized many different fields [21]. By understanding the mechanisms of materials construction, it can allow scientists to design and develop the desired structure, morphology, porosity, and overall physicochemical characteristics of the materials with the aim to boost the photo-stability and photo-efficiency.

In materials science, self-supported materials are described as the hierarchical porous materials constructed via free-standing or self-assembly interconnected open-framework either in 1D, 2D or 3D structures that could unveil more active sites, offer beneficial mass transfer conduits and better interaction at the interior surface [2, 10, 11, 21]. Progresses in self-supported chemistry are always followed by the new growth of molecular structures with uniform and tunable nanoarchitecture as the unique materials are characterized by their specific recipes for controlling the fabrication of different kind of nanostructures [18]. Porous materials with several structure sizes, such as micro-, meso-, and macroporous can perform different roles due to their varying pore size which also results in improved specific surface areas and the manifestation of versatile materials [17]. The versatility of this material could be used as exceptional catalytic supports, super-active photoelectrodes, and the avoidance of catalytic deactivation. However, to fabricate simple and ligand-free self-supported materials at the microscopic level is still a considerable challenge.

Therefore, in this chapter, several fabrication methods are introduced for the construction of self-supported photocatalytic materials such as electrodeposition, vacuum deposition, hydrothermal and solvothermal, dip-coating infiltration, sol-gel, and sputtering. The brief explanation of the fabrication methods and its basic principles are also discussed with respect to their improved physicochemical characteristics and performances. Different kind of self-supported materials would vary in terms of its mechanism of formation, the interaction between the interacting media and instruments which produced different shapes ranging from wires, cubic, flower, cocoon, and rod-like structures. The recent applications of this material are also briefly reviewed especially as efficient photocatalyst or photoelectrode for wastewater treatment, PEC water-splitting for hydrogen evolution, and CO_2 reduction.

2 Fabrication and Modification of Self-support Nanostructure Photocatalyst/Photoelectrode

The most common method to synthesize self-supported nanostructure is to produce materials which have an ability to forms as a free-standing material either in one-dimensional (1D), two-dimensional (2D) or three-dimensional (3D) hierarchical structures with or without a template. It can be synthesized via various methods, such as electrochemical deposition, chemical vapor deposition, hydrothermal/ solvothermal method, sol-gel method, sputtering, and thermal solid-state reaction.

On the other hand, several strategies have been adopted in the modification of self-supported nanostructure photocatalyst/photoelectrode to improve photocatalytic performance such as metal doping, non-metal doping, loading with a noble metal, and graphene derivative coupling. In this section, the fabrication methods, as well as their modification on the self-support nanostructure, are summarized.

2.1 Electrodeposition

Electrochemical deposition (ECD) and electrodeposition (ED) are an example for coating photoelectrode in low temperature, controllable, low cost and more interestingly the simple procedure which make these methods are selected either in small or large scales in various applications [23].

Xu et al. prepared highly ordered Cu micro-cone arrays (Cu-MCA) electrodeposited with Cu_2O nanostructure as the photocathode for photoelectrochemical water splitting [53]. It was reported the 3D Cu_2O/Cu/MCA structure has higher photocurrent as compared to Cu_2/Cu photoelectrode. The detailed morphology of the photoelectrodes can be seen in the SEM images. Figure 1a shows a rough surface of Cu cone, which was investigated to be {111} orientation. The electrodeposited Cu_2O grains were about 500 nm to 1 μm in size with a cubic morphology and exposed {100} facets. Each Cu_2O exposed more facet when grew on the Cu-MCA, which expanded the surface range leading to the 2.17 times higher in photocurrent as compared to which grown on the Cu foil (Fig. 1b). In this study, the bare Cu_2O/Cu-MCA was further covered with a thin and uniform protection layer by atomic layer deposition (ALD) to conserve the material from photo-corrosion (Fig. 1d).

With the same method, pure ZnO and ZnO coated with TiO_2 have been successfully fabricated as a three-dimensional (3D) self-supporting material and

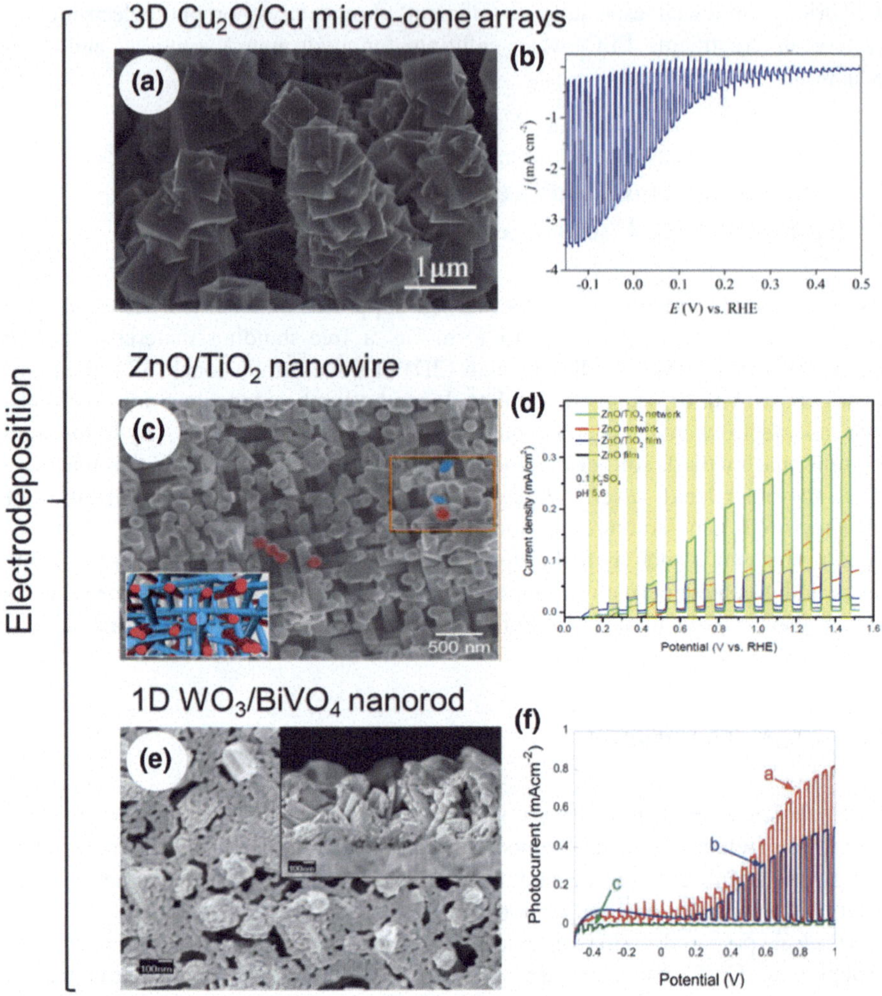

Fig. 1 Electrodeposition method to prepare various type of nanostructures. **a** Scanning electron micrographs of the bare Cu_2O/Cu-MCA. **b** The j-V plots tested from positive to negative potential range for the bare Cu_2O/Cu-*MCA* (*Reprinted with permission from* [53]. *Copyright 2016 The Electrochemical Society*). **c** High magnification SEM image nanowire network of ZnO/TiO_2, The inset in (**c**) shows a 3D illustration of the top view of the network with wires tilted (blue) and perpendicular (red) with respect to the substrate. **d** Linear sweep voltammograms of the bare ZnO, and TiO_2-coated ZnO nanowire networks and films (*Reprinted with permission from* [27]. *Copyright 2018 MDPI*). **e** Top and cross-sectional (inset) views of WO_3/$BiVO_4$ nanorod heterojunction films. **f** Current potential (with respect to normal hydrogen electrode NHE scale) plots (*Reprinted with permission from* [44]. *Copyright 2011 American Chemical Society*)

functionalized to be applied as photoanode for photoelectrochemical application [27]. The synthesis of 3D hierarchical structure provides the interconnection between different directions of wire, which improve the stability of the material. The 3D self-supporting material was further coated with TiO_2 to improve the stability during measurement. The homogeneous growth is shown in Fig. 1c over a $\sim 100 \times 120$ μm area while in the inset, tilted and vertical wires are marked in blue and red, respectively. The advantage of this network structure is that the overlap of wires is mainly limited to junctions, keeping the one-dimensionality of the structure. Figure 1d presents the current density of both 3D nanowire network samples (ZnO and ZnO/TiO_2). By raising the potential from ~ 0.1 and ~ 0.35 mA/cm^2 at 1.5 V versus RHE, the higher photocurrents were achieved. Further, the TiO_2 layer also helps to enhance the photocurrent and act as the corrosion preservation for the nanowire network and film. The mechanism can be explained by the additional electron-hole pairs generates from ZnO in the n-type TiO_2, reducing the electron-hole recombination where the contribution is from the ZnO–TiO_2 interface which improves charge separation, thereby reducing the probability of electron-hole recombination.

Su et al. introduced the heterojunction photoanode of WO_3/$BiVO_4$ for photoelectrochemical measurement by demonstrated the nanostructure coordination [44]. However, the resistive defeat in the structure is due to the defect in WO_3-NWs lead to the low photocurrent, which less than 1 mA cm^{-2}. Therefore, the demand to produce more conductive, high aspect ratio, and uniform free standing WO_3 nanostructures are important to boost the efficiency of WO_3/$BiVO_4$ heterojunction photoanodes. Meanwhile, Kitamori et al. demonstrated the fabrication of WO_3-NRs prepared by Glancing Angle Deposition (GLAD) and ensuing Electrochemical Deposition (ED) of $BiVO_4$ and CoPi. Scanning electron microscopy (SEM) studies of the developed WO_3-NRs and WO_3-NRs/$BiVO_4$ reveals the nanorods structure in Fig. 1e are well detached and homogenous with the length of 2.5 μm and the average diameter ranging from 200 to 300 nm along the nanorods. In the study, the photocurrent density, as shown in Fig. 1f increased from 0.5 to 5.45 mA cm^{-2}. The suitable band gap of 2.4 eV of bismuth vanadate ($BiVO_4$) indicates that material is relevant for water splitting, but due to poor electron transport properties, cobalt-phosphate (CoPi) was used as co-catalysts to enhance the kinetics of oxygen evolution reaction (OER). This study also reported the optimization of WO_3-NRs/$BiVO_4$ + CoPi photoanode recorded the highest photocurrent of 6.72 mA cm^{-2} at 1.23 V RHE, which approached 90% of the theoretical limit value.

2.2 *Chemical Vapor Deposition*

Chemical vapor deposition (CVD) is a method in which gas molecules are heated and decomposed on substrates growing as a thin film (so-called the template-assisted method). There are numerous types of CVD method which include thermal CVD, plasma CVD with the hot or cold wall. The method selection is depending on the required structure to obtain, material quality, and also the type of precursor involved [31]. CVD process is a standard method to coat thin film involving metals, insulator, and semiconductor. For the first time it was introduced by Li et al. by established the synthesis of single-layer graphene on Cu foil with methane and hydrogen gases flow at 1000 °C, and the demonstrated method has been successfully synthesized the single-crystalline graphene. It has been proven that CVD method is suitable for the growth of many other nanosheets such as h-BN [41, 43], silicine [46], insulators (e.g., In_2Se_3 and Bi_2Se_3) [32, 41] and metal carbides [8].

Warwick et al. successfully reported single-phase hematite (α-Fe_2O_3) nanopyramid which deposited on FTO substrates via single-step PE-CVD at temperatures of 300 and 400 °C, followed by thermal treatment in air at 650 °C. As shown in Fig. 2a, the porous morphology of dendrites grown vertically aligned with respect to the substrate. Such nanostructures may facilitate charge carrier and enhance the rate of charge transport, consequently allowing photogenerated holes to reach the interface with the electrolyte [17] efficiently. The presence of similar vertical arrays enhanced the light harvesting properties of the nanomaterial by introducing for light trapping and scattering phenomena [29]. Taking into account the importance of absorption efficiency, the factor may be a good criterion for photoelectrochemical performance. The photoefficiency as in Fig. 2b for a sample of 400 °C was read at $j \approx 680\ \mu A\ cm^{-2}$ at 1.50 V is higher than bare Fe_2O_3 and also higher than doped or functionalized Fe_2O_3.

Flat dense structures of WO_3 at 325 °C and nano-needle structured WO_3 at 375 °C using an aerosol-assisted chemical vapor deposition (AACVD) method has been reported by Kafiza et al. where the effect of increasing precursor volume is produced a series of the flat samples (F $\sim$ 300 nm) with nanoneedles grown on top (N $\sim$ 2.3, 4.6 and 6.2 μm) (Fig. 2c). As a result, both the length and width of the nano-needles also increased when the volume of precursor increased. Meanwhile, at the higher deposition volumes, the tips of nano-needles start to merge. This nanoneedle structure shows the photocurrent of 1.24 mA $cm^{-2,}$ which is $\sim$25% of the theoretical limit. This structure has five times higher photocatalytic currents than what was observed in flat materials, as shown in Fig. 2d.

Ma et al. synthesized vertical graphene (VG) nanosheets by a one-step copper-assisted plasma-enhanced chemical vapor deposition (PECVD) which successfully grown at a temperature below 500 °C on glass substrates [22]. The advantage of low operating temperature is the potential to be extended for industrial scale applications due to the simple process yet cost effectiveness. By varying the growth times from 1 to 12 min, the growth of VG films on glass substrates is studied, as shown in Fig. 2e. A thin layer is grown on the glass surface within the first 1 min, but no obvious VG nanosheets can be observed. The thin layer act as a buffer layer, to create a connection between substrate and VG nanosheets which confirmed by a scratch made by tweezers, as shown in the inset. When growth time reached 4 min, a large amount of VG nanosheets start to emerge on the buffer layer

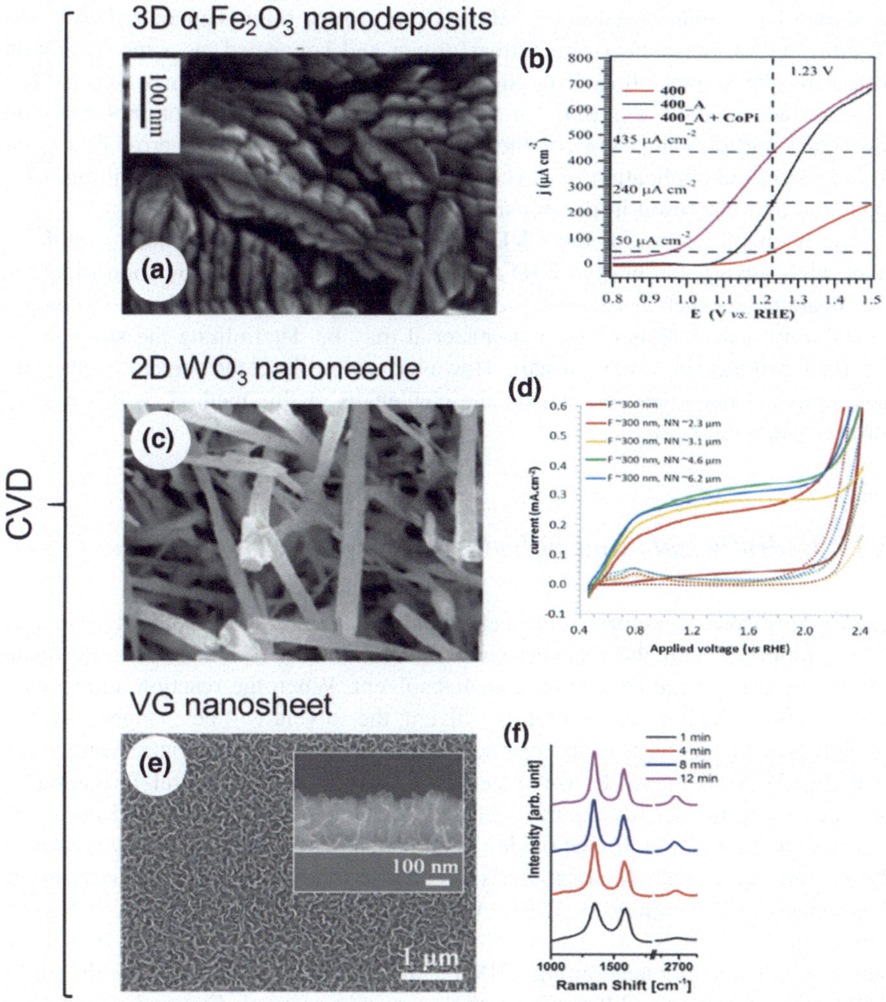

Fig. 2 Chemical vapor deposition method to prepare various type of nanostructures. **a** FE-SEM images for α-Fe_2O_3 nanodeposits grown at 400 °C, before thermal treatment. **b** Photocurrent for as-prepared and annealed Fe_2O_3 specimens deposited at 400 °C eventually functionalized with Co-Pi cocatalyst (*Reprinted with permission from* [50]. *Copyright 2015 American Chemical Society*). **c** SEM images of select WO_3 photoanodes at and 50,000 magnification with average width about 1.4 μm. **d** A series of nano-needle photoanodes with constant flat layer thickness beneath (*Reprinted with permission from* [12]. *Copyright 2017 American Chemical Society*) **e** SEM of VG on glass substrate and with copper catalyst. **f** Raman spectra of VG nanosheets grown with assistance of a copper catalyst for different growth times (*Reprinted with permission from* [22]. *Copyright 2015 Springer*)

as shown by Raman spectra (Fig. 2f). As the growth time is increased to 8 and 12 min, the VG nanosheets are further grown and connected, as a result, is with increasing the growth time, it is affecting the height of the VG nanosheets. The large surface area of VG films in the vertical systems can accommodate more functional particles thus the conductivity, and optical properties provide a good advantage in the application of developing photoelectrode for water splitting, CO_2 reduction and also used in dye-sensitized solar cells.

The main preference of the CVD process is the ability to produce a uniform layer, high-quality and purity of 2D nanomaterials. Interestingly, by controlling the CVD parameters such as catalyst, precursor and temperature, different morphology, crystallinity, and defects of the nanomaterial may be determining the structure of the final product in CVD growth. However, controlling the stoichiometry and defects in 2D nanomaterials remain the challenges in this method, which require further optimization studies.

2.3 Hydrothermal and Solvothermal

The hydro/solvothermal process is a chemical route synthesis in a closed system by using autoclave with the reaction temperature between room temperature up to 240 °C by using aqueous or non-aqueous solvent. When the reaction temperature reached above the boiling point of the solvent, the solvent will be changed to vapor in high pressure to start up the reaction and crystallize the materials. More interestingly, the way to synthesize materials is similar to the natural formation of precious crystals such as quartz crystal which formed in our earth. Lina et al. reported the fabrication of zinc oxide (ZnO) nanostructured films directly grown on the Zn foil via hydrothermal synthesis to enhance the photocatalytic activity in the degradation of Rhodamine B (RhB) under UV light irradiation [20]. By fixing the temperature to 150 °C for 2 h reaction time, using the different type of solvent which is hexamethylenetetramine (HMT), methylamine or the mixed solution of KOH and zinc nitrate, different morphology was observed. Figure 3a shows the SEM image of the porous nanosheet obtained from the HMT, aligned nanotube arrays in methylamine and the nanoflower film were produced from a mixed solution of KOH and zinc nitrate. Among all, the porous nanosheet- based ZnO film has been recognized by having a large surface range leads to the higher contact area of ZnO with RhB and UV light absorption. Therefore, it increased the photocatalytic activity for the degradation of dyes, as shown in Fig. 3b.

In other works, via a facile self-assembly method, continued with hydrothermal using autoclave, Jiang et al. successfully fabricated self-supported oxygen-doped carbon nitride aerogel, without using any harmful solvents or cross-linking agents [10]. This environmentally friendly method was varying reaction time and fixed the temperature in order to observe the changes in the morphology of the

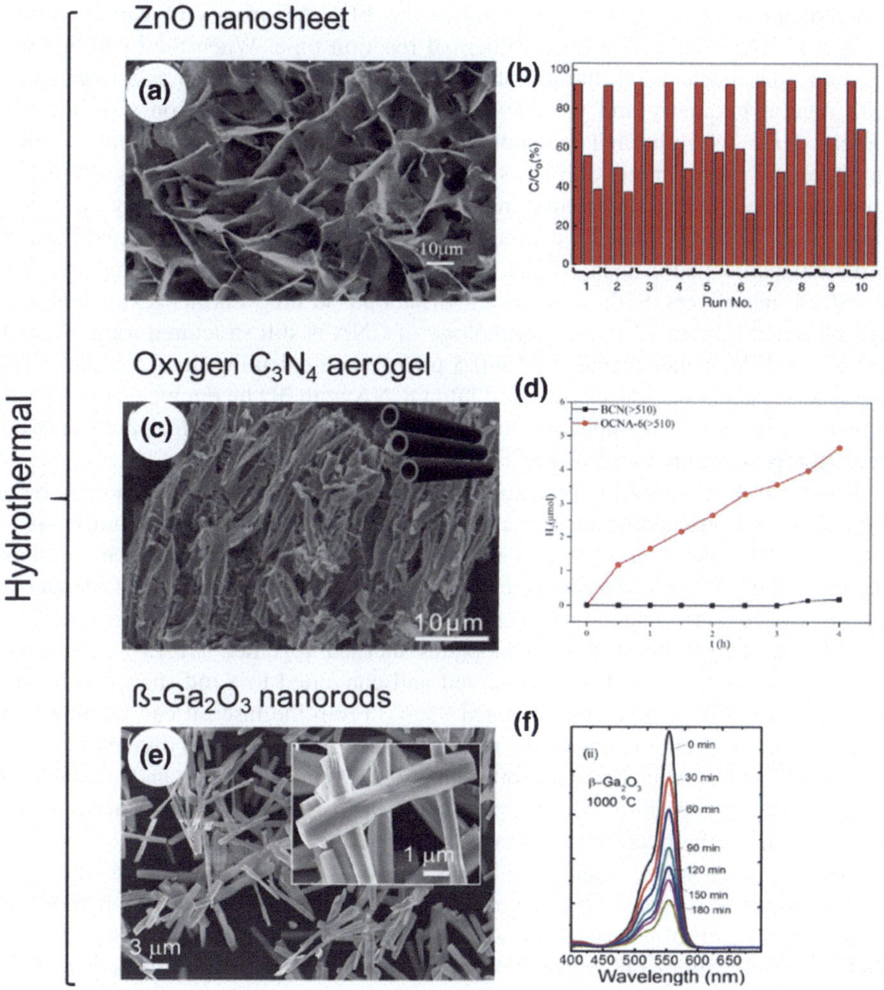

Fig. 3 Hydrothermal method to prepare various type of nanostructures. **a** SEM images of the as-prepared nanostructured ZnO films formed in HMT solution. **b** The photocatalytic decomposition of RhB for 10 cycles for 0 min, 60 min and 120 min, respectively (*Reprinted with permission from* [20]. *Copyright 2016 Elsevier*). **c** FESEM image of OCNA at 6 h reaction. **d** Hydrogen evolution reaction of BCN and OCNA-6 $\lambda > 510$ nm (*Reprinted with permission from* [10]. *Copyright 2018 Elsevier*). **e** GaOOH nanostructures from hydrothermal method temperature 95 °C for 5 h. **f** Absorbance spectra of the RhB solution at different UV irradiation times in the presence of β-Ga_2O_3 nanorods (*Reprinted with permission from* [36]. *Copyright 2015 Springer*)

nanostructured material. Figure 3c shows the FESEM images of carbon nitrite aerogel (CNA) with different hydrothermal reaction time. When the hydrothermal reaction time reached 3 h, the morphology of CNA was observed as a large sheet with regular geometry and the thickness is in the range of micron size. Further increased the hydrothermal reaction time to 5 h, both large sheet and rod-like structure can be observed, and the surface of the sample becomes rough. Finally, after extended the hydrothermal reaction time 6 h, the morphology of CNA changed to the trunk-shaped rod structure with a diameter of about 1–3 μm. Further structural investigation using TEM on 6 h present the trunk-shaped rod is composed of nanofibers with a diameter of about 30 nm. Further extending the hydrothermal time to 12 h, the morphology of CNA is still structured trunk-shaped rod but with a smaller diameter (0.3–0.5 μm). When calcining the CNA under N_2 atmosphere, CNA was re-polymerized into OCNA with 3D hierarchical structure as shown in Fig. 3c. The sample also has the potential to be applied for photocatalytic hydrogen production, as shown in Fig. 3d.

Reddy et al. reported the preparation of Gallium oxide (Ga_2O_3) nanorods by a facile hydrothermal synthesis at various temperatures starting from room temperature to 95 °C, and their structures and photocatalytic properties were observed as shown in Fig. 3f [36]. As shown in Fig. 3e, the cocoon shaped GaOOH nanostructures were constructed at room temperature. When the temperature increased to 50 °C, the multilayer of small nanoplates merged together. At 75 °C, stacked structures of multi-layered were observed and converted to a rod-shaped structure nanorods when the temperature reached 95 °C. From the inset, it can be observed when the temperature increased; the nanorods have a possibility to divided into two rods, which is a smaller size. Generally, in order to control the size and architecture of GaOOH nanostructures, the temperature is one of the parameters needs to be adjusted during the hydrothermal process.

On the other hand, Yang et al. have discussed the comparative study of self-supported graphene-TiO_2 nanostructures prepared from a combination of hydrothermal and sol-gel method towards the photocatalytic behavior. The effect of TiO_2 dimensionalities either 1D, 2D, or 3D, and their coupling with graphene on interfacial charge transfer also has been reported [54]. Coupling of TiO_2 with graphene derivative has improved the charge separation, which influences the photocatalytic activity. The research also found that TiO_2 nanomaterial dimensionalities also plays a crucial aspect role in regulating the unique charge-transfer properties with graphene, having great potential to be exploited as composite photocatalyst.

Hydrothermal or solvothermal condition is a promising method to crystallized material in a controllable atmosphere. The low-cost process which acquires low temperature and straightforward method are among the benefits of these processes where it is favorable during the selection of fabrication method in high scale application. On the other hand, it is difficult to explain the growth mechanism due to the reactions are taking place in the sealed container, the autoclave. Furthermore, the method is quite responsive to different parameters such as temperature, the concentration of solvents and precursors, etc. Therefore, it makes the method challenging when the repetitions are required during the observation.

2.4 *Other Methods*

Solid state reaction is a heating process in high temperature in a certain period to enhance the kinetic reaction within the material. Through thermal solid state reaction, Yang et al. fabricated 3D macroporous $ZnO/ZnFe_2O_4$ for high efficient light trapping and charge collection in photoelectrochemical water splitting by employing dip coating infiltration [56] as shown in Fig. 4a. PEC activities of $ZnO/ZnFe_2O_4$ show primarily improved about 6.2 times photoconversion efficiency compare to pristine ZnO even though without co-catalyst (Fig. 4b), the stability reached until 10 h measurement. The enhancement attributed from high charge separation efficiency in binder-free heterojunction interfaces in the 3D nanostructures. The advantage of this reaction is the direct contact of reacting material, which will minimize the use of heat. However, slow kinetics reaction is the drawbacks, which cause a longer time for the reaction process.

Sol-gel (SG) is a common coating processing method for metal oxide which grows small molecules from the solid product. Theoretically, a colloidal solution (sol) will be used to create an integrated network (gel) in the SG method. Free-standing TiO_2 nanowires with photocurrent generation 2.58 $\mu A\ cm^{-2}$ is much higher than TiO_2 nanoparticle, which read at 0.09 $\mu A\ cm^{-2}$ due to an increase in charge carrier transport and light harvesting properties (Fig. 4c–d). Interestingly, the photoelectrode coating by the SG method has efficient charge separation and excellent surface properties. However, it takes a longer time in the fabrication method for the formation of gel, and thermal treatment is required after the process to produce the material with good mechanical properties.

Another method such as sputtering is the process of coating substrate in a vacuum chamber that contains inert gas to produce a thin film of material on the substrate surface. Girardi et al. used Direct Current (DC) magnetron sputtering method to grow cobalt oxide (Co_3O_4) nanopetals in silicon (Fig. 4e) and compared the performance in photocurrent with different nanostructures for application in water remediation [7] (Fig. 4f). The results showed that the photocurrent increased proportionally with the organic content of the solution. Through the different thermal treatments, the different nanostructures were observed.

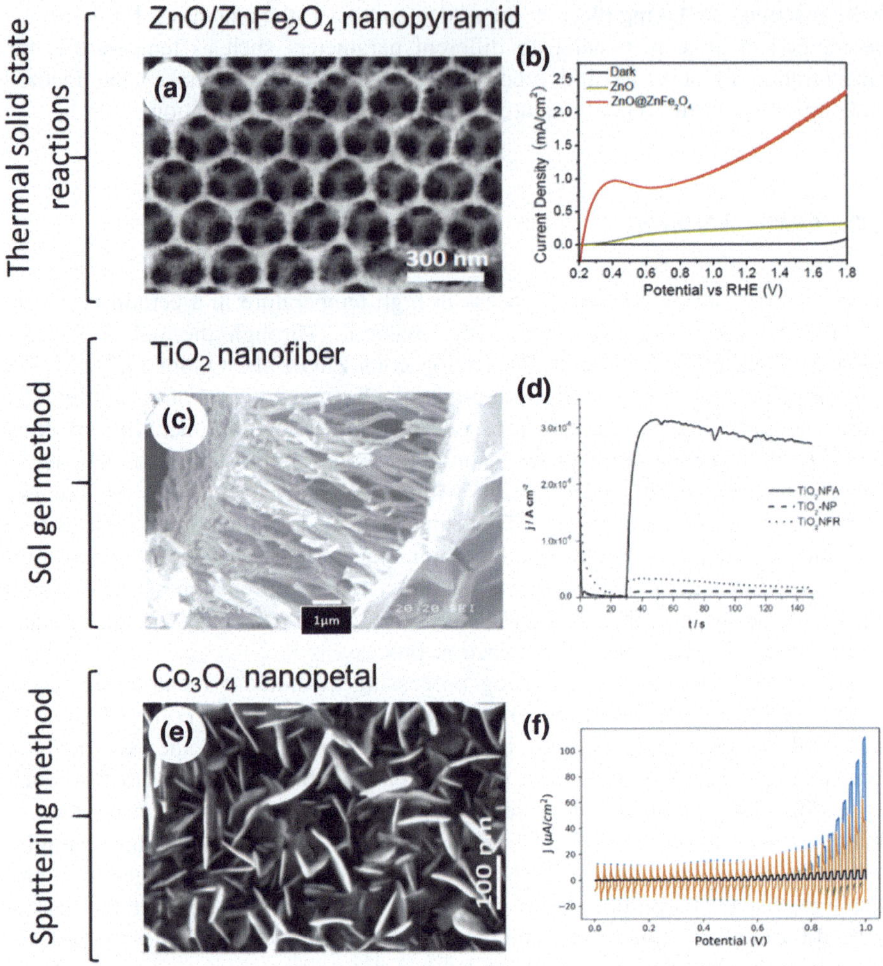

Fig. 4 Other methods to prepare various type of nanostructures (i) Thermal solid-state reactions to synthesis $ZnO/ZnFe_2O_4$ nanopyramid. **a** SEM images of $ZnO/ZnFe_2O_4$ sample prepared with 60 dip coating cycles. **b** Photocurrent measured at 0.400 V versus Ag/AgCl KCl sat. in 0.1 M phosphate buffer solution pH 7, under UV light irradiation (*Reprinted with permission from* [56]. *Copyright 2017 Royal Society of Chemistry*), (ii) Sol-gel method to prepare TiO_2 nanofiber. **c** Lateral view of SEM images of the nanofibres constituting a TiO_2 NFA photoanode. **d** Photocurrent measured at 0.400 V versus Ag/AgCl KCl sat. in 0.1 M phosphate buffer solution pH 7, under UV light irradiation, (*Reprinted with permission from* [28]. *Copyright 2015 IOP Publishing*) (iii) Sputtering method to prepare Co_3O_4 nanopetal. **e** FESEM top-view images of the samples PETAL. **f** LSV with chopped illumination of the sample PETAL in sodium sulphate before (orange) and after (blue) the addition of glucose (*Reprinted with permission from* [7]. *Copyright 2019 MDPI*)

3 Recent Application of Self-support Nanostructure Photocatalyst/Photoelectrode

3.1 Efficient Wastewater Treatment with Self-supported Photocatalyst

The rapid increase of population has contributed to the rapid expansion of industrialization and urbanization; clean water has been an enormous demand; unfortunately, limited sources throughout the world. One of the most promising options for treating wastewater is the exploitation of semiconductor nanostructured photocatalyst owing to its enormous potential and high efficiency to discard organic pollutant and pathogenic bacteria with the aid of sunlight [45, 55].

Briefly, the wastewater treatment through photocatalysis system involves photo-harvester of a catalyst made of semiconductor materials in suspended form and several active species in bulk liquid to carry out pollutant degradation. The ideal photolysis occurs via a sequential redox reactions as follows; (1) the semiconductor is irradiated by photon energy larger than its band gap ($hv > E_g$), an electron (e^-) bound at the valence band (VB) is photoexcited to the conduction band (CB), creating a hole (h^+) at the VB. (2) Accordingly, the photogenerated e^- are transported to the photocatalytic site of the semiconductor to interact with complementary charge species in bulk liquid.

Generally, there are three main ionic species involved in the photocatalytic reaction; the $*OH$, h^+, and $*O_2^-$, represents hydroxyl radical, proton, and superoxide radical. In aqueous solution, $*OH$ is essentially an oxidant in the photocatalytic degradation of pollutant. In aqueous solution, $*OH$ radicals can be generated via two pathways; (1) spontaneous oxidation of H_2O and OH^- by h^+ to generate $^*OH$ radicals, (2) The evolution of O_2 is reduced by the photoexcited e^- to produce $^*O_2^-$ radicals, which then accepting h^+ to form $^*OOH$ radicals and finally decomposed to yield $^*OH$ radicals [26].

Tuning semiconductor photocatalyst into nanostructure privileges photocatalytic reaction owing to the large photocatalytic active area and purposely to decouple the mismatch between the inherent carrier diffusion length and the depth of light absorption proximal to the band gap, a prevailing issue among semiconductor photocatalyst [48]. For example, a 3D $ZnIn_2S_4$/PVDF–poly(MMA-co-MAA)(ZIS/Polymer heterostructure) composite mats were designed using sequential electrospinning technique and hydrothermal process (Fig. 5a) [33]. With 50 wt% loading of ZIS on the polymer, a four folds improvement of methyl orange (MO) degradation (C/C_o) over ZIS powder was achieved (Fig. 5b) (Table 1), which might be benefitted from encouraging adsorption of radical and dye molecules.

In another study, a remarkable photocatalytic degradation of perfluorooctanoic acid (PFOA) was achieved using a novel sheaf-like structure of Ga_2O_3 (Fig. 5c) which synthesized through polyvinyl alcohol (PVA)-assisted hydrothermal technique with subsequent calcination [39]. The rate constant of photocatalytic degradation of PFOA in pure water under UV illumination is 4.85 h^{-1}, which correspond

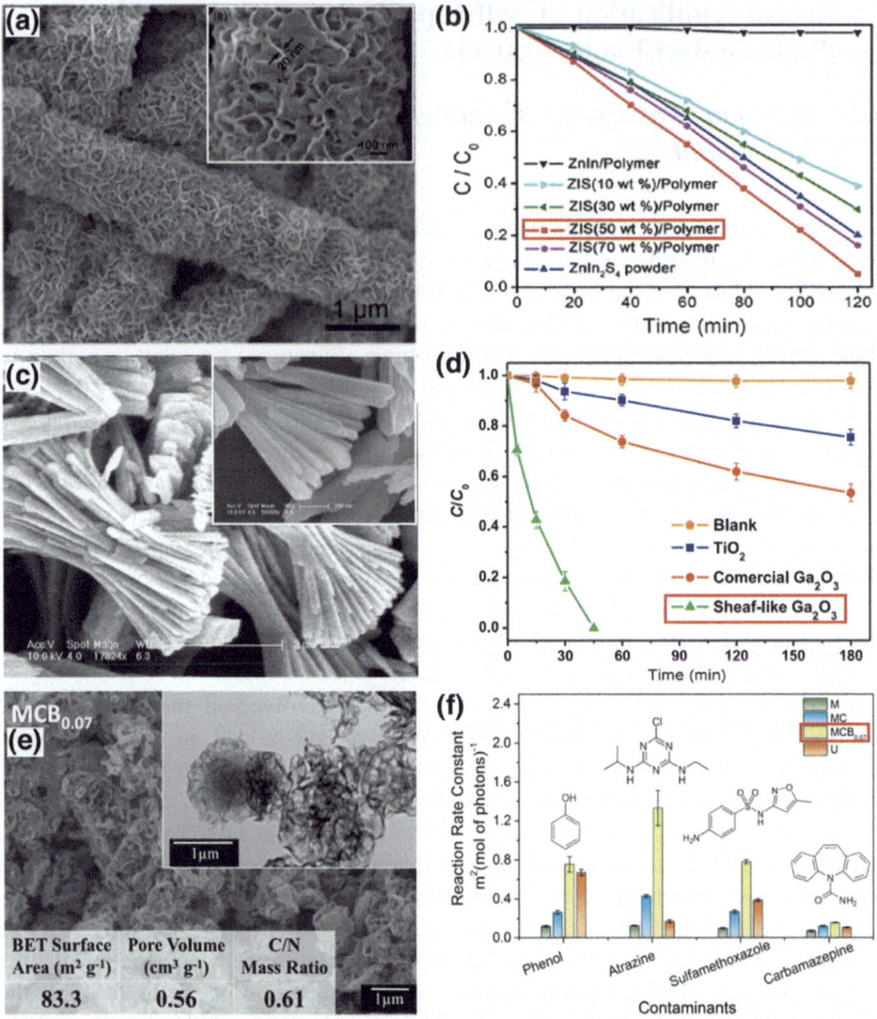

Fig. 5 The FESEM and photocatalytic activity of wastewater treatment using self-supported photocatalyst. **a** Low magnification SEM images of electrospun ZIS (50 wt%)/Polymer mat, inset; high magnification. **b** Photocatalytic degradation of MO under visible light irradiation in the presence of ZIS (50 wt%)/Polymer (red box) composites under visible light irradiation ($\lambda > 420$ nm) (*Reprinted with permission from* [33]. *Copyright 2012 American Chemical Society*). FESEM images of as-synthesized. **c** Shelf-like β-Ga_2O_3, inset; magnified image. **d** Time dependence of PFOA in pure water under UV irradiation with sheaf-like Ga_2O_3 (red box) (*Reprinted with permission from* [39]. *Copyright 2013 Elsevier*). **e** SEM (insets TEM) images of g-C_3N_4 (MCB0.07), inset tables show the surface area, pore volume, and bulk carbon to nitrogen (C/N) mass ratio of the sample. **f** Photocatalytic degradation rate constants of phenol, atrazine, sulfamethoxazole, and carbamazepine with g-C_3N_4 (MCB0.07) (red box) under simulated visible sunlight irradiation ($\lambda > 400$ nm, Xenon lamp). (*Reprinted with permission from* [58]. *Copyright 2016 American Chemical Society*)

Table 1 The properties and the photoactivity of photocatalyst or photoelectrode for water treatment application

Photocatalyst	Synthesis method	Optical properties and band gap	BET surface area	Pollutant	Degradation efficiency or reaction rate	Ref.
3D $ZnIn_2S_4$/PVDF–Poly (MMA-coMAA)	Electrospinning + hydrothermal	Maximum absorption band at 553 nm	59.2 $m^2\ g^{-1}$	Methyl orange	C/C_o 0.05 after 2 h	[33]
β-Ga_2O_3 (shelf like nanostructure)	PVA-assisted hydrothermal	High efficiency of trace PFOA removal at 185 nm vacuum UV irradiation	Surface area: 36.1 m^2g^{-1} Nanopores (2–4 nm and 8 nm)	Perfluoro-octanoic acid (PFOA)	4.85 h^{-1}	[39]
Carbon doped g-C_3N_4	Thermal polycondensation	The band gap of Carbon-doped g-C_3N_4 at N_1 nitrogen site is 2.57 eV (theoretical calculation)	83.3 m^2g^{-1}	Phenol	0.8 m^2 (mol of photons)$^{-1}$	[58]

to 16 and 44 folds greater than that obtained by commercial Ga_2O_3 and P25, respectively (Fig. 5d) (Table 1). The improved photocatalytic activity is attributed to a high specific surface area (36.1 m^2/g), enormous nanopores 2–8 nm, and close interaction between Ga_2O_3 and PFOA. It has been known that the decomposition of PFOA is constrained by the interference of organic matter [39], which was not the case for photocatalytic degradation of PFOA in secondary effluent using sheaf-like Ga_2O_3.

A visible-light-responsive photocatalyst of Graphitic carbon nitride (g-C_3N_4), has been the material of choice for photocatalytic water treatment study [25, 26, 58]. However, structural defects reduce conductivity and localized excitons, which causes a passive interfacial electron transfer reaction and hampering the conversion of adsorbed active oxygen radicals, such as hydroxide radical (*OH) [19]. Therefore, for instance, by doping the supramolecule-based g-C_3N_4 with carbon (Fig. 5e), the reaction rate of photodegradation of phenol and persistent organic micropollutant was enhanced by 2.3–10.5 folds relative to melamine-based g-C_3N_4 (Fig. 5f). This result is the manifestation of improved charge separation and transportation across the semiconductor-liquid interface.

3.2 Enhanced PEC Water-Splitting Reaction with Highly Photoactive Self-supported Photocatalyst/Photoelectrode

Relative to photocatalytic wastewater treatment, the photocatalytic and photoelectrochemical (PEC) water-splitting reaction is slightly more straightforward, where H_2O molecule is only split to H_2 and O_2. In a similar way, an illuminated semiconductor ($hv > E_g$) produces photoexcited electrons at the CB and photogenerated holes at the VB. In an ideal condition, these photogenerated carriers reside at semiconductor/electrolyte surface can interact with complementary charges from bulk liquid to drive water redox reaction. Water-lysis can be realized should the band gap of the semiconductor straddles the water redox potential, which is +1.23 and +0 V versus NHE for hydrogen evolution reaction (HER) and oxygen evolution reaction (OER), respectively at pH 0.

An immobilized photocatalyst, bearing a photoelectrode, is an alternative for hydrogen evolution reaction (HER) via PEC water-splitting reaction and yet more efficient than water photocatalysis reaction owing to facile gas separation and lower production cost. In the PEC reaction system, there is a number of configurations that have been broadly explained [24, 47, 52]. For a quick background, the use of an *n*-type semiconductor photoelectrode in which the population of electrons is higher than holes; the oxygen evolution takes place by accepting protons; $4OH^- + 4h^+ \rightarrow O_2 + 2H_2O$. The photoexcited electrons are migrated to counter electrode through an external circuit to drive hydrogen evolution via proton reduction; $4H_2O + 4e^- \rightarrow 2H_2 + 4OH^-$. Contrastingly, the use of *p*-type semiconductor works reversibly. In practice, the efficiency of HER and OER can be both improved when an external bias is applied.

The fabrication of nanostructured photocatalyst or photoelectrode for PEC water splitting also aids in improving PEC activity as proven by many studies [1, 16, 37, 38, 51]. The elucidation on the improved PEC activity via nanostructuring for wastewater treatment is also applicable to explain the attributes of improved HER activity. Unlike other photocatalytic reaction, the attained photocurrent generation in ideal PEC-water splitting condition is the mirror of HER activity with photolysis ratio of water 2:1 (H_2:O_2). Hence, the reported maximum photocurrent generation is considered the reflection of theoretical hydrogen production and the photoactivity of the photoelectrode.

In recent years, hematite (α-Fe_2O_3) as photoanode has received considerable attention for PEC water-splitting application due to its chemical stability and facile morphological modulation [4, 30, 57]. However, its conductivity and photoactivity in pristine thin film form are poor. There were many enhancement efforts to improve the low conductivity of hematite; thermal annealing is one of them being the most straight-forward strategy [13, 14, 59]. Literally, it is responsible for the high efficiency of hematite as photoanode [15, 34, 49]. Although, in particular, high annealing temperature above 700 °C is favorable for an enhanced photocurrent generation owing to the increase in crystallinity [42], the nanostructured hematite experienced severe agglomeration. Through an alternative approach of hybrid microwave annealing, the nanostructured hematite was synthesized (Fig. 6a) which demonstrated 2 folds higher photocurrent generation than that conventional thermal-annealed photoanode (Fig. 6b) (Table 2). The presented PEC performance is the result of intact nanostructured hematite during hybrid microwave annealing (HMA) process (Fig. 6a) [16].

WO_3 is another proven-stable metal-oxide semiconductor under PEC condition, but it suffers from passive charge transfer and rapid charge recombination [29]. As an effort to alleviate the problem, a hierarchical porous WO_3 was synthesized on tungsten (W) substrate through ultrafast laser processing with subsequent thermal heating (Fig. 6c) [1]. The PEC measurement revealed that the hierarchical porous WO_3 produced maximum photocurrent density of 1.2 mA cm^{-2} at 1.0 V versus Ag/AgCl, correspond to nearly 2 folds higher than that WO_3 NW owing to improved charge separation (Fig. 6d) [1]. The hierarchical porous WO_3 also showcased an excellent photostability during long-term PEC water-splitting operation (Table 2).

A remarkably high photocurrent generation was obtained using highly dense Cu_2O NW photocathode fabricated using a modified wet chemical oxidation method (WCOM) (Fig. 6e) [38]. The WCOM method, although offer simple and highly scalable fabrication technique, but the interference of Cu_2O micro flower (MF) hinders the photocurrent generation [37, 38]. With the addition of NH_3 into the WCOM medium, a highly dense NW array was produced which translated into photocurrent generation of −5.45 mA cm^{-2} at −0.6 V versus Ag/AgCl, correspond to 2.7 folds improvement compared to Cu_2O MF (Fig. 6f) (Table 2). The enhanced photocurrent generation with NW as compared to MF is suggested due to higher carrier density, lower charge transfer resistant, larger surface active area, and better radial charge diffusion [38].

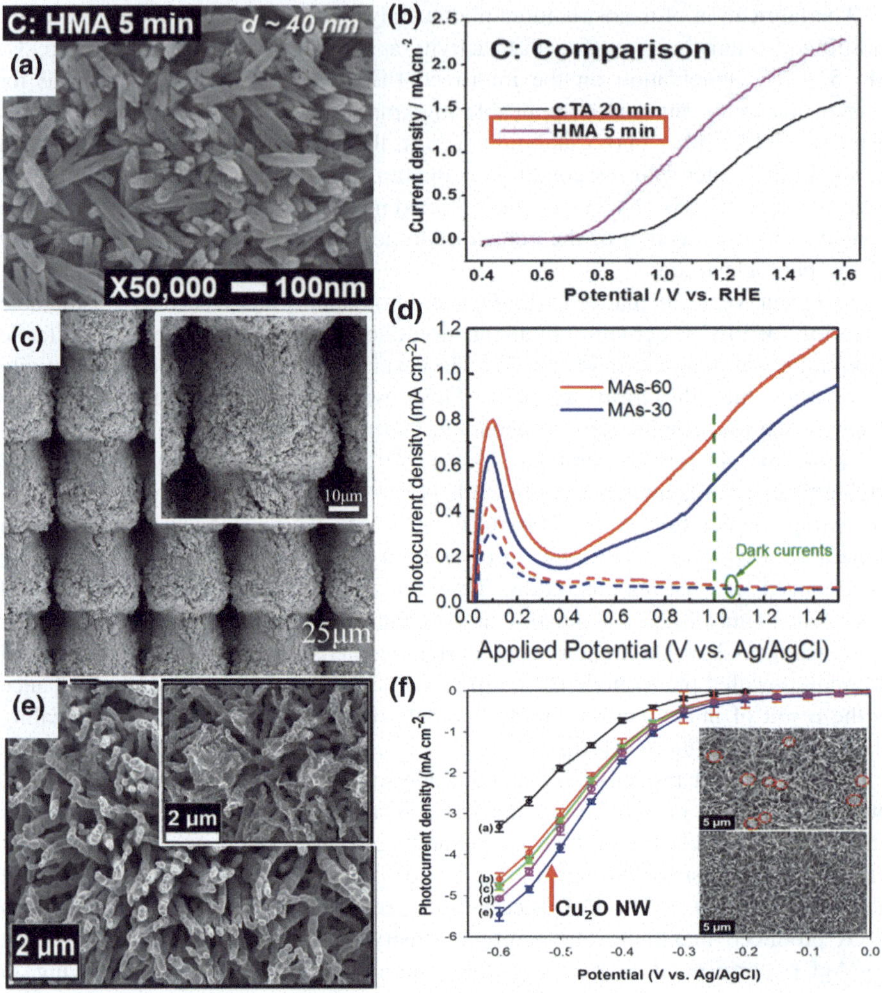

Fig. 6 The FESEM images of self-supported photocatalyst/photoelectrode and their corresponding photoactivities. **a** Hematite thin film after 5 min of hybrid microwave annealing (HMA). **b** The photocurrent density of hematite thin film photoanodes treated by HMA (red box) (*Reprinted with permission from* [16]. *Copyright 2015 American Chemical Society*). **c** Hierarchical porous WO_3 nanoparticle aggregates, inset; magnified image. **d** Photocurrent density of micro-nano structured WO_3 measured through LSV (red box) (*Reprinted with permission from* [1]. *Copyright 2016 American Chemical Society*). **e** Highly dense and uniform Cu_2O NW, inset: the interference of unfavorable Cu_2O microflower (MF). **f** Photocurrent density of Cu_2O NW and Cu_2O MF, inset: The distribution of MF and NW on Cu_2O film (*Reprinted with permission from* [38]. *Copyright 2018 Elsevier*)

Table 2 Various type of self-supported nanostructured photocatalyst/photoelectrode with their corresponding photoactivities for PEC water-splitting application

Type of photoelectrode	Fabrication method	Electrolyte	Photocurrent density (mA cm^{-2}) at V versus Ag/AgCl	Stability J/J_O (%) at V versus Ag/AgCl	Ref.
Nanostructure-preserved Fe_2O_3	Hybrid microwave annealing (HMA)	1 M NaOH (pH = 13.6)	2.25 mA cm^{-2} at 0.6 V	N/A	[16]
3D self-supporting WO_3 micro-nano architectures	Ultrafast laser processing hybrid with thermal oxidation	0.1 M H_2SO_4 (pH ≈ 0.7)	1.2 mA cm^{-2} at 1.0 V	97% after 120 min	[1]
Highly dense Cu_2O NW	Wet chemical oxidation method and annealing	0.5 M Na_2SO_4 (pH = 6.8)	−5.45 mA cm^{-2} at 0.6 V	N/A	[38]

3.3 Reduction of CO_2 to Valuable Products with Highly Photoactive Self-supported Photocatalyst or Photoelectrode

In response to the ascending amount of CO_2 in the atmosphere and continuous global demand of fuel, progressive research has been geared to the photochemical and more intense to the photoelectrochemical reduction of CO_2 (PECRC) to fuel approach [5, 9, 40]. Despite other established CO_2 conversion approach such as carbon capture-reuse, capture-storage, CO_2 sequestration, and electrochemical reduction of CO_2, these techniques inflict a high cost and a large amount of energy consumption to run the conversion process. Important to note that, although the conversion of CO_2 to fuel via photoelectrochemical means imposes an immense fundamental and technical challenges, but the potential benefit is advantageous.

Similar photoelectrode configuration, photocell design and technical parameters as for PEC water-splitting can be applied, except for the saturation of electrolyte with CO_2 that differ. The conversion products can be harnessed at two phases, which are aqueous and gas phases. The possible generated products in the aqueous phase are formic acid, formaldehyde, methanol, ethanol, isopropanol, and acetone, whereas carbon monoxide, methane, and ethylene are among the products evolved in gas phases. Potential conversion products derived from electrochemical reduction of CO_2 with their corresponding thermodynamic potential is tabulated in Table 3.

Table 3 The potential product produced from PEC reduction of CO_2 and their corresponding redox potential

Reaction	E_0 (V) versus NHE at pH 7
$2H^+ + 2e^- \rightarrow H_2$	−0.41
$CO_2 + e^- \rightarrow CO_2^{-}\bullet$	−1.90
$CO_2 + 2H^+ + 2e^- \rightarrow HCO_2H$	−0.61
$CO_2 + 2H^+ + 2e^- \rightarrow CO + H_2O$	−0.53
$CO_2 + 4H^+ + 4e^- \rightarrow C + 2H_2O$	−0.20
$CO_2 + 4H^+ + 4e^- \rightarrow HCHO + H_2O$	−0.48
$CO_2 + 6H^+ + 6e^- \rightarrow CH_3OH + H_2O$	−0.38
$CO_2 + 8H^+ + 8e^- \rightarrow CH_4 + 2H_2O$	−0.24
$2CO_2 + 8H_2O + 12e^- \rightarrow C_2H_4 + 12OH^-$	−0.34
$2CO_2 + 9H_2O + 12e^- \rightarrow C_2H_5OH + 12OH^-$	−0.33
$3CO_2 + 13H_2O + 18e^- \rightarrow C_3H_7OH + 18OH^-$	−0.32

The engineering of nanostructured photocatalyst or photoelectrode material for PEC CO_2 reduction not only reflected on its photoactivity, but the selectivity of fuel generation as proven by recent studies [5, 6, 9, 35, 40]. Product selectivity is not solely governed by the morphological differences, but it is also influenced by the bias potential, reaction time, type, and pH of the electrolyte.

For example, the selectivity of formate in PEC reduction of CO_2 was catalyzed by the unique hierarchical Co_3O_4 electrode [9]. A detailed study found that the hierarchical structure is consist of single-crystalline microflower and thin nanopetals atop 1-D rhombus nanorod (Fig. 7a). The synergistic of both hierarchical structures have contributed to the outstanding photocatalytic and PECRC owing to the large surface-active area for CO_2 and HCO_3 surface adsorption (Fig. 7b) (Table 4). Research further elucidated that the surface comprised of Co^{3+} derived from the (121) facet augmented the electrochemically active surface area, induces enhanced PEC responses of CO_2 and HCO_3. With respect to product selectivity, de Brito et al. proposed that, based on experimental finding, the generation of CH_3OH is possible under PEC at an early period of reaction, in $NaHCO_3/Na_2CO_2$ buffer solution, pH 8, at +0.2 V, with UV-Vis illumination [3].

In another development, the deposition of Cu nanoparticles (NPs) on Co_3O_4 nanotube (NT) arrays has considerably increased the yield and highly selective to formate production reaching 100% Faradaic efficiency (FE) upon PEC reduction of CO_2 (Fig. 7c) [40]. Within 8 h of PEC reaction, 6.75 mmol L^{-1} cm^{-2} of formate was significantly produced, which correspond to 1.5 folds enhancement than that obtained by bare Co_3O_4 NTs (Fig. 7d) (Table 4). The improved PEC performance is contributed by the minimization of charge transfer resistance across nanotube layer straight to substrate underneath, favorable interface band structure between Co_3O_4 NTs and Cu NPs, the large surface area of active sites, and high light reflection/scattering which mainly benefited from the metal-decorated upright nanotube structure.

Through a simple, reproducible modified TiO_2 sol-gel fabrication technique, self-assembly of TiO_2 microsphere which composed of P25 TiO_2 nanoparticles was

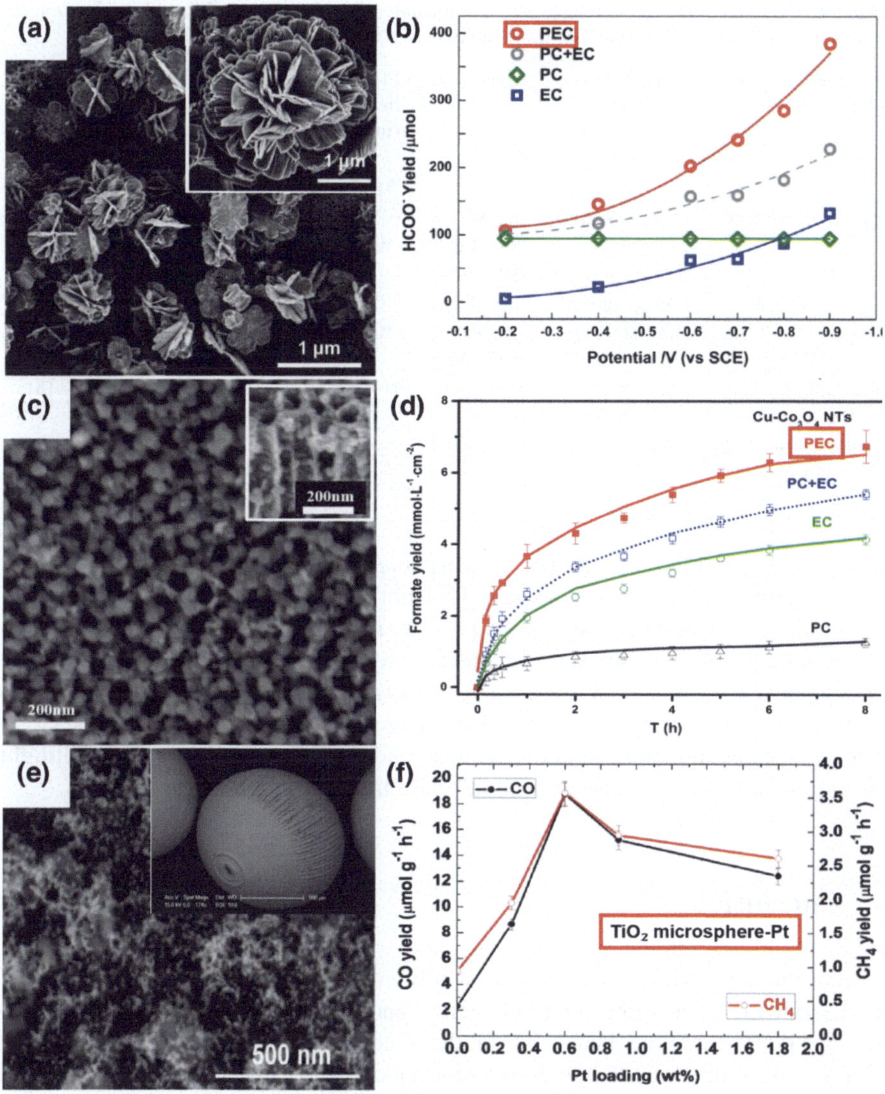

Fig. 7 SEM images of developed self-supported photoelectrodes or photocatalysts for the reduction of CO_2 and their photoactivity. **a** SEM image of HA-Co_3O_4, inset: upper layer microflower morphology. **b** The yield of formate at different negative voltage under PEC condition (red box) (*Reprinted with permission from* [9]. *Copyright 2013 American Chemical Society*). **c** SEM images (top view) of anodic layer of the Cu-Co_3O_4 NTs, Inset; side views. **d** The CO and CH_4 yields as photocatalyzed by the TiO_2 microsphere/Pt (red box) (photocurrent density of Cu-Co_3O_4 NTs in CO_2 saturated condition (red box) (*Reprinted with permission from* [40]. *Copyright 2015 American Chemical Society*). **e** SEM images of TiO_2 spheres, inset; magnified image of the sample. **f** Yields of the CO and CH_4 produced during 24 h of photodriven CO_2 reduction on TiO_2 sphere catalysts with various Pt loadings *Reprinted with permission from* [5]. *Copyright 2014 American Chemical Society*)

Table 4 Different types of self-supported photocatalyst or photoelectrodes with their photoactivities and converted products from the reduction of CO_2

Type of photoelectrode	Fabrication method	Electrolyte	Photocurrent density ($mA\ cm^{-2}$) at V versus Ag/AgCl	CO_2 reduction product	Ref.
Co_3O_4 microflower	Solvothermal synthesis	0.1 M Na_2SO_4 (CO_2 sat.)	−0.5; −0.90 V	Formate 384.8 ± 7.4 μmol High selectivity	[9]
Co_3O_4 NT/Cu	Anodization/ Cu ED	0.1 M Na_2SO_4 (CO_2 sat.)	−0.122; −0.83 V	Formate 6.75 mM cm^{-2} High selectivity	[40]
TiO_2 microspheres with hierarchical nanostructure (with platinum loading)	Modified TiO_2 sol-gel approach	Deionized water (CO_2 sat.)	N/A	Methane 19 μmol g^{-1} h^{-1}	[5]

synthesized via (Fig. 7e) [5]. The microsphere was constructed with hierarchical porosity consist of larger mesopores (12.4 nm) and smaller mesopores (2.3 nm). A significant enhancement in converting CO_2 to CH_4 with Pt-loaded TiO_2 microsphere through photocatalysis was achieved, presenting 4 to 5 folds production rate as compared to commercial Degussa P25 TiO_2 (Fig. 7f). The enhanced photocatalytic activity is mainly attributed to the larger mesoporous volume, thus larger surface area, and hierarchical porosity, which is advantageous to increase multilight reflection/scattering and promotes rapid mass transport.

4 Conclusion

Several known methods can be successfully employed to examine self-supported materials with outstanding photo-efficiency and photo-stability. Different fabrication methods might be beneficial to the data obtained from different kind of materials used in which the phase transformation of the materials can be observed with control of the growth and nucleation processes. The direct fabrication method is preferable due to the formation of free-standing materials either in 1D, 2D, or 3D nanoarchitectures with or without a template. The chemically induced self-supported strategy displayed enhanced photocatalytic and photoelectric conversion efficiency than their counterparts. The boosted performance of the hierarchical materials can be ascribed to their utmost potential capability in adsorbing more substrate molecules due to larger surface areas, porosity, effective transport and fast diffusion of electrolyte in the rich micro–mesoporous conduits and prolonging the lifespan of electron for effective migration. Furthermore, the unique structures of the materials which benefit the activity by simultaneously enhancing light utilization via the multiple reflections within the organized open framework providing better reaction site accessibility.

References

1. Cai, M., Fan, P., Long, J., Han, J., Lin, Y., Zhang, H., Zhong, M.: Large-scale tunable 3D Self-supporting WO_3 micro-nano architectures as direct photoanodes for efficient photoelectrochemical water splitting. ACS Appl. Mater. Interfaces. **9**, 17856–17864 (2017)
2. Chen, X., Shi, R., Chen, Q., Zhang, Z., Jiang, W., Zhu, Y., Zhang, T.: Three-dimensional porous g-C_3N_4 for highly efficient photocatalytic overall water splitting. Nano Energy **59**, 644–650 (2019)
3. de Brito, J.F., Araujo, A.R., Rajeshwar, K., Zanoni, M.V.B.: Photoelectrochemical reduction of CO_2 on Cu/Cu_2O films: Product distribution and pH effects. Chem. Eng. J. **264**, 302–309 (2015)
4. Dias, P., Vilanova, A., Lopes, T., Andrade, L., Mendes, A.: Extremely stable bare hematite photoanode for solar water splitting. Nano Energy **23**, 70–79 (2016)
5. Fang, B., Bonakdarpour, A., Reilly, K., Xing, Y., Taghipour, F., Wilkinson, D.P.: Large-scale synthesis of TiO_2 microspheres with hierarchical nanostructure for highly efficient photodriven reduction of CO_2 to CH_4. ACS Appl. Mater. Interfaces. **6**, 15488–15498 (2014)
6. Ghadimkhani, G., de Tacconi, N.R., Chanmanee, W., Janaky, C., Rajeshwar, K.: Efficient solar photoelectrosynthesis of methanol from carbon dioxide using hybrid CuO–Cu_2O semiconductor nanorod arrays. Chem. Commun. **49**, 1297 (2013)
7. Girardi, L., Bardini, L., Michieli, N., Kalinic, B., Maurizio, C., Rizzi, G.A., Mattei, G.: Co_3O_4 nanopetals on si as photoanodes for the oxidation of organics. Surfaces **2**, 41–53 (2019)
8. Gogotsi, Y.: Transition metal carbides go 2D. Nat. Mater. **14**, 1079–1080 (2015)
9. Huang, X., Cao, T., Liu, M., Zhao, G.: Synergistic photoelectrochemical synthesis of formate from CO2 on 12Γ hierarchical Co3O4. J. Phys. Chem. C **117**, 26432–26440 (2013)
10. Jiang, W., Ruan, Q., Xie, J., Chen, X., Zhu, Y., Tang, J.: Oxygen-doped carbon nitride aerogel: a self-supported photocatalyst for solar-to-chemical energy conversion. Appl. Catal. B Environ. **236**, 428–435 (2018)
11. Jiao, W., Cheng, Y., Zhang, J., Che, R.: Self-assembled 3D hierarchical copper hydroxyphosphate modified by the oxidation of copper foil as a recyclable, wide wavelength photocatalyst. Langmuir **33**, 13649–13656 (2017)
12. Kafizas, A., Francàs, L., Sotelo-vazquez, C., Ling, M., Li, Y., Mccafferty, L., Blackman, C.S., Darr, J.A., Parkin, I.P.: Optimizing the activity of nano-needle structured WO_3 photoanodes for solar water splitting–direct synthesis via chemical vapor deposition. J. Phys. Chem. **121**, 5983–5993 (2017)
13. Kanan, M.W., Surendranath, Y., Nocera, D.G.: Cobalt–phosphate oxygen-evolving compound. Chem. Soc. Rev. **38**, 109–114 (2009)
14. Kim, J.Y., Jang, J., Youn, D.H., Magesh, G., Lee, J.S.: A stable and efficient hematite photoanode in a neutral electrolyte for solar water splitting: towards stability engineering. Adv. Energy Mater. **4**, 1400476 (2014)
15. Kim, J.Y., Magesh, G., Youn, D.H., Jang, J., Kubota, J., Domen, K., Lee, J.S.: Single-crystalline, wormlike hematite photoanodes for efficient solar water splitting. Sci. Rep. **3**, 2681 (2013)
16. Kim, J.Y., Youn, D.H., Kim, J.H., Kim, H.G., Lee, J.S.: Nanostructure-preserved hematite thin film for efficient solar water splitting. ACS Appl. Mater. Interfaces **7**, 14123–14129 (2015)

17. Kimura, T., Sato, S., Kataoka, K., Morikawa, T., Nakamura, D.: Self-assembled single-crystalline GaN having a bimodal meso/macropore structure to enhance photoabsorption and photocatalytic reactions. ACS Appl. Mater. Interfaces. **11**, 4233–4241 (2019)
18. Li, B., Li, W., Li, H., Wu, L.: Ionic complexes of metal oxide clusters for versatile self-assemblies. Acc. Chem. Res. **50**, 1391–1399 (2017)
19. Li, Z., Luan, Y., Qu, Y., Jing, L.: Modification strategies with inorganic acids for efficient photocatalysts by promoting the adsorption of O_2. ACS Appl. Mater. Interfaces **7**, 22727–22740 (2015)
20. Lina, W., Yingying, Z., Xiaoyun, L., Wenjun, D., Weihua, T., Benyong, C., Chaorong, L., Xiao, L., Tierui, Z., Wei, X.: Nanostructured porous ZnO film with enhanced photocatalytic activity. Thin Solid Films **519**, 5673–5678 (2011)
21. Lotito, V., Zambelli, T.: Approaches to self-assembly of colloidal monolayers: a guide for nanotechnologists. Adv. Colloid Interface Sci. **246**, 217–274 (2017)
22. Ma, Y., Jang, H., Kim, S.J., Pang, C., Chae, H.: Copper-assisted direct growth of vertical graphene nanosheets on glass substrates by low-temperature plasma-enhanced chemical vapour deposition process. Nanoscale Res. Lett. **10**, 308 (2015)
23. Medina-Ramos, J., Pupillo, R.C., Keane, T.P., Dimeglio, J.L., Rosenthal, J.: Efficient conversion of CO_2 to CO using tin and other inexpensive and easily prepared post-transition metal catalysts. J. Am. Chem. Soc. **137**, 5021–5027 (2015)
24. Minggu, L.J., Wan Daud, W.R., Kassim, M.B.: An overview of photocells and photoreactors for photoelectrochemical water splitting. Int. J. Hydrogen Energy **35**, 5233–5244 (2010)
25. Mohamed, M.A., Jaafar, J.M., Zain, M.F., Minggu, L.J., Kassim, M.B., Salehmin, M.N.I., Rosmi, M.S., Salleh, W.N., Othman, M.H.D.: Concurrent growth, structural and photocatalytic properties of hybridized C, N co-doped TiO_2 mixed phase over g-C_3N_4 nanostructured. Scr. Mater. **142**, 143–147 (2018)
26. Mohamed, M.A., Zain, M.F., Minggu, L.J., Kassim, M.B., Saidina Amin, N.A., Salleh, W.N., Salehmin, M.N.I., Md Nasir, M.F., Mohd Hir, Z.A.: Constructing bio-templated 3D porous microtubular C-doped g-C_3N_4 with tunable band structure and enhanced charge carrier separation. Appl. Catal. B Environ. **236**, 265–279 (2018)
27. Movsesyan, L., Maijenburg, A., Goethals, N., Sigle, W., Spende, A., Yang, F., Kaiser, B., Jaegermann, W., Park, S., Mul, G., Trautmann, C., Toimil-Molares, M.: ZnO Nanowire networks as photoanode model systems for photoelectrochemical applications. Nanomaterials **8**, 693 (2018)
28. Ongaro, M., Mardegan, A., Nanotech, V., Stortini, A.M., Signoretto, M.: Arrays of templated TiO_2 nanofibres as improved photoanodes for water splitting under visible light. Nanotechnology **26**, 165404 (2015)
29. Park, S., Jin, H., Woo, C., Jo, H., Sik, S., Won, S., Kee, H., Lee, S., Kim, D., Sun, K.: Sn self-doped Fe_2O_3 nanobranch arrays supported on a transparent, conductive SnO_2 trunk to improve photoelectrochemical water oxidation. Int. J. Hydrogen Energy **39**, 16459–16467 (2014)
30. Peerakiatkhajohn, P., Yun, J., Chen, H., Lyu, M., Butburee, T., Wang, L.: Stable hematite nanosheet photoanodes for enhanced photoelectrochemical water splitting. Adv. Mater. **28**, 6405–6410 (2016)
31. Pei, S., Cheng, H.M.: The reduction of graphene oxide. Carbon **50**, 3210–3228 (2012)
32. Peng, H., Liu, Z.: Controlled synthesis of topological insulator nanoplate arrays on mica. J. Am. Chem. Soc. **134**, 6132–6135 (2012)
33. Peng, S., Zhu, P., Mhaisalkar, S.G., Ramakrishna, S.: Self-supporting three-dimensional $ZnIn_2S_4$/PVDF–Poly(MMA-co-MAA) composite mats with hierarchical nanostructures for high photocatalytic activity. J. Phys. Chem. C **116**, 13849–13857 (2012)
34. Phuan, Y.W., Chong, M.N., Zhu, T., Yong, S., Chan, E.S.: Effects of annealing temperature on the physicochemical, optical and photoelectrochemical properties of nanostructured hematite thin films prepared via electrodeposition method. Mater. Res. Bull. **69**, 71–77 (2015)

35. Rajeshwar, K., de Tacconi, N.R., Ghadimkhani, G., Chanmanee, W., Janáky, C.: Tailoring copper oxide semiconductor nanorod arrays for photoelectrochemical reduction of carbon dioxide to methanol. Chem. Phys. Chem. **14**, 2251–2259 (2013)
36. Reddy, L.S., Ko, Y.H., Yu, J.S.: Hydrothermal synthesis and photocatalytic property of β-Ga_2O_3 nanorods. Nanoscale Res. Lett. **10**, 364 (2015)
37. Salehmin, M.N.I., Jeffery Minggu, L., Arifin, K., Mohamad Yunus, R., Mohamed, M.A., Kassim, M.B.: Recent advances on state-of-the-art copper (I/II) oxide as photoelectrode for solar green fuel generation: challenges and mitigation strategies. Appl. Catal. A Gen. **582**, 117104 (2019)
38. Salehmin, M.N.I., Jeffery Minggu, L., Mark-Lee, W.F., Mohamed, M.A., Arifin, K., Jumali, M.H.H., Kassim, M.B.: Highly photoactive Cu_2O nanowire film prepared with modified scalable synthesis method for enhanced photoelectrochemical performance. Sol. Energy Mater. Sol. Cells **182**, 237–245 (2018)
39. Shao, T., Zhang, P., Jin, L., Li, Z.: Photocatalytic decomposition of perfluorooctanoic acid in pure water and sewage water by nanostructured gallium oxide. Appl. Catal. B Environ. **142–143**, 654–661 (2013)
40. Shen, Q., Chen, Z., Huang, X., Liu, M., Zhao, G.: High-yield and selective photoelectrocatalytic reduction of CO_2 to formate by metallic copper decorated Co_3O_4 nanotube arrays. Environ. Sci. Technol. **49**, 5828–5835 (2015)
41. Shi, Y., Hamsen, C., Jia, X., Kim, K.K., Reina, A., Hofmann, M., Hsu, A.L., Zhang, K., Li, H., Juang, Z., Dresselhaus, M.S., Li, L., Kong, J.: Synthesis of few-layer hexagonal boron nitride thin film by chemical vapor deposition. Nano Lett. **10**, 4134–4139 (2010)
42. Sivula, K., Zboril, R., Le Formal, F., Robert, R., Weidenkaff, A., Tucek, J., Frydrych, J., Grätzel, M.: Photoelectrochemical water splitting with mesoporous hematite prepared by a solution-based colloidal approach. J. Am. Chem. Soc. **132**, 7436–7444 (2010)
43. Song, L., Ci, L., Lu, H., Sorokin, P.B., Jin, C., Ni, J., Kvashnin, A.G., Kvashnin, D.G., Lou, J., Yakobson, B.I., Ajayan, P.M.: Large Scale growth and characterization of atomic hexagonal boron nitride layers. Nano Lett. **10**, 3209–3215 (2010)
44. Su, J., Guo, L., Bao, N., Grimes, C.A.: Nanostructured $WO_3/BiVO_4$ heterojunction films for efficient photoelectrochemical water splitting. Nano Lett. **11**, 1928–1933 (2011)
45. Tanveer, M., Tezcanli Guyer, G.: Solar assisted photo degradation of wastewater by compound parabolic collectors: review of design and operational parameters. Renew. Sustain. Energy Rev. **24**, 534–543 (2013)
46. Tao, L., Cinquanta, E., Chiappe, D., Grazianetti, C., Fanciulli, M., Dubey, M., Molle, A., Akinwande, D.: Silicene field-effect transistors operating at room temperature. Nat. Nanotechnol. **10**, 227–231 (2015)
47. Vilanova, A., Lopes, T., Spenke, C., Wullenkord, M., Mendes, A.: Optimized photoelectrochemical tandem cell for solar water splitting. Energy Storage Mater. **13**, 175–188 (2018)
48. Walter, M.G., Warren, E.L., McKone, J.R., Boettcher, S.W., Mi, Q., Santori, E.A., Lewis, N. S.: Solar water splitting cells. Chem. Rev. **110**, 6446–6473 (2010)
49. Wang, L., Lee, C.-Y., Mazare, A., Lee, K., Müller, J., Spiecker, E., Schmuki, P.: Enhancing the water splitting efficiency of Sn-doped hematite nanoflakes by flame annealing. Chem. A Eur. J. **20**, 77–82 (2014)
50. Warwick, M.E.A., Kaunisto, K., Barreca, D., Carraro, G., Gasparotto, A., Maccato, C., Bontempi, E., Sada, C., Ruoko, T., Turner, S., Van Tendeloo, G.: Vapor phase processing of α-Fe_2O_3 photoelectrodes for water splitting: an insight into the structure/property interplay. ACS Appl. Mater. Interfaces **7**, 8667–8676 (2015)
51. Wu, M., Wang, Y., Xu, Y., Ming, J., Zhou, M., Xu, R., Fu, Q., Lei, Y.: Self-supported Bi_2MoO_6 nanowall for photoelectrochemical water splitting. ACS Appl. Mater. Interfaces **9**, 23647–23653 (2017)
52. Xing, Z., Zong, X., Pan, J., Wang, L.: On the engineering part of solar hydrogen production from water splitting: photoreactor design. Chem. Eng. Sci. **104**, 125–146 (2013)

53. Xu, Q., Qian, X., Qu, Y., Hang, T., Zhang, P., Li, M., Gao, L.: Electrodeposition of Cu_2O nanostructure on 3D Cu micro-cone arrays as photocathode for photoelectrochemical water reduction. J. Electrochem. Soc. **163**, 976–981 (2016)
54. Yang, J., Wen, Z., Shen, X., Dai, J., Li, Y., Li, Y.: A comparative study on the photocatalytic behavior of graphene-TiO_2 nanostructures: Effect of TiO_2 dimensionality on interfacial charge transfer. Chem. Eng. J. **334**, 907–921 (2018)
55. Yang, L., Dong, S., Sun, J.-H., Feng, J., Wu, Q., Sun, S.: Microwave-assisted preparation, characterization and photocatalytic properties of a dumbbell-shaped ZnO photocatalyst. J. Hazard. Mater. **179**, 438–443 (2010)
56. Yang, T., Xue, J., Tan, H., Xie, A., Li, S., Yan, W., Shen, Y.: Highly Ordered ZnO/$ZnFe_2O_4$ inverse opals with binder-free heterojunction interfaces for high-performance photoelectrochemical water splitting. J. Mater. Chem. A **6**, 1210–1218 (2018)
57. Yu, Q., Meng, X., Wang, T., Li, P., Ye, J.: Hematite films decorated with nanostructured ferric oxyhydroxide as photoanodes for efficient and stable photoelectrochemical water splitting. Adv. Funct. Mater. **25**, 2686–2692 (2015)
58. Zheng, Q., Durkin, D.P., Elenewski, J.E., Sun, Y., Banek, N.A., Hua, L., Chen, H., Wagner, M.J., Zhang, W., Shuai, D.: Visible-light-responsive graphitic carbon nitride: rational design and photocatalytic applications for water treatment. Environ. Sci. Technol. **50**, 12938–12948 (2016)
59. Zhong, D.K., Cornuz, M., Sivula, K., Grätzel, M., Gamelin, D.R.: Photo-assisted electrodeposition of cobalt–phosphate (Co–Pi) catalyst on hematite photoanodes for solar water oxidation. Energy Environ. Sci. **4**, 1759 (2011)

Surface-Enhanced Raman Scattering Substrates: Fabrication, Properties, and Applications

Sajan D. George

Abstract Recent advancements in nanotechnology largely enabled fabrication of plasmonic nanostructures of desired structural features and substantially improved the sensitivity and selectivity of the conventional optical sensing techniques. The plasmonic nanostructure mitigates the limitation of weak scattering cross-section in Raman spectroscopy via electromagnetic as well as chemical enhancement mechanism. The plasmonic nanostructure combined with the Raman spectroscopy technique, popularly known surface-enhanced Raman scattering spectroscopy, has been now established as an effective tool for molecular finger printing of analyte molecule and find applications diverse areas, ranging from biosensors to art. This chapter explains the mechanism behind the surface-enhanced Raman scattering spectroscopy with an emphasis on the factors contributing towards the enhancement in the Raman signal. Further, an account of the difference between conventional and surface enhanced Raman spectroscopy is presented. The role of hot spots and the rationale behind the choice of metal nanoparticles for surface-enhanced Raman scattering substrates is described. In addition, various approaches adopted for the fabrication of substrates in 1D, 2D, and 3D is explained in detail. A detailed account of a few emerging areas wherein this technique finds applications is also given in the chapter.

Keywords Surface-enhanced Raman scattering · Substrate fabrication · SERS applications

S. D. George (✉)
Centre for Applied Nanosciences, Manipal Academy of Higher Education,
Manipal, Karnataka 576104, India
e-mail: sajan.george@manipal.edu

S. D. George
Department of Atomic and Molecular Physics, Manipal Academy of Higher Education,
Manipal, Karnataka 576104, India

Inamuddin et al. (eds.), *Self-standing Substrates*,
Engineering Materials, https://doi.org/10.1007/978-3-030-29522-6_3

1 Introduction

The quest for developing sensitive and specific platforms for analysing the molecules at a single molecule level has been strongly triggered by development in the area of nanotechnology [1–3]. Recent years have been witnessing considerable progress in nanotechnology-enabled chemical and biosensors, particularly for the detection of toxic chemicals, biomarkers, pollutants, and hazardous materials [4–6]. A wide variety of analytical techniques, namely mass spectroscopy, electrochemical analyser, optical techniques, etc., are being used for the accurate identification of the ultra-low concentration of the analyte molecules. Among the various sensing approaches employed, optical techniques are emerging as a popular candidate due to its advantages such as rapid detection of very low concentration of analyte molecules and its non-contact nature. Raman spectroscopy is considered as an ideal optical analytical tool as it can fingerprint the molecule of interest and can provide information about the constituents, the symmetry, and the environment. The Raman effect relies on the inelastic scattering of photons arising from the interaction with vibrational oscillations of the bonds in a molecule that spectrally manifests as a set of satellite peaks to either side of central incident photon frequency [7, 8]. The higher photon energy spectral signatures on the central Rayleigh scattering is known as anti-Stokes scattering, and that spectral signature on the lower energy side of the Rayleigh scattering is known as Stoke shifted radiation (schematically illustrated in Fig. 1).

In general, the Raman shift is expressed in units of cm^{-1}, and the corresponding spectrum is called the Raman spectrum. Despite the potential in molecular sensing and obtaining molecular structural information with the high spatial resolution achieved through diffraction limited wavelength, the practical applications of this technique are largely limited even now by the extraordinarily weak Raman scattering cross-section. The typical differential Raman scattering cross section is around 10^{-27} cm^2 $molecule^{-1}$ for dye molecules with conjugated systems and even lower ($\sim 10^{-30}$ cm^2 $molecule^{-1}$) for non-resonant compounds. The scattering cross-section of the Raman effect is much lower as compared to other spectroscopic techniques such as UV ($\sim 10^{-18}$ cm^2 $moleule^{-1}$), IR ($\sim 10^{-21}$ cm^2 $moleule^{-1}$), and

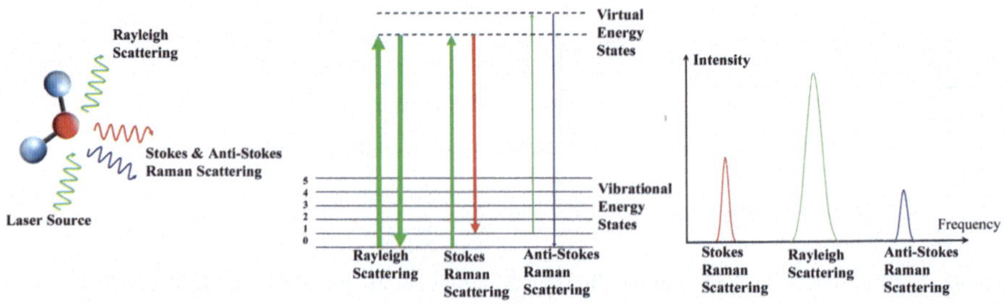

Fig. 1 Schematic of Rayleigh and Raman scattering

fluorescence ($\sim 10^{-24}$–10^{-14} cm^2 moleule^{-1}) [9]. The ultra-low scattering cross section of Raman effect largely limit the real-life applications of this technique.

In 1974, Fleischmann, Hendra, and McQuillan accidentally identified the intense Raman signal from pyridine molecules placed onto a roughened silver surface and speculated that the large Raman intensities arise from the adsorption of pyridine molecules onto the substrate [10]. In 1977, detailed studies carried out independently by Jearmaire and Van Duyne along with Albrecht and Creighton, showed that Raman signal arises from the increase of scattering cross-section of the adsorbed pyridine molecules [11]. Later it is proposed that charge transfer between the molecule is responsible for the experimentally observed Raman signal enhancement [12]. Though non-conclusive yet, it has been shown that the experimentally observed enhancement in Raman signal arises from the plasmonic optical properties of the substrate (electromagnetic enhancement) and molecule-substrate interaction (chemical enhancement). This effect is now popularly known as surface-enhanced Raman scattering/spectroscopy (SERS) [13–15]. As the cause of these effects is independent to each other, they have a multiplicative effect on the experimentally observed Raman signal strength (as illustrated in Fig. 2). The presence of the substrate can alter the available local electromagnetic field intensity for the analyte molecule, possibly can change the symmetry and intensity of the Raman transitions, and can lead to the enhancement in scattered radiation that includes Raman signature.

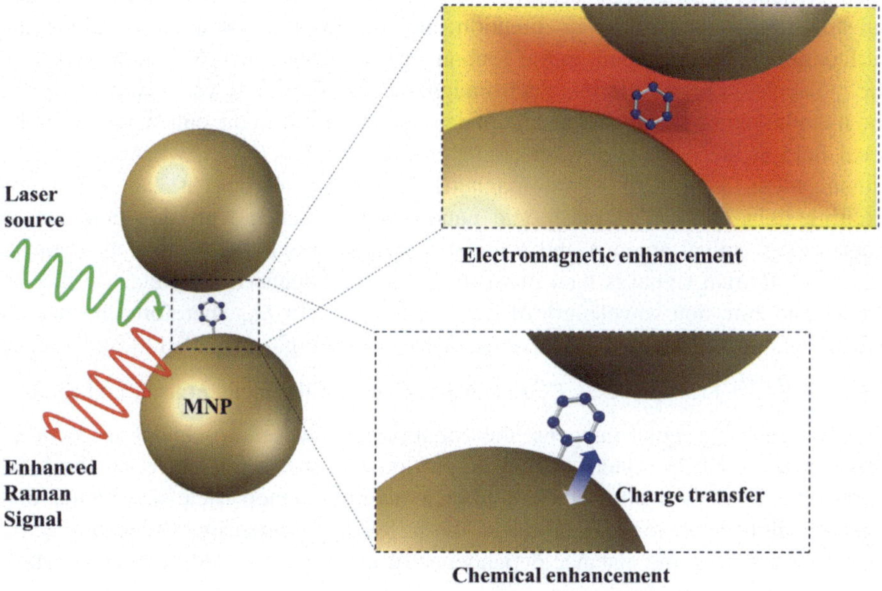

Fig. 2 Schematic of electromagnetic and chemical enhancement mechanism of Raman scattering

The sensing of SERS can be done label-free (direct) or labelled (indirect) schemes wherein the former directly fingerprint the molecule of interest, whereas the latter utilize SERS tags for the signal detection [16–19]. Though the direct SERS scheme can simultaneously map multiple targets, often the spectrum depends upon the chemical structure, orientation, and conformation of the molecule in the complex medium. The indirect SERS scheme offers advantages such as multiplexing capability based on narrow peak widths and high signal stability. However, while multiplexing, the spectral overlap has to be taken into account.

2 Electromagnetic Enhancement Contribution to SERS

The principle of electromagnetic enhancement is based on the magnification of the electric field due to the excitation of localized surface plasmon resonances (collective oscillations of the electrons) of the substrate material. Therefore, the enhancement depends largely on the substrate physical properties as well as on its interaction with the exciting electromagnetic radiation. When the wavelength of the incident light meets the plasmonic resonance condition of the metallic nanoparticles or nanostructures, the surface plasmon polaritons can be excited. It induces a strong electromagnetic field in the metallic nanostructures with evanescent character. However, SERS signal strength not only depends upon the increase in excitation field strength but also depend upon the strength of scattering photon [20, 21]. The enhancement in scattering is achieved when the plasmonic oscillations of the SERS substrate are perpendicular in direction, and the in-plane oscillations (along the surface) does not contribute to the Raman signal enhancement (only adsorption of the molecules) [22, 23]. In the electromagnetic contribution towards Raman signal, the metallic nanoparticles act as an optical antenna result in the enhancement in the local field at the excitation wavelength (λ_e); $E_{loc}(\lambda_e) = En_1 E_0$, where En_1 is the enhancement factor of the electromagnetic field in the nearfield at λ_e and E_0 is the exciting light at λ_e. The plasmonic nanoparticles excited with this field act as transmitting optical antenna and transfer the Raman signal from near field to far field. The Raman signal is then proportional to the enhanced local electric field at the Raman emission wavelength of λ_{em}; $E_{loc}(\lambda_{em}) = En_2 E_0$. Therefore, the overall SERS signal enhancement depends upon "exciting" and "emitting" fields; $En_{SERS} \alpha \left[\frac{E_{loc}(\lambda_e)}{E_0}\right]^2 \left[\frac{E_{loc}(\lambda_{em})}{E_0}\right]^2 = E_{n1}^2 E_{n2}^2$. A resonance of the incident laser and Stoke's Raman scattering signal happens, the enhancement factors En_1 and En_2 becomes equal, and the SERS signal will be proportional to the fourth power of the local electric field. As the local electric field depends upon nanoparticle size, shape, and distance, all these factors can influence the SERS signal intensity. The local electric field depends upon the distance dependence of the molecule, and the metal surface varies as $\left(1 + \frac{r}{a}\right)^{-3}$ where a is the nanoparticle radius, and the corresponding SERS signal varies as $\left(1 + \frac{r}{a}\right)^{-12}$ [24] (as depicted in Fig. 3).

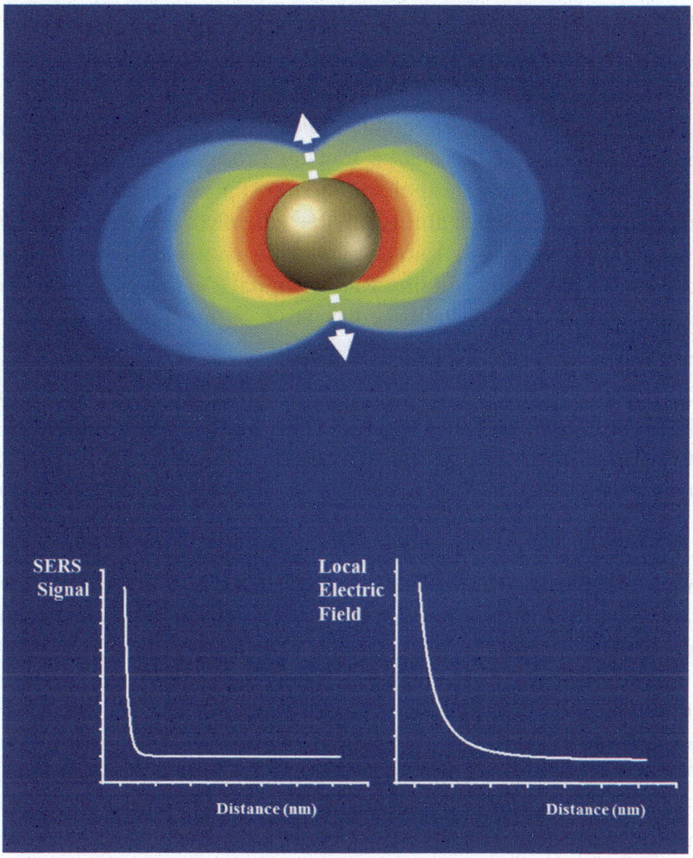

Fig. 3 Schematic of electric field distribution around the single nanoparticle and distance dependent variation of the local electric field strength and SERS signal

Many times, the field from single nanoparticle cannot create sufficient local electric field strength so that researchers are frequently bringing two or more nanoparticles together. When two nanoparticles are spatially brought together, the localized electric field between the particles (hot spots) is very high and lead to considerable enhancement in the Raman signal. When two nanoparticles of diameter D and an interparticle distance of d are placed in a uniform electric field E_0 polarized along the dimer axis, the local field between the gap can be $E_{loc} = E_0 \frac{(D+d)}{d}$ so that enhancement factor becomes $\left(1 + \frac{D}{d}\right)$ Depending upon the distance between the particles, the enhancement factor due to electromagnetic contribution is typically reported to be 10^7–10^{11}.

3 Chemical Enhancement Contribution to SERS

The chemical enhancement contribution to the SERS signal originates from many factors such as the interaction between the molecules and nanoparticle in the ground state, resonant excitation of the charge transfer process between the nanoparticle and the molecules and excitation of electronic transfer within the molecule [25–28]. The modification of the polarizability of the adsorbed molecule onto the SERS substrate implies the change in Raman scattering cross-section of the vibrational modes and hence of the corresponding Raman signal. As is well known, the Raman cross section is proportional to the first derivative of the polarizability with respect to the vibrational normal mode (Q_k): $P_{\text{Raman}} \propto \sigma_k \propto \frac{\partial \alpha(Q_{k,\omega})}{\partial Q_k}$. Besides, the shifted and broadened electronic levels of the adsorbate molecules can lead to the resonance Raman effect. Depending upon the adsorbate, the charge transfer can occur between the LUMO or HOMO layer and the metal surface. For example, in the case of p-aminothiophenol on silver surface, the charge transfer occurs from the metal state near the Fermi level to the lower unoccupied molecular orbital (LUMO) whereas in the case of the aniline-metal interaction, the charge transfer occurs from the highest occupied molecular orbital (HOMO) of the adsorbed molecules to the LUMO of the metal clusters [29]. Owing to the requirement of direct adsorption of the molecules to the metal surface, the molecules with high chemisorption onto the metal surface normally exhibit maximum chemical enhancement in SERS signal. In the physisorption, the substrate-molecule interaction is mediated via van der Waal force and the enthalpy change is small enough (less than −25 kJ/mol) to break the chemical bonds and possibly cause only electronic cloud distortion. On the other hand, the chemisorption mediated via chemical bond have greater enthalpy change (less than −40 kJ/mol) and causes changes in the electronic structure of the molecules and lead to the appearance of vibrational modes forbidden in an isolated state [30]. Typically, the enhancement factor due to the chemical mechanism is in the range of 10^1–10^3.

4 SERS Is Simply an Enhancement of the Raman Effect?

Though Raman and SERS spectra are similar, they are not exactly the same. As compared to the Raman signal, the SERS spectrum can show many spectroscopically forbidden Raman modes as the selection rules can be varied due to the electric field gradient. The molecules in the proximity of the strong localized evanescent wave from the metal surface are affected by the field gradient and can lead to the breaking down of conventional selection rules applicable for Raman spectroscopy. Such forbidden electronic transitions induced by the electric field gradient are reported to observe in the form of resonance Raman bands in the SERS spectra of carbon nanotubes. In such cases, the surface selection rules determine the Raman bands. As explained earlier, the enhancement of various Raman active modes depends upon the in-plane and perpendicular field components and due to the dominance of the Raman tensor diagonal components with Cartesian coordinates parallel to the surface normal. Thus, only dynamic dipoles perpendicular to the surface contribute to the signal and parallel components—if any—are normally weak in the spectrum. In addition, the coupling rate, which is a measure of the interaction of the plasmon modes with the molecular excitonic resonances, depend upon the oscillator strength of the molecular resonance and the mode volume of the confined fields. The approximation of weak coupling in which both resonances can be treated independently fails when the coupling rate approaches the dephasing rate of two resonances. In such a condition, the energy is exchanged coherently between the plasmon and exciton so that hybrid resonances get formed and the transition from the weak to the strong coupling regime is experimentally observed as Rabi splitting or Fano-like resonances in the scattering spectrum [31, 32]. The interaction of the plasmons of the nanoparticles results in hybridized plasmon modes with low energy "bonding" and high energy "anti-bonding" plasmon modes which correspond to the symmetric and asymmetric coupling of the modes, as in the case of the molecular orbital theory of the atomic orbitals.

5 HOT Spots

It is the spatial region at the metallic substrate surface where the local fields are extremely high as compared to the excitation field, and Kleiman defined hot spot as "*a junction or close interaction of two or more plasmonic objects where at least one object has a small radius of curvature on the nm scale*" [33]. In general, the field increase around a single particle with high curvature may not be as high as a junction. When two nanoparticles are brought together, the peak position of the extinction spectra may get altered due to dipole-dipole coupling and normally lead to the red shift in the extinction spectra. Aside large enhancement factor for SERS signal with the hot decrease in spot distance as schematically shown in Fig. 4 [34], the shift in peak position wavelength as compared to the isolated particle depends upon the nanoparticle distance.

Moreover, some additional peaks can arise from the higher order coupling between the two nanospheres (Refer Fig. 6.19 of the Ref. [22]). The enhancement factor strongly depends upon the distance between the particles as it increases from 10^6 to 10^{11} for a change in distance from 10 to 1 nm [30]. The increase in hot spots and consequent enhancement in Raman signal due to the aggregation of more nanoparticles as compared to the two nanoparticle brought together (dimer configuration) is found to have a less pronounced effect, with a maximum increase of an order of magnitude or so. Moreover, the reduction in the distance between the nanoparticle to an extent $g < 1$ nm can lead to the quantum mechanical effects like quantum tunneling. When $g < 2$ nm, the enhancement is averaged over the whole dimer and produce 300 times less enhancement as compared to the hot spot. However, for a given interparticle distance, the shape of the particle also plays an

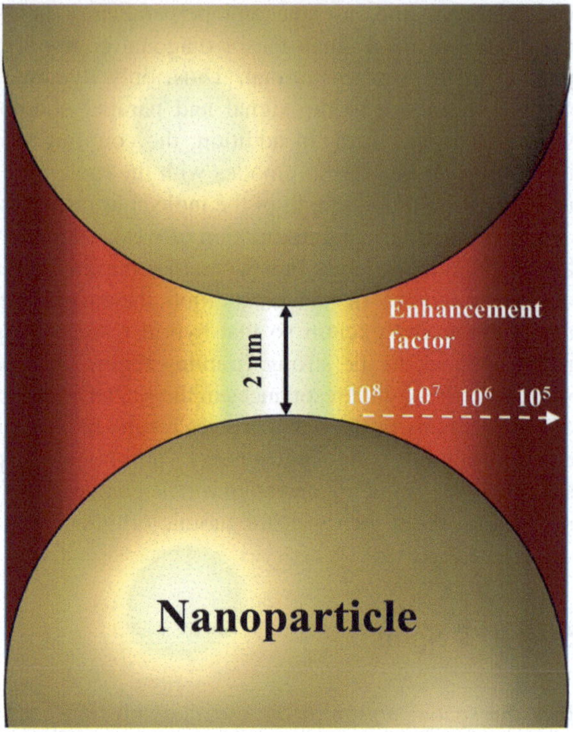

Fig. 4 Inter nanoparticle-distance dependent variation of the enhancement factor in SERS signal

important role in signal enhancement. For example, when the nanoparticle distance is in the order of 1–2 nm, the cube-shaped particles exhibit two modes of local surface plasmon resonance in dependence of the curvature of the cube corners. Aside spherical and cube nanoparticles, nanorods, and star or triangular shaped particles are also of paramount interest to fabricate SERS substrates [33]. It is worthwhile to note here that there is no correlation between the extinction coefficient of the single nanoparticle and the SERS signal enhancement at the hotspot. Most of the reported SERS hot spots are point-like, one-dimensional linear, or two-dimensional planar geometries that can produce high SERS signal. However, the laser focus volume is 3-D and thus the aforementioned hot spot geometries under-utilizing the available confocal volume for the SERS signal. Thus, the recent efforts are made to fabricate 3-D hot spots [35–37]. However, 3D hot spot fabrication has to overcome the challenges such as the decrease in electric field enhancement (i) due to quantum tunnelling when the interparticle distance is less than 1 nm (ii) charge exchange phenomena occurring when the particles touch each other (iii) electrostatic shielding due to the equipotential energy created due to the solid contact. Nevertheless, many successful efforts are reported recently to fabricate hot spots in 3D space.

Despite the progress in fabricating hot spots in a controlled manner, only a small portion of the entire SERS substrate can be considered as a hot spot. In a study, it has been shown that 35% of the surface of the Ag sphere and from 0.6% of the surface of the Ag dimer contribute towards the 80% of the total SERS signal [38]. Moreover, the efficacy of SERS drops drastically over a few nanometres away from the hotspot. As explained earlier, smaller the distance between the particles, higher the enhancement in the signal but comes at the cost of reduced probability of finding the hotspot. Thus, the molecules can easily diffuse out of the hotspot region and cause significant variation in the Raman signal. Thus, the focus should be to fabricate a hot surface rather than hotspots.

6 Why Metals Are Preferred for SERS?

In the quasi-static approximation, when excited with an external electric field, the electric field generated in a sphere can be obtained from the expression

$$E_{in} = \frac{3\varepsilon_d}{\varepsilon(\omega) + 2\varepsilon_d} E_0 \tag{1}$$

Here $\varepsilon(\omega) = \varepsilon'(\omega) + i\varepsilon''(\omega)$ is the complex dielectric constant of the material that forms the nanoparticle and ε_d is the dielectric constant of the medium in which nanoparticle is immersed. The real part of the dielectric constant represents the strength of the polarization, whereas the imaginary part accounts for the dissipation process in the material under the electric field. Conduction electron losses are due to electron-electron scattering, electron-phonon scattering, and scattering of electrons at the surface and at the grain boundaries or with lattice defects. The interband transitions account for bound electron losses in which absorption of photons promotes the electrons to higher empty bands. In the resonance condition, when the denominator of Eq. (1) is close to zero, i.e. when $\varepsilon'(\omega) = -2\varepsilon_d$, leads to the following expression

$$E_{in} = \frac{3\varepsilon_d}{i\varepsilon''(\omega)} E_0 \tag{2}$$

It imposes restrictions for the dielectric constant of the nanoparticle material such as the real part has to be negative and imaginary part should be small to maximize the electric field according to the Eq. (2). These conditions are often satisfied for metals in the optical and NIR frequency range. The popular Drude model that treat conduction electrons as free electron gas provides an expression for dielectric constant as

$$\varepsilon(\omega) = 1 - \frac{\omega_p^2}{\omega(1 + i\Gamma)} \tag{3}$$

Here, the plasma frequency for conduction electrons with density n and effective optical mass m^* can be obtained from the expression $\omega_p = \sqrt{\frac{ne^2}{m^* \varepsilon_0}}$. Γ is the total damping rate. Thus, when $\omega_p > \omega$, $\varepsilon'(\omega) < 0$ and when, $\omega_p < \omega$, $\varepsilon'(\omega) > 0$ An estimation of the plasma frequency of the many metals shows that they fall into the UV range of the electromagnetic spectrum, for examples it is 8.7 eV (142 nm) for copper, 8.9 eV (139 nm) for gold, 9.2 eV (134 nm) for silver and 12.7 eV (98 nm) for aluminium [39]. Thus these metal are better candidates to satisfy the resonance condition of Eq. (1) in the visible and infrared spectral range of electromagnetic spectrum as compared to dielectrics ($\varepsilon'(\omega) \sim 1 - 10$). However, the neglected intraband transition plays a critical role in the optical response of the metallic nanoparticles, especially when they overlap with the plasmonic excitation band. In the case of silver nanoparticles, the intraband transitions are at 4 eV (310 nm) which is far away from the plasmonic resonance absorption band ($\sim$420 nm for aqueous dispersion) [40]. On the other hand, the for gold and copper, the intraband transitions occur at 2 eV (620 nm) and thus partially coupled with a plasmonic band that falls in the 2–2.5 eV and causes the redshift and broadening of the plasmonic resonance of gold and copper [39, 40]. Owing to the dependence of the real term of dielectric constant on the polarizability and the imaginary term on the losses, the quality factor for a spherical nanoparticle can be defined as

$$Q_{LSPR}(\omega) = \frac{-\varepsilon'(\omega)}{\varepsilon''(\omega)} \tag{4}$$

The negative value of dielectric constant for a wide range of frequencies ensures that the metals can be used for visible or infrared part of the electromagnetic spectrum. Among the metals, silver shows the lowest losses as the intraband transitions have higher energy as compared to other metals. Aluminium is a good substrate in the UV range where it exhibits negative $\varepsilon'(\omega)$ and low loss $\varepsilon''(\omega)$. Nevertheless, silver and gold are widely used in SERS studies as they are air stable, and other metals like copper and aluminium are more reactive. The spectral position of the surface plasmons can thus be tailored by changing the surrounding medium, nanoparticle shape, and size. As compared to the nanoparticles, in the case of metallic surfaces, propagating surface plasmon polaritons (PSPP) mode can be found at boundaries between the dielectric and metal conductor (as depicted in Fig. 5).

When the incident wavelength matches with plasmon resonance condition, a strong electromagnetic field with evanescent wave character is induced on the metallic nanostructure. This field enables the squared field incident intensity of the Raman modes of molecules adsorbed to the substrate, and Raman scattering occurs in all spatial directions due to Hertzian dipole nature of the molecule [41]. It is pertinent to note here that recently, other materials liked doped semiconductors, oxide, chalcogenides are being explored as an alternative plasmonic substrate for SERS studies [9, 42–47].

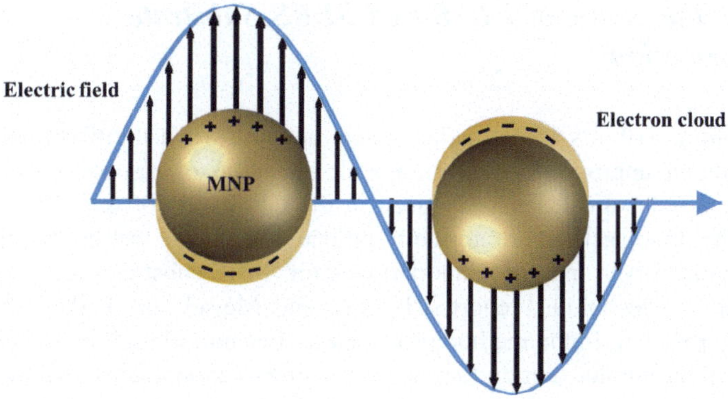

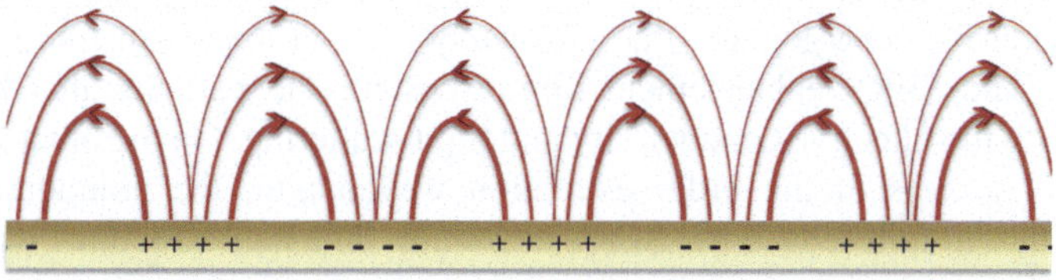

Fig. 5 Schematic of localized surface plasmons around a single nanoparticle (top) and the propagating surface plasmon polaritons at the metal-dielectric interface (bottom)

7 SERS Substrate Fabrication Approaches

The fabrication of nanostructures for SERS substrates can be done using the bottom up or top down approaches as well as template method. It has been well established that the plasmonic field around a metallic nanoparticle can contribute towards the Raman signal, tailoring of the fluorescence emission of a nearby fluorophore, altering the quantum efficiency of the fluorophore by controlling the radiative and non-radiative channel of deexcitation of an excited molecule, etc. [48–55]. Typical requirements for a good SERS substrate are: high reproducibility in Raman signal from different parts of the substrate and between the substrates, high enhancement factor, good shelf life, large area with low-cost of production, chemical inertness, good adsorption of the molecule of interest to the substrate, no photochemistry upon excitation with laser, and low or little toxicity. Though there are several techniques existing in the literature to fabricate the SERS substrates, a few of the commonly used approaches are outlined below.

7.1 Metallic Nanoparticle-Based SERS Substrate Fabrication

Owing to the possibility to act as an optical antenna when the plasmonic field is excited with an appropriate electromagnetic radiation of resonance wavelength, metal nanoparticles are extensively used in the fabrication of the SERS substrate [56–60]. The nanoparticles are normally prepared using the wet chemical process, and the popular silver and gold nanoparticles for SERS substrates are prepared via citrate reduction technique, reported by Lee and Meisel and Turkevich method, respectively [61, 62]. In the method proposed by Lee and Meisel, $NaBH_4$ is used to reduce the silver nitrate solution to prepare the silver nanoparticles, and the addition of salts could yield aggregated metal nanoparticles. On the other hand, in the Turkevich method, reduction of chloroauric acid by citrate yield gold nanoparticles. By controlling the experimental conditions like temperature, ionic strength, and reactant concentrations, it is possible to adjust the particle size, shape, aggregation properties, surface charge, etc., that ultimately play an important role in the SERS efficiency. The SERS enhancement diminishes for the particles that have a low radius due to the effective conductivity and light scattering. On the other hand, when the particle size is in the order excitation wavelength, the nonradiative modes dominate that in turn reduce the SERS efficiency. The nanostructures with different morphologies such as nanorods, nanotriangle, nanocubes, nanocages, nanostars, nanoflower, etc., have been employed to fabricate the SERS substrates [63–67]. The studies imply that the increased sharpness, roughness, and plasmonic hot spot density contribute positively towards the SERS signal enhancement. In general, the SERS efficiency scales as follows: nanoparticles < aggregated nanoparticles < nanotriangles ≪ nanostars and the difference arise from the difference in hot spots per particle. Besides the Lee-Meisel protocol, many other approaches such a reduction of silver nitrate by reducing agents such as sodium borohydride [68–70], hydroxylamine hydrochloride [71] etc. have been reported to synthesize the silver nanoparticles. Recently, bimetallic nanoparticles consisting of two metals, for example, Ag and Au, with varying shape, size, and structure are being used as opposed to the conventional monometallic nanoparticles [72–76]. The advantage of using the nanoparticles for SERS substrate fabrication is that it allows surface modification through chemisorption of the structurally modified ligands On the other hand, the aggregation of metal nanoparticles often raises a big challenge in the reproducibility of SERS spectra. Moreover, analyte solution must be mixed with metal nanoparticle solution to obtain SERS and thus limit sampling in some applications, such as in examining adsorbates on non-SERS active surfaces. The sampling requirement has been addressed by shell isolated metal nanoparticles by Tian group recently [77]. Additionally, the particles with a plasmonic particle as the core and the dielectric material as a shell are emerging as a promising candidate due to increased stability, the controlled distance between the particles and thus hot spot size [78–82]. In this case, for example, colloidal gold nanoparticles are prepared first and then mixed with pH-activated silicate. The Raman reporters can be incorporated at this step to create SERS labels. The tetraethylorthosilicate (TEOS) and ammonia can be used to grow the silica shell of desired thickness. The SERS efficiency depends upon the clustering degrees (i.e., monomers, dimers, etc.), and it is found that trimer exhibits maximum SERS signal enhancement. The metal nanoparticles can be irreversibly immobilized onto the planar surface to fabricate SERS substrates such as glass by using derivatized silanes as the immobilizing agent. Metal nanoparticle assisted fabrication approaches to involve various techniques. Self-assembly of the nanoparticles that include chemical attachment of metal nanoparticles to solid substrates, electrostatic interaction mediated attachment of metal nanoparticles onto the solid substrate, capillary force driven force metal nanoparticles onto the substrate, direct transfer of pre-assembled metal nanoparticles, etc. are a few commonly adopted approaches for the SERS fabrication [83–85]. In the chemical attachment, the surface is modified with functional groups so that one of its functional group enables the anchoring of the molecule and other functional groups facilitate the binding of the nanoparticle to the substrate. In the case of the glass substrate, first, the surface will be functionalized with an amine of thiol group via surface polymerization procedure and then immerse in metal nanoparticle suspension for fabricating SERS substrate. The chemical functional groups of the substrate, incubation time between the nanoparticle and functionalized substrate, metal nanoparticle concentration, and the surface adsorbents on the metal nanoparticles ultimately determine the enhancement factor in this case. The self-assembly of metal nanoparticles via electrostatic attraction of polymers or biomolecules are exploited to fabricate SERS substrates. Though the reproducibility of such substrate is questionable, the studies show that the substrates prepared via this route provide a better sample to sample variation. The capillary force present during the evaporation of the liquid is also exploited for assembling the nanoparticles to make SERS substrates. Following the Langmuir-Blodgett technique, the metal nanoparticle layer at the fluid-fluid interface can be transferred to a solid surface to fabricate SERS substrate. Of late, the laminar flow inside a microfluidic channel and the consequent concentration gradient of chemicals created has been explored to mix the metallic nanoparticles with the samples and for SERS measurements [86–90].

Another promising emerging approach is to utilize the liquid-infused porous surface-enhanced Raman scattering substrates (SLIPSERS). Herein, a slipper liquid-infused porous surface consists of a film lubricating fluid lock in place by a micro/nanoporous substrate that creates a smooth and stable surface that eliminates the pinning of the liquid contact line (Fig. 6). The evaporation of metal nanofluid along with analyte molecules without depinning the contact line enables the clustering of the nanoparticles and the molecules to obtain the SERS signal [91].

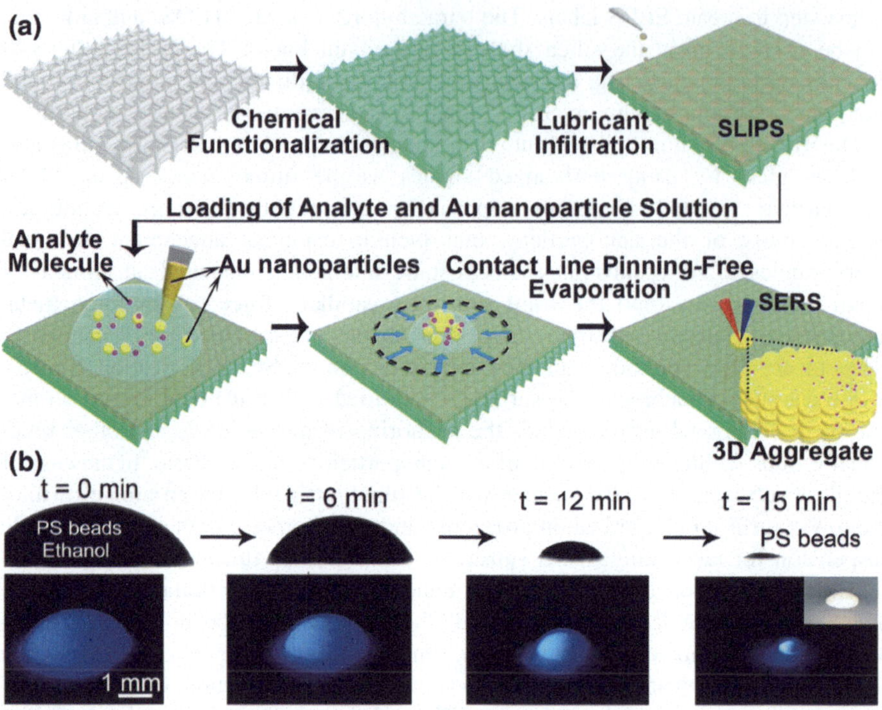

Fig. 6 Concept of SLIPSERS. **a** Schematic illustration showing the concept of SLIPSERS. Any analyte can be enriched to form a small aggregate, regardless of its interactions with the Au nanoparticles, facilitating subsequent SERS detections. **b** Visualization of the analyte enrichment process on SLIPS using luminescent PS spheres dispersed in ethanol. (Inset) Optical image of the enriched PS sphere aggregate. Reproduced Fig. 1 with permission from PNAS 113(2), 268–273 (2016)

7.2 Fabrication of the Nanostructures on the Substrate Via Lithography

Most of the nanoparticle-based routes fall into the bottom-up approach to fabricate the SERS substrates wherein respective chemical, biochemical, or electrochemical reactions determine the SERS structures. These approaches often fail to fabricate the structures of desired properties to cross-validate the results obtained through theoretical investigations. The top-down approach wherein the fabrication of plasmonic structures of desired features are considered to be a favorable alternative, especially for highly specific applications like the bio-molecule analysis. In this context, various nanolithographic techniques are being used to fabricate the plasmonic nanostructures for SERS studies.

One of the popular nanolithographic technique is the electron beam lithography wherein 10–50 keV electron beams are scanned across radiation sensitive polymer (making them either less or more soluble in the developer solution) coated on solid SiO_x/Si wafers. The electron beam exposed or non-exposed region can selectively be etched off from the substrate so as to fabricate the structures of dimensions as small as 20 nm [92–96]. The deposition of metal makes these substrates as plasmonic-active. The commonly used electroresist is polymer poly(methyl-methacrylate), which is a positive electroresist. A typical schematic procedure adopted for SERS substrate fabrication using electron beam lithography is shown Fig. 7. As illustrated in Fig. 7, the metallic coating can be done after carrying out reactive ion etching of the electroresist or even before the etching process [94]. The fabrication of continuous film proven to provide better SERS efficiency as compared to the discrete metal nanoparticles on the surface due to the presence of several sharp surfaces. The electron beam lithography technique offers the advantage of fabricating ordered homogenous structures with well-controlled interparticle distance. This technique allows fabrication of differently shaped nanostructures such as nanodisks, nanoholes, aperiodic nanostructures, etc., for the potential application in the field of SERS [15].

An average enhancement factor of 10^9 has been reported for the substrates fabricated via oblique angle deposition technique combined electron beam lithography technique [97]. Therein, the metal nanoparticle dimers are separated by distance ~ 1 nm achieved via double mask technique by depositing two metallic layers at two different angles. Instead of an electron beam, the focussed ion beam is also used to lithographically fabricate the SERS substrates, especially for fabricating the nanohole arrays in a plasmonic substrates. For this normally focussed ion beam machine with Ga^+ ions accelerated at a high voltage of 30 keV is focussed on a plasmonic substrate so that ion beam mills out layers of the surface to create holes with a spatial resolution as good as 10 nm [98, 99]. However, electron or ion beam lithography technique is unfavorable for large area fabrication and cost-effective approach. Moreover, fabrication of substrates of gaps less than 10 nm routinely is difficult via this technique.

Laser-based lithographic techniques are emerging as an efficient tool for large area fabrication of SERS substrates on different kinds of substrates [100, 101]. In extreme ultraviolet interference lithography, instead of using an electron beam, the pattern is written on the photoresist using the interference of coherent synchrotron light beams operating at 1.35 nm [102–105]. The direct laser patterning via ablating the substrates using picosecond and femtosecond laser pulses is emerging as green and rapid means of fabricating large area SERS substrates (as schematically shown in Fig. 8).

The usage of femtosecond laser offers advantages such as minimum heat affected zone, little residue, no usage of additional chemicals, etc. The capability of direct laser patterning for fabricating SERS substrates has been successfully demonstrated in the case of semiconductors, metal, and polymers [106, 107]. The potential of the high-intensity radiation to in situ reduce the metallic nanoparticles also been exploited to fabricate the SERS substrates.

Nanosphere lithography is one of the widely used techniques for fabricating SERS substrate as it is inexpensive and provide inherently parallel, high throughput

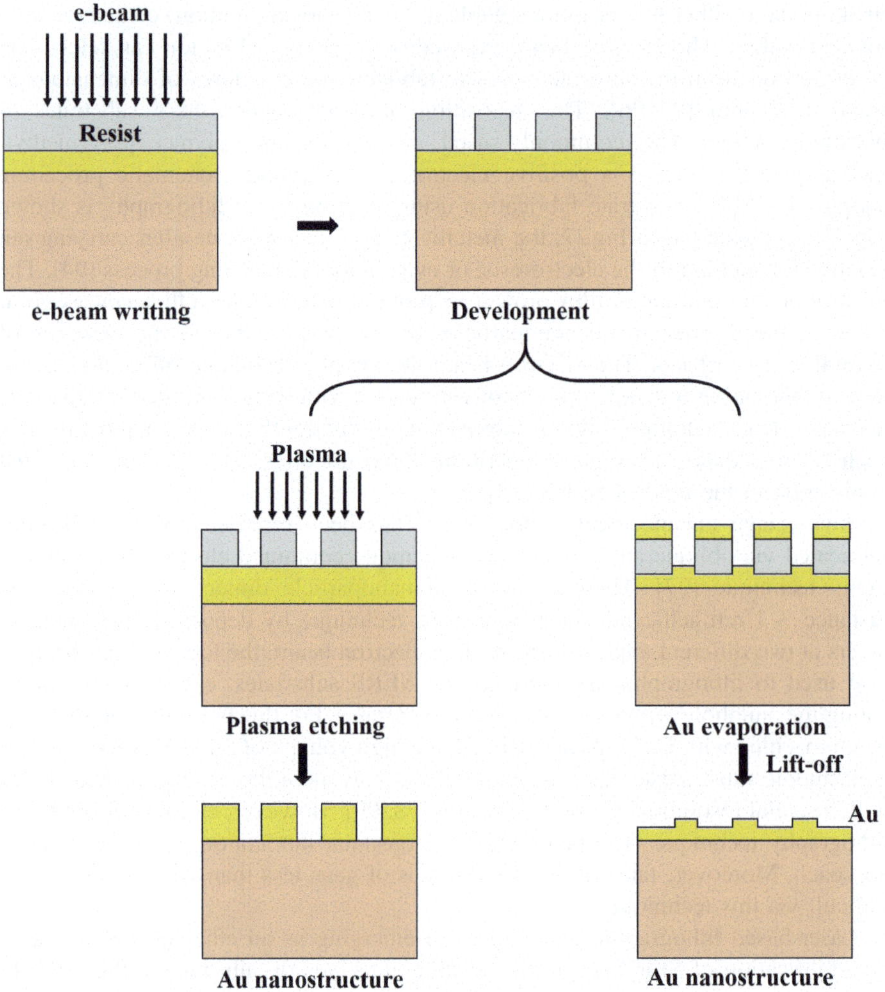

Fig. 7 Schematic of electron beam lithography technique to create plasmonic nanostructures

substrates [108–111]. The basic procedure for the fabrication of SERS substrates using this technique is as follows: Initially, the substrate is drop coated with nanosphere SiO_2 particles to self-assemble to form a mask on clean conductive ITO substrate or evaporated metal substrate over the glass. Following it, a metal layer of controlled thickness is deposited through the mask by physical vapor deposition or electrochemical deposition. It may lead fabrication of three different kinds of SRES substrates. The direct physical vapor deposition onto the nanospheres create plasmonic metal over nanosphere surface (FON), and the removal of nanospheres via sonication or reactive ion etching create the triangular footprint. The removal of spheres following the electrochemical deposition create a thin film of hexagonal

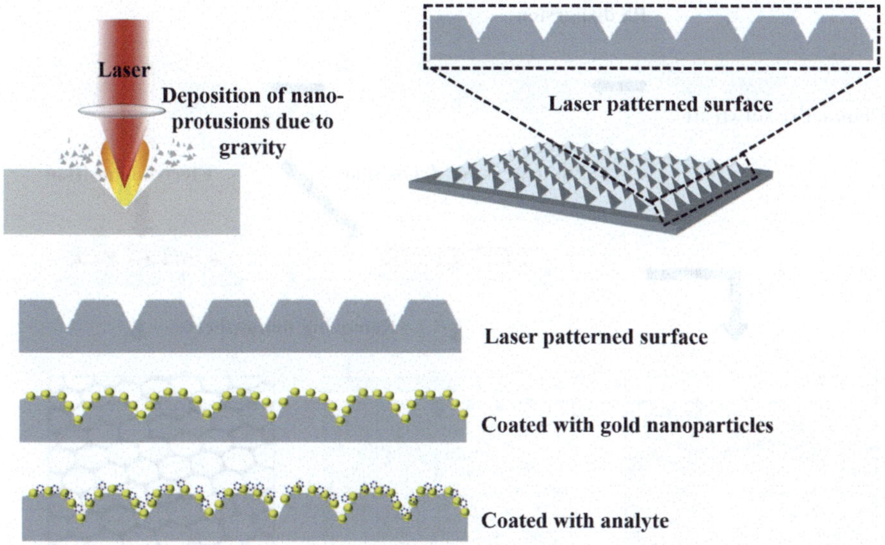

Fig. 8 Schematic of direct laser based lithography to fabricate SERS substrate

array uniform nanosphere voids. The structural parameters of the SERS substrate can be tailored to match the excitation wavelength by controlling the metal coating film thickness and the nanosphere radius (as schematically shown in Fig. 9).

In combination with soft lithography, nanosphere lithography has been used for fabricating flexible SERS substrates. In one approach, a soft elastomer like polydimethylsiloxane (PDMS) is poured onto the template made by the self-assembly of nanospheres. After curing at the appropriate temperature, the peeled off bowl-shaped PDMS is further coated with metallic silver films by sputtering of fabricating SERS [112, 113]. A similar replication procedure via soft lithography can be done from a laser patterned surface followed by in situ reduction of the metallic nanoparticles, as shown in Fig. 10.

Such substrates are reported to provide a femtomolar limit of detection of analyte rhodamine 6G molecules [114]. The nanoporous alumina made by the electrochemical method is also shown to be an efficient SERS substrate after depositing with a plasmonic metallic layer [115–119]. Herein, by controlling the applied voltage, type of acid, and temperature, it is possible to control the pore parameters such as diameter and depth. It is possible to obtain the metal nanorods or nanotubes by the dissolution of the alumina layer in an electrodeposited metal layer onto the alumina. Additionally, the nanoporous alumina can be used as a master structure for replicating nanopillars onto the flexible polymers to create SERS substrates [120].

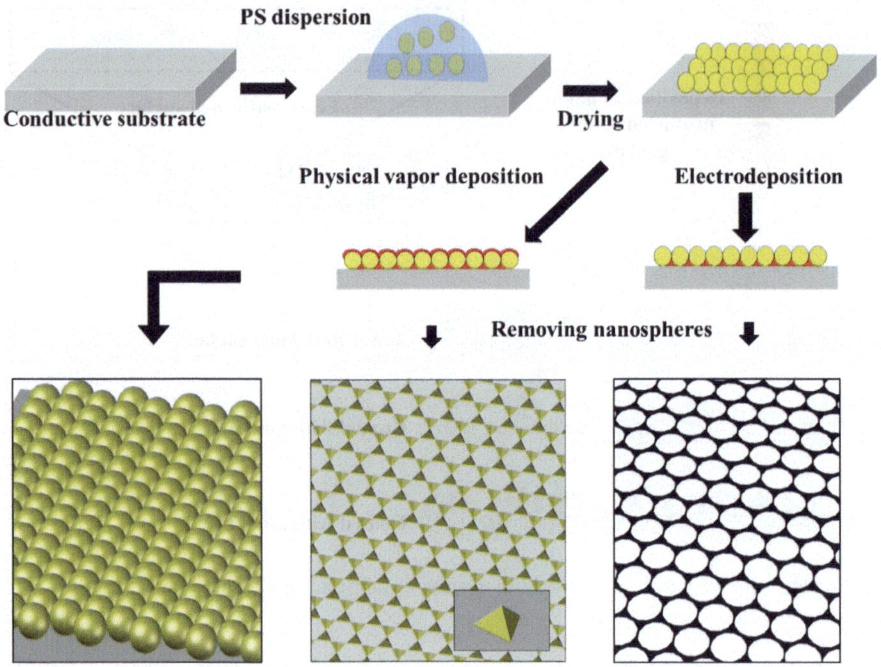

Fig. 9 Schematic of the nanosphere lithography technique to fabricate three differently ordered SERS substrates

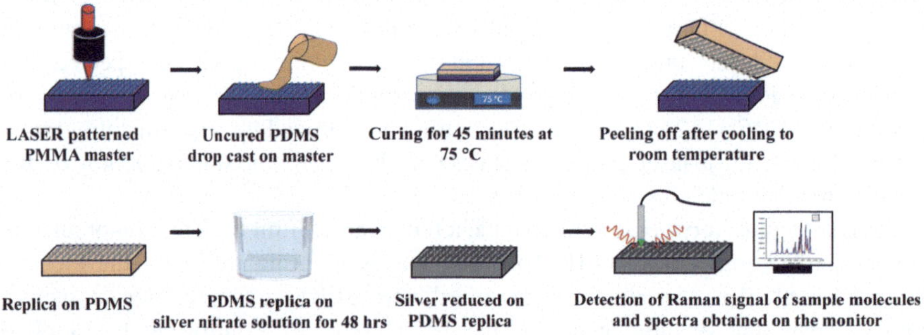

Fig. 10 Schematic of SERS substrate fabrication from laser patterned surface on soft elastomer via soft-lithography followed by in situ metal nanoparticle reduction

7.3 2D SERS Substrates

Owing to inherent atomically flat planar nature of the surface with no dangling bonds, 2-D materials like graphene, hexagonal boron nitride, and molybdenum disulfide are explored as a promising candidate for Raman substrate and target capture [121–123]. These 2D materials contribute towards the chemical enhancement factor of the Raman signal via ground state charge transfer mechanism [124–127]. The ground state charge transfers between the 2D substrate and the analyte molecule does not require chemical bond formation, but it can modify the electronic distribution of analyte molecules. In the case of widely used graphene, the ground state charge transfer occurs easily as it has abundant π electrons on its surface and has a continuous energy band. As illustrated in the Fig. 11, instead of excitation of the molecule to the virtual energy level and subsequent relaxation to the vibrational subs-structure, the graphene electrons are involved in Raman scattering process via an electron-phonon coupling.

The vibrational modes involving the lone pair or π electrons have stronger coupling with the graphene and thus exhibit the highest enhancement in Raman signal when these 2D materials incorporated along with plasmonic structures can provide a combined effect of high local field enhancement and chemical enhancement [128–130]. Traditionally, the 2D materials are prepared by physical or chemical exfoliation or chemical vapor deposition. These chemically vapor deposited 2D graphene, or MoS_2 can be transferred to the metal surface to fabricate hybrid SERS substrates. However, to ensure the 2D conformity over the SERS substrates, the nanopyramid or nanocone structures for the metallic structures are preferred, though some ripples are unavoidable. Alternatively, metallic nanoparticles can be coated onto the exfoliated 2D materials to fabricate the hybrid SERS substrates. Another approach is to sandwich the Ag nanoparticles between the layers of graphene oxide and reduced graphene oxide. The hybrid SERS substrates are found to be efficient in detecting a single molecule of analytes like Rh6G and lysozyme and provide a good detection limit of 10^{-9} M for molecules with very low Raman scattering cross section such as dopamine and serotonin. Other materials like black phosphorous, MXenes also recently using in the fabrication of 2D SERS substrates [42].

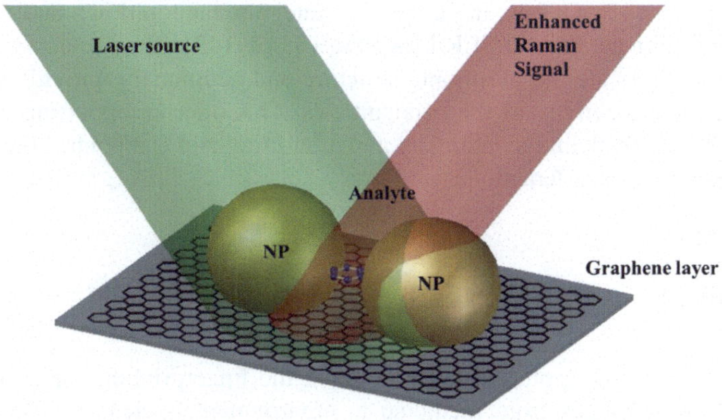

Fig. 11 Schematic of plasmonic nanoparticle integrated 2D-SERS substrate

7.4 3D SERS Substrates

The prerequisite of the tight focusing of incident light on the correct plane to achieve optimal SERS signal makes them a less favorable candidate for on-site applications and lead to the development of 3D SERS platforms. Extension from the 2D substrate to 3D platform increases the overall surface area available for the molecules, especially in the third dimension. Though the hotspot number in 3D structures is expected to be more than the 2D surface, the number of hot spots per nanoparticle may not always be greater as compared to 2D nanoparticles. Moreover, beyond the skin depth of the excitation radiation, the 3D architecture has little influence on the SERS signal [131–133]. Nevertheless, more hotspots in terms of absolute number in a 3D space can contribute positively towards the SERS signal. The 3D SERS substrates can be fabricated through bottom-up as well as a top-down approach. In the bottom-up approach, the chemical properties of the nanoparticle are exploited to self-organize into larger assemblies to from 3D-SERS substrates. Other bottom-up approaches, such as template metal deposition, electroless deposition, and direct chemical growth, are also explored for the fabrication of 3D-SERS substrates. On the other hand, the top-down approach utilizes a lithographic technique like photolithography to fabricate 3D SERS substrates. The self-organized differently shaped 3D plasmonic clusters can be fabricated by the usage of organic molecules or nucleic acid as linking element. The behavior and morphology of the superstructure and the SERS signal enhancement largely depend upon the shape of the clustering unit. Many particles like Ag polyhedral, core-shell Au-Ag nanorods are being used plasmonic particle to fabricate highly efficient reproducible 3D-SERS substrates [134, 135]. The highly symmetric and strongly coupled 3D plasmonic structures can be obtained through DNA driven self-assembly. The usage of DNA facilitates well control over the inter-nanoparticle distance [136–138]. In the case of a top-down approach, aside from photolithography, other techniques such as optical lithography, deep reaction ion etching, electron beam lithography, ion beam lithography, etc., can be used to 3D SERS structures [139–141]. Of late, the trend is to integrate the top-down and bottom-up approach to fabricate the 3D-SERS substrates [142, 143]. The top-down approach can precisely fabricate the plasmonic structure and additionally provide the active sites for the self-assembly of the nanoparticles. This offer an opportunity to fabricate hot spots at predefine locations in a regular array and can control the near and far field nanoparticle interactions.

8 Applications of SERS Substrates

The possible sensing applications by utilizing the fingerprinting capability of the SERS technique ranges from medicine to environmental sciences. The advancement in the field of photonics and nanotechnology-enabled the realization of affordable, highly sensitive portable Raman spectrometers so as to extend the applications of this technique to many unexplored areas. As there exists a large quantum of focussed reviews on the application of SERS technique in diverse areas, herein, the effort is to delve into a few interesting and emerging applications. The readers are strongly encouraged to refer to the recent reviews for a more in-depth understanding of applications. As mentioned earlier, the choice of direct or indirect SERS technique largely relies on the complexity of the matrix that the Raman technique probe for the signal. The direct approach commonly employed for the molecules with high Raman cross-section like small molecules, explosives, contaminants, etc. The indirect approach requires tagging moieties to selectively recognize and capture the molecule of interest (for e.g., via antigen-antibody interaction). Irrespective of the method is direct or indirect, many times, the matrix effect arises in the complex mixture of biological fluids (blood, serum, saliva) or from phospholipids of cell membranes. A few of the potential areas wherein the SERS technique find applications are outlined here.

8.1 Biosensing and Bioimaging Applications

The widespread application of SERS technique for biosensing and bioimaging stems from the fact that highly photostable Raman targets have sharp spectral bands and it can be multiplexed and provide optimal contrast upon near-infrared excitation and thus devoid autofluorescence from other biological samples such as blood, tissue, urine, etc. Identification of multiplexed biomarkers via non-invasive Raman technique will greatly facilitate clinicians in accurate diagnosis and monitoring the progress of treatment [144]. However, the detection of the ultra-low concentration of possible biomarkers present in the body fluids demands the utilization of SERS technique for the measurement with high sensitivity and specificity. Label-free SERS technique for bio-analysis often suffers from lack of specificity originating from the similarity in vibrational peaks of the biomolecules. In the labeled SERS technique, the specific target is detected through the probe and the ligand molecule on SERS nanotags. The specific detection of the biomolecules is normally carried out using the dyes or the fluorophores. On the other hand, the SERS nanotags comprise of Raman signal-enhancing plasmonic nanoparticle, Raman labels or reporter linked to the plasmonic particle surface for identification, the shell around the nanoparticle for the protection and target recognizing ligand. The choice of Raman reporter has to satisfy the conditions like high Raman scattering cross-section, a minimum number of Raman bands for multiplexing, high photostability, surface groups that bind the metal nanoparticle surface with target molecules. Commercially available dyes such as BODIPY, Cy5.5 are typical examples of Raman reporters. The protective layer is normally made from biomolecules like bovine serum albumin (BSA) or polymers like PEG or glasses like silica. Depending upon the specific cases, the corresponding antibodies can be used as ligand. The SERS technique is commonly used to detect the pH in microcellular environments. For example, high glucose metabolism rate and poor vascular perfusion lead to the variation in intercellular pH value during the progress of cancer [145, 146]. It is reported that the signal intensity of the Raman reporter (para-mercapto-benzoic acid (pMBA)) on Au nanostar substrate at 1700 cm^{-1} is decreased for a variation in pH from 5 to 9 with a concomitant increase in intensity at 1014, 1136, and 1390 cm^{-1} [147]. Kneipp group did extensive work on exploring the application of SERS technique as a pH probe [148, 149]. The immunoassay platform of SERS technique to detect the protein biomarkers has been explored in a microfluidic environment [150]. The potential of this technique to detect nucleic acid using the SERS based lateral flow and SERS nanotags via ligase chain reaction has been explored [151]. Besides, the capability of the SERS technique to detect other biologically relevant samples such as glucose, lipids, pathogens, etc. are also demonstrated successfully [152–156]. Aside from the biosensing applications, of late SERS technique has been used for in vitro and in vivo bioimaging. The in vitro studies are carried out for biomarkers using nanotags in tissue samples as well as in cell lines. The cancer biomarkers-EGFR, CD 44, TGFβRII in a breast cancer model is detected by using biocompatible SERS nanotags in a multiplexed platform [157]. In another approach, NIR responsive gold nanostars with resonant dyes are multiplexed for SERS bioimaging. In another interesting work, new folate-targeted SERS tags are employed for selective bioimaging of FR overexpressed cancer cells [158]. The strategy employed therein is to cover the plasmonic gold nanoparticle with Raman active azide derivatives to increase the number of molecules to conjugate with folate cyclooctyne derivatives by the copper-free click relation. Such SERS nanotags can be selectively bound to the FR positive cancer cells and can be exploited for distinguishing between FR positive and FR negative cells. Along with the advancement in microscopic techniques, the SERS technique is now being explored for in vivo imaging of small animals like zebrafish. Raman mapping of the microinjected SERS tags in a zebrafish embryo at once cell stage is carried out at different five stages ranging from 6 to 96 h post-fertilization to evaluate the distribution and different types of tissues [159]. Owing to the potential of probing at the cellular level optically, this technique can be a good alternative for magnetic resonance imaging (MRI).

8.2 Food Safety Evaluation

The development of a reliable and rapid technique is the demand of consumer to probe the safety and quality of the food product. The fingerprinting capability and non-contact nature of the SERS technique make them a favorable tool to address the requirements of the food industry. However, the identification of trace level of a target analyte in a complex matrix like food is a challenging task as the Raman signature of the target is often masked by the other molecules present in the matrix

[160]. To circumvent these, target specific capture mediators such as molecularly imprinted polymers (MIPs) like plastic antibodies, and aptamers like nucleic acid antibodies are employed. The functionalization of SERS substrates or the complex formation between the target molecule and capture molecules are utilized for the target analyte molecule. The direct-antibody SERS technique proven its capability to chemical hazards present in food samples. For example, the immunomagnetic separation route is shown to detect 1 $mgml^{-1}$ ovalbumin in milk samples within 20 min in a two-step process. Using a one-step approach, a limit of detection of 0.79×10^{-3} mmoL/l is reported for melamine in milk within a test time ~20 min [161]. Apart from the direct antibody-SERS method, indirect antibody SERS method is also extensively used for food analysis. The strategy employed here is to first deposit the antibody onto magnetic beads or a substrate and then immobilize the antibodies to selectively bind the target antigen and then finally combine the tags with the captured antigens to form a usual sandwich structure. It can be isolated from the complex solution by applying a magnetic field or washing. In the direct aptamer-SERS, the reaction is carried out between aptamer solution is mixed with Ag dendritic colloidal substrate so as to form Ag-thiol binding connections. The thiolated Ag dendrites can be attached to 6-mercaptohexanol to achieve specific binding to detect four pesticides, including profenofos, phorate, isocarbophos, and omethoate in apple juice [162]. Later the two setup method is utilized for detecting low level or ricin in orange juice and milk, Bacillus anthracis spores in orange juice, etc. [163, 164]. In the indirect aptamer-SERS technique the following strategies are usually used: target sequence can be captured by its aptamer and immobilize on the substrate, or utilize the existing target sequence with particular reaction so that reporter is free for binding with metal nanoparticles or explore the change in certain variation in SERS signal that arises from the change in position or conformation of the reporter when the reporter comes close to active surface that relies on the presence of target sequence and specific reaction. On the other hand, molecular imprinting technique that generates template shaped holes in polymer matrices with a memory of the template molecules that can be applied in molecular capturing and recognition can be utilized for detection of low-molecular-weight analytes. Owing to the capability of the selective separation of the target analyte and the specific concentration can be achieved simultaneously, this technique can improve the SERS sensitivity and selectivity. The success of MIP has been demonstrated for trace level molecule or organic compound detection in complex samples, for example, Sudan-I (1 ppm) in paprika powder, nicotine, artificial 4-mercaptobenzoic acid in an artificial sample, etc. [165–167]. Trace levels of various hazardous chemicals in different kinds of food materials such as fruits, vegetables, meat, dairy products, fishery products, etc. are detected using the SERS technique. However, further, work in hotspot engineering and multiplexed Raman reporters are required to develop this technique as a routine analyte tool in the food industry.

8.3 SERS Technique for Environmental Pollutants

Aside from the prerequisites like high sensitivity and specificity and reproducibility of conventional measurements techniques, SERS technique for environmental pollutant measurement additionally demands portability too [168]. To improve the sensitivity of the SERS technique for pollutant measurements, the hot spot engineering, and sample pre-concentration techniques are adopted. In addition to employing indirect SERS techniques with tags to increase the selectivity, the SERS technique is coupled with capillary chromatography, thin layer chromatography, and ion-chromatography to distinguish different analyte components in a mixture [162, 169]. The lithographically patterned plasmonic nanostructures can ensure the reproducibility of the SERS based environment pollutant detectors. The advancement in electronics along with the utilization of optical fiber for coupling of optical radiation and detection facilitates the miniaturization of SERS devices for field applications. The development of paper-based SERS substrates in conjunction with portable Raman spectrometer largely facilitate the on-site monitoring of environmental pollutants. The direct probing of vibrational energies in the SERS techniques makes them a favorable tool to detect the organic pollutants existing in the environment. Many organic pollutants like pesticides (e.g., thiram, parathion, malachite green), PAHs and PCBs (Fluoranthene, Fluorine, Pyrene, Naphthalene), other organic pollutants (ciprofloxacin, furadantin, endosulfan) are efficiently detected using the SERS technique [169]. Apart from organic pollutants, the SERS technique is extensively used to detect heavy metals. The oxy-metal ions (As^{5+}, As^{3+}, UO_2^{2+}) can be detected directly by the SERS technique by fingerprinting them. The monatomic metal ions (e.g., Hg^{2+}) indirectly detected by SERS technique suing the "turn off" or "turn on" mechanism. In the turn-off sensor, SERS technique probe the quenching in Raman signature of a dye molecule (rhodamine B) adsorbed onto the SERS substrate in the presence of Hg^{2+}. On the other hand, the "turn on" sensor relies on the strong and specific binding of Hg^{2+} with thymine-thymine base pairs. Similar sensors are developed for Pb^{2+}, Cd^{2+}, Cu^{2+}, and $Cr^{3+,}$ etc. [169] Additionally, this technique is also used to detect the pathogens present in the environment by probing the chemical composition at the pathogen membrane.

8.4 SERS Technique in Forensic Science

Forensic science demand sensitive and selective investigation of different kinds of samples, ranging from solid cocaine to paint, to aid the crime investigation. To meet this requirement, the SERS technique with mentioned advantages promises to be an ideal candidate. The biological applications mentioned for the SERS techniques equally well applicable for the biologically relevant samples like DNA, RNA, urea, blood, etc. obtained from the crime scene [170]. With the utilization of statistical tools like principal component analysis (PCA), the antigen-antibody interactions are analyzed for different ABO blood groups to classify them with 96–98% accuracy by utilizing the 1311–1345 cm^{-1} band of the Raman spectrum [171]. The application of the micro-SERS to detect cathinone designer drugs in the range of 10^{-10}–10^{-12} M is also demonstrated [172]. Aside these, different materials like cocaine, methamphetamine, and morphine is detected at low concentrations (13 ng/ml for morphine, 4.6 ng/ml for cocaine, and 4.5 ng/nl for methamphetamine) is detected using the SERS technique for the optimized gold and silver particle sizes of 50 and 60 nm, respectively with borate as most efficient capping agent [173]. In an interesting work, Xu et al. used FlexBrite, a plasmonic nano-mushroom substrate, to detect up to a concentration of 0.5 mg/ml methamphetamine in water [174]. In recent work, Dong et al. by exploiting the potential of a statistical tool like support vector machine, a portable dynamic SERS system is developed to detect methamphetamine and 3,4-methylenedioxy methamphetamine in 2μL of a human urine sample [175]. The potential of this technique to detect explosive like trinitrotoluene (TNT) has also been demonstrated using a silicon chip sensor in a portable SERS device [176]. Additionally, the potential of this technique to identify other explosive and gunshot residue are investigated by other research groups [177]. Even though the SERS technique is a well-established approach to fingerprinting the molecule of interest, the application of this technique in forensic science is still in infancy.

8.5 SERS in Art and Archaeology

The non-destructive nature and small sample size requirement make the SERS technique as a preferred tool in archaeology and conservation science. Along with the advancement in photonics (CCDs, laser sources) and its integration with microscope technologies facilitate the efficient SERS measurements at dimensions as small as 1 μM, which is a typical size order of pigment grains. In spite of a large amount of literature available on the application of SERS technique to analyze synthetic dyes, the studies on museum objects are still rare. A SERS library of natural dyes is essential to compare the data obtained from the works of arts and historical artifacts for the positive analysis [178]. Additionally, computer controlled classification algorithm is needed to match peaks of the sample of interest with the data bank to make the process rapid and trustworthy. Due to the complexation of metals with the dyes used routinely in painting and textiles, pre-treatment of samples in strong acids and alkali is often employed before the SERS analysis. However, such treatments pose the challenge of disruption of the matrix and can lead to the reduced interaction between the dye of interest and SERS substrate. To mitigate these issues, gentle extraction and non-extractive in situ hydrolysis methodologies are proposed. In 1987, Guineau and Guichard reported the usage of SERS technique to record alizarin and extracts of the plant dye madder and related anthraquinoid derivatives from 8th century textile samples [178]. A revival of this field had happened in 2004 when three independent works reported on the development of analytical protocols for the identification of synthetic alizarin, madder root extracts and lakes, a semi-quantitative analysis of purpurin using aggregated silver nanoparticles and adsorption and acidic behavior studies on silver nanoparticles [179–181]. To get further insight into the application of SERS in art and archeology, authors are encouraged to read a review article by F. Pozzi and M. Leona and the references therein [178]. In an interesting case study presented by A. Cesaratto et al. carried out SERS analysis of aniline dyes in purpole paper cut-outs by Henri Matise and in a 19th century silk fabric, and characterization of xanthene dyes and lakes in a selection of pinke Matisse cut-outs, in Vincent Van Gogh's Irises and in Japanese woodblock prints [182]. Their studies reveal that aniline dyes on fabric require pre-sample treatment and treatment with a nitric acid solution substantially enhance the SERS signal of xanthene dyes and lakes display. Additionally, for the first time, the authors demonstrated eosin Y and phloxine in actual works of art. In the Japanese woodblock prints, the authors characterized eosin-carmine binary mixtures. Owing to the scarcity of the sample, the SERS technique is still to be established as a preferred tool in the area of art and archaeology.

9 Conclusion and Future Outlook

Ever since the early works in the field of SERS technique in the 1970s, the technique has now emerged as an efficient as a sensing and imaging tool in diverse areas, ranging from biology to archaeology. The large enhancement in Raman signature in the SERS technique is attributed to "electromagnetic" as well "chemical" or a multiplicative effect of both. This chapter unravelled the contribution of these mechanisms towards the SERS enhancement and critical role of hotspots in SERS studies. Additionally, the reason behind the most preferred choice of metal nanoparticles is discussed in detail. The chapter accounts for the various strategies adopted for fabricating the SERS substrates from nanoparticles and nanostructures via bottom-up, top-down, and template assisted approaches. Additionally, the recent trends in the 2D and 3D SERS substrates are discussed in detail. Though the applications of SERS are wide spread and difficult to comprehend in a chapter, an over view of a few applications like biosensing and bioimaging, food safety evaluation, environmental pollutant detection, forensic science, art, and archaeology wherein the direct and indirect SERS techniques are employed is explained.

The advancement in nanoscale plasmonic and non-plasmonic particles with hot-spots engineered in the 1D, 2D, and 3D could pave the way for the development of SERS based portable Raman devices that can find applications in daily life, e.g. food safety, pathogens in a hospital environment, drug abuse, etc. The biggest challenge of reproducibility and long term stability of SERS substrates can be mitigated with the advancement in nanotechnology. In addition, the development of miniaturization in device technologies may further facilitate the application of this technique in hitherto unexplored areas. Aside these, novel strategies to direct and confine the analyte molecules into electromagnetic hotspots via approaches like molecular coupling, a surface grating of analyte molecules, superhydrophobic, superoleophobic or SLIP surfaces, may enable the ultra-low concentration of the molecule of interest. In addition, by conjugating with stimuli-responsive polymers, the on-demand SERS detection can also be made possible. With the development of hybrid SERS techniques, photonics and nanotechnology, many of the existing challenges are expected to address in the coming decade.

Acknowledgements The author acknowledge the Manipal Academy of Higher Education for the support through Dr. TMA Pai Endowment Chair in Applied Nanosciences. He is also grateful to Mr. Aravind M and Ms. Alina Peethan for their support in creating the illustrations shown in the chapter. Support from Prof. Santhosh Chidangil of Department of Atomic and Molecular Physics, Manipal Academy of Higher Education is greatly appreciated.

References

1. Liang, F., Guo, Y., Hou, S., Quan, Q.: Photonic-plasmonic hybrid single-molecule nanosensor measures the effect of fluorescent labels on DNA-protein dynamics. Sci. Adv. **3** (5), e1602991 (2017)
2. Mejía-Salazar, J., Oliveira Jr, O.N.: Plasmonic biosensing: focus review. Chem. Rev. **118** (20), 10617–10625 (2018)
3. Taylor, A.B., Zijlstra, P.: Single-molecule plasmon sensing: current status and future prospects. ACS Sens. **2**(8), 1103–1122 (2017)
4. Willner, M.R., Vikesland, P.J.: Nanomaterial enabled sensors for environmental contaminants. J. Nanobiotechnol. **16**(1), 95 (2018)
5. Ye, D., Zuo, X., Fan, C.: DNA nanotechnology-enabled interfacial engineering for biosensor development. Ann. Rev. Anal. Chem. **11**, 171–195 (2018)
6. Yunus, I.S., Harwin, Kurniawan, A., Adityawarman, D., Indarto, A.: Nanotechnologies in water and air pollution treatment. Environ. Technol. Rev. **1**(1), 136–148 (2012)
7. Butler, H.J., Ashton, L., Bird, B., Cinque, G., Curtis, K., Dorney, J., Esmonde-White, K., Fullwood, N.J., Gardner, B., Martin-Hirsch, P.L.: Using Raman spectroscopy to characterize biological materials. Nat. Protocols **11**(4), 664 (2016)
8. Czamara, K., Majzner, K., Pacia, M. Z., Kochan, K., Kaczor, A., Baranska, M.: Raman spectroscopy of lipids: a review. J. Raman Spectrosc. **46**(1), 4–20 (2015)
9. Demirel, G., Usta, H., Yilmaz, M., Celik, M., Alidagi, H.A., Buyukserin, F.: Surface-enhanced Raman spectroscopy (SERS): an adventure from plasmonic metals to organic semiconductors as SERS platforms. J. Mater. Chem. C **6**(20), 5314–5335 (2018)
10. Fleischmann, M., Hendra, P.J., McQuillan, A.J.: Raman spectra of pyridine adsorbed at a silver electrode. Chem. Phys. Lett. **26**(2), 163–166 (1974)
11. Jeanmaire, D.L., Van Duyne, R.P.: Surface Raman spectroelectrochemistry: Part I. Heterocyclic, aromatic, and aliphatic amines adsorbed on the anodized silver electrode. J. Electroanal. Chem. Interfacial Electrochem. **84**(1), 1–20 (1977)
12. Otto, A.: The 'chemical'(electronic) contribution to surface-enhanced Raman scattering. J. Raman Spectrosc. Int. J. Orig. Work Aspects Raman Spectrosc. Includ. Higher Order Process. Brill. Rayleigh Scatt. **36**(6–7), 497–509 (2005)

13. Yamamoto, Y.S., Ishikawa, M., Ozaki, Y., Itoh, T.: Fundamental studies on enhancement and blinking mechanism of surface-enhanced Raman scattering (SERS) and basic applications of SERS biological sensing. Front. Phys. **9**(1), 31–46 (2014)
14. Stiles, P.L., Dieringer, J.A., Shah, N.C., Van Duyne, R.P.: Surface-enhanced Raman spectroscopy. Annu. Rev. Anal. Chem. **1**, 601–626 (2008)
15. Fan, M., Andrade, G.F., Brolo, A.G.: A review on the fabrication of substrates for surface enhanced Raman spectroscopy and their applications in analytical chemistry. Anal. Chimica Acta **693**(1–2), 7–25 (2011)
16. Ngo, H.T., Wang, H.-N., Fales, A.M., Vo-Dinh, T.: Plasmonic SERS biosensing nanochips for DNA detection. Anal. Bioanal. Chem. **408**(7), 1773–1781 (2016)
17. Tsoutsi, D., Sanles-Sobrido, M., Cabot, A., Gil, P.-R.: Common aspects influencing the translocation of SERS to biomedicine. Current Med. Chem. **25**(35), 4638–4652 (2018)
18. Torres-Nunez, A., Faulds, K., Graham, D., Alvarez-Puebla, R., Guerrini, L.: Silver colloids as plasmonic substrates for direct label-free surface-enhanced Raman scattering analysis of DNA. Analyst **141**(17), 5170–5180 (2016)
19. Garcia-Rico, E., Alvarez-Puebla, R.A., Guerrini, L.: Direct surface-enhanced Raman scattering (SERS) spectroscopy of nucleic acids: from fundamental studies to real-life applications. Chem. Soc. Rev. **47**(13), 4909–4923 (2018)
20. Campion, A., Kambhampati, P.: Surface-enhanced Raman scattering. Chem. Soc. Rev. **27** (4), 241–250 (1998)
21. Ding, S.-Y., You, E.-M., Tian, Z.-Q., Moskovits, M.: Electromagnetic theories of surface-enhanced Raman spectroscopy. Chem. Soc. Rev. **46**(13), 4042–4076 (2017)
22. Le Ru, E., Etchegoin, P.: Principles of Surface-Enhanced Raman Spectroscopy: and related plasmonic effects. Elsevier (2008)
23. Chalmers, J.M., Edwards, H.G., Hargreaves, M.D.: Infrared and Raman Spectroscopy in Forensic Science. Wiley (2012)
24. Zong, C., Xu, M., Xu, L.-J., Wei, T., Ma, X., Zheng, X.-S., Hu, R., Ren, B.: Surface-enhanced Raman spectroscopy for bioanalysis: reliability and challenges. Chem. Rev. **118**(10), 4946–4980 (2018)
25. Persson, B.N.J., Zhao, K., Zhang, Z.: Chemical contribution to surface-enhanced Raman scattering. Phys. Rev. Lett. **96**(20), 207401 (2006)
26. Valley, N., Greeneltch, N., Van Duyne, R.P., Schatz, G.C.: A look at the origin and magnitude of the chemical contribution to the enhancement mechanism of surface-enhanced Raman spectroscopy (SERS): theory and experiment. J. Phys. Chem. Lett. **4**(16), 2599–2604 (2013)
27. Zayak, A., Hu, Y., Choo, H., Bokor, J., Cabrini, S., Schuck, P., Neaton, J.: Chemical Raman enhancement of organic adsorbates on metal surfaces. Physical Rev. Lett. **106**(8), 083003 (2011)
28. Moskovits, M.: Surface-enhanced Raman spectroscopy: a brief retrospective. J. Raman Spectrosc. Int. J. Orig. Work Aspects Raman Spectrosc. Includ. Higher Order Process. Brill. Rayleigh Scatt. **36**(6–7), 485–496 (2005)
29. Lee, H.K., Lee, Y.H., Koh, C.S.L., Phan-Quang, G.C., Han, X., Lay, C.L., Sim, H.Y.F., Kao, Y.-C., An, Q., Ling, X.Y.: Designing surface-enhanced Raman scattering (SERS) platforms beyond hotspot engineering: emerging opportunities in analyte manipulations and hybrid materials. Chem. Soc. Rev. **48**(3), 731–756 (2019)
30. Pilot, R., Signorini, R., Fabris, L.: Surface-enhanced raman spectroscopy: principles, substrates, and applications. In: Metal Nanoparticles and Clusters, pp. 89–164. Springer (2018)
31. Ye, J., Wen, F., Sobhani, H., Lassiter, J.B., Van Dorpe, P., Nordlander, P., Halas, N.J.: Plasmonic nanoclusters: near field properties of the Fano resonance interrogated with SERS. Nano Lett. **12**(3), 1660–1667 (2012)
32. Postaci, S., Yildiz, B.C., Bek, A., Tasgin, M.E.: Silent enhancement of SERS signal without increasing hot spot intensities. Nanophotonics **7**(10), 1687–1695 (2018)

33. Kleinman, S.L., Frontiera, R.R., Henry, A.-I., Dieringer, J.A., Van Duyne, R.P.: Creating, characterizing, and controlling chemistry with SERS hot spots. Phys. Chem. Chem. Phys. **15** (1), 21–36 (2013)
34. Petryayeva, E., Krull, U.J.: Localized surface plasmon resonance: nanostructures, bioassays and biosensing—a review. Anal. Chimica Acta **706**(1), 8–24 (2011)
35. Liu, H., Yang, Z., Meng, L., Sun, Y., Wang, J., Yang, L., Liu, J., Tian, Z.: Three-dimensional and time-ordered surface-enhanced Raman scattering hotspot matrix. J. Am. Chem. Soc. **136**(14), 5332–5341 (2014)
36. Sun, Y., Han, Z., Liu, H., He, S., Yang, L., Liu, J.: Three-dimensional hotspots in evaporating nanoparticle sols for ultrahigh Raman scattering: solid–liquid interface effects. Nanoscale **7**(15), 6619–6626 (2015)
37. Wang, H., Fang, J., Xu, J., Wang, F., Sun, B., He, S., Sun, G., Liu, H.: A hanging plasmonic droplet: three-dimensional SERS hotspots for a highly sensitive multiplex detection of amino acids. Analyst **140**(9), 2973–2978 (2015)
38. Le Ru, E.C., Etchegoin, P.G.: Single-molecule surface-enhanced Raman spectroscopy. Ann. Rev. Phys. Chem. **63**, 65–87 (2012)
39. West, P.R., Ishii, S., Naik, G.V., Emani, N.K., Shalaev, V.M., Boltasseva, A.: Searching for better plasmonic materials. Laser Photonics Rev. **4**(6), 795-808 (2010)
40. Cottancin, E., Celep, G., Lermé, J., Pellarin, M., Huntzinger, J., Vialle, J., Broyer, M.: Optical properties of noble metal clusters as a function of the size: comparison between experiments and a semi-quantal theory. Theor. Chem. Acc. **116**(4–5), 514-523 (2006)
41. Cialla, D., Pollok, S., Steinbrücker, C., Weber, K., Popp, J.: SERS-based detection of biomolecules. Nanophotonics **3**(6), 383–411 (2014)
42. Karthick Kannan, P., Shankar, P., Blackman, C., Chung, C.H.: Recent advances in 2D inorganic nanomaterials for SERS sensing. Adv. Mater. 1803432 (2019)
43. Jayram, N.D., Aishwarya, D., Sonia, S., Mangalaraj, D., Kumar, P.S., Rao, G.M.: Analysis on superhydrophobic silver decorated copper Oxide nanostructured thin films for SERS studies. J. Colloid Interface Sci. **477**, 209–219 (2016)
44. Cong, S., Yuan, Y., Chen, Z., Hou, J., Yang, M., Su, Y., Zhang, Y., Li, L., Li, Q., Geng, F.: Noble metal-comparable SERS enhancement from semiconducting metal oxides by making oxygen vacancies. Nat. Commun. **6**, 7800 (2015)
45. Yu, Z., Yun, F.F., Wang, Y., Yao, L., Dou, S., Liu, K., Jiang, L., Wang, X.: Desert Beetle-inspired superwettable patterned surfaces for water harvesting. Small **13**(36), 1701403 (2017)
46. Kim, N.-J., Kim, J., Park, J.-B., Kim, H., Yi, G.-C., Yoon, S.: Direct observation of quantum tunnelling charge transfers between molecules and semiconductors for SERS. Nanoscale **11** (1), 45–49 (2019)
47. Wang, X., Shi, W., She, G., Mu, L.: Surface-enhanced Raman scattering (SERS) on transition metal and semiconductor nanostructures. Phys. Chem. Chem. Phys. **14**(17), 5891–5901 (2012)
48. Kini, S., Ganiga, V., Kulkarni, S.D., Chidangil, S., George, S.D.: Sensitive detection of mercury using the fluorescence resonance energy transfer between CdTe/CdS quantum dots and Rhodamine 6G. J. Nanoparticle Res. **20**(9), 232 (2018)
49. Basheer, N.S., Kumar, B.R., Kurian, A., George, S.D.: Thermal lens probing of distant dependent fluorescence quenching of Rhodamine 6G by silver nanoparticles. J. Lumin. **137**, 225–229 (2013)
50. Kumar, B.R., Basheer, N.S., Kurian, A., George, S.D.: Thermal-lens study on the distance-dependent energy transfer from Rhodamine 6G to gold nanoparticles. Int. J. Thermophys. **34**(10), 1982–1992 (2013)
51. John, J., Thomas, L., Kurian, A., Nampoori, V., George, S.D.: Role of decoration method of gold nanoparticles on the thermal and optical properties of mesoporous silica-Rhodamine 6G hybrids. Microporous Mesoporous Mater. **244**, 171–179 (2017)

52. John, J., Thomas, L., Kurian, A., George, S.D.: Modulating fluorescence quantum yield of highly concentrated fluorescein using differently shaped green synthesized gold nanoparticles. J. Lumin. **172**, 39–46 (2016)
53. John, J., Thomas, L., George, N.A., Kurian, A., George, S.D.: Tailoring of optical properties of fluorescein using green synthesized gold nanoparticles. Phys. Chem. Chem. Phys. **17**(24), 15813–15821 (2015)
54. Kumar, B.R., Basheer, N.S., Kurian, A., George, S.D.: Study of concentration-dependent quantum yield of Rhodamine 6G by gold nanoparticles using thermal-lens technique. Appl. Phys. B **115**(3), 335–342 (2014)
55. Basheer, N.S., Kumar, B.R., Kurian, A., George, S.D.: Silver nanoparticle size–dependent measurement of quantum efficiency of Rhodamine 6G. Appl. Phys. B **113**(4), 581–587 (2013)
56. Lane, L.A., Qian, X., Nie, S.: SERS nanoparticles in medicine: from label-free detection to spectroscopic tagging. Chem. Rev. **115**(19), 10489–10529 (2015)
57. Cardinal, M.F., Vander Ende, E., Hackler, R.A., McAnally, M.O., Stair, P.C., Schatz, G.C., Van Duyne, R.P.: Expanding applications of SERS through versatile nanomaterials engineering. Chem. Soc. Rev. **46**(13), 3886–3903 (2017)
58. Cao, Y., Zhang, J., Yang, Y., Huang, Z., Long, N.V., Fu, C.: Engineering of SERS substrates based on noble metal nanomaterials for chemical and biomedical applications. Appl. Spectrosc. Rev. **50**(6), 499–525 (2015)
59. Si, S., Liang, W., Sun, Y., Huang, J., Ma, W., Liang, Z., Bao, Q., Jiang, L.: Facile fabrication of high-density sub-1-nm gaps from Au nanoparticle monolayers as reproducible SERS substrates. Adv. Funct. Mater. **26**(44), 8137–8145 (2016)
60. Reguera, J., Langer, J., de Aberasturi, D.J., Liz-Marzán, L.M.: Anisotropic metal nanoparticles for surface enhanced Raman scattering. Chem. Soc. Rev. **46**(13), 3866–3885 (2017)
61. Turkevich, J., Stevenson, P.C., Hillier, J.: A study of the nucleation and growth processes in the synthesis of colloidal gold. Discuss. Faraday Soc. **11**, 55–75 (1951)
62. Lee, P., Meisel, D.: Adsorption and surface-enhanced Raman of dyes on silver and gold sols. J. Phys. Chem. **86**(17), 3391–3395 (1982)
63. Benz, F., Chikkaraddy, R., Salmon, A., Ohadi, H., De Nijs, B., Mertens, J., Carnegie, C., Bowman, R.W., Baumberg, J.J.: SERS of individual nanoparticles on a mirror: size does matter, but so does shape. J. Phys. Chem. Lett. **7**(12), 2264–2269 (2016)
64. Quester, K., Avalos-Borja, M., Vilchis-Nestor, A.R., Camacho-López, M.A., Castro-Longoria, E.: SERS properties of different sized and shaped gold nanoparticles biosynthesized under different environmental conditions by Neurospora crassa extract. PloS one **8**(10), e77486 (2013)
65. Nguyen, B.H., Nguyen, V.H., Tran, H.N.: Rich variety of substrates for surface enhanced Raman spectroscopy. Adv. Natl. Sci. Nanosci. Nanotechnol. **7**(3), 033001 (2016)
66. Solís, D.M., Taboada, J.M., Obelleiro, F., Liz-Marzán, L.M., García de Abajo, F.J.: Optimization of nanoparticle-based SERS substrates through large-scale realistic simulations. ACS Photon.**4**(2), 329–337 (2017)
67. Betz, J.F., Wei, W.Y., Cheng, Y., White, I.M., Rubloff, G.W.: Simple SERS substrates: powerful, portable, and full of potential. Phys. Chem. Chem. Phys. **16**(6), 2224–2239 (2014)
68. Mavani, K., Shah, M.: Synthesis of silver nanoparticles by using sodium borohydride as a reducing agent. Int. J. Eng. Res. Technol. **2**(3), 1–5 (2013)
69. Song, K.C., Lee, S.M., Park, T.S., Lee, B.S.: Preparation of colloidal silver nanoparticles by chemical reduction method. Korean J. Chem. Eng. **26**(1), 153–155 (2009)
70. Mulfinger, L., Solomon, S.D., Bahadory, M., Jeyarajasingam, A.V., Rutkowsky, S.A., Boritz, C.: Synthesis and study of silver nanoparticles. J. Chem. Edu. **84**(2), 322 (2007)
71. Leopold, N., Lendl, B.: A new method for fast preparation of highly surface-enhanced Raman scattering (SERS) active silver colloids at room temperature by reduction of silver nitrate with hydroxylamine hydrochloride. J. Phys. Chem. B **107**(24), 5723–5727 (2003)

72. Sun, D., Zhang, G., Jiang, X., Huang, J., Jing, X., Zheng, Y., He, J., Li, Q.: Biogenic flower-shaped Au–Pd nanoparticles: synthesis, SERS detection and catalysis towards benzyl alcohol oxidation. J. Mater. Chem. A **2**(6), 1767–1773 (2014)
73. Hunyadi, S.E., Murphy, C.J.: Bimetallic silver–gold nanowires: fabrication and use in surface-enhanced Raman scattering. J. Mater. Chem. **16**(40), 3929–3935 (2006)
74. Fan, M., Lai, F.-J., Chou, H.-L., Lu, W.-T., Hwang, B.-J., Brolo, A.G.: Surface-enhanced Raman scattering (SERS) from Au: Ag bimetallic nanoparticles: the effect of the molecular probe. Chem. Sci. **4**(1), 509–515 (2013)
75. Muniz-Miranda, M., Gellini, C., Canton, P., Marsili, P., Giorgetti, E.: SERS and catalytically active Ag/Pd nanoparticles obtained by combining laser ablation and galvanic replacement. J. Alloy. Compd. **615**, S352–S356 (2014)
76. Yang, Y., Shi, J., Kawamura, G., Nogami, M.: Preparation of Au–Ag, Ag–Au core–shell bimetallic nanoparticles for surface-enhanced Raman scattering. Scr. Mater. **58**(10), 862–865 (2008)
77. Li, J.F., Huang, Y.F., Ding, Y., Yang, Z.L., Li, S.B., Zhou, X.S., Fan, F.R., Zhang, W., Zhou, Z.Y., Ren, B.: Shell-isolated nanoparticle-enhanced Raman spectroscopy. Nature **464** (7287), 392 (2010)
78. Guo, P., Sikdar, D., Huang, X., Si, K.J., Xiong, W., Gong, S., Yap, L.W., Premaratne, M., Cheng, W.: Plasmonic core–shell nanoparticles for SERS detection of the pesticide thiram: size-and shape-dependent Raman enhancement. Nanoscale **7**(7), 2862–2868 (2015)
79. Xie, W., Herrmann, C., Kömpe, K., Haase, M., Schlücker, S.: Synthesis of bifunctional Au/Pt/Au core/shell nanoraspberries for in situ SERS monitoring of platinum-catalyzed reactions. J. Am. Chem. Soc. **133**(48), 19302–19305 (2011)
80. Shanthil, M., Thomas, R., Swathi, R., George Thomas, K.: Ag@ SiO_2 core–shell nanostructures: distance-dependent plasmon coupling and SERS investigation. J. Phys. Chem. Lett. **3**(11), 1459–1464 (2012)
81. Yang, Y., Liu, J., Fu, Z.-W., Qin, D.: Galvanic replacement-free deposition of Au on Ag for core–shell nanocubes with enhanced chemical stability and SERS activity. J. Am. Chem. Soc. **136**(23), 8153–8156 (2014)
82. Pande, S., Ghosh, S.K., Praharaj, S., Panigrahi, S., Basu, S., Jana, S., Pal, A., Tsukuda, T., Pal, T.: Synthesis of normal and inverted gold–silver core–shell architectures in β-cyclodextrin and their applications in SERS. J. Phys. Chem. C **111**(29), 10806–10813 (2007)
83. Wang, J., Zhu, T., Tang, M., Cai, S., Liu, Z.: Fabricating surface enhanced Raman scattering (SERS)-active substrates by assembling colloidal Au nanoparticles with self-assembled monolayers. Jpn. J. Appl. Phys. **35**, L1381–L1384 (1996)
84. Hu, X., Cheng, W., Wang, T., Wang, Y., Wang, E., Dong, S.: Fabrication, characterization, and application in SERS of self-assembled polyelectrolyte-gold nanorod multilayered films. J. Phys. Chem. B **109**(41), 19385–19389 (2005)
85. Chen, J., Guo, L., Qiu, B., Lin, Z., Wang, T.: Application of ordered nanoparticle self-assemblies in surface-enhanced spectroscopy. Mater. Chem. Front. **2**(5), 835–860 (2018)
86. Pu, H., Xiao, W., Sun, D.-W.: SERS-microfluidic systems: a potential platform for rapid analysis of food contaminants. Trends Food Sci. Technol. **70**, 114–126 (2017)
87. Kant, K., Abalde-Cela, S.: Surface-enhanced Raman scattering spectroscopy and microfluidics: towards ultrasensitive label-free sensing. Biosensors **8**(3), 62 (2018)
88. Willner, M.R., McMillan, K.S., Graham, D., Vikesland, P.J., Zagnoni, M.: Surface-enhanced Raman scattering based microfluidics for single-cell analysis. Anal. Chem. **90**(20), 12004–12010 (2018)
89. Wu, Y., Jiang, Y., Zheng, X., Jia, S., Zhu, Z., Ren, B., Ma, H.: Facile fabrication of microfluidic surface-enhanced Raman scattering devices via lift-up lithography. R. Soc. Open Sci. **5**(4), 172034 (2018)
90. Jahn, I., Žukovskaja, O., Zheng, X.-S., Weber, K., Bocklitz, T., Cialla-May, D., Popp, J: Surface-enhanced Raman spectroscopy and microfluidic platforms: challenges, solutions and potential applications. Analyst **142**(7), 1022–1047 (2017)

91. Yang, S., Dai, X., Stogin, B.B., Wong, T.-S.: Ultrasensitive surface-enhanced Raman scattering detection in common fluids. Proc. Natl. Acad. Sci. **113**(2), 268–273 (2016)
92. Marquestaut, N., Martin, A., Talaga, D., Servant, L., Ravaine, S., Reculusa, S., Bassani, D. M., Gillies, E., Lagugné-Labarthet, F.: Raman enhancement of azobenzene monolayers on substrates prepared by Langmuir–Blodgett deposition and electron-beam lithography techniques. Langmuir **24**(19), 11313–11321 (2008)
93. Abu Hatab, N.A., Oran, J.M., Sepaniak, M.: Surface-enhanced Raman spectroscopy substrates created via electron beam lithography and nanotransfer printing. ACS Nano **2**(2), 377–385 (2008)
94. Kahl, M., Voges, E., Kostrewa, S., Viets, C., Hill, W.: Periodically structured metallic substrates for SERS. Sens. Actuators B: Chem. **51**(1–3), 285–291 (1998)
95. Petti, L., Capasso, R., Rippa, M., Pannico, M., La Manna, P., Peluso, G., Calarco, A., Bobeico, E., Musto, P.: A plasmonic nanostructure fabricated by electron beam lithography as a sensitive and highly homogeneous SERS substrate for bio-sensing applications. Vib. Spectrosc. **82**, 22–30 (2016)
96. Yue, W., Wang, Z., Yang, Y., Chen, L., Syed, A., Wong, K., Wang, X.: Electron-beam lithography of gold nanostructures for surface-enhanced Raman scattering. J. Micromech. Microeng. **22**(12), 125007 (2012)
97. Yu, Q., Guan, P., Qin, D., Golden, G., Wallace, P.M.: Inverted size-dependence of surface-enhanced Raman scattering on gold nanohole and nanodisk arrays. Nano Lett. **8**(7), 1923–1928 (2008)
98. Brolo, A.G., Gordon, R., Leathem, B., Kavanagh, K.L.: Surface plasmon sensor based on the enhanced light transmission through arrays of nanoholes in gold films. Langmuir **20**(12), 4813–4815 (2004)
99. Sivashanmugan, K., Liao, J.-D., You, J.-W., Wu, C.-L.: Focused-ion-beam-fabricated Au/Ag multilayered nanorod array as SERS-active substrate for virus strain detection. Sens. Actuators B: Chem. **181**, 361–367 (2013)
100. Sree Satya Bharati, M., Byram, C., Soma, V.R.: Femtosecond laser fabricated Ag@ Au and Cu@ Au alloy nanoparticles for surface enhanced Raman spectroscopy based trace explosives detection. Front. Phys. **6**, 28 (2018)
101. Quilis, N.G., Lequeux, M., Venugopalan, P., Khan, I., Knoll, W., Boujday, S., de La Chapelle, M.L., Dostalek, J.: Tunable laser interference lithography preparation of plasmonic nanoparticle arrays tailored for SERS. Nanoscale **10**(21), 10268–10276 (2018)
102. Solak, H.H., David, C., Gobrecht, J., Golovkina, V., Cerrina, F., Kim, S.O., Nealey, P.: Sub-50 nm period patterns with EUV interference lithography. Microelectron. Eng. **67**, 56–62 (2003)
103. Zhang, P., Yang, S., Wang, L., Zhao, J., Zhu, Z., Liu, B., Zhong, J., Sun, X.: Large-scale uniform Au nanodisk arrays fabricated via x-ray interference lithography for reproducible and sensitive SERS substrate. Nanotechnology **25**(24), 245301 (2014)
104. Hwang, J., Yang, M.: Sensitive and reproducible gold SERS sensor based on interference lithography and electrophoretic deposition. Sensors **18**(11), 4076 (2018)
105. Jha, S.K., Ekinci, Y., Agio, M., Löffler, J.F.: Towards deep-UV surface-enhanced resonance Raman spectroscopy of explosives: ultrasensitive, real-time and reproducible detection of TNT. Analyst **140**(16), 5671–5677 (2015)
106. Izquierdo-Lorenzo, I., Jradi, S., Adam, P.-M.: Direct laser writing of random au nanoparticle three-dimensional structures for highly reproducible micro-SERS measurements. RSC Adv. **4**(8), 4128–4133 (2014)
107. Byram, C., Moram, S.S.B., Shaik, A.K., Soma, V.R.: Versatile gold based SERS substrates fabricated by ultrafast laser ablation for sensing picric acid and ammonium nitrate. Chem. Phys. Lett. **685**, 103–107 (2017)
108. Dick, L.A., McFarland, A.D., Haynes, C.L., Van Duyne, R.P.: Metal film over nanosphere (MFON) electrodes for surface-enhanced Raman spectroscopy (SERS): improvements in surface nanostructure stability and suppression of irreversible loss. J. Phys. Chem. B **106**(4), 853–860 (2002)

109. Haynes, C.L., Van Duyne, R.P.: Nanosphere Lithography: A Versatile Nanofabrication Tool for Studies of Size-Dependent Nanoparticle Optics. ACS Publications (2001)
110. Quero, G., Zito, G., Managò, S., Galeotti, F., Pisco, M., De Luca, A., Cusano, A.: Nanosphere lithography on fiber: towards engineered lab-on-fiber SERS optrodes. Sensors **18**(3), 680 (2018)
111. Wallace, G.Q., Tabatabaei, M., Lagugné-Labarthet, F.: Towards attomolar detection using a surface-enhanced Raman spectroscopy platform fabricated by nanosphere lithography. Can. J. Chem. **92**(1), 1–8 (2013)
112. Kahraman, M., Daggumati, P., Kurtulus, O., Seker, E., Wachsmann-Hogiu, S.: Fabrication and characterization of flexible and tunable plasmonic nanostructures. Sci. Rep. **3**, 3396 (2013)
113. Lee, C., Robertson, C.S., Nguyen, A.H., Kahraman, M., Wachsmann-Hogiu, S.: Thickness of a metallic film, in addition to its roughness, plays a significant role in SERS activity. Sci. Rep. **5**, 11644 (2015)
114. George, J.E., Unnikrishnan, V., Mathur, D., Chidangil, S., George, S.D.: Flexible superhydrophobic SERS substrates fabricated by in situ reduction of Ag on femtosecond laser-written hierarchical surfaces. Sens. Actuators B: Chem. **272**, 485–493 (2018)
115. Ji, N., Ruan, W., Wang, C., Lu, Z., Zhao, B.: Fabrication of silver decorated anodic aluminum oxide substrate and its optical properties on surface-enhanced Raman scattering and thin film interference. Langmuir **25**(19), 11869–11873 (2009)
116. Kassu, A., Farley, C., Sharma, A., Kim, W., Guo, J.: Effect of pore size and film thickness on gold-coated nanoporous anodic aluminum oxide substrates for surface-enhanced Raman scattering sensor. Sensors **15**(12), 29924–29937 (2015)
117. Tran, B., Nam, N., Son, S., Lee, N.: Nanoporous anodic aluminum oxide internalized with gold nanoparticles for on-chip PCR and direct detection by surface-enhanced Raman scattering. Analyst **143**(4), 808–812 (2018)
118. Terekhov, S., Mojzes, P., Kachan, S., Mukhurov, N., Zhvavyi, S., Panarin, A.Y., Khodasevich, I., Orlovich, V., Thorel, A., Grillon, F.: A comparative study of surface-enhanced Raman scattering from silver-coated anodic aluminum oxide and porous silicon. J. Raman Spectrosc. **42**(1), 12–20 (2011)
119. Toccafondi, C., La Rocca, R., Scarpellini, A., Salerno, M., Das, G., Dante, S.: Thin nanoporous alumina-based SERS platform for single cell sensing. Appl. Surf. Sci. **351**, 738–745 (2015)
120. Zhang, C., Yi, P., Peng, L., Lai, X., Chen, J., Huang, M., Ni, J.: Continuous fabrication of nanostructure arrays for flexible surface enhanced Raman scattering substrate. Sci. Rep. **7**, 39814 (2017)
121. Sun, L., Hu, H., Zhan, D., Yan, J., Liu, L., Teguh, J.S., Yeow, E.K., Lee, P.S., Shen, Z.: Plasma modified MoS_2 nanoflakes for surface enhanced Raman scattering. Small **10**(6), 1090–1095 (2014)
122. Lu, X., Luo, X., Zhang, J., Quek, S.Y., Xiong, Q.: Lattice vibrations and Raman scattering in two-dimensional layered materials beyond graphene. Nano Res. **9**(12), 3559–3597 (2016)
123. Zhang, S., Zhang, N., Zhao, Y., Cheng, T., Li, X., Feng, R., Xu, H., Liu, Z., Zhang, J., Tong, L.: Spotting the differences in two-dimensional materials–the Raman scattering perspective. Chem. Soc. Rev. **47**(9), 3217–3240 (2018)
124. Ling, X., Zhang, J.: First-layer effect in graphene-enhanced Raman scattering. Small **6**(18), 2020–2025 (2010)
125. Ling, X., Moura, L., Pimenta, M.A., Zhang, J.: Charge-transfer mechanism in graphene-enhanced Raman scattering. J. Phys. Chem. C **116**(47), 25112–25118 (2012)
126. Ling, X., Xie, L., Fang, Y., Xu, H., Zhang, H., Kong, J., Dresselhaus, M.S., Zhang, J., Liu, Z.: Can graphene be used as a substrate for Raman enhancement? Nano Lett. **10**(2), 553–561 (2009)
127. Ling, X., Fang, W., Lee, Y.-H., Araujo, P.T., Zhang, X., Rodriguez-Nieva, J.F., Lin, Y., Zhang, J., Kong, J., Dresselhaus, M.S.: Raman enhancement effect on two-dimensional layered materials: graphene, h-BN and MoS_2. Nano Lett. **14**(6), 3033–3040 (2014)

128. Kim, N.-Y., Leem, Y.-C., Hong, S.-H., Park, J.-H., Yim, S.-Y.: Ultrasensitive and stable plasmonic surface-enhanced Raman scattering substrates covered with atomically thin monolayers: effect of the insulating property. ACS Appl. Mater. Interfaces (2019)
129. Xia, M.: 2D materials-coated plasmonic structures for SERS applications. Coatings **8**(4), 137 (2018)
130. Yi, N., Zhang, C., Song, Q., Xiao, S.: A hybrid system with highly enhanced graphene SERS for rapid and tag-free tumor cells detection. Sci. Rep. **6**, 25134 (2016)
131. Jahn, M., Patze, S., Hidi, I.J., Knipper, R., Radu, A.I., Mühlig, A., Yüksel, S., Peksa, V., Weber, K., Mayerhöfer, T.: Plasmonic nanostructures for surface enhanced spectroscopic methods. Analyst **141**(3), 756–793 (2016)
132. Lee, S., Choi, I.: Fabrication Strategies of 3D Plasmonic Structures for SERS. BioChip J. **13** (1), 30–42 (2019)
133. Liu, H., Yang, L., Liu, J.: Three-dimensional SERS hot spots for chemical sensing: Towards developing a practical analyzer. TrAC Trends Anal. Chem. **80**, 364-372 (2016)
134. Gómez-Graña, S., Pérez-Juste, J., Alvarez-Puebla, R.A., Guerrero-Martínez, A., Liz-Marzán, L.M.: Self-assembly of Au@ Ag nanorods mediated by Gemini surfactants for highly efficient SERS-active supercrystals. Adv. Opt. Mater. **1**(7), 477–481 (2013)
135. Henzie, J., Andrews, S.C., Ling, X.Y., Li, Z., Yang, P.: Oriented assembly of polyhedral plasmonic nanoparticle clusters. Proc. Natl. Acad. Sci. **110**(17), 6640–6645 (2013)
136. Lim, D.-K., Jeon, K.-S., Hwang, J.-H., Kim, H., Kwon, S., Suh, Y.D., Nam, J.-M.: Highly uniform and reproducible surface-enhanced Raman scattering from DNA-tailorable nanoparticles with 1-nm interior gap. Nat. Nanotechnol. **6**(7), 452 (2011)
137. Lim, D.-K., Jeon, K.-S., Kim, H.M., Nam, J.-M., Suh, Y.D.: Nanogap-engineerable Raman-active nanodumbbells for single-molecule detection. Nat. Mater. **9**(1), 60 (2010)
138. Barrow, S.J., Wei, X., Baldauf, J.S., Funston, A.M., Mulvaney, P.: The surface plasmon modes of self-assembled gold nanocrystals. Nat. Commun. **3**, 1275 (2012)
139. Chirumamilla, M., Toma, A., Gopalakrishnan, A., Das, G., Zaccaria, R.P., Krahne, R., Rondanina, E., Leoncini, M., Liberale, C., De Angelis, F.: 3D nanostar dimers with a sub-10-nm gap for single-/few-molecule surface-enhanced Raman scattering. Adv. Mater. **26** (15), 2353–2358 (2014)
140. Cinel, N.A., Bütün, S., Ertaş, G., Özbay, E.: 'Fairy Chimney'-shaped tandem metamaterials as double resonance SERS substrates. Small **9**(4), 531–537 (2013)
141. Qian, C., Ni, C., Yu, W., Wu, W., Mao, H., Wang, Y., Xu, J.: Highly-ordered, 3D petal-like array for surface-enhanced Raman scattering. Small **7**(13), 1801–1806 (2011)
142. Yan, B., Thubagere, A., Premasiri, W.R., Ziegler, L.D., Dal Negro, L., Reinhard, B.M.: Engineered SERS substrates with multiscale signal enhancement: nanoparticle cluster arrays. ACS Nano **3**(5), 1190–1202 (2009)
143. Alba, M., Pazos-Perez, N., Vaz, B., Formentin, P., Tebbe, M., Correa-Duarte, M.A., Granero, P., Ferré-Borrull, J., Alvarez, R., Pallares, J.: Macroscale plasmonic substrates for highly sensitive surface-enhanced Raman scattering. Angew. Chem. Int. Ed. **52**(25), 6459–6463 (2013)
144. Zhang, W., Jiang, L., Piper, J.A., Wang, Y.: SERS nanotags and their applications in biosensing and bioimaging. J. Anal. Test. **2**(1), 26–44 (2018)
145. Hashim, A.I., Zhang, X., Wojtkowiak, J.W., Martinez, G.V., Gillies, R.J.: Imaging pH and metastasis. NMR Biomed. **24**(6), 582–591 (2011)
146. Tannock, I.F., Rotin, D.: Acid pH in tumors and its potential for therapeutic exploitation. Cancer Res. **49**(16), 4373–4384 (1989)
147. Liu, Y., Yuan, H., Fales, A.M., Vo-Dinh, T.: pH-sensing nanostar probe using surface-enhanced Raman scattering (SERS): Theoretical and experimental studies. J. Raman Spectrosc. **44**(7), 980–986 (2013)
148. Kneipp, J., Kneipp, H., Wittig, B., Kneipp, K.: Following the dynamics of pH in endosomes of live cells with SERS nanosensors. J. Phys. Chem. C **114**(16), 7421–7426 (2010)

149. Gühlke, M., Heiner, Z., Kneipp, J.: Combined near-infrared excited SEHRS and SERS spectra of pH sensors using silver nanostructures. Phys. Chem. Chem. Phys. **17**(39), 26093–26100 (2015)
150. Wang, Y., Rauf, S., Grewal, Y.S., Spadafora, L.J., Shiddiky, M.J., Cangelosi, G.A., Schlücker, S., Trau, M.: Duplex microfluidic SERS detection of pathogen antigens with nanoyeast single-chain variable fragments. Anal. Chem. **86**(19), 9930–9938 (2014)
151. Wang, X., Choi, N., Cheng, Z., Ko, J., Chen, L., Choo, J.: Simultaneous detection of dual nucleic acids using a SERS-based lateral flow assay biosensor. Anal. Chem. **89**(2), 1163–1169 (2016)
152. Wang, Y., Ravindranath, S., Irudayaraj, J.: Separation and detection of multiple pathogens in a food matrix by magnetic SERS nanoprobes. Anal. Bioanal. Chem. **399**(3), 1271–1278 (2011)
153. Levin, C.S., Kundu, J., Janesko, B.G., Scuseria, G.E., Raphael, R.M., Halas, N.: Interactions of ibuprofen with hybrid lipid bilayers probed by complementary surface-enhanced vibrational spectroscopies. J. Phys. Chem. B **112**(45), 14168–14175 (2008)
154. Šimáková, P., Kočišová, E., Procházka, M.: Sensitive Raman spectroscopy of lipids based on drop deposition using DCDR and SERS. J. Raman Spectrosc. **44**(11), 1479–1482 (2013)
155. Kong, K.V., Lam, Z., Lau, W.K.O., Leong, W.K., Olivo, M.: A transition metal carbonyl probe for use in a highly specific and sensitive SERS-based assay for glucose. J. Am. Chem. Soc. **135**(48), 18028–18031 (2013)
156. Shafer-Peltier, K.E., Haynes, C.L., Glucksberg, M.R., Van Duyne, R.P.: Toward a glucose biosensor based on surface-enhanced Raman scattering. J. Am. Chem. Soc. **125**(2), 588–593 (2003)
157. Dinish, U., Balasundaram, G., Chang, Y.-T., Olivo, M.: Actively targeted in vivo multiplex detection of intrinsic cancer biomarkers using biocompatible SERS nanotags. Sci. Rep. **4**, 4075 (2014)
158. Yuan, H., Liu, Y., Fales, A.M., Li, Y.L., Liu, J., Vo-Dinh, T.: Quantitative surface-enhanced resonant Raman scattering multiplexing of biocompatible gold nanostars for in vitro and ex vivo detection. Anal. Chem. **85**(1), 208–212 (2012)
159. Wang, Y., Seebald, J.L., Szeto, D.P., Irudayaraj, J.: Biocompatibility and biodistribution of surface-enhanced Raman scattering nanoprobes in zebrafish embryos: in vivo and multiplex imaging. ACS Nano **4**(7), 4039–4053 (2010)
160. Yaseen, T., Pu, H., Sun, D.-W.: Functionalization techniques for improving SERS substrates and their applications in food safety evaluation: a review of recent research trends. Trends Food Sci. Technol. **72**, 162–174 (2018)
161. Hu, Y., Feng, S., Gao, F., Li-Chan, E.C., Grant, E., Lu, X.: Detection of melamine in milk using molecularly imprinted polymers–surface enhanced Raman spectroscopy. Food Chem. **176**, 123–129 (2015)
162. Pang, S., Labuza, T.P., He, L.: Development of a single aptamer-based surface enhanced Raman scattering method for rapid detection of multiple pesticides. Analyst **139**(8), 1895–1901 (2014)
163. He, L., Deen, B.D., Pagel, A.H., Diez-Gonzalez, F., Labuza, T.P: Concentration, detection and discrimination of Bacillus anthracis spores in orange juice using aptamer based surface enhanced Raman spectroscopy. Analyst **138**(6), 1657–1659 (2013)
164. He, L., Deen, B., Rodda, T., Ronningen, I., Blasius, T., Haynes, C., Diez-Gonzalez, F., Labuza, T.P.: Rapid detection of ricin in milk using immunomagnetic separation combined with surface-enhanced Raman spectroscopy. J. Food Sci. **76**(5), N49-N53 (2011)
165. Kamra, T., Chaudhary, S., Xu, C., Montelius, L., Schnadt, J., Ye, L.: Covalent immobilization of molecularly imprinted polymer nanoparticles on a gold surface using carbodiimide coupling for chemical sensing. J. Colloid Interface Sci. **461**, 1–8 (2016)
166. Kamra, T., Zhou, T., Montelius, L., Schnadt, J., Ye, L.: Implementation of molecularly imprinted polymer beads for surface enhanced Raman detection. Anal. Chem. **87**(10), 5056–5061 (2015)

167. Chang, L., Ding, Y., Li, X.: Surface molecular imprinting onto silver microspheres for surface enhanc24 June 2013ed Raman scattering applications. Biosens. Bioelectron. **50**, 106–110 (2013)
168. Sarfo, D.K., Izake, E.L., O'Mullane, A.P., Ayoko, G.A.: Fabrication of nanostructured SERS substrates on conductive solid platforms for environmental application. Crit. Rev. Environ. Sci. Technol. 1–36 (2019)
169. Li, D., Qu, L., Zhai, W., Xue, J., Fossey, J.S., Long, Y.: Facile on-site detection of substituted aromatic pollutants in water using thin layer chromatography combined with surface-enhanced Raman spectroscopy. Environ. Sci. Technol. **45**(9), 4046–4052 (2011)
170. Fikiet, M.A., Khandasammy, S.R., Mistek, E., Ahmed, Y., Halámková, L., Bueno, J., Lednev, I.K.: Surface enhanced Raman spectroscopy: a review of recent applications in forensic science. Spectrochim. Acta Part A: Mol. Biomol. Spectrosc. **197**, 255-260 (2018)
171. Kamińska, A., Kowalska, A., Albrycht, P., Witkowska, E., Waluk, J.: ABO blood groups' antigen–antibody interactions studied using SERS spectroscopy: towards blood typing. Anal. Methods **8**(7), 1463–1472 (2016)
172. Halouzka, V., Halouzkova, B., Jirovsky, D., Hemzal, D., Ondra, P., Siranidi, E., Kontos, A. G., Falaras, P., Hrbac, J.: Copper nanowire coated carbon fibers as efficient substrates for detecting designer drugs using SERS. Talanta **165**, 384–390 (2017)
173. Kline, N.D., Tripathi, A., Mirsafavi, R., Pardoe, I., Moskovits, M., Meinhart, C., Guicheteau, J.A., Christesen, S.D., Fountain III, A.W.: Optimization of surface-enhanced Raman spectroscopy conditions for implementation into a microfluidic device for drug detection. Anal. Chem. **88**(21), 10513–10522 (2016)
174. Xu, Z., Jiang, J., Wang, X., Han, K., Ameen, A., Khan, I., Chang, T.-W., Liu, G.L.: Large-area, uniform and low-cost dual-mode plasmonic naked-eye colorimetry and SERS sensor with handheld Raman spectrometer. Nanoscale **8**(11), 6162–6172 (2016)
175. Dong, R., Weng, S., Yang, L., Liu, J.: Detection and direct readout of drugs in human urine using dynamic surface-enhanced Raman spectroscopy and support vector machines. Anal. Chem. **87**(5), 2937–2944 (2015)
176. Chen, N., Ding, P., Shi, Y., Jin, T., Su, Y., Wang, H., He, Y.: Portable and reliable surface-enhanced Raman scattering silicon chip for signal-on detection of trace trinitrotoluene explosive in real systems. Anal. Chem. **89**(9), 5072–5078 (2017)
177. López-López, M., Merk, V., García-Ruiz, C., Kneipp, J.: Surface-enhanced Raman spectroscopy for the analysis of smokeless gunpowders and macroscopic gunshot residues. Anal. Bioanal. Chem. **408**(18), 4965–4973 (2016)
178. Pozzi, F., Leona, M.: Surface-enhanced Raman spectroscopy in art and archaeology. J. Raman Spectrosc. **47**(1), 67–77 (2016)
179. Shadi, I.T., Chowdhry, B.Z., Snowden, M.J., Withnall, R.: Semi-quantitative analysis of alizarin and purpurin by surface-enhanced resonance Raman spectroscopy (SERRS) using silver colloids. J. Raman Spectrosc. **35**(8–9), 800–807 (2004)
180. Canamares, M., Garcia-Ramos, J., Domingo, C., Sanchez-Cortes, S.: Surface-enhanced Raman scattering study of the adsorption of the anthraquinone pigment alizarin on Ag nanoparticles. J. Raman Spectrosc. **35**(11), 921–927 (2004)
181. Chen, K., Leona, M., Vo-Dinh, K.C., Yan, F., Wabuyele, M.B., Vo-Dinh, T.: Application of surface-enhanced Raman scattering (SERS) for the identification of anthraquinone dyes used in works of art. J. Raman Spectrosc. Int. J. Orig. Work Aspects Raman Spectrosc. Includ. Higher Order Process. Brill. Rayleigh Scatt. **37**(4), 520–527 (2006)
182. Pozzi, F., Cesaratto, A., Leona, F.: Recent advances on the analysis of polychrome works of art: SERS of synthetic colorants and their mixtures with natural dyes. Front. Chem. **7**, 105 (2019)

Ultrafiltration Membrane for Water Treatment

Mohd Hafiz Dzarfan Othman, Mohd Ridhwan Adam, Mohammad Arif Budiman Pauzan, Siti Khadijah Hubadillah, Mukhlis A. Rahman and Juhana Jaafar

Abstract This chapter discusses the utilization of ultrafiltration membrane (UF) in water treatment system. The UF membrane overview including the type, preparation and characterization is concisely reviewed. The two major types of membrane namely polymeric and ceramic membranes are subdivided into two distinguished subchapters focusing on their fabrication and physicochemical properties. Additionally, the main converge of this chapter is the application of these UF membranes on multidisciplinary industries such as textile, dairy, beverages, microelectronics, petrochemical, cosmetic and pharmaceutical, and few others. The advantages and limitations of UF carefully addressed in their respective subchapters. At the end of this chapter, an attempt is also made to show the future direction of the UF membrane towards the advance membrane technology system such as membrane distillation, membrane contactor and many others.

Keywords UF membrane · Water treatment · Polymeric membrane · Ceramic membrane · Membrane fabrication

M. H. D. Othman (✉) · M. R. Adam · M. A. B. Pauzan · S. K. Hubadillah · M. A. Rahman · J. Jaafar
Advanced Membrane Technology Research Centre (AMTEC), School of Chemical and Energy Engineering, Faculty of Engineering, Universiti Teknologi Malaysia, 81310 UTM Johor Bahru, Johor, Malaysia
e-mail: hafiz@petroleum.utm.my

M. R. Adam
e-mail: ridhwan_adam@yahoo.com

M. A. B. Pauzan
e-mail: mabudiman90@yahoo.com

S. K. Hubadillah
e-mail: ctkhadeeja@gmail.com

M. A. Rahman
e-mail: mukhlis@petroluem.utm.my

J. Jaafar
e-mail: juhana@petroleum.utm.my

Inamuddin et al. (eds.), *Self-standing Substrates*,
Engineering Materials, https://doi.org/10.1007/978-3-030-29522-6_4

1 Introduction and History

Membrane technologies have been extensively used in wastewater treatment owing to their precious properties such as cost effective, ease of fabrication and simple operation system. Accordingly, membrane technologies are divided into four types based on their pore sizes which are microfiltration (MF), ultrafiltration (UF), nanofiltration (NF) and reverse osmosis (RO) [6, 34, 61]. In term of rejection performance, NF and RO showed excellent performances due to their smaller pore size. Unfortunately, both of these membranes show drawbacks of having low flux, needing another pretreatment and high pressure consumption. It should be mentioned here that UF membrane offers advantages of high flux, simple operation and low pressure consumption. The average pore diameter for UF membranes is 10–1000 Å. In early 1900s, Bechhold pioneering the fabrication of UF membrane by prepared it from nitro cellulose [8]. Later in 1918, Zsigmondy and Bachmann explained in detail in their publication on how to modify the MF membrane into UF membrane [70]. In 1922, they got their invention patented in US patent by exposing a thin coating of a nitrocellulose solution [71]. The most highlighted breakthrough of UF membrane was in 1963, in which the first cellulose acetate membrane prepared by Loeb and Sourirajan [45]. Three years later, Abcor installed the first commercial UF membrane to recover electrocoat paint from automobile paint shop rinse water [6]. Interestingly, UF membrane received widely attention and started its commercialization with various configuration such as spiral wound and hollow fibre. Figure 1 shows the flow of the development of UF membranes.

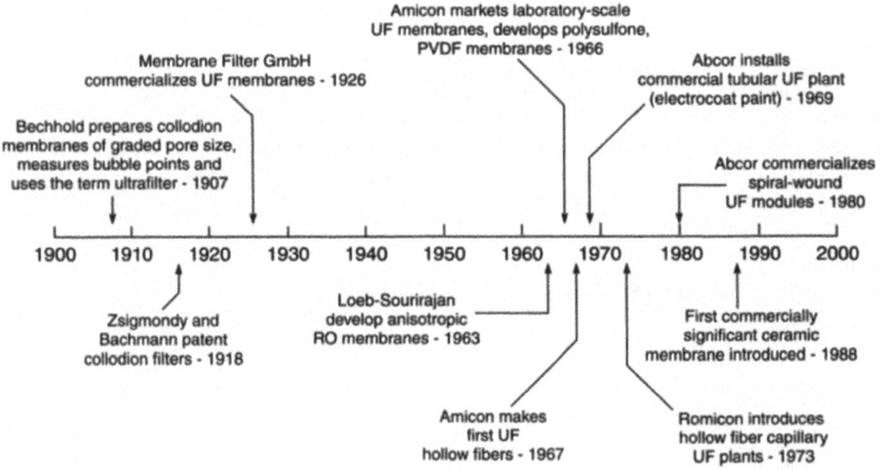

Fig. 1 Milestones in the development of UF membranes. From Baker [6]

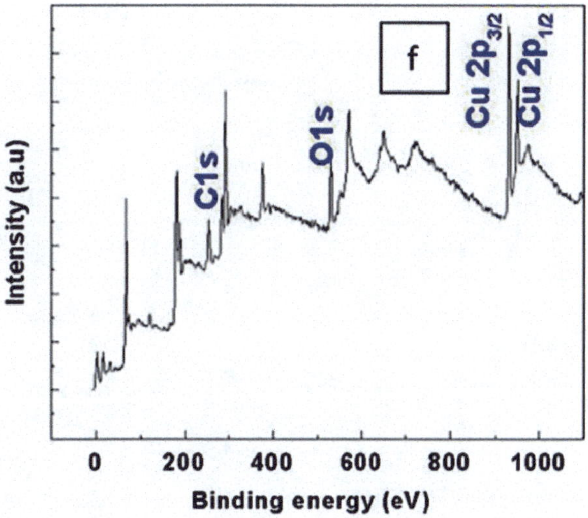

Fig. 2 XPS analysis of UF ceramic membrane derived from clay-alumina surface and CuO. From Choudhury et al. [12]

Nowadays, there has been an increased interest in development of UF membrane, not only towards polymeric membrane type but also towards ceramic membrane. In 1983, Tsapyuk et al. introduced the first UF ceramic membrane by forming a layer of cellulose acetate on porous tubular ceramic membrane for treating wastewater containing colloidal solutions of gelatin and silver [62]. A recent study on UF membrane has been investigated by investigating ceramic membrane incorporating with copper oxide nanoparticles through dip coating for chromium (VI) removal [12]. A detailed XPS analysis (Fig. 2) revealed the binding properties of the CuO and clay-alumina ceramic membrane surface and the rejection mechanism of chromium (VI).

Herein, it is also interesting to note that the application of UF ceramic membranes have been widely applied as photocatalytic membrane [3, 59]. For example, a research team from Dr. Othman's lab has been successfully prepared a low-cost ceramic membrane derived from kaolin. The ceramic membrane was then modified with TiO_2 via hydrothermal method for decolorization of reactive black 5 [51]. Figure 3 shows the SEM images of the prepared membrane and proved a well-dispersed TiO_2 nanorods (Fig. 3) have been obtained. Accordingly, the prepared membrane induced high water permeation of 165 L/h m^2bar and photocatalytic activity of 80.3% under UV irradiation.

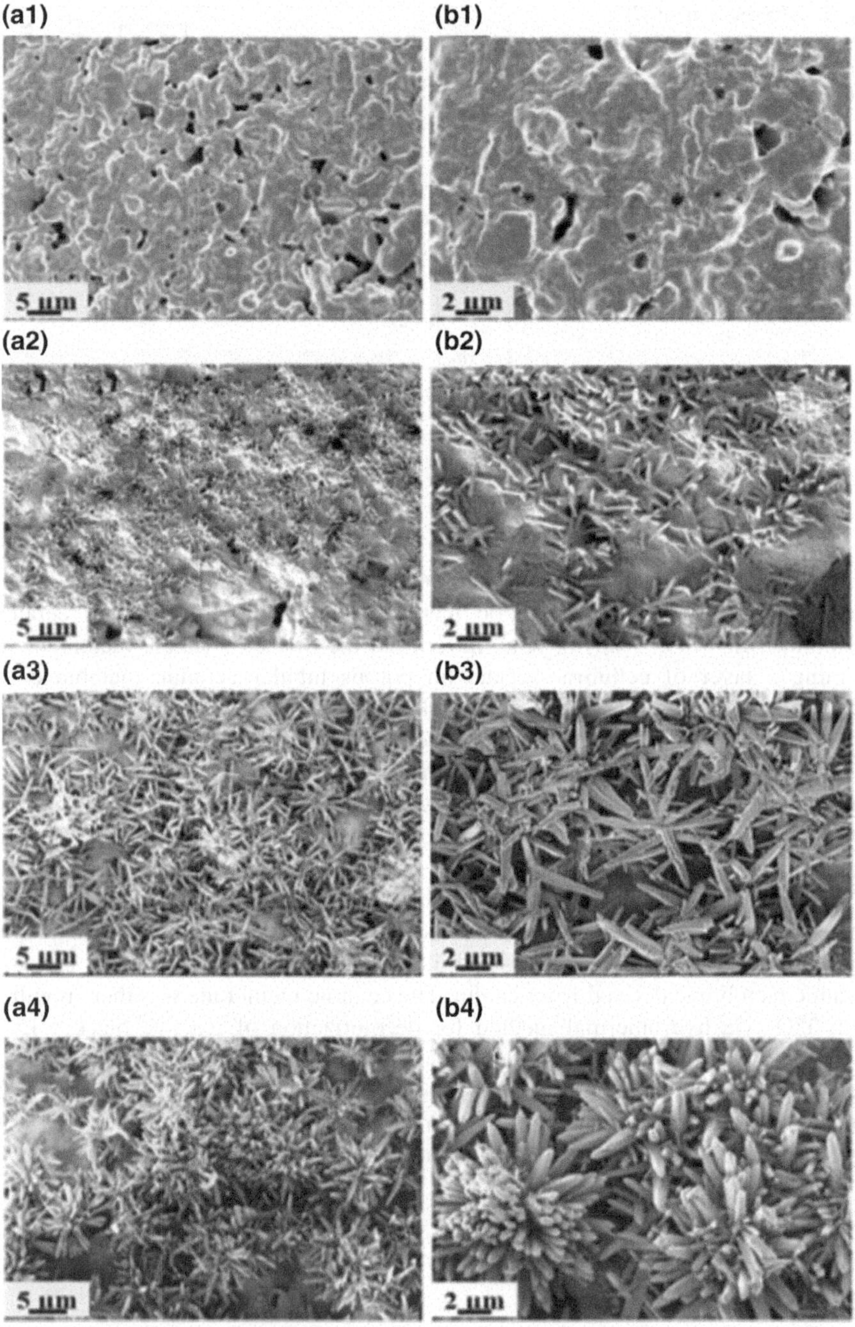

Fig. 3 Morphology images of well-dispersed of TiO_2 nanorods on kaolin-ceramic membrane at **a** 2000x and **b** 5000x of (1) kaolin support and kaolin/TNR membrane that prepared at (2) 2 h, (3) 6 h, and (4) 10 h of reaction time. From Mohtor et al. [51]

2 UF Membrane Type, Characteristics and Preparation

2.1 Polymeric Membrane

Polymeric membranes is known as the major source of membrane that commercially available in the market due to its low cost and abundance of production [66]. Unlike the MF range process, the UF membranes are mainly operating via sieving process and rather in a wider separation range to that of MF membranes. The pore size of the membranes generally between 0.01 and 0.1 μm, are capable of removing particles, colloids, viruses and pathogens as well as allows the permeability of microsolutes with molecular weight (MW) of less than 300 through the membrane [20]. Although these UF membranes are capable of removing those species, the selectivity of the membrane is basically based on the surface charge and size difference of the components to be separated, besides the properties of the membranes itself and its hydrodynamic conditions.

There are numerous types of polymers have been used for the fabrication of UF membranes and few have become an emerged selection as a leading choice for potable reuse applications. The two major considerations that determining the potable recycle of the membrane technologies are the properties of the material that composed the membrane and the mechanisms of the membrane formation. The materials selection of the membranes is mainly depending on the physicochemical properties of the polymer including distribution of the membrane pore size, porosity, wetting vulnerability, polymer flexibility, cost, mechanical strength, durability, stability and chemical resistance [64]. Apart of that, the other advantageous properties of the polymer that associated to the fabrication of the membrane are namely as low tortuosity besides the surface properties including the surface charge that mainly attributed to the rejection performance. Additionally, there are few characteristics that principally influence the membrane performance by enhancing the membrane regeneration and fouling recovery resulted from the weakly adhering materials, resistance towards aggressive cleaning agents as well as low surface roughness [67].

Of all the polymeric materials chosen for the membranes production, there are few reported and widely used attributable to their superior properties contributable to the high performance of the membranes. Table 1 summarizes a list of frequently used polymeric membranes with their respective properties [2, 52, 67]. As been aforementioned earlier, the selection of such polymers is primarily due to their physicochemical properties contributing to the high performance of the fabricated membranes.

There were variety of polymeric membrane have been utilized for the potable reuse membrane facilities. One of its kind is the polypropylene membrane which controls the largest potable reuse membrane in the world. The higher mechanical strength accompanied with the high resistivity towards many solvents has attributed to this successfulness. Over the time, the industries have now shifted to PVDF as a material of choice. This could be possibly due to the improved membrane lifespan,

Table 1 Summary of common polymeric materials used for the potable reuse membranes

Polymer	Acronym	Application	Advantages	Disadvantages
Polyacrylonitrile	PAN	MF/UF	Rigid, highly permeable, narrow pore size distribution	Low mechanical strength
Polyethylene	PE	MF/UF	Oxidant tolerant, low cost, high resistance to organic solvent	Poor fouling resistance, weak thermal properties
Polyvinylidene fluoride	PVDF	MF/UF	Chlorine resistance, oxidant tolerant	Broad pore size distribution
Polypropylene	PP	MF/UF	Reasonable mechanical strength, great organic solvents resistance	Poor resistance to fouling, poor oxidant tolerant
Polysulfone	PSU	MF/UF	High mechanical strength, good chemical resistant	Considerably low hydrophilicity
Polyether sulfone	PES	MF/UF	Robust, highly permeable, small pore size distribution	Considerably low hydrophilicity
Polyvinyl chloride	PVC	MF/UF	–	Weak thermal steadiness, poor oxidant tolerant
Polyamide	PA	MF/UF	Narrow pore size, high rejection, selectivity	Relatively, impermeable, dense structure

reduced capital cost relatable to the production and declined nominal pore size as compared to that of polypropylene [44]. On the other hand, while the new materials along with advantageous operation continue to develop, the contemplation of the polymer constituents of the current facilities can involve a substantial principal cost for the system retrofit because the low pressure of the UF membrane systems have normally not homogeneous to the common platform and thus would permit the latest product materials to be installed promptly to the system with prior modification. Therefore, these UF membrane systems are unique and proposed in a variation of shapes in both submerged and pressurized patterns [63].

Apart from the materials selection, the fabrication of the polymeric UF membranes plays a vital role that consequently determine the performance of the membranes. There are many techniques attempted by the manufacturers in fabricating the membranes. These fabrication techniques are mainly depending more on the membrane materials rather than the class of membrane (i.e. MF, UF, etc.). The most common techniques used for the membrane fabrication are phase inversion and electrospinning process. Notably, phase inversion techniques are varied including non-solvent induced phase separation (NIPS), vapor induced phase inversion (VIPS), thermal phase separation and controlled evaporation. On the other hand, more complicated approaches are utilized for the composite-type membranes especially those acquired the surface modifications [65]. These included the film casting, hydrothermal synthesis process, dip coating, surface adsorption, spray

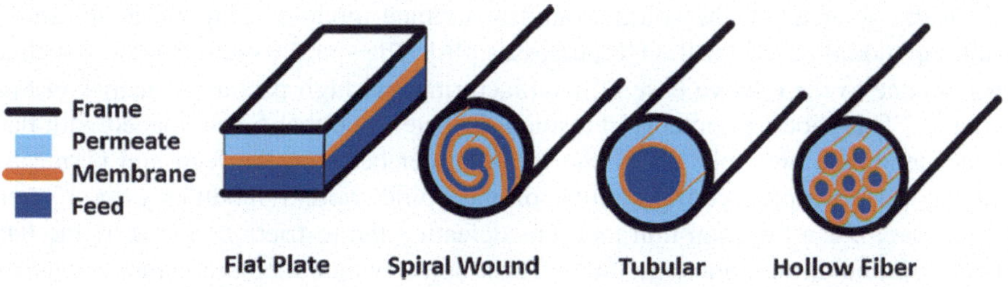

Fig. 4 Schematic diagrams of the most common membrane modules in varies applications. From Loo et al. [46], Crowder and Gooding [14]

coating, sputtering and etching, aerosol deposition and layer-by-layer deposition [41]. Besides that, the chemical process is also attempted to modify the membranes chemically, depending on the applications. These techniques included radical grafting, chemical coupling and hydrophilization with plasma.

The membranes package in reactors and modules has become a vital step in developing the membrane filtration technology. The module forms frequently occupy the great limitations on the membrane constituents attributable to the existing fabrication techniques. Therefore, a broad range of membrane components and configurations have been established that well-matched to a variation of appliances. Figure 4 depicts the four conventional polymeric membrane modules namely flat sheet, tubular, spiral wound and hollow fibre configurations.

Of all the membrane configurations, flat sheet and the uncommonly used tubular membrane modules with diameter of 1–3 cm have a lower packing density and thus lead to the higher costs. The stacked membrane modules of flat sheet configure are constructed to bear an elevated pressure that can overload of 100 bar. However, their applications are only restricted to cutting-edge water treatment process attributable to the low surface area per unit volume densities. Extensively, these membranes are exposed to severe membrane fouling and thus possessed low treatment efficiencies [1]. Alternatively, these modules have largely replaced by the spiral wound and hollow fibre configurations for reuse applications and water treatment processes.

Unlike the flat sheet counterpart, hollow fibre components may comprise of quite a few or several thousands of hollow fibre membranes of various lumen dimension of 0.5–1.5 mm in diameter. This feature would subsequently enhance the surface area per unit volume thus, mostly preferred over other membrane configurations. Additionally, hollow fibre configuration fundamentally has a favourable mass transfer coefficient and higher packing densities and thus making it suitable for the UF processes [29]. Moreover, the filtration in hollow fibre can be achieved from inside-out or vice versa. This could be attributable to the removal of foulants by backwashing process. Unfortunately, most of the polymeric hollow fibre membranes are limited to the applications to pressures below that 4 bar, even after the surface modifications [44].

On the other hand, the spiral wound membrane module is known as the most common module used for the UF processes [46]. The spiral wound design is such a success due to the relative ease of manufacturing and high packing densities of the modules. This module contains of a small diameter tube that firmly loaded with flat sheet membrane and split up by the mesh spacer between the feed and permeate outlets. The compact configurations of the spiral wound modules offer higher surface area for a filtration unit and thus defeating the restrictions faced by the flat sheet and tubular membrane modules. These spiral wound membrane modules have been found and reported to be successfully removing traditional contaminants from the feed, and rather emerging contaminants from wastewater sources [21, 47].

2.2 Ceramic Membrane

Ceramic membrane has been known to be the counter for polymeric membrane [28, 34]. Normally, ceramic membrane is fabricated from inorganic materials such as silica, titania, alumina and zirconia. The membrane mostly has been used for filtration due to its permselective barrier which acts as a fine sieve especially to be used in water treatment process. Porosity is the main characteristic possessed by ceramic membrane which make it able to be used in waste water filtration which can be divided into porous and dense together with other criteria including pore size, membrane's thickness and surface porosity. The summary of membranes type is shown in Table 2 used in filtration according to its pore size and porosity.

Divided into four layers kind of porosity, categorization of the membrane will be choose based on the pore size and the type of filtration as well as the contaminants or pollutants that will be used according to its application.

Basically, there are four layers possessed by ceramic membrane where the performance of the membrane was taken for consideration in term of separation and permeability properties. These layers are characterized as macroporous (the biggest particle size), followed by mesoporous, microporous and dense (the smallest particle size). The mechanism of the ceramic membrane can be described in Fig. 5 where the lowest layer act as a mechanical support for the membrane, the top layer is the place for the separation and is bridged by a middle layer which compromising by two or more mesoporous structure with varied of pore size.

Table 2 Type of ceramic membrane porosity. From Li [43]

Type	Pore size (nm)	Mechanism	Application
Macroporous	>50	Sieving	UF, MF
Mesoporous	2–50	Knudsen diffusion	UF, MF, gas separation
Microporous	<2	Micropore diffusion	Gas separation
Dense	–	Diffusion	Gas separation, reaction

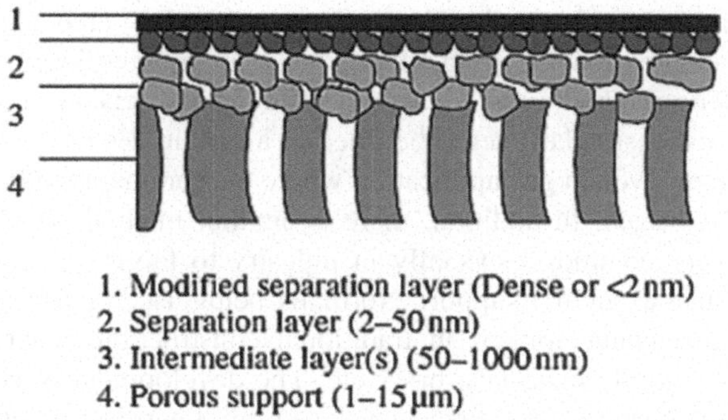

Fig. 5 Schematic diagram of ceramic membrane. From Hubadillah et al. [34]

Depending on its application, silica, titania, alumina and zirconia or even the combination or derivative of these materials are the most commonly used raw ceramic materials for ceramic membrane. Recently, alumina has been the most commonly commercialized materials for ceramic membrane owing to the excellent structure both chemically as well as thermally of the material. However, due to its high cost and some drawbacks, the emphasising of low cost and alternative materials for ceramic materials has drawn more interest. Alternative materials from clay such as kaolin, bauxite, bentonite, ball clay, sagger clay and fire clay has been proposed due to its excellent performance similar to conventional ceramic membrane materials [32, 33, 35]. Other than that, zeolite also has been proposed as starting materials for ceramic membrane due to its adsorption and ion exchange capacity properties [34].

Filtration of membrane can be defined as a technique which utilized the use of strong, thin and semi-permeable membrane for separation process. Due to gradient pressure or driving force, the dissolved components and suspended solids in the water or waste water will be halted from pass through the membrane. MF comprises of 0.1–10 μm considered larger compared to UF which utilized to remove large colloids, cells, viruses and microbes. On the hand, the size of UF which comprises of 0.001–0.1 μm pores useful for removing smaller colloids, other macromolecules from various media such as protein. The separation by these membranes were accomplished by the size of the pore size where the smaller pore size membrane will reject large solutes while smaller solutes such as water will flow through the pore of the membrane [48].

Currently, most UF are in the shape of pressurized, multi-bore and spiral wound hollow fibre membrane depending on its application [24]. The fabrication process of ceramic membrane normally involved few stages which are (1) preparation of suspension; (2) spinning or pressing of suspension and lastly (3) final sintering.

Commonly, slip casting has been the well-known used method to fabricate ceramic membrane. Not only requires a very long time to cast, the wall thickness of the membrane is also hard to control which eventually lead to unstable thickness and very thick. Another casting method is tape casting where the method was adapted from casting of polymeric membrane involved pressure press technique. The method uses a casting knife which acts as moving carrier for the powder suspension later let to be dried in a certain desired area. This method suitable to be employed in gas application where the ceramic membrane was used for oxygen and hydrogen. In addition, there is another method which is extrusion method which very common especially in industry to fabricate ceramic or inorganic membrane such as tubular support. Normally being used in large scale production, this method also being applied in traditional construction materials such as porcelain, tiles, electronic materials, brick etc. The development of sol-gel method of ceramic UF membranes was first introduced by Leenaars et al. [42]. Sol-gel technique is advantageous in term of small pore-size as this technique allows controlling the pore-size hence make it the most important technique in ceramic membranes fabrication (Figs. 6 and 7).

Membrane morphology of ceramic membrane plays one huge factor in its separation performance. Associated with its shape, distribution, density, pore size, and the membrane's surface which determine the membrane's category to be either dense or porous. Depending on its application such as to be used for gas or liquid separation, it is very crucial for manufacturers to be able to specify the properties of the ceramic membrane. For gas separation application, dense ceramic membrane is suitable as this membrane possess gas tightness, crystal structures and permeation characteristics. Unlike porous ceramic membranes which suitable to be used for liquid application, few parameters need to be considered such as the particle or pore size, packing, surface area etc. which is very important to be taken prior to manufacturing. Therefore, it is essential to analysis the characterization and properties of the membranes according to its application.

A membrane's pore structure will determine how the solutes will pass through, while the membrane surface determines the uptake of these solutes and potential for fouling. Many membranes are asymmetric, so a complete set of information about each membrane layer is very important. From those studies, it is remarkable the significant influence of the solute–solute and solute–membrane interactions on the final performance of the filtration process. However, most of the studies on UF membranes involve the use of tight UF membranes [43].

The used of ceramic membrane for UF has been quite popular among researchers in membrane technology due to its excellent chemical and physical properties. Many applications have been used and described in detail in followed subtopics. Basically, ceramic membranes for UF requires the pore size of the membrane within range more than 50 nm to be taken place. However, it is hard to be applied in actual projects, due to its shortages of high cost, great investment, poor stain- resistant capacity, and high energy consumption. Researchers have utilized different methods to overcome these problems, concluding optimized structural design, adding turbulence promoter, and coupling with other treating

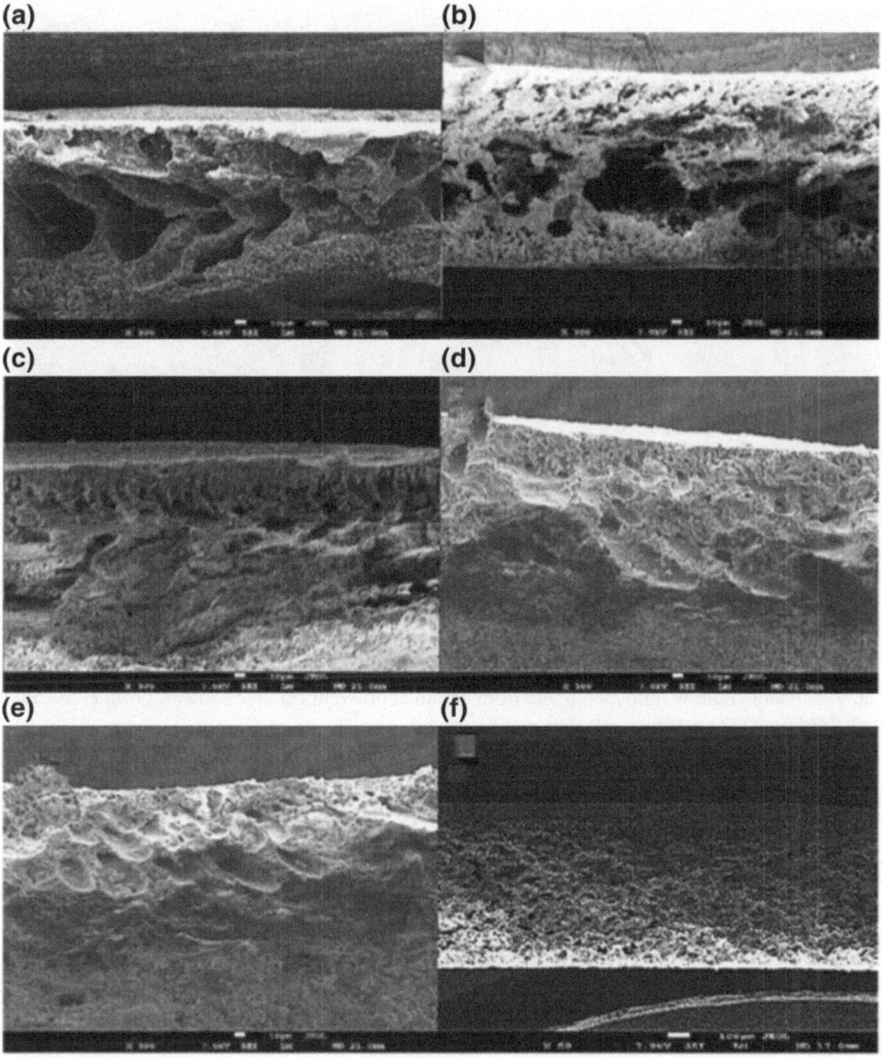

Fig. 6 FESEM image of ceramic membrane showing different pore size. From Hubadillah et al. [31]

processes. On the other hand, membrane fouling is still the most difficult problem to solve. Thus, it is necessary to enrich the theoretical aspects of ceramic membrane process by investigating the membrane fouling control and cleaning procedure.

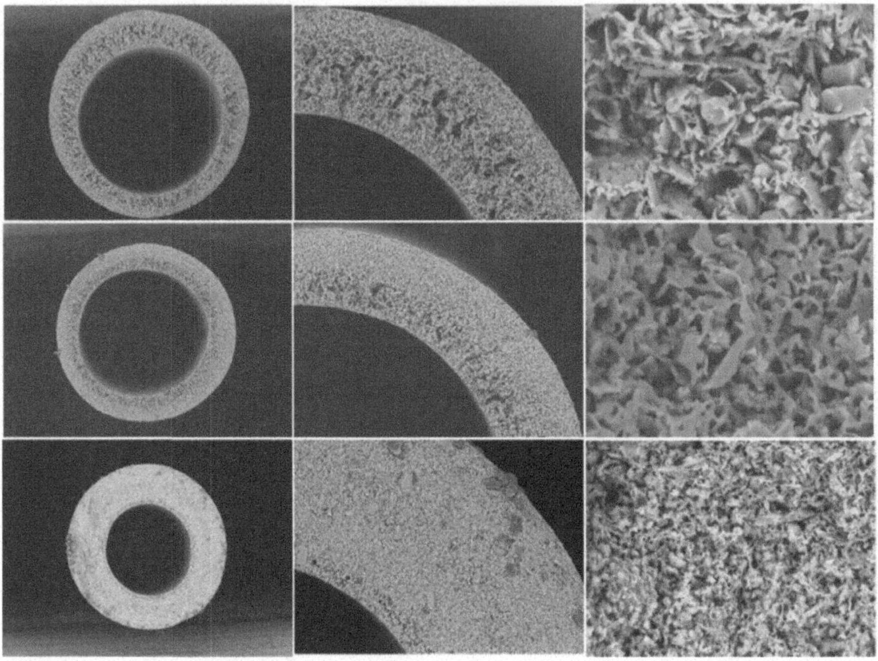

Fig. 7 Ceramic hollow fibre membrane from kaolin at different ceramic content. From Hubadillah et al. [36]

3 Application of UF Membrane in Industrial Influent

3.1 Textile Industry

Another remarkable problem related to wastewater is textile industry due to the production of dyes. Salts and toxic contaminants in its wastewater. This is due to the various activities done by textile industry such as printing, bleaching and dyeing that used colors and other chemicals. Among all technologies, NF membranes were found to be effective in treating this wastewater because of its precious behaviour of having synergistic effects of steric hindrance exclusion and electrostatic repulsion [9]. Unfortunately, NF membranes is negatively charged membrane in which having strong electrostatic repulsion that prohibited the salt particles contained in the wastewater from being treated [19]. Due to this, the usage of NF membrane in the textile wastewater has been replaced with UF membrane. Table 3 summarizes the investigation of UF membrane towards wastewater treatment from textile industry.

Table 3 Summary of UF membranes application towards textile wastewater treatment

Type of membrane	Method of fabrication	Flux (L/m^2h)	Dye rejection (%)	References
Ceramic membrane	Electrodialysis using platinized titanium	220–280	~95	[40]
Ceramic membrane	Commercial ceramic membrane with different pore size	>200	~80	[17]
Polymeric membrane	Phase inversion process of dope solution consisting PVDF and SAN	20–100	57.9–98.4	[58]
Ceramic membrane	Dip coating and ultrasonic treatment	>20	85–98	[10]
Polymeric membrane	Dip-coating and cross-linking process	3.06	97.7	[4]

3.2 *Dairy Industry*

The membrane technology has transfigured along with the diary industries. The membranes have been utilized for many purposes in this field including the calibration of the main constituents of the milk for fitting with the new product, extending the shelf life of the dairy products in the absence of heat treatment as well as enhancing the yield and quality of the dairy food productions. Additionally, the membranes have also used for concentrating, purifying and fractionating the valuable milk proteins in its natural states [39].

The milk is considered as an essential and complete food for the human diet especially for the vegetarians. It is also known as the ideal liquid for the membrane separation due to its compositions. The utilization of the membrane technologies for the diary production has been started since the 1960s and currently has become the second largest industries utilizing the membranes behind the wastewater treatment process [57]. It is reported that several hundreds of thousands square meters of membranes (400,000 m^2) is currently used in these industries [56]. From that figure, about 2/3 of the membranes set up in these industries is utilized for the whey treatment and the remaining 1/3 for the milk. The application of the membrane technology in the dairy industries have been significantly boosted with the establishment of the superior substances namely polyamide, polysulfone and cellulose acetate accompanied by the technological processes of UF. Figure 8 shows the milk component indicated by size and suitable membrane processes.

There are numerous types of membranes with diverse properties presented in the market and frequently utilized in the dairy industries. These membrane filtrations have been utilized in many areas of the industries including the whey processing, milk protein processing, shelf life of milk, cheese industry, demineralization or desalting as well as fractionation of milk fat. The extended shelf life of milk products has been applied in a manner of reducing the microbial activities thus, prolong the lifetime under the chilled surroundings [54]. On the other hand, the

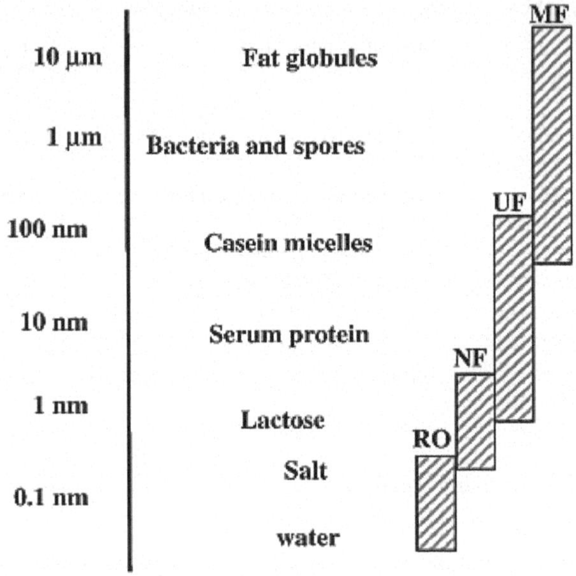

Fig. 8 The milk components due to size and appropriate membrane processes for the treatment. From Daufin et al. [15]

membrane technology in cheese industry used for the concentration of milk by a factor of 1.2–2.0 times and increases the casein: protein ratio thus, reducing the requirement of the processing equipment as well as enhancing the yielding of cheese production [26].

The application of the membrane technology could be beneficial in these industries due to a wide range of advantages offer by this technology. The membrane filtration is capable of producing a desired quality of product via non-thermal environmentally friendly approaches that possibly minimizes the adverse effect rise from the temperature. The effect of denaturation of proteins, changes in phase as well as sensory attributed by the heat treatment could be avoided using the membrane technology. Additionally, the membranes can be utilized for the removal of the undesirable constituents namely microorganisms, sediments or drugs that consequently have a adverse effect of the product quality. Furthermore, there were reported that these membrane technology have used the fusion of membranes relatively to that a single membrane systems [7].

3.3 Beverage Industry

Over the past few years, the growth of beverage industry can be seen is likely increased especially in juice manufacturing. The search to develop the best method to keep the highest percentage of originality of its original source of food such as fruit juice so that the original flavour of the juice can be retained. Converting from lab scale to manufacturing or pilot scale has been one of the most challenging things encountered by many manufacturing plants. Not only problem may arise during the process, another problem could occur after the process which is fruit-waste-containing water. Membrane technology has been discovered could minimize these problems especially UF or MF membrane, however, here comes another problem which is membrane fouling which could hinder the expansion on the use of the membranes in pilot plants.

The utilization of UF is not solely for filtration, instead, fractionation and fixation also can be considered to be another usage of UF. Clara et al. [13] stated that UF can be exploited for fractionation of milk to generate cheese where lactose and solved salts containing in the permeate and proteins, fat and some insoluble salts containing in the retentate part. It has been discovered that skimmed milk produced by UF contain high concentration of calcium and protein which is very vital in dairy industry application. On the other hand, in fruit juice application where UF has been used for clarification of the desired juice depending on the molecular weight cut-off. Instead of collecting from the permeate part, the juice was collected from the retentate part. UF also has been used in order to keep the juice from any miscellaneous such as bacteria, undesired proteins and polysaccharides which polluting the clarity of the juice.

Membrane technology has been considered one important technology to be developed in beverage industry specially to separate varieties of wastes from beverage industries such as fruit juices, wines and some vegetables juices wastes. UF has been known to allow the separation process from retentate containing the concentrated fibrous or pulp and a permeate containing solution after clarification of free from microorganisms. Not only in the process in waste water treatments, UF has been used to clarify fruit juice to prevent pasteurized fraction. In addition, UF also brings huge impact in beverage industry where suspended solid separated from clear final products such as in liqueur, clear fruit juice and carbonated soft drinks has been considered to possess negative impact to the quality of the products [16]. Study by Galaverna et al. [22] stated that apart from separation process, UF was also used in the recovery of bioactive components from fruit juice. In the study, the bioactive components from depectinized kiwi fruit juice were mostly recovered from the process of clarification using UF.

Mohamad et al. [50], has summarised some studies by several researches by using ceramic membranes in UF process. The polymeric membranes that has been used in UF process include polysulfone, polyethersulfone, regenerated and acetate cellulose. However, during cleaning process by using strong chemicals, the membranes tend to break which leads to membrane fouling due to chemical degradation in long term period which shortened the lifespan of the membranes. Due to this drawback, ceramic membranes specifically to be made to be used in UF process has drawn much interests among researchers due to its great performance in term of chemically stable in long term period, thermally and mechanically. UF ceramic membranes which comprised of alumina oxide, zirconia oxide and titania oxide and some supported by these materials or carbon substructure were made according to its MWCO to identify the membrane compatibility to be used in apple juice separation/clarification process. From the observation, the reduction of

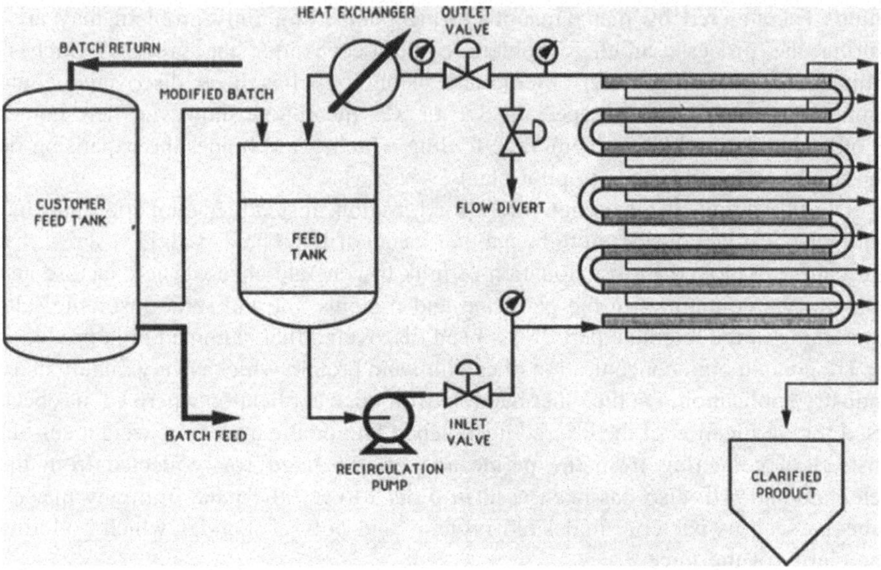

Fig. 9 Schematic of a UF fruit juice in plant. From Cheryan and Alvarez [11]

permeate flux over time up was due to the formation of layer on the surface of the membranes because of the retained solids juice which inhibited a smooth trans-membrane process. Also, at pressure below 300 kPa, the membranes also shown some improvement in term of reducing the fouling resistance. On the other hands, ceramic membranes also have been reported showing a good performance to separate protein and lactose where almost PR ~80% content of high protein and LR ~7% content of low lactose were retained considering high permeate flux at approximately 40 L/m^2 h (Fig. 9).

According to Girard and Fukumoto [25], compounds present in fruit juices are mostly made up of organic and amino acids, sugars, pigments, vitamins, phenolic compounds, essential oils, nucleotides, minerals, esters, alcohols and aldehydes. Low molecular size of these molecules which is <1 kDa together with pore size of the membrane was observed to influence the enzyme depectinization. During this process, pectin was broken down making components of MW below 10 kDa increased. Due to huge range of MW of pectin, which is about 10–500 kDa, these large molecules could disperse which later formed a solid suspensions or colloids called polysaccharides. This colloid not only will produce unclear solution to the juice, it also endures large particles and hence allowing transmembrane with UF membrane to take place.

Another fouling on membrane was also examined by Saha et al. [55]. By using polymeric membrane of PS/PES, the fouling on the membrane caused by polysaccharides in sugarcane juice was examined and it has been found that, the large particles size of polysaccharide contributed to the major fouling of the

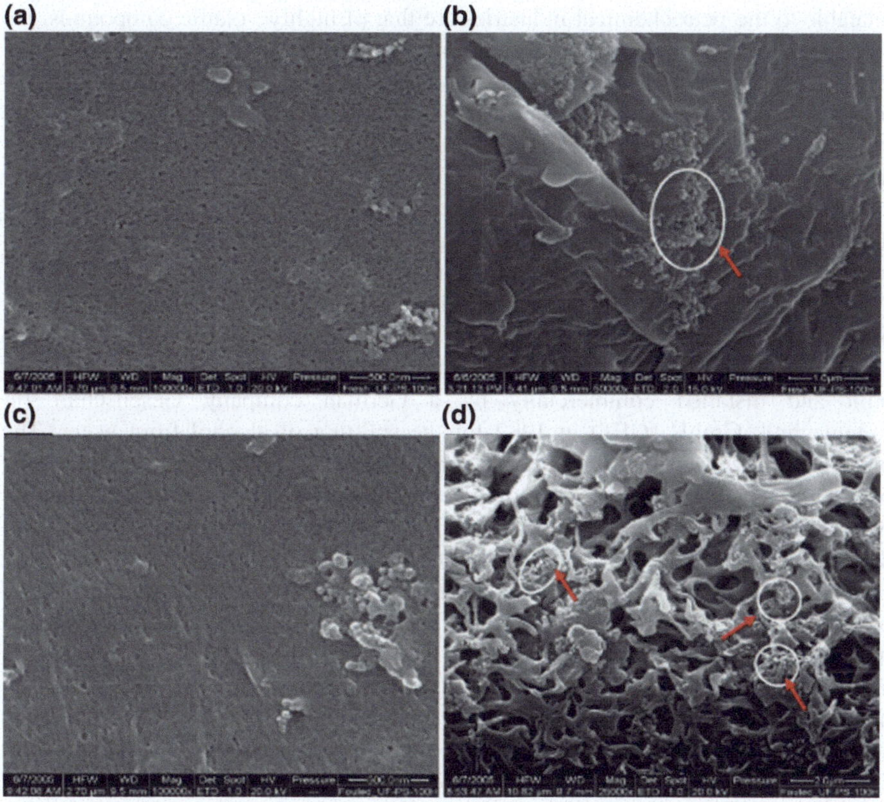

Fig. 10 SEM images of UF-PS-100H membrane: **a** pristine (surface), **b** pristine (cross-section), **c** polysaccharide fouled (surface) and **d** polysaccharide fouled (cross-section). From Saha et al. [55]

membrane. Apparently, arabinogalactan, lipids and some phenolics components have become the major contributor for membrane fouling in which this researcher used a method to modify the membrane's surface by grafting a thin layer of polymer made up from hydrogel to overcome this problem. This modified approach literally has been the best option proposed in order to solve the problem of membrane fouling due to large molecules such as polysaccharides (Fig. 10).

3.4 Petrochemical Industries

Unlike the other applications, the utilization of the membrane technology in petrochemical industries is mainly aimed for the separation and recovery process of several chemicals attained in the petrochemical fields. Since most of the chemicals relatable to the petrochemical industries are that of highly volatile compounds, the UF membranes for these processes are principally using the pervaporation approaches. Compared to the conventional approaches, membranes can possibly offer easy-operate, low maintenance and simple process option. Additionally, the membrane modules can be added and improved to achieve the anticipated separation process [53].

The separation of light solvent and the dewaxing process of solvent are among the common focus of the usage of membrane system in this area. The waste stream of petrochemical process may contain aromatic amines and/or phenolic compounds. Pervaporation filtration could possibly offer the separation of solutions or mixture of components that close boiling points. This process is normally difficult to be attained by other means including distillation. The first pervaporation plant was built and installed commercially by a German company Gesellschaft für Trenntechnik GmbH (GFT) in 1982 for the splitting of alcohol from water. The plant has utilized the polyvinyl alcohol as composite membrane since it is more permeable to water than the alcohol. The membrane was capable of reducing the water content of feed ethanol from ~10 to <1% as water permeate. Figure 11 depicts the flow scheme of the GFT plant for the ethanol recovery.

On the other hand, membrane technology has also been used for the recovery of phenol besides the polymer adsorption, solvent extraction and activated carbon utilization [27]. The membrane aromatic recovery system (MARS) is fairly innovative procedure for the reclamation of aromatic acids and bases. In this procedure, the aromatics are selectively eradicated from the wastewater into the stripping solution. The process normally used a tubular silicon rubber membrane with 500 μm wall thickness. For the recovery of the aromatic acids such as phenol, the stripping solution is preserved at basic condition using NaOH whilst the acidic stripping solution of HCl is used for the reclamation of aromatic bases such as aniline.

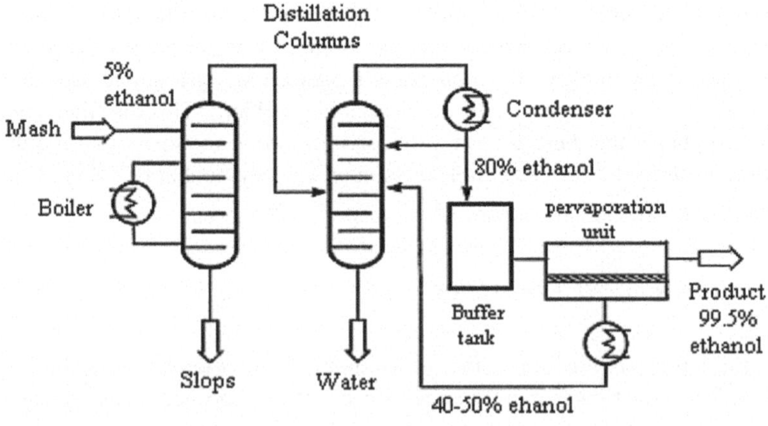

Fig. 11 The flow scheme of GFT plant for ethanol recovery process. From Baker [5]

Therefore, it can be clearly seen that the reaction and separation procedure possessed by the membrane play a crucial role in achieving the desired product of the petrochemical industries. Membrane that works in very low energy, do not require any heating-cooling system and do not produced any environmental pollution. The recovery of the small amount component from the stream using membrane can be achieved without the requirement of any substantial additional energy costs. Additionally, the system usually composed of low capital cost due to compact size, low power consumption and modular configuration will be subsequently reducing the production cost.

4 Application of UF Membrane in Industrial Effluent

4.1 Cosmetic and Pharmaceutical Industries

Recently, quality issue in wastewater especially from industry correspond to the risen of personal care products (PPCPs) and endocrine disrupting compounds, pharmaceuticals (EDC) has been concerning to fresh water supply. EDC mainly can be found in pharmaceutical products and can be considered as one of micropollutants in water or wastewater especially from industries. The removal of EDC/PPCPs to water stream from cosmetic and pharmaceutical industries one of the greatest micropollutants that need to be taken seriously.

According to Clara et al. [13], normally, the concentration of EDC in water was measured from nanogram per litre (ng/L) to micrograms per litre (μg/L) which be later specified the size of membrane should be used in treating the wastewater also according to its molecular weight cut-off (MWCO). Study by this researcher shown that, during wastewater treatment process, some substances such as carbamazepine which is one type of EDC, was not really removed by using membrane bioreactor during the process. This may due to the size of the molecules which is considered large and the membrane operated was not suitable to be used in the process. Hence, solution to this matter would the use of dense membrane with UF in reverse or forward osmosis process to enhance the power consumption of the membrane reactor.

Another study by Yoon et al. [68] has indicated that hydrophobic adsorption has been the biggest factor to assist high retention in a NF and UF dead-end stirred-reactor measurements. The results presented that less polar retention exhibit high volatile and more hydrophobic behaviour and vice versa. The membrane used for the removal of EDC/PPCPs has shown NF has better performance rather than UF in term of retention due to the size of the EDC which can be considered large. The pore size of the UF was investigated and it has shown that performance of the membrane was affected by the pore size. Other factors including source of waste water, electrostatic repulsion, feed and retentate solution, as well as size molecules which may affect the hydrophobicity adsorption of the membrane (Fig. 12).

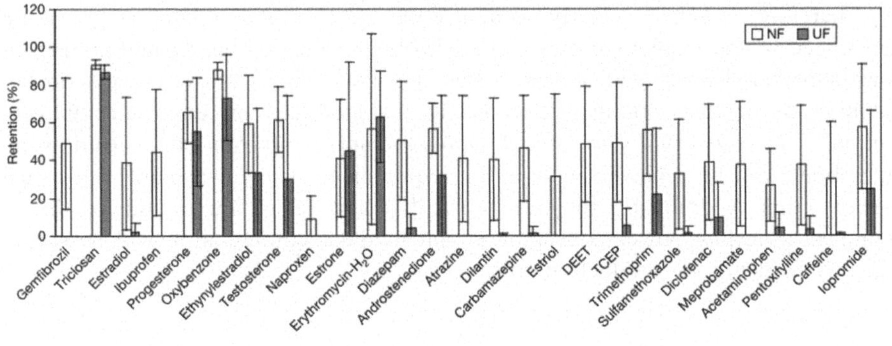

Fig. 12 Comparison on average percentage retention of EDC/PPCPs by the NF and UF membranes. From Yoon et al. [69]

Another study was done by Yoon et al. [69] using the same membrane and reactor in order to investigate EDC/PPCPs removal from various effluents sources. Rejection of retention by size of membrane hugely affected the removal especially once the operation has reached a stable or steady state. With the enhance of hydrophobic adsorption, it was found that based on the result of average of 0 to <2.5 ng/cm^2 adsorbed mass of EDC/PPCPs was calculated.

On the other hand, Taheran et al. [60] has pointed that estimation on these parameters including pore size, porosity, hydrophobicity, molecular size, weight and charge of the molecules in the feed, as well as morphology of the membranes should be included in order to remove pharmaceutically active compounds (PhAC) from water by using separation method. Salt rejection is one of the keys in membrane separation and to ensure an efficient removal of PhACs. Normally, MWCO of MF and UF membrane could be more than thousands Dalton where range for most of MWCO molecules only in between 200 and 800 Daltons. Hence, the adsorption capacity of the membrane should be evaluated for better performance to avoid any failure in separation process. Apart from membrane separation process which involved UF and NF, adsorption should be taken as another alternative due to the size of PhACs which hindering the used of membrane separation method (Fig. 13).

4.2 *Food Processing Industries*

Another remarkable problem related to wastewater is from food processing industries. This wastewater produced from the equipment cleaning process where all the contaminants flow out together into the drainage. Accordingly, wastewater from food industries contains very high BOD and protein, thus was automatically polluted the environment when pumped out from the industries.

A recent study by Miron et al. towards the treatment of protein-based solution have been investigated with ceramic membrane technology [49]. In the study,

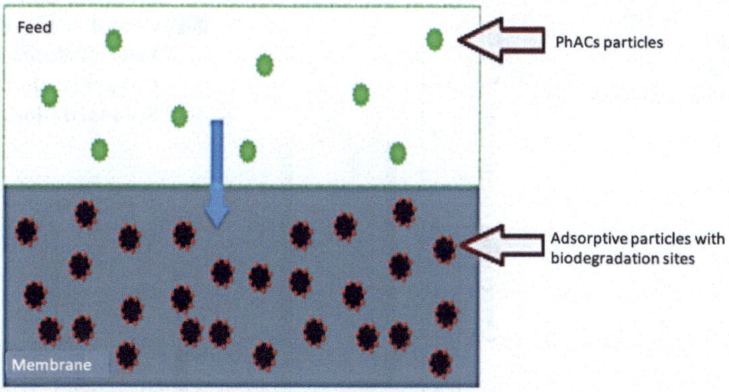

Fig. 13 Proposed mechanism of PhACs removal from wastewater. From Taheran et al. [60]

ceramic membrane was prepared and treated via hydrothermal process with partially irreversible cleaning operations. As a result, 100% lysosome rejection was successfully obtained. To be noted, HPLC test of the protein sample showed changes in its peaks during permeation as well as retention. For example, a peak with value of 0.4 min indicating a hydrodynamic radius of protein was shown during retention process, showing that the protein was denatured after the process. Figure 14 shows the chromatograms of retentate samples obtained during filtration test 2 of lysozyme using prepared UF membrane.

Later, another recent study by Ding et al. was towards protein treatment in detail together with membrane fouling problem [18]. A filtration system with 76 mm diameter stirres ultrafiltration cell and conducted by a pressure of 0.1 MPa. Accordingly, the performance was conducted by using 250 mg/L BSA solution as protein-based solution. As a result, the regression results for flux decline using the four fouling models and are shown in Fig. 15. Interestingly, the results proved that the corresponding correlation coefficient (R^2) values regarding 20 mg/L BSA were

Fig. 14 Chromatograms of retentate samples taken at different pressures. From Miron et al. [49]

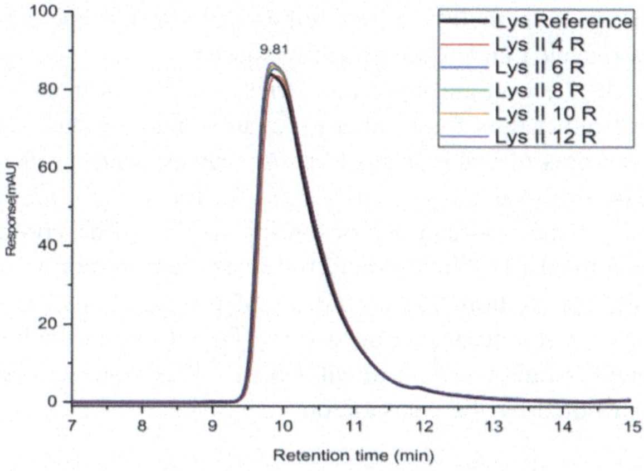

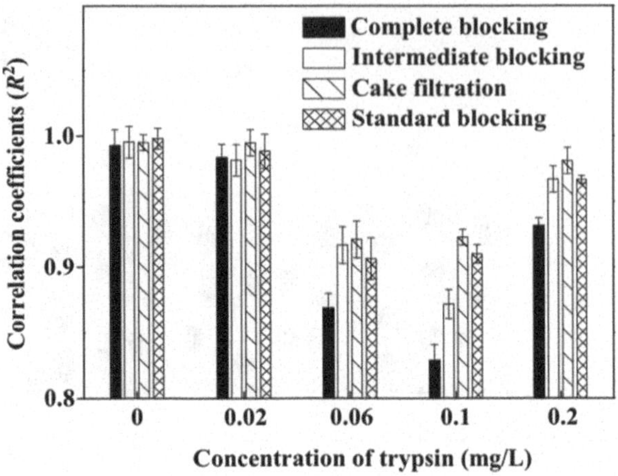

Fig. 15 R^2 of the four fouling models for bovine serum albumin (BSA) filtration with trypsin

0.993, 0.995, 0.995, and 0.998 for complete blocking, intermediate blocking, cake filtration, and standard blocking, respectively. All these R^2 values fit the flux decline well, indicating that there was a possibility of the four fouling mechanisms (pore blocking, adsorption fouling, cake layer and combined fouling) having occurred. Of the fouling models, complete blocking deviated farther from the flux decline than the other three model equations did.

4.3 Iron and Steel Industries

The implementation of major resources efficiency by minimizing the freshwater intake and emission has become vital due to the stringent environmental regulations and awareness amongst the environment activist across the globe. Due to that, the industries have become the most pressured entities regulates by this sanction worldwide [38]. One of its kind is the steel industries that use a large amount of water. Nonetheless, a little amount of water is spent as the largest part of the water is recycled and restored to the source.

In the iron and steel industries, water is utilized not only for cooling purposes, but rather uses for another procedures namely dust scrubbing and descaling. Thus, all forms of water (from fresh to seawater) are used in this steelmaking processes. The freshwater is primarily used for the straight and secondary cooling processes, whilst the seawater is consuming for one-time cooling upon the antifouling pre-treatment [37]. In average, the water consumption for an assimilated steelmaking process is about 28.6 m^3/tonne of produced steel whereas the water release is of 25.3 m^3/tonne of produced steel. Therefore, the inclusive water usage per tonne of steel is rather low of about 1.6 m^3. This water consumption and losses is mainly attributed by the evaporation.

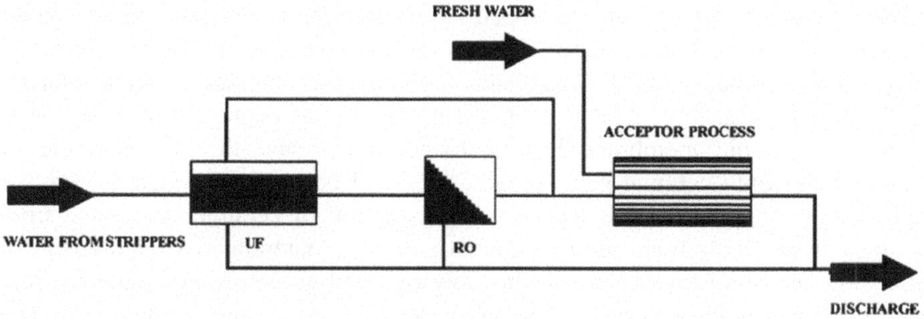

Fig. 16 The integrated membrane systems for wastewater reclamation in iron and steel industries. From Huang et al. [30]

The water resource management is a vital challenge to be monitored to increase the sustainability of the manufacture cycle as the freshwater accessibility and quality become the foremost interests. The increase in water recycling upon the cooling and cleaning processes can decline the water quality. Upon the cooling and desalination of this water, the escalate of salt content in the water rotation system attributable to the evaporation will not merely possess the ecological harmfulness, but rather undesirably disrupt the factory equipment due to corrosion.

Both conventional and advanced wastewater treatments such as chemical precipitation/clarification in the combination with flocculation have been attempted in the focus of producing high-grade water and thus, high proficiency of water reutilizing. Similarly, there were studies reported that membrane technology have been carried out in many industries for the treatment and reclaim of wastewater [23]. Additionally, the employment of UF membranes has efficiently removed the tiny colloidal suspended solid organic matter thus, causing in the premium water for iron and steel industries purposes. This subsequently reduced the cost of operation for the industries [30]. Figure 16 illustrates the integrated membrane systems for wastewater reclamation from iron and steel industries.

5 Conclusion and Future Direction

In summary, UF membranes considered to be one the viable technology to treat wastewater due to possessed precious behaviour include low pressure, high flux, cost-effective investment and operational simplicity. UF membrane induced smaller pore size ranging from 10 to 1000 Å. Accordingly, it was found that UF membranes has an ability to treat various industrial wastewater such as textile, diary and food processing industries. Interestingly, the development of UF membrane is not only referred to polymeric membrane, but also ceramic membrane. Compared to polymeric membrane, ceramic membrane showed better performance especially in term of its ability that can endure harsh condition. Due to this, development of UF ceramic membrane has been developed by incorporating nanoparticles on membrane surface or membrane matrix through coating and dope mixing, respectively. It was also reported that UF membranes have become attention in photocatalytic application to treat dye or other recalcitrant compounds contained in wastewater. However, ceramic membrane has micro-size pores and the incorporation of nanoparticles onto ceramic membrane surface will be a great challenge. Another remarkable encounter is that the cost of production of ceramic membrane from alumina is relatively high due to expensive starting materials and requirement of high sintering temperature. Due to this, low cost and green ceramic material from clays and wastes have been used to fabricate low cost ceramic membranes. The most prominent thing here is that these green materials possessed the presence of natural photocatalytic material such as TiO_2 in small amount. Therefore, further research and development of dual-layer or triple-layer ceramic membrane that prepared in single step is an interesting topic to be studied towards photocatalytic process of dyes in wastewater.

References

1. Adham, S., Chiu, K.-P., Lehman, G.: Optimization of Membrane Treatment for Direct and Clarified Water Filtration. American Water Works Association, USA (2006)
2. Alspach, B., Adham, S., Cooke, T., Delphos, P., Garcia-Aleman, J., Jacangelo, J.: Microfiltration and ultrafiltration membranes for drinking water. J. Am. Water Works Assoc. **100**, 84–97 (2008)
3. Athanasekou, C.P., Moustakas, N.G., Morales-Torres, S., Pastrana-Martínez, L.M., Figueiredo, J.L., Faria, J.L., Silva, A.M.T., Dona-Rodriguez, J.M., Romanos, G.E., Falaras, P.: Ceramic photocatalytic membranes for water filtration under UV and visible light. Appl. Catal. B Environ. **178**, 12–19 (2015)
4. Babu, J., Murthy, Z.V.P.: Treatment of textile dyes containing wastewaters with PES/PVA thin film composite nanofiltration membranes. Sep. Purif. Technol. **183**, 66–72 (2017)
5. Baker, R.W.: Membrane Technology. Wiley, USA (2000)
6. Baker, R.W.: Membrane Technology and Applications. Wiley, USA (2012)
7. Balannec, B., Vourch, M., Rabiller-Baudry, M., Chaufer, B.: Comparative study of different nanofiltration and reverse osmosis membranes for dairy effluent treatment by dead-end filtration. Sep. Purif. Technol. **42**, 195–200 (2005)
8. Bechhol, H.: Kolloidstudien mit der Filtrationsmethode. Z. Physik. Chem. **60**, 257–318 (1907)
9. Bejaoui, I., Mnif, A., Hamrouni, B.: Performance of reverse osmosis and nanofiltration in the removal of fluoride from model water and metal packaging industrial effluent. Sep. Sci. Technol. **49**, 1135–1145 (2014)
10. Bouazizi, A., Breida, M., Achiou, B., Ouammou, M., Calvo, J.I., Aaddane, A., Younssi, S.A.: Removal of dyes by a new nano–TiO_2 ultrafiltration membrane deposited on low-cost support prepared from natural Moroccan bentonite. Appl. Clay Sci. **149**, 127–135 (2017)
11. Cheryan, M., Alvarez, J.R.: Food and beverage industry applications. Membr. Sci. Technol. (USA, Elsevier) (1995)
12. Choudhury, P., Mondal, P., Majumdar, S., Saha, S., Sahoo, G.C.: Preparation of ceramic ultrafiltration membrane using green synthesized CuO nanoparticles for chromium (VI) removal and optimization by response surface methodology. J. Clean. Prod. **203**, 511–520 (2018)

13. Clara, M., Strenn, B., Gans, O., Martinez, E., Kreuzinger, N., Kroiss, H.: Removal of selected pharmaceuticals, fragrances and endocrine disrupting compounds in a membrane bioreactor and conventional wastewater treatment plants. Water Res. **39**, 4797–4807 (2005)
14. Crowder, M.L., Gooding, C.H.: Spiral wound, hollow fiber membrane modules: a new approach to higher mass transfer efficiency. J. Membr. Sci. **137**, 17–29 (1997)
15. Daufin, G., Escudier, J.P., Carrere, H., Berot, S., Fillaudeau, L., Decloux, M.: Recent and emerging applications of membrane processes in the food and dairy industry. Food Bioprod. Process. **79**, 89–102 (2001)
16. De Barros, S., Andrade, C., Mendes, E., Peres, L.: Study of fouling mechanism in pineapple juice clarification by ultrafiltration. J. Membr. Sci. **215**, 213–224 (2003)
17. Dilaver, M., Hocaoğlu, S.M., Soydemir, G., Dursun, M., Keskinler, B., Koyuncu, İ., Ağtaş, M.: Hot wastewater recovery by using ceramic membrane ultrafiltration and its reusability in textile industry. J. Clean. Prod. **171**, 220–233 (2018)
18. Ding, Y., Ma, B., Liu, H., Qu, J.: Effects of protein properties on ultrafiltration membrane fouling performance in water treatment. J. Environ. Sci. **77**, 273–281 (2018)
19. Dow, N., Villalobos Garcia, J., Niadoo, L., Milne, N., Zhang, J., Gray, S., Duke, M.: Demonstration of membrane distillation on textile waste water: assessment of long term performance, membrane cleaning and waste heat integration. Environ. Sci. Water Res. Technol. **3**, 433–449 (2017)
20. Fane, A.G., Wang, R., Jia, Y.: Membrane technology: past, present and future. In: Wang, L. K., Chen, J.P., Hung, Y.T., Shammas, N.K. (eds.) Membrane and Desalination Technologies. Humana Press, Totowa (2011)
21. Fujioka, T., Khan, S.J., McDonald, J.A., Roux, A., Poussade, Y., Drewes, J.E., Nghiem, L.D.: Modelling the rejection of N-nitrosamines by a spiral-wound reverse osmosis system: mathematical model development and validation. J. Membr. Sci. **454**, 212–219 (2014)
22. Galaverna, G., Di Silvestro, G., Cassano, A., Sforza, S., Dossena, A., Drioli, E., Marchelli, R.: A new integrated membrane process for the production of concentrated blood orange juice: effect on bioactive compounds and antioxidant activity. Food Chem. **106**, 1021–1030 (2008)
23. Galiana-Aleixandre, M., Iborra-Clar, A., Bes-Piá, B., Mendoza-Roca, J., Cuartas-Uribe, B., Iborra-Clar, M.: Nanofiltration for sulfate removal and water reuse of the pickling and tanning processes in a tannery. Desalination **179**, 307–313 (2005)
24. Gille, D., Czolkoss, W.: Ultrafiltration with multi-bore membranes as seawater pre-treatment. Desalination **182**, 301–307 (2005)
25. Girard, B., Fukumoto, L.R.: Membrane processing of fruit juices and beverages: a review. Crit. Rev. Food Sci. Nutr. **40**(2), 91–157 (2000)
26. Guinee, T., O'kennedy, B., Kelly, P.: Effect of milk protein standardization using different methods on the composition and yields of Cheddar cheese. J. Dairy Sci. **89**, 468–482 (2006)
27. Han, S., Ferreira, F.C., Livingston, A.: Membrane aromatic recovery system (MARS)—a new membrane process for the recovery of phenols from wastewaters. J. Membr. Sci. **188**, 219–233 (2001)
28. Harun, Z., Hubadillah, S.K., Hasan, S., Yunos, M.Z.: Effect of thermodynamic properties on porosity of ceramic membrane prepared by phase inversion. Appl. Mech. Mater. **575**, 31–35 (2014)
29. Hernández, S., Lei, S., Rong, W., Ormsbee, L., Bhattacharyya, D.: Functionalization of flat sheet and hollow fiber microfiltration membranes for water applications. ACS Sustain. Chem. Eng. **4**, 907–918 (2016)
30. Huang, X.F., Ling, J., Xu, J.C., Feng, Y., Li, G.M.: Advanced treatment of wastewater from an iron and steel enterprise by a constructed wetland/ultrafiltration/reverse osmosis process. Desalination **269**, 41–49 (2011)
31. Hubadillah, S.K., Harun, Z., Othman, M.H.D., Ismail, A.F., Gani, P.: Effect of kaolin particle size and loading on the characteristics of kaolin ceramic support prepared via phase inversion technique. J. Asian Ceram. Soc. **4**, 164–177 (2016)

32. Hubadillah, S.K., Othman, M.H.D., Harun, Z., Ismail, A.F., Iwamoto, Y., Honda, S., Rahman, M.A., Jaafar, J., Gani, P., Mohd Sokri, M.N.: Effect of fabrication parameters on physical properties of metakaolin-based ceramic hollow fibre membrane (CHFM). Ceram. Int. **42**, 15547–15558 (2016)
33. Hubadillah, S.K., Othman, M.H.D., Ismail, A.F., Rahman, M.A., Jaafar, J., Iwamoto, Y., Honda, S., Dzahir, M.I.H.M., Yusop, M.Z.M.: Fabrication of low cost, green silica based ceramic hollow fibre membrane prepared from waste rice husk for water filtration application. Ceram. Int. **44**, 10498–10509 (2018)
34. Hubadillah, S.K., Othman, M.H.D., Matsuura, T., Ismail, A.F., Rahman, M.A., Harun, Z., Jaafar, J., Nomura, M.: Fabrications and applications of low cost ceramic membrane from kaolin: a comprehensive review. Ceram. Int. **44**, 4538–4560 (2018)
35. Hubadillah, S.K., Othman, M.H.D., Matsuura, T., Rahman, M.A., Jaafar, J., Ismail, A.F., Amin, S.Z.M.: Green silica-based ceramic hollow fiber membrane for seawater desalination via direct contact membrane distillation. Sep. Sci. Technol. **205**, 22–31 (2018)
36. Hubadillah, S.K., Othman, M.H.D., Rahman, M.A., Ismail, A.F., Jaafar, J.: Preparation and characterization of inexpensive kaolin hollow fibre membrane (KHFM) prepared using phase inversion/sintering technique for the efficient separation of real oily wastewater. Arab. J. Chem. (2018)
37. Jamalludin, M.R., Harun, Z., Hubadillah, S.K., Basri, H., Ismail, A.F., Othman, M.H.D., Shohur, M.F., Yunos, M.Z.: Antifouling polysulfone membranes blended with green SiO2 from rice husk ash (RHA) for humic acid separation. Chem. Eng. Res. Des. **114**, 268–279 (2016)
38. Klemeš, J.J., Varbanov, P.S., Kravanja, Z.: Recent developments in process integration. Chem. Eng. Res. Des. **91**, 2037–2053 (2013)
39. Kumar, P., Sharma, N., Ranjan, R., Kumar, S., Bhat, Z., Jeong, D.K.: Perspective of membrane technology in dairy industry: a review. Asian-Australas. J. Anim. Sci. **26**, 1347 (2013)
40. Lafi, R., Gzara, L., Lajimi, R.H., Hafiane, A.: Treatment of textile wastewater by a hybrid ultrafiltration/electrodialysis process. Chem. Eng. Process. Process Intens. **132**, 105–113 (2018)
41. Lalia, B.S., Kochkodan, V., Hashaikeh, R., Hilal, N.: A review on membrane fabrication: structure, properties and performance relationship. Desalination **326**, 77–95 (2013)
42. Leenaars, A., Keizer, K., Burggraaf, A.: The preparation and characterization of alumina membranes with ultra-fine pores. J. Mater. Sci. **19**, 1077–1088 (1984)
43. Li, K.: Ceramic Membranes for Separation and Reaction. Wiley, USA (2007)
44. Liu, F., Hashim, N.A., Liu, Y., Abed, M.R.M., Li, K.: Progress in the production and modification of PVDF membranes. J. Membr. Sci. **375**, 1–27 (2011)
45. Loeb, S., Sourirajan, S.: Sea water demineralization by means of an osmotic membrane. In: Saline Water Conversion-II. American Chemical Society (1963)
46. Loo, S.L., Fane, A.G., Krantz, W.B., Lim, T.T.: Emergency water supply: a review of potential technologies and selection criteria. Water Res. **46**, 3125–3151 (2012)
47. Madaeni, S.S., Koocheki, S.: Application of taguchi method in the optimization of wastewater treatment using spiral-wound reverse osmosis element. Chem. Eng. J. **119**, 37–44 (2006)
48. Miller, D.J., Dreyer, D.R., Bielawski, C.W., Paul, D.R., Freeman, B.D.: Surface modification of water purification membranes. Angew. Chem. Int. Ed. **56**, 4662–4711 (2016)
49. Miron, S.M., Dutournié, P., Thabet, K., Ponche, A.: Filtration of protein-based solutions with ceramic ultrafiltration membrane. Study of selectivity, adsorption, and protein denaturation. C. R. Chim. (2018)
50. Mohamad, A.W., Ching, Y.N., Gen, H.N.: Ultrafiltration in food processing industry: review on application, membrane fouling, and fouling control. Food Bioprocess Technol. **5**, 1143–1156 (2012)
51. Mohtor, N.H., Othman, M.H.D., Bakar, S.A., Kurniawan, T.A., Dzinun, H., Norddin, M.N.A. M., Rajis, Z.: Synthesis of nanostructured titanium dioxide layer onto kaolin hollow fibre membrane via hydrothermal method for decolourisation of reactive black 5. Chemosphere **208**, 595–605 (2018)

52. Pearce, G.K.: UF/MF Membrane Water Treatment: Principles and Design. Water Treatment Academy Bangkok (2011)
53. Ravanchi, M.T., Kaghazchi, T., Kargari, A.: Application of membrane separation processes in petrochemical industry: a review. Desalination **235**, 199–244 (2009)
54. Rysstad, G., Kolstad, J.: Extended shelf life milk—advances in technology. Int. J. Dairy Technol. **59**, 85–96 (2006)
55. Saha, N., Balakrishnan, M., Ulbricht, M.: Sugarcane juice ultrafiltration: FTIR and SEM analysis of polysaccharide fouling. J. Membr. Sci. **306**, 287–297 (2007)
56. Saxena, A., Tripathi, B.P., Kumar, M., Shahi, V.K.: Membrane-based techniques for the separation and purification of proteins: an overview. Adv. Colloid Interface Sci. **145**, 1–22 (2009)
57. Shi, X., Tal, G., Hankins, N.P., Gitis, V.: Fouling and cleaning of ultrafiltration membranes: a review. J. Water Process Eng. **1**, 121–138 (2014)
58. Srivastava, H.P., Arthanareeswaran, G., Anantharaman, N., Starov, V.M.: Performance of modified poly(vinylidene fluoride) membrane for textile wastewater ultrafiltration. Desalination **282**, 87–94 (2011)
59. Szymański, K., Morawski, A.W., Mozia, S.: Humic acids removal in a photocatalytic membrane reactor with a ceramic UF membrane. Chem. Eng. J. **305**, 19–27 (2016)
60. Taheran, M., Brar, S.K., Verma, M., Surampalli, R.Y., Zhang, T.C., Valéro, J.R.: Membrane processes for removal of pharmaceutically active compounds (PhACs) from water and wastewaters. Sci. Total Environ. **547**, 60–77 (2016)
61. Tamime, A.Y.: Membrane Processing: Dairy and Beverage Applications. Wiley, USA (2012)
62. Tsapyuk, E.A., Medvedev, M.I., Badekha, V.P., Kucheruk, D.D.: Treating wastewater from the production of light-sensitive photographic and cinematographic materials and the treatment of the resulting wastes. J. Water Chem. Technol. **5**, 67–72 (1983)
63. Turken, T., Sengur-Tasdemir, R., Koseoglu-Imer, D.Y., Koyuncu, I.: Determination of filtration performances of nanocomposite hollow fiber membranes with silver nanoparticles. Environ. Eng. Sci. **32**, 656–665 (2015)
64. Van Der Bruggen, B., Vandecasteele, C., Van Gestel, T., Doyen, W., Leysen, R.: A review of pressure-driven membrane processes in wastewater treatment and drinking water production. Environ. Prog. **22**, 46–56 (2004)
65. Vrouwenvelder, J.S., Graf Von Der Schulenburg, D.A., Kruithof, J.C., Johns, M.L., Van Loosdrecht, M.C.M.: Biofouling of spiral-wound nanofiltration and reverse osmosis membranes: a feed spacer problem. Water Res. **43**, 583–594 (2009)
66. Warsinger, D.M., Chakraborty, S., Tow, E.W., Plumlee, M.H., Bellona, C., Loutatidou, S., Karimi, L., Mikelonis, A.M., Achilli, A., Ghassemi, A., Padhye, L.P., Snyder, S.A., Curcio, S., Vecitis, C.D., Arafat, H.A., Lienhard, J.H.: A review of polymeric membranes and processes for potable water reuse. Prog. Polym. Sci. **81**, 209–237 (2018)
67. Wilf, M., Tech, T.: Membrane Types and Factors Affecting Membrane Performance. National Water Research Institute, Fountain Valley, CA (2008)
68. Yoon, Y., Westerhoff, P., Snyder, S.A., Wert, E.C.: Nanofiltration and ultrafiltration of endocrine disrupting compounds, pharmaceuticals and personal care products. J. Membrane Sci. **270**, 88–100 (2006)
69. Yoon, Y., Westerhoff, P., Snyder, S.A., Wert, E.C., Yoon, J.: Removal of endocrine disrupting compounds and pharmaceuticals by nanofiltration and ultrafiltration membranes. Desalination **202**, 16–23 (2007)
70. Zsigmondy, R., Bachmann, W.: Über neue Filter. Z. Anorg. Allg. Chem. **103**, 119–128 (1918)
71. Zsigmondy, R., Wilhelm, B.: Filter and method of producing same. United State Patent Application (1916)

Conducting Polymer Membranes and Their Applications

Ibrahim Alarifi, Brijesh Prasad and Mohammad Kashif Uddin

Abstract This book chapter presents the information on conducting polymers and their applications. Conducting polymers are the class of polymeric material which was discovered in 1977, and since then these become the exciting topic of new research. Their preparation methods, advantages, applications have been studied extensively. The excellent properties of conductive polymers have enabled them to be used in and as a sensor, energy storage devices, solar cells, fuel cells, lithium ion batteries, supercapacitors, microwave absorption, electrorheological fluids, light emitting diode and separation membrane. Adsorption is the most commonly used method for water treatment because of its numerous advantages. The results show that the conductive polymers have effective adsorptive properties towards various heavy metal ions and thus can be applied for the remediation of toxic pollutants from wastewater.

Keywords Conductive membrane · Ion sensor · Adsorption · Gas sensors · Stress sensors

1 Introduction

Synthetic polymers act as artificial materials for substituting limited metals resources. The conventional polymers generally resist the flow of charge or so-called electrical conduction, and in some cases, they also act as the dielectric

I. Alarifi (✉)
Department of Mechanical and Industrial Engineering, College of Engineering, Majmaah University, Al-Majmaah 11952, Saudi Arabia
e-mail: i.alarifi@mu.edu.sa

B. Prasad
Department of Mechanical Engineering, Graphic Era Deemed to be University, Dehradun 248001, India

M. K. Uddin
Basic Engineering Sciences Department, Majmaah University, Al-Majmaah 11952, Saudi Arabia

Inamuddin et al. (eds.), *Self-standing Substrates*, Engineering Materials, https://doi.org/10.1007/978-3-030-29522-6_5

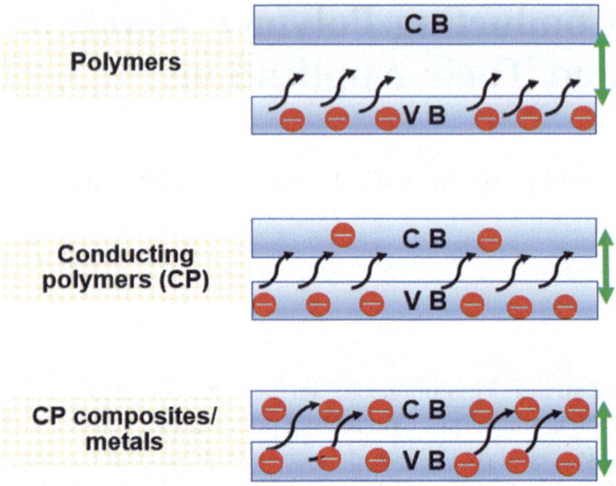

Fig. 1 Energy band distribution in polymers, CP and CP composites

medium or as insulators. The forms of conventional polymers are plastics, fibers, rubber, etc. They can be tailored accordingly by different processing techniques as per the need of applications. Conducting polymers (CP) are a new class of polymers which have unique properties of electrical conduction and optical properties. Currently, the research is focused on developing nano size CP materials of high flexibility, easy transportation, and low cost. Nanotechnology uses materials in a range of 1–100 nm and is developed by different fabrication techniques showing a physical and chemical modification in chemical structures of the CP nanomaterials [1–3] (Fig. 1).

Developing and assembling of nanostructures in the material needs lots of patience, effort, strategies, accuracy, and precision to render their functional attachment for combining nano building units. These nano units include nanoparticles, nano-sphere, hollow nano-sphere, nanotubes, nanofibers, nanopatterns and nanostructures of core-shell. CP nanomaterials have raised the interest of scientists because of their versatile properties of mechanical, electrical, magnetic, optical, chemical and physical performance. The ability to tune properties had developed new methods of fabrication of organic, inorganic, carbon, metallic, and other fillers with the polymer composites by different chemical and physical methods [4–12].

Conducting polymer composites (CPC) are the burning topics for researchers and have the potential of revolutionizing the automobile, aircraft safety study, human health, soft electronics, sports, water treatment, and many others. They are used as resistive sensors, strain sensors, governors, actuators for automobile applications, while in biomedical applications they are used for human health monitoring devices such as electromagnetic interface shielding (EMI), artificial nerves, biosensors, and soft actuators. They are also used for energy storage devices like solar cells, supercapacitors, lithium-ion batteries light emitting diodes, etc. Polymer membranes are also being used for the separation of the ions and filtration of pharmaceutical compounds from the water by playing the role as a filter. This MEMS-based device bought a revolution in the development of human health-related devices which were soft, flexible, cost-effective and biocompatible. The strong biocompatible property of conducting polymer composites make them highly appreciated material for the application of bioimplants, soft limbs, actuators, and biosensors [13]. A. M. Saheb and his team developed the GSTP1 probe model for DNA sequencing by the conducting polymer sheets using Polypyrrole (PPy). This probe helped in the diagnosis of prostate cancer and the diagnosis of tumor and its behavior [14, 15]. Koshla et al. described the fabricated conducting polymers by doping the nanoparticles in the polymer matrix for the development of nano-micro flexible systems [16]. Polymers consist of molecules having one or more ionic groups; mostly there are anionic groups which are also known as macroions. The movement of macroions ionic charge is from the starting monomer to the end of the chain and has the ability to attract more groups to it. With all these advantages CP have some limitations like CP are non-soluble in the common solvents because of their high glass transition temperature above burning point and therefore, cannot be processed again as thermoplastics [17].

Lots of efforts have been made by the researcher's scientist to develop advance CP materials by different nanotechnology techniques to form nanocrystals using polymers, metals, magnetic materials and semiconductors [18–23].

1.1 Conductive Polymer Background

Polymers are present in natural and synthetic forms, but due to limitations of natural resources synthetic polymers have attracted the world as they can be developed as per need and can be tailored respectively. Conducting polymers with nano-material can lead to the development of smart composite materials having high tunable mechanical and surface modification possibilities. The development of CPs started with the Polypyrrole derivatives having low resistivity of 1 Ω in the year 1963 by Bolto and Weiss [24]. Polyacetylene [$(CH)_2$] developed by Shirakawa et al. in the year 1970 which laid the foundation of new conductive materials. In 1971 Shirakawa with his team developed various conductive polymers [25–28].

Polyacetylene was one of the CP which showed conductive behavior after that efforts have been done for enhancing the conductive properties of the CPs and received high attention in the field of engineering and science communities. Lots of work was done in 1970–1990 for understanding the basic behavior of conducting polymers which provided a new edge for the researchers with lots of applications in the field of conducting polymers. The research culminated by the discovery of electrically conductive polymer in 2000 by Alan J. Heeger and his team for which they were awarded the Nobel Prize in the year 2000 [29–31]. Conducting devices have also been used for the MEMS application by S. Takamatsu et al. They developed MEMS-based devices by making use of conducting polymer and developed biochemical sensor, touch sensor and electrochromic pixel display [32].

With the development of intrinsically conducting polymers which were doped with the electrically conductive dopant and conjugated double bond at the backbone of the polymer chain, the conducting polymers are being explored for their unusual electronic charge response in different applications [33, 34]. These conductive membranes contain conjugated π electrons on the backbone of the polymer chain. The conjugated π electrons are delocalized from the backbone to move in the conduction band in result this conduction band showed the same behavior of conduction as metals. But this system is very unstable to the alteration of the bond and gives rise to the formation of an energy gap in the electric charge spectrum [35–45].

To overcome this problem of formation of the energy gap, dopants were introduced inside the polymer in the form of conductive fillers for raising the conductivity which carried the charges as extra electrons and stabilized the instability of the backbone by donating and accepting of electrons [26, 46].

Because of which, they exhibited some special properties such as the transmission of low energy, high electron affinity with low potential of ionization. When the potential was applied across the conductive polymer charges, flow from inside to out or out to inside depending on the dopant and mobility of the charge which led to disturbance of ions in the polymer backbone. Both p-type and n-type dopants can be used as conductive fillers. Some of them are Br^-, Cl^-, NO_3^- etc. [47–53] (Fig. 2).

Conducting composite polymer membranes field is becoming inter-disciplinary in the field of polymer membrane research. They are being used for various applications of electrical conduction in minor scale for the purpose of sensors and actuators in the field of technology and polymer membrane science. EAP's has been identified as an important factor for membrane development because of their unique property of reversible redox and electroactive behavior with ease of processing. EAPs are used for conducting membranes as they have very good barrier properties with the conductive fillers for the conductions of charge which can be tuned as per need by controlling the oxidation state of the polymer, versatility in synthesis, thin films, and conductive substrates. In EAP, the conductive mass transfer characteristics, chemical tenability, and response are present to stimuli with remarkable electrochemical stability [54–57].

The electrical conductivity of various ranges can be obtained in the conductive composites simply by selecting favorable dopants like carbon-based nanofillers, metal oxides, graphene, etc. This property can be used for the conduction of charge in the insulator interface by maintaining the thermal stability and mechanical stability with flexibility. They can widely be used for the applications of Light

Fig. 2 Schematic diagram of conducting composite

Emitting Diode (LED) lightning and charge storage capacitor. Conductive polymers have an advantage over the metal conductors as they are light in weight, flexible, can be designed for sophisticated shapes and have nearly the same mechanical properties which can be tuned as per requirement. Polymers having a large number of ionic compounds having a nonpolar chain backbone are called ionomers. These types of materials are very much helpful for preparing the coating and the transparent packaging material. They are also very interesting used as the ion exchange membranes. Polymers having a large number of ionic groups with water solubility property are known as polyelectrolytes [58–60].

1.2 *Conductive Polymers and Their Mechanism of Conductivity*

CP are organic materials which consist of conjugated π electrons on their backbone. The π electrons delocalize from the backbone and move freely to construct an electrical path for charge carriers through the unsaturated backbone. CP consist of a backbone consisting of double or triple bonds in a molecule. These molecules are separated by a single bond, along which some sharing of electrons occurs. The π bonds in p orbitals of conductive polymer overlap each other allowing the electrons to be delocalized and move freely in the bound atom space [61, 62].

To enhance the conductivity of the polymer, distinct fillers generally having a negative charge are introduced in the polymer to form a conductive polymer [63]. Polythiophene (PT), Polypyrrole (PPy), polyaniline (PANI) and Polyacetylene (PA) are the widely used CP [64–67].

Figure 3 shows a summary of CP and their chemical structure. The conductivity of Polypropylene (PPy) is nearly 7.5×10^3 S cm^{-1} [68] which is because of the P-type conduction present in the polymer backbone chain with the movement of cations and anions [69–71].

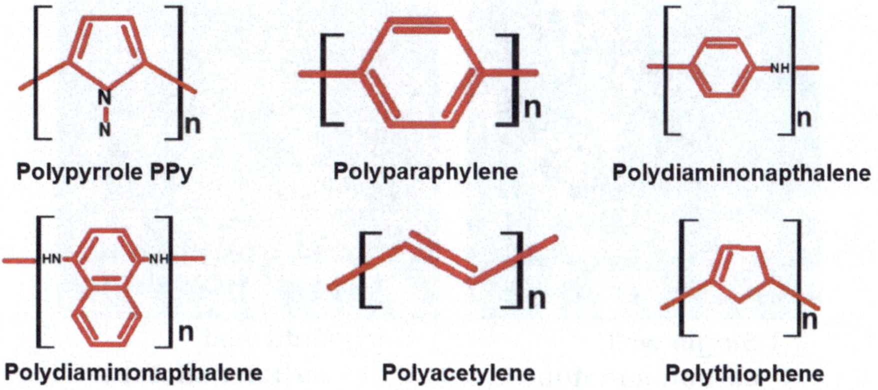

Fig. 3 Conducting polymers chemical structure

Conducting polymers have the affinity to conjugate with the double bonds of different pollutants like chlorophenol [72], mercury [73], lead [74], silver [75] phenols and organic solutes [76].

Specific proportion of the dopants used with the base polymer to form the conductive substrate, membrane or film to achieve the required conductivity [77], by introducing dopant, charge carriers are varied which create a potential difference inside the composite, due to which flow of electrons takes place in the backbone of the polymer by delocalizing the electrons and forming of new chains of polarons and polarons in the crystal lattice [78, 79].

Small amounts of dopant near to threshold point gives good sensitivity, for achieving higher conductivity the fillers are added in higher concentration with low sensitivity. Small dopants like ions and large dopants like sodium polystyrene sulfonate depending on the molecular size are used for the fabrication of CP. Smaller dopants reenter in the polymer and leave the system on regular usage, while the larger dopants work for a longer time without leaching and are more electro stable in comparison to small dopants [80–82] (Fig. 4).

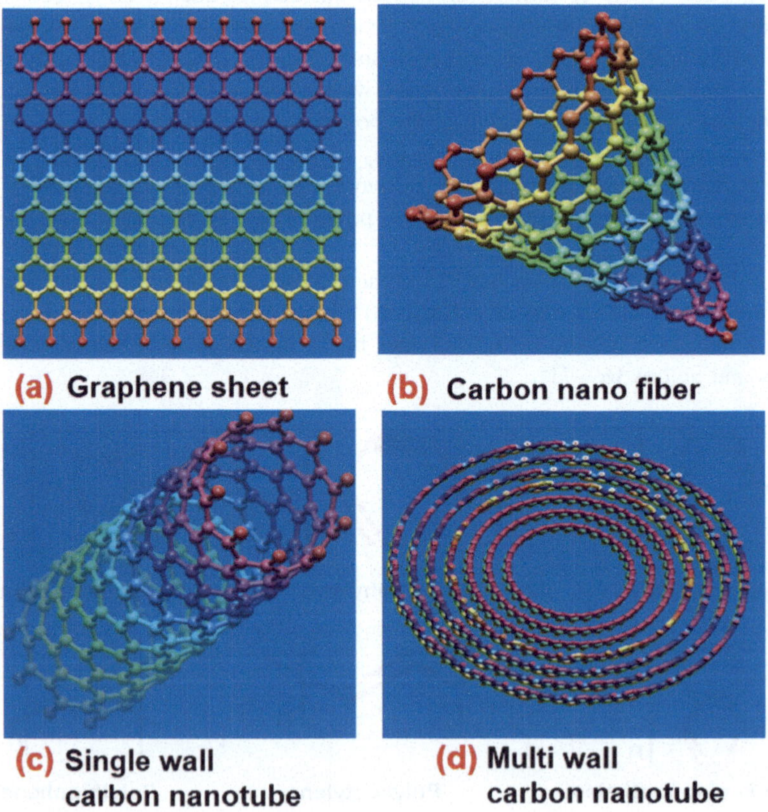

Fig. 4 Conducting fillers structure of graphene, CNF, CNT, and MWCNT for CP

1.3 Synthesis of Conducting Polymers

These conductive polymers can be categorized based on their ionic behavior as cationic which were obtained from commonly used electrochemical polymerization and chemical oxidation [83, 84]. In addition, polymerization and condensation are the common chemical methods used for the development of conductive polymers. In oxidation, the electrons are removed from the backbone which leads to the formation of cationic radical and at the time of the reduction, electrons are added to the backbone of the polymer chain leading to the formation of anionic radicals. These anionic radicals aware much stable than the cationic radicals [85].

Generally, two methods have been previously proposed for the development of conductive polymer nanocomposite: one is chemical polymerization which is mostly used for the mass production of 20 nm thick films but is less eco-friendly [86], and the other is electrochemical polymerization which is used for the preparation of thin films. Electrical conductivity is imparted to the polymer membrane which behaves as an insulator by introducing the conductive fillers [87]. Presently, the nanofillers like carbon black, carbon nanotubes, graphene, reduced graphene oxides, carbon nanofibers, silver, and other metals are used with the polymer matrix to form the conductive channel for the flow of charge [88–97].

The conductivity of the conductive polymer composite is governed by the dispersion state, intrinsic properties, aspect ratio, geometry, fabrication technique and interaction of the filler with the base polymer. Therefore, defining the filler content with the polymer is of great importance. The conductivity of the composite suddenly starts to rise very high, above the threshold point, by introducing the low quantities of filler elements as per the percolation threshold theory. At percolation threshold point, fillers concentration is much enough to form a conductive network without any barriers thought out the composite. It has also been observed during the study that with the increase of aspect ratio that is the length to diameter ratio, there was a decrease in percolation threshold point in the composite [98–100].

1.3.1 Fabrication Techniques Conductive Polymer Composites

Solution Mixing Techniques

Solution mixing is the most widely and easy technique used for the preparation of thin conductive membranes. It facilitates conductive fillers such as CNT, CNF, RGO and graphene sheets. For it, the polymer is dissolved in a suitable solvent for complete dispersion with regroups stirring and heating. On the other hand, the filler is dispersed in the same solvent used for polymer dispersion with the help of sonicator followed by stirring [101, 102] (Fig. 5).

Then the polymer solution and the filler solution are mixed with a magnetic stirrer and heated to form a homogenous solution at a specific temperature for embedding the conductive filler completely in the polymer matrix. The

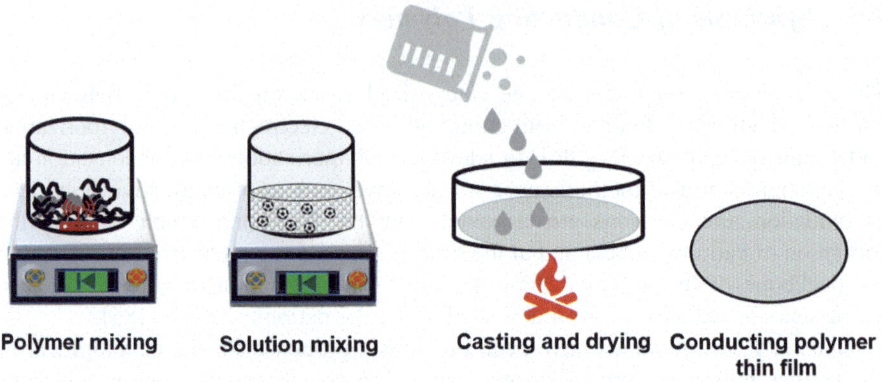

Fig. 5 Development of conducting polymer composite by solvent casting technique

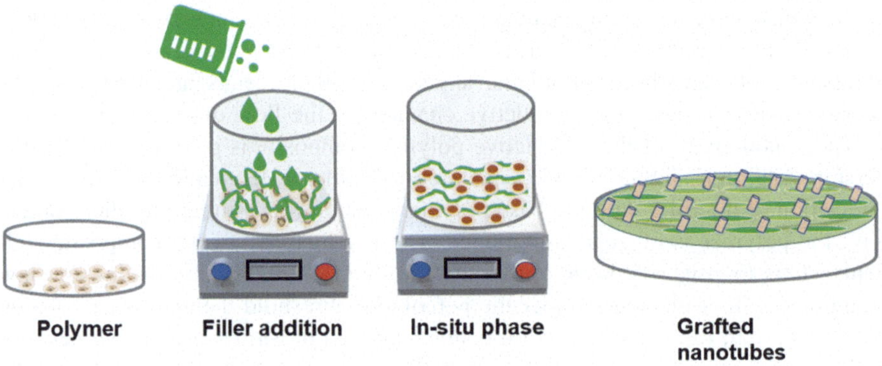

Fig. 6 Synthesis of conducting polymer by in situ technique

homogeneous mixture obtained by the complete mixing is finally cast over the patterns of different design for solvent evaporation. After the evaporation of the solvent, the thin membrane is peeled off and is used for characterization by preparing them in different geometric shapes. The researchers for obtaining different composites have used water and organic solvents [103, 104].

In situ Fabrication Technique

For the in situ polymerization technique, the filler is allowed to be swollen in an aqueous monomer solution, after that the catalyst is diffused to initiate the polymerization by radiation or heating [105] (Fig. 6).

In this process, covalent bonds are formed in between the filler and the polymer matrix to develop CNT/polymer composites. This process is generally used for the

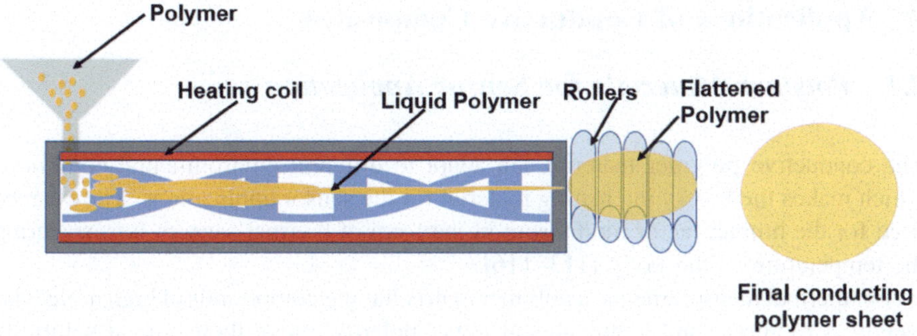

Fig. 7 Schematic diagram of a melt processing technique

high level of filler dispersion in the matrix to form highly conductive membranes [106] (Fig. 7).

Melt Processing

Melt promising method is an attractive fabrication technique because there is low usage of solvents which were evaporated in above mentioned other techniques and were little harmful to the environment. In this process, the polymer and the filler are mixed in the molten state [107]. The polymer is mechanically mixed with the fillers at high temperatures by applying the conventional methods of injection molding and extrusion [108].

This is generally preferred for mass production in industries at a faster rate with simplicity; this method is used for the polymers which show unstable behavior in other fabrication techniques of in situ and solution mixing [109].

Latex Technology of Fabrication

To obtain electrically conductive composites apart from the methods mentioned above, latex technology of fabrication is also used for incorporating the fillers in the polymer matrix. Latex technology is highly appreciated for the development of uniformly dispersed homogeneous fillers in the composite. This process is easy and helps in upscaling of the process. There are three main steps in this technique: the first formation of the colloidal solution by making use of filler, secondly mixing of the colloidal solution with the latex polymer and third is lyophilization (drying) of the colloidal solution to form composite [110].

For this process, polymers can be artificially bought to form polymer latex using emulsion polymerization and filler that can disperse in aqueous solution are used. Here, the formation of the three-dimensional framework takes place by incorporating the conductive filler in a highly viscous matrix of the polymer [111, 112].

2 Applications of Conductive Composites

2.1 Polymer Materials for Sensor Application

The conductive polymer material can adapt to different environmental conditions which makes them very interesting material for the sensor applications. They can be used for the human factor monitoring in the form of thermal sensors for measuring the temperature of the body [113–116].

For thermal measurements, a polymer matrix having compounds of high molecular weight is required, and aggregates at lower polarity above their critical solubility temperature are considered. N-Polyacrylamide (NNPAM), N-Isopropylacrylamide (NIPAM), N-Tert-Butylacrylamide (NTBAM), N-Isopropylmethacrylamide (NIPMAM), are the homopolymers used for the development of thermal sensors [117, 118].

Tensile sensors which can be used for measuring the change in strain caused by the deformation and stretching of muscles are used for detection of harmful vapors in dangerous working conditions of tunnels, chemical laboratories. These sensors can also be used for detecting the acidic and basic behavior by playing the role of pH meter. In water treatment, they can be used for detecting the pollutants in the industrial wastewater by detecting the chemical level of the discharged pollutants [119, 120] (Fig. 8).

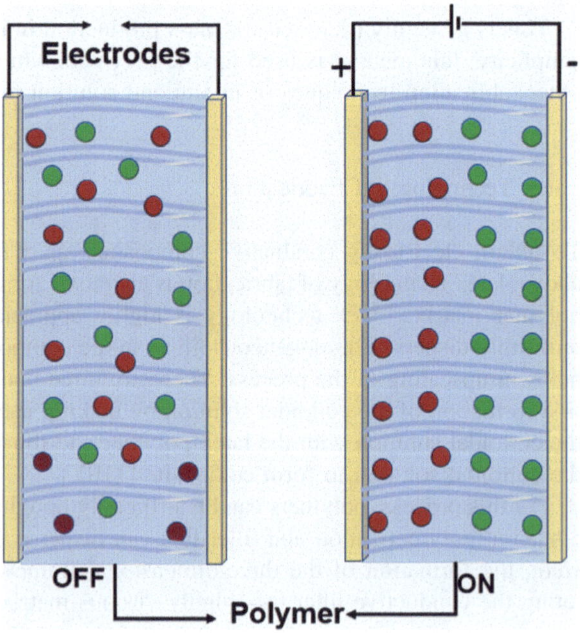

Fig. 8 Structure of conducting polymer with ion displacement on deformation

2.1.1 Ion Sensor

Ion sensors are the conductive polymer sensors, in which the ions play the role of conductive charge carriers and form a conductive network. When the analyte comes in contact with the conductive component, interaction takes place, and this phenomenon is measured in the form of electrical signals. Ion selective electrode sensors (ISEs) is one the example of an ion sensor used for measuring the specific ions in the presence of various solutions, despite other dissolved ions presents in the solution [121].

The conductive membrane was developed using polyaniline having organic phosphate bis [4-(1,1,3,3-tetramethyl butyl)]. Phenylphosphoric acid over which electrodes of polyaniline- (PANI) were applied. This conductive membrane was used for calcium ion detection [122]. Different types of polymers used for the application of ion sensors are poly (vinyl chloride) (PVC) [123], poly(3,4-ethylene dioxythiophene) (PEDOT) [124], and poly (ethylene glycol) (PEG) [125]. These materials allow faster generation of signals with stabilization on injecting of the analyte.

2.1.2 pH Sensor

Our nature is composed of different types of materials and day to day changes caused by the chemical reactions occurring due to different processes in nature, which play an important role in determining the final result. Therefore, determining the acidic and basic behavior of substance becomes very important and simply can be determined by the pH scale on different levels. pH works on the principle of strain change which occurs due to swelling of the polymer membrane when bought in contact with the analyte. It causes changes in the level of pH as per the behavior of the analyte. This stress change brings the change in resistance and in the result which changes electrical impulse is obtained as output [126]. Maintaining pH can improve the properties of the substance used in chemistry, biochemistry and for environmental science. Poly (acrylic acid) (PAA), poly (vinyl alcohol) (PVA) [127], Methylene-bis-acrylamide, acrylamide, polyaniline (PANI) [128], 2-Dimethylaminoethyl methacrylate, Poly(p-phenylenediamine) (PPV) and poly (2-hydroxyethyl methacrylate) [129, 130] are some of the materials used for the development of pH sensors [131, 132].

Leiner et al. in his study used a polyester film which was deposited with a hydrogel having ion permeability made of polyurethane (PU) having aminoethyl cellulose fiber and 1-hydroxypyrrole-3,6,8- trisulphochlorane for detecting the pH of the blood. The developed conductive membrane changed the conductivity up to a measurable stretch when in contact with the hydrogen ions which leads to the measurement of pH [133, 134].

2.1.3 Gas Sensors

Gas sensors indicate the presence of a gas of particular type in the environment. PVC having functional groups which are active in the structure have been used by various researchers for developing different types of gas sensors [135].

Gas sensors can also be used to determine the moisture by calculating the relative humidity of the environment in medicinal industries, households, factories, etc. During this phenomenon, the electrical conductivity of the material is changed by the presence of water vapor which can be adsorbed by the surface of the sensor. These sensors contain hydrophilic functional groups such as –COOH, $–N^{+}(R)_3Cl$ and $–SO_3H$ [136–138].

2.1.4 Stress Sensors

Stress sensors are a kind of smart sensors which works on photoluminescence phenomenon, i.e., arising of different color on deformation by application of different types of forces like shearing, compression tensile and stretching [139, 140]. Polymers used for the development of stress sensors are low density polyethylene (LDPE) [141], oligo (p-phenylene vinylene), [142] and poly (methyl acrylate) (PMA) [143].

Biomedical Applications

Conducting polymers are being used for the various biomedical applications. They are used as biosensors, electrode materials for neural networks, bio actuators, tissue scaffolds, healing of wounds, drug delivery systems, neural prostheses, soft muscles and for regulating cell functions [144–146].

The CP works on the principle of the electromagnetic field which is used for regulating the biological process for interacting the cells within the tissues. They have a major advantage of showing electrical behavior with biocompatibility which is very much important for the biological systems [147]. However, CP's also have some drawbacks as they show low mechanical strength individually (Fig. 9).

Therefore, they need to be integrated with some flexible matrix material such as polymers which generally show insulating behavior, resulting in flexible conductive composites in the form of conductive polymer membranes, polymer gels, and thin sheets. Several approaches have been used for tailoring the mechanical properties and topographical properties using various functionalizing nanomaterials such as carbon fibers, nanotubes by micro-fabrication techniques. Ultrathin membranes have been produced by inorganic materials such as metal nanoparticles, silicon, graphene ceramic nanoparticles [148–151].

Tissue Engineering

Fatal diseases such as cancer, diabetics, and organ failures are increasing these days due to change in lifestyle and food habits. Therefore, the researchers are focusing on the development of artificial organs which can mimic the original organs and can replace the infected part [152–154]. These artificial organs are supported by the

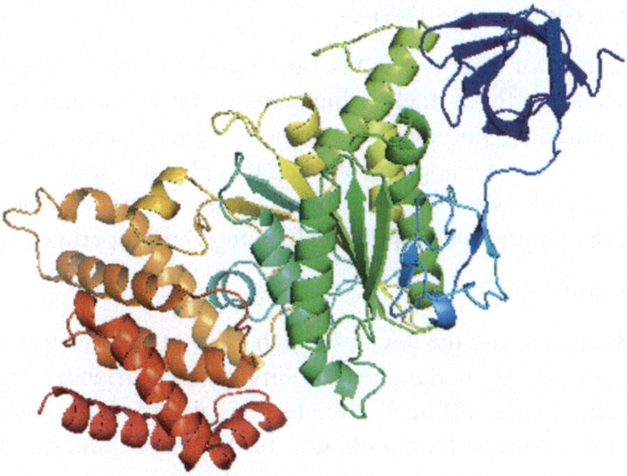

Fig. 9 Structure of enzymes used with CP used for tissue engineering

conductive polymers which can provide adequate power to run the function of organs. Stem cells and the gene delivery systems are some of the examples of the regenerative medicines developed by tissue engineering. The capability of governing and manipulating the self-engineered micro assemblies at nanoscale confirm about the macroscopic changes occurring in the nanomaterials. With the advancement in the neuropathology, cell biology and neurophysiology, it is possible to gain the neuroprotection of tissues by regenerating the dead parts with the help of the CP integrated micro machines [151, 152].

By the tissue engineering process, the desired biocompatible material can be achieved which can provide the required mechanical, electrical and degradation characteristics for the fabricated synthetic organ systems. Methods are being searched for developing electrical, chemical, physical, topological and biological stimuli in CP based systems for regeneration of the nerve [151]. But there are also some constraints of the conductive polymer for biomaterials such as degradability and mechanical characteristics for processing as complex structures in 3D. PPy is the most prominent conductive polymer used for the nerve regeneration system applications but lacks mechanical stability as a biomaterial [153, 154].

Therefore, this limitation drove the search of hybrid materials which can be incorporated in the host polymer matrix encompassing the desired electrical characteristics of the conductive polymers for charge transportation in biomedical materials. Some of the other material used for this purpose are poly (caprolactone fumarate), poly (ethyl cyanoacrylate) and chitosan [155, 156].

Recent advancement in the conductive polymer is of conductive behavior hydrogel. Another approach of developing the conducting polymers is by fixing the carbon-based nanofillers like CNTs in the host polymer. However, the CNTs showed toxic behavior in vivo test because of their functional grouping ability, size, and incorporation of high concentration [157–159].

In addition to its glial cell obtained in PNS is a well-known stem cell used for the nerve regeneration applications.

Energy Storage Devices

Storing of Energy has become a global concern because of the limited quantity of available fossil fuels which are to be exhausted sooner or later. Therefore, the conducting polymers are looked as the substitute for the new compact energy storage device applications and can play a significant role. CP can be used for other applications for charge storing such as batteries, solar cells, fuel cells, nanogenerators, supercapacitors by enhancing the properties [160–162].

Solar Cells

Solar cells are the devices which convert solar energy directly into electrical energy. Solar energy is the cheapest environment-friendly renewable resource or so-called green energy [163]. But the use of solar energy is limited because of the high prices of the photovoltaic modules and lower efficiencies of the solar panels. Improving the photovoltaic absorption with the minimum loss with minimum cost is a major concern [164].

Polymers are considered as a suitable material for developing cost-effective organic polymer photovoltaic devices. These devices are light, low cost, and have easy fabrication techniques. The photovoltaic cell works on the acceptor and donor approach. The PV polymer cell consists of two major constituents, one electron acceptor and another electron donor. Conjugated polymers such as polyfluorene (PF), polyphenylene vinylene (PPV), polythiophene (PT) and some low bandgap materials like thiophene, fluorine, and pyrazine used as the donor constituents [165–167].

Poly(3-hexylthiophene) (P_3HT) is the widely used conjugated polymer for developing photovoltaic cell [168]. CP infiltrated with conductive metal particles, carbon fillers such as CNT, CNF, CB to form conductive composites show a unique property of hole transportation and light absorption leading to photovoltaic efficiencies enhancement [169–173]. Nanotubes of TiO_2 is considered as a good candidate for the development of solar cells as they provide high electron migration pathways [174].

Fuel Cells

The demand for fuel cells has increased in electric vehicles because of their ease of transportation, low cost, and low-cost fabrication techniques. Fuel cells directly convert chemical energy by the electrochemical process into electrical energy [175, 176]. CP developed by the electrochemical process is a suitable candidate for solar cell applications. CP with porous structures is highly appreciated for PV applications as they can store charge and help in smooth transportation of the costs. CP is very much used for fuel cells as they can store energy in the polymer matrix and can be directly used by the device. CP made by the graphene is considered as a suitable candidate for the fuel cell applications [177, 178].

Lithium-ion Batteries

Conventionally cadmium and nickel-based rechargeable batteries were used but had limited charge storage capacity and less durable. While, lithium-ion batteries are lighter having more charge storage capacity, highly durable, eco-friendly and rechargeable, CP plays an important role in battery electrodes as they have a high specific capacity and show good cyclic performance [179, 180].

Supercapacitors

The urgent demand for supercapacitors has arisen due to the high demand for hybrid electric vehicles, portable electronic devices, and uninterruptible eco-friendly power supplies. Supercapacitors are developed as the modified electrode materials and so-called as the electrochemical capacitors. They have a long life cycle, works on the simple principle of charging/discharging, high charge prorogation and show pulse power supply [181, 182].

These supercapacitors can store hundred to thousand farads charge and are developed from a carbon having a higher surface area. Supercapacitors show higher specific power in comparison to lithium-ion batteries. Different types of materials are used for the electrode of supercapacitors such as metal oxides, carbon oxides, and conducting polymers. Supercapacitors are classified into three types: (i) transition metal oxides (ii) high surface carbons and (iii) conducting polymers. CP show high specific capacitance but poor cyclic stability [183, 184] (Fig. 10).

The conducting polymer nanocomposites can improve poor cyclic stability. Different types of carbon compounds are used for the electrode material such as CNT, CNF, CB, activated carbon graphene and its oxides because of low cost, ease of fabrication, controllable porosity, ease of functionalization variable forms (powder, sheets, fibers, tubes, and monoliths). Carbon-based composites and nanocomposites show high power capabilities but lower specific energy due to sorption and desorption of ions at a high-speed rate. PPy/AC has been used as the electrode material. Further PPy/CNF has been studied by Kim et al. by developing the composite with a thickness of 5–10 nm [185–189] by electrochemical method and developed symmetric, asymmetric PANI-LiPF6 electrodes and carbon electrodes. The supercapacitor with mesoporous carbon electrode using active carbon

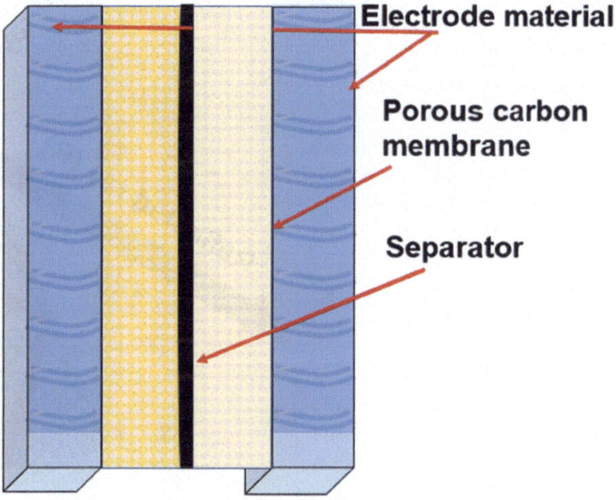

Fig. 10 Schematic diagram of supercapacitor developed by porous carbon membrane

showed higher electrochemical efficiency than the redox type electrodes. Active carbon electrode increased the voltage, and the electrode of polymer helped in maintaining the redox reaction for the longer duration [190].

EMI and Microwave Absorption

Electromagnetic interference is an important issue which is caused by the interference of the different frequencies and is important for the electronic devices for biomedical applications, wireless communications, navigation, space technologies, and living beings. Therefore, shielding of the devices is a major issue which can be controlled by a coating of a layer of material to stop or reduce the chances of interference. Conducting polymers and their composites showing high electrical conductivity are suitable material for the applications of electromagnetic shielding [191–194] (Fig. 11).

Electrorheological Fluids

Electrorheological (ER) fluids are used in clutch systems, hydraulic brakes, valves, and dampers. These fluids show high damping properties under external stimuli's. ER fluids consist of high dielectric property particles and low conductivity components. Due to this property, the electric field can be changed in very short timings. Conducting polymers consist of the polarized particles which show high environmental stability [195–202].

Light Emitting Diode (LED)

Organic and inorganic light emitting diodes has always been a subject of study among researchers. Introduction of polymers in the field of LED to the development

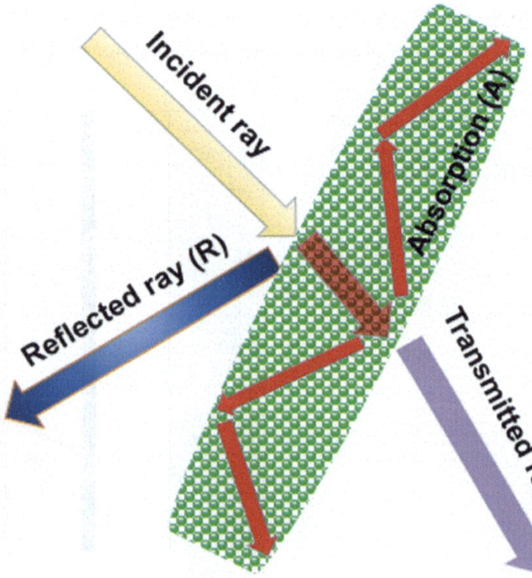

Fig. 11 Schematic diagram of EMI shielding by conductive polymer composites

of PLED. These PLEDS show a fast response, cover a larger area, high process time and can operate with low voltages. Different techniques have been used for enhancing the optical and electrical properties of the PLEDs by producing a change in the chemical structure of the composite. Presently, the commercial leads are present in only two colors that are orange and green made by poly (Phenylenevinylenes) (PPVs). However, they also exhibit the red, blue and green the Venot et al. studied the PLED and produced plead using ITO/TPD-Si_2/PFO/Ca/Al and also studied about the continuity and charged injection at the layer interfaces [203, 204].

Poly (2-methoxy-5-(2 ethyl-hexoxy)-1, 4-phenylenevinylene) (MEH-PPV) is highly used for producing the red-orange colour in PLEDs. Polythiophene (PT), poly (2-methoxy-5-(2 ethyl-hexoxy)-1,4-phenylene-vinylene), poly(9,9-dioctylfluorene) (PFO), poly{[3-hexylthiophene]- CO_3-[2-(p-cyano-phenoxy) ethyl] thiophene (PTOPhCN), Polyfluorene (PF) and its derivative poly (9, 9-di-Nhexylfluorene) (PDHF), for producing blue light are some other polymers used for producing thin films with excellent chemical stability and solubility [205–209].

The PLEDs can be integrated into the thin film transistors TFT which is a very tedious task for developing flat displays.

2.2 *Conducting Membrane for Wastewater Treatment*

Conducting membranes are used as a cathode for filtration of organic matter of high concentration and are very effective in solving the fouling problem with endowing (quality) of wastewater by advanced functionalities of the conductive membrane [210]. The membrane used as a cathode can work under 5 V with very less fouling and good effluent [211]. On the application of electrical potential, these conductive membranes demonstrated high anti-scaling and extremely high anti-biofouling properties [212–215].

These electrically conductive membranes have also been used as electro catalytic bases for transforming the liquid contaminants electrochemically [216]. This class of membrane is also used for treating the industrial foulants for stripping the paints from metals.

Conducting polymer composites (CPC) are becoming the prime choice as a material for water treatment. If we look at the last year of research dealing with environmental issues, CPC is commonly used to treat several contaminants from polluted water. Adsorption method is the most popular and easy among the available techniques for wastewater treatment [217–224]. There are some review papers which highlighted the importance of CNPs in various forms for water treatment [225, 226].

For the adsorption process, the nanomaterials acted as an excellent source to adsorb various harmful aqueous pollutants [227, 228]. Various forms of nanomaterials such as nanoparticles, nanotubes, nanofibers, nanopowder, quantum dots, fullerenes, nanostructured colloids and semiconductors, nanocomposites, etc. have been proved their effectiveness in stopping the water damage. Nanoadsorbents based on conducting polymer nanocomposites are the recent attractive materials for water purification. A recent detailed review article mentioned the use of polyaniline-based nanoadsorbents and their derivatives in heavy metals and dyes removal [229]. Mallakpour and Behranvand in 2017 utilized the recycled PET bottle waste in the synthesis of MWCNT-ZnO quantum dot nanocomposites using mixing method for cadmium (II) ion removal [230]. Organodisulfide polymer modified magnetic microspheres were recently prepared for the adsorption of different heavy metals, i.e. Hg(II), Pb(II) and Cd(II) from salinity wastewater. The excellent adsorption capacity and regeneration studies confirmed the potential of magnetic nanoparticles in wastewater treatment [231]. Polypyrrole and polyvinyl alcohol polymers were also used recently for the uptake of trivalent arsenic by incorporated it with Fe3O4 nanoparticle via in situ polymerization [232]. It was found in this study that pH 12, 30 min time and 0.1 g of the synthesized nanocomposite were suitable for 100% As(III) removal from contaminated water. The ability of polyaniline nanofibers was also tested for heavy metal adsorption [233]. The nanofibers were synthesized by the combination of electrospinning and in situ polymerization methods. The synthesized nanofibers obtained the high adsorption capacity of 111.11 mg/g for Pb(II) and 833.33 mg/g for Cu(II) at 7.5 pH. It was found that the adsorption followed Langmuir isotherm and the pseudo-second-order kinetic model while the intra-particle diffusion controlled the adsorption process. The removal of heavy metals was occurred by electrostatic attraction mechanism.

A novel super magnetic nanocomposite of self-doped polyaniline was prepared by the coprecipitation method, and its adsorption potential was successfully investigated for the potential removal of Cu(II), Cd(II), Co(II), and Cr(III) metal ions [234]. The conductivity study of the nano-adsorbent showed the great enhancement of conductivity which can be applied in superconductive applications. The single and multiple metal ion adsorption experiments were conducted to remove Pb (II), Zn (II) and Cu (II) by polyaniline/TiO2 composite [235]. It was found that TiO2 has special adsorption affinity towards zinc and poor affinity towards copper.

3 Conclusions

The present book chapter confirms that today's research has a prime focus on conducting polymers which have numerous applications and uses. The wide variety of nanoadsorbents based on conducting polymers have been considered for the removal of various heavy metals. The results confirmed the potential of polymeric adsorbents in removing toxic pollutants from water and wastewater. Conducting polymers have the affinity to conjugate with the double bonds of different pollutants. The synthesis procedure with different surfactants and polymers is an ongoing subject of industrial research and interest to the academic field. The conductive polymers were obtained from commonly used electrochemical polymerization, chemical oxidation, and other common chemical methods such as polymerization and condensation. The conductive polymer material can adapt to different environmental conditions which makes them very interesting material for the ion, pH, gas, and stress sensor applications. Conducting polymers are being successfully used for the various biomedical, tissue engineering, electrical, energy, electromagnetic and environmental applications.

Acknowledgements The authors would like to thank Deanship of Scientific Research at Majmaah University for supporting and would like to thank the College of Engineering for motivation and support.

References

1. Zhao, Q.Q., Boxman, A., Chowdhry, U.: Nanotechnology in the chemical industry–opportunities and challenges. J. Nanopart. Res. **5**(5–6), 567–572 (2003)
2. Lazzari, M., López-Quintela, M.A.: Block copolymers as a tool for nanomaterial fabrication. Adv. Mater. **15**(19), 1583–1594 (2003)
3. Rao, C.N.R., Cheetham, A.K.: Science and technology of nanomaterials: current status and future prospects. J. Mater. Chem. **11**(12), 2887–2894 (2001)
4. Li, W., Sha, X., Dong, W., Wang, Z.: Synthesis of stable hollow silica microspheres with mesoporous shell in nonionic W/O emulsion. Chem. Commun. **20**, 2434–2435 (2002)
5. Li, M., Mann, S.: Emergent nanostructures: water-induced mesoscale transformation of surfactant-stabilized amorphous calcium carbonate nanoparticles in reverse microemulsions. Adv. Func. Mater. **12**(11–12), 773–779 (2002)
6. Barbe, C.J., Graf, R., Finnie, K.S., Blackford, M., Trautman, R., Bartlett, J.R.: Sol-gel synthesis of complex titanates in micro-emulsions. J. Sol-Gel. Sci. Technol. **26**(1–3), 457–462 (2003)
7. Jang, J.: Conducting polymer nanomaterials and their applications. In: Emissive Materials Nanomaterials, pp. 189–260. Springer, Berlin, Heidelberg (2006)
8. Anderson, M., Osterlund, L., Ljungstrom, S., Palmgvist, A.: J. Phys. Chem. B **9**, 241 (2002)
9. Mori, Y., Okastu, Y., Tsujimoto, Y.: Titanium dioxide nanoparticles produced in water-in-oil emulsion. J. Nanopart. Res. **3**(2–3), 219–225 (2001)
10. Shchukin, D.G., Sukhorukov, G.B.: Nanoparticle synthesis in engineered organic nanoscale reactors. Adv. Mater. **16**(8), 671–682 (2004)
11. Parak, W.J., Gerion, D., Pellegrino, T., Zanchet, D., Micheel, C., Williams, S.C., Boudreau, R., Le Gros, M.A., Larabell, C.A., Alivisatos, A.P.: Biological applications of colloidal nanocrystals. Nanotechnology **14**(7), R15 (2003)
12. Rao, C.R., Kulkarni, G.U., Thomas, P.J., Edwards, P.P.: Metal nanoparticles and their assemblies. Chem. Soc. Rev. **29**(1), 27–35 (2000)
13. Rubehn, B., Bosman, C., Oostenveld, R., Fries, P., Stieglitz, T.: A MEMS-based flexible multichannel ECoG-electrode array. J. Neural Eng. **6**(3), 036003 (2009)
14. Saheb, A.H., Leon, M., Josowicz, M.: Electrochemical assay of GSTP1-related DNA sequence for prostate cancer screening. Electrochem. Soc. Interface **21**(3–4), 71–75 (2012)
15. Ates, M., Karazehir, T., Sezai Sarac, A.: Conducting polymers and their applications. Curr. Phys. Chem. **2**(3), 224–240 (2012)

16. Khosla, A., Gray, B.L.: Micropatternable multifunctional nanocomposite polymers for flexible soft NEMS and MEMS applications. ECS Trans. **45**(3), 477–494 (2012)
17. McCullough, R.D.: The chemistry of conducting polythiophenes. Adv. Mater. **10**(2), 93–116 (1998)
18. Pileni, M.P.: Fabrication and physical properties of self-organized silver nanocrystals. Pure Appl. Chem. **72**(1–2), 53–65 (2000)
19. Ohde, H., Hunt, F., Wai, C.M.: Synthesis of silver and copper nanoparticles in a water-in-supercritical-carbon dioxide microemulsion. Chem. Mater. **13**(11), 4130–4135 (2001)
20. Hirai, T., Saito, T., Komasawa, I.: Stabilization of CdS nanoparticles immobilized on thiol-modified polystyrene particles by encapsulation with polythiourethane. J. Phys. Chem. B **105**(40), 9711–9714 (2001)
21. Dutta, P., Fendler, J.H.: Preparation of cadmium sulfide nanoparticles in self-reproducing reversed micelles. J. Colloid Interface Sci. **247**(1), 47–53 (2002)
22. Chen, D.H., Wu, S.H.: Synthesis of nickel nanoparticles in water-in-oil microemulsions. Chem. Mater. **12**(5), 1354–1360 (2000)
23. Capek, I.: Preparation of metal nanoparticles in water-in-oil (w/o) microemulsions. Adv. Colloid Interface Sci. **110**(1–2), 49–74 (2004)
24. Bolto, B.A., McNeill, R., Weiss, D.E.: Electronic conduction in polymers. III. Electronic properties of polypyrrole. Aust. J. Chem. **16**(6), 1090–1103 (1963)
25. Shirakawa, H., Ikeda, S.: Infrared spectra of poly (acetylene). Polym. J. **2**(2), 231 (1971)
26. Chiang, C.K., Fincher Jr., C.R., Park, Y.W., Heeger, A.J., Shirakawa, H., Louis, E.J., Gau, S.C., MacDiarmid, A.G.: Electrical conductivity in doped polyacetylene. Phys. Rev. Lett. **39** (17), 1098 (1977)
27. Chiang, C.K., Druy, M.A., Gau, S.C., Heeger, A.J., Louis, E.J., MacDiarmid, A.G., Park, Y. W., Shirakawa, H.: Synthesis of highly conducting films of derivatives of polyacetylene, (CH)x. J. Am. Chem. Soc. **100**(3), 1013–1015 (1978)
28. Shirakawa, H., Louis, E.J., MacDiarmid, A.G., Chiang, C.K., Heeger, A.J.: Synthesis of electrically conducting organic polymers: halogen derivatives of polyacetylene, (CH)x. J. Chem. Soc., Chem. Commun. **16**, 578–580 (1977)
29. Heeger, A.J.: Polyaniline with surfactant counterions: conducting polymer materials which are processible in the conducting form. Synth. Met. **57**(1), 3471–3482 (1993)
30. Alarifi, I., Alharbi, A., Khan, W., Swindle, A., Asmatulu, R.: Thermal, electrical and surface hydrophobic properties of electrospun polyacrylonitrile nanofibers for structural health monitoring. Materials **8**(10), 7017–7031 (2015)
31. Chiang, J.C., MacDiarmid, A.G.: 'Polyaniline': protonic acid doping of the emeraldine form to the metallic regime. Synth. Metals **13**(1–3), 193–205 (1986)
32. Bernards, D.A., Macaya, D.J., Nikolou, M., DeFranco, J.A., Takamatsu, S., Malliaras, G.G.: Enzymatic sensing with organic electrochemical transistors. J. Mater. Chem. **18**(1), 116–120 (2008)
33. Kivelson, S., Heeger, A.J.: Intrinsic conductivity of conducting polymers. Synth. Metals **22** (4), 371–384 (1988)
34. Heywang, G., Jonas, F.: Poly(alkylenedioxythiophene)s—new, very stable conducting polymers. Adv. Mater. **4**(2), 116–118 (1992)
35. Wang, Y., Jing, X.: Intrinsically conducting polymers for electromagnetic interference shielding. Polym. Adv. Technol. **16**(4), 344–351 (2005)
36. Hatchett, D.W., Josowicz, M.: Composites of intrinsically conducting polymers as sensing nanomaterials. Chem. Rev. **108**(2), 746–769 (2008)
37. Menon, A.K., Wolfe, R.M., Kommandur, S., Yee, S.K.: Progress in Nickel-coordinated polymers as intrinsically conducting n-type thermoelectric materials. Adv. Electron. Mater. 1800884 (2019)
38. Keplinger, T., Wang, X., Burgert, I.: Nanofibrillated cellulose composites and wood derived scaffolds for functional materials. J. Mater. Chem. A **7**(7), 2981–2992 (2019)

39. Chiang, C.K., Gau, S.C., Fincher Jr., C.R., Park, Y.W., MacDiarmid, A.G., Heeger, A.J.: Polyacetylene, (CH)x: n-type and p-type doping and compensation. Appl. Phys. Lett. **33**(1), 18–20 (1978)
40. Shirakawa, H., Ito, T., Ikeda, S.: Electrical properties of polyacetylene with various cis-trans compositions. Die Makromolekulare Chemie: Macromol. Chem. Phys. **179**(6), 1565–1573 (1978)
41. Rabolt, J.F., Clarke, T.C., Street, G.B.: Vibronic intensity enhancement in the infrared spectra of heavily doped (CH)x and (CD)x. J. Chem. Phys. **71**(11), 4614–4619 (1979)
42. Fincher Jr., C.R., Ozaki, M., Heeger, A.J., MacDiarmid, A.G.: Donor and acceptor states in lightly doped polyacetylene, (CH)x. Phys. Rev. B **19**(8), 4140 (1979)
43. Harada, I., Furukawa, Y., Tasumi, M., Shirakawa, H., Ikeda, S.: Spectroscopic studies on doped polyacetylene and β-carotene. J. Chem. Phys. **73**(10), 4746–4757 (1980)
44. Heeger, A.J.: Technical feature-semiconducting and metallic polymers: the fourth generation of polymeric materials 2001 mrs spring meeting plenary presentation. MRS Bull. Mater. Res. Soc. **26**(11), 900–904 (2001)
45. Percec, V., Glodde, M., Bera, T.K., Miura, Y., Shiyanovskaya, I., Singer, K.D., Balagurusamy, V.S.K., Heiney, P.A., Schnell, I., Rapp, A., Spiess, H.W.: Self-organization of supramolecular helical dendrimers into complex electronic materials. Nature **419**(6905), 384 (2002)
46. Farajikhah, S., Van Amber, R., Sayyar, S., Shafei, S., Fay, C.D., Beirne, S., Javadi, M., Wang, X., Innis, P.C., Paull, B., Wallace, G.G.: Processable thermally conductive polyurethane composite fibers. Macromol. Mater. Eng. **304**(3), 1800542 (2019)
47. Guimard, N.K., Gomez, N., Schmidt, C.E.: Conducting polymers in biomedical engineering. Prog. Polym. Sci. **32**(8–9), 876–921 (2007)
48. Guo, B., Glavas, L., Albertsson, A.C.: Biodegradable and electrically conducting polymers for biomedical applications. Prog. Polym. Sci. **38**(9), 1263–1286 (2013)
49. Balint, R., Cassidy, N.J., Cartmell, S.H.: Conductive polymers: Towards a smart biomaterial for tissue engineering. Acta Biomater. **10**(6), 2341–2353 (2014)
50. Huang, Z.B., Yin, G.F., Liao, X.M., Gu, J.W.: Conducting polypyrrole in tissue engineering applications. Front. Mater. Sci. **8**(1), 39–45 (2014)
51. Altaf, F., Gill, R., Batool, R., Drexler, M., Alamgir, F., Abbas, G., Jacob, K.: Proton conductivity and methanol permeability study of polymer electrolyte membranes with range of functionalized clay content for fuel cell application. Eur. Polym. J. **110**, 155–167 (2019)
52. Yelilarasi, A., Anbarasan, R., Manikandan, K.M.: Electrical conductivity studies on the nanocomposites of Poly (aniline) with various initiator and oxide nanoparticles. Vacuum **163**, 172–175 (2019)
53. Bendrea, A.D., Cianga, L., Cianga, I.: Progress in the field of conducting polymers for tissue engineering applications. J. Biomater. Appl. **26**(1), 3–84 (2011)
54. Bar-Cohen, Y., Bar-Cohen, Y.: Electroactive Polymer (EAP) Actuators as Artificial Muscles: Reality, Potential, and Challenges, vol. 136, pp. 1–765. SPIE Press, Bellingham, WA (2004)
55. Bar-Cohen, Y., Zhang, Q.: Electroactive polymer actuators and sensors. MRS Bull. **33**(3), 173–181 (2008)
56. Carpi, F., Smela, E. (eds.): Biomedical Applications of Electroactive Polymer Actuators. Wiley (2009)
57. Kim, K.J., Tadokoro, S.: Electroactive polymers for robotic applications. Artif. Muscles Sens. **23**, 291 (2007)
58. Unal, F.A., Burhan, H., Karakus, S., Karaelioglu, G., Sen, F.: Ionic polymer-metal composite actuators operable in dry conditions. In: Ionic Polymer Metal Composites for Sensors and Actuators, pp. 149–159. Springer, Cham (2019)
59. Mukai, K., Asaka, K., Sugino, T., Kiyohara, K., Takeuchi, I., Terasawa, N., Futaba, D.N., Hata, K., Fukushima, T., Aida, T.: Highly conductive sheets from millimeter-long single-walled carbon nanotubes and ionic liquids: application to fast-moving, low-voltage electromechanical actuators operable in air. Adv. Mater. **21**(16), 1582–1585 (2009)

60. Bar-Cohen, Y.: Electroactive polymers: current capabilities and challenges. In: Smart Structures and Materials 2002: Electroactive Polymer Actuators and Devices (EAPAD), vol. 4695, pp. 1–8. International Society for Optics and Photonics (2002)
61. Thakur, M.: A class of conducting polymers having nonconjugated backbones. Macromolecules **21**(3), 661–664 (1988)
62. Bay, L., Jacobsen, T., Skaarup, S., West, K.: Mechanism of actuation in conducting polymers: osmotic expansion. J. Phys. Chem. B **105**(36), 8492–8497 (2001)
63. Scrosati, B. (ed.): Applications of electroactive polymers (vol. 75). Chapman & Hall, London (1993)
64. Ateh, D.D., Navsaria, H.A., Vadgama, P.: Polypyrrole-based conducting polymers and interactions with biological tissues. J. R. Soc. Interface **3**(11), 741–752 (2006)
65. Wei, Z., Zhang, Z., Wan, M.: Formation mechanism of self-assembled polyaniline micro/nanotubes. Langmuir **18**(3), 917–921 (2002)
66. Chu, J., Lu, D., Wu, B., Wang, X., Gong, M., Zhang, R., Xiong, S.: Synthesis and electrochromic properties of conducting polymers: polyaniline directly grown on fluorine-doped tin oxide substrate via hydrothermal techniques. Sol. Energy Mater. Sol. Cells **177**, 70–74 (2018)
67. Edmonds, M.S.: Understanding the effects of structural modification on the electrical and optical properties of polymer-acid-templated conducting polymers. Doctoral dissertation, Princeton University (2018)
68. Dai, L.: Intelligent Macromolecules for Smart Devices: From Materials Synthesis to Device Applications. Springer Science & Business Media (2004)
69. Elabd, Y.A.: Highly Conductive Anion Exchange Block Copolymers. Texas Engineering Experiment Station College Station, College Station (2018)
70. Meek, K.M., Sun, R., Willis, C., Elabd, Y.A.: Hydroxide conducting polymerized ionic liquid pentablock terpolymer anion exchange membranes with methylpyrrolidinium cations. Polymer **134**, 221–226 (2018)
71. Liu, F.H., Lin, C.X., Hu, E.N., Yang, Q., Zhang, Q.G., Zhu, A.M., Liu, Q.L.: Anion exchange membranes with well-developed conductive channels: Effect of the functional groups. J. Membr. Sci. **564**, 298–307 (2018)
72. Bagheri, H., Mohammadi, A., Salemi, A.: On-line trace enrichment of phenolic compounds from water using a pyrrole-based polymer as the solid-phase extraction sorbent coupled with high-performance liquid chromatography. Anal. Chim. Acta **513**(2), 445–449 (2004)
73. Krishna, M.B., Karunasagar, D., Rao, S.V., Arunachalam, J.: Preconcentration and speciation of inorganic and methyl mercury in waters using polyaniline and gold trap-CVAAS. Talanta **68**(2), 329–335 (2005)
74. Huang, M.R., Peng, Q.Y., Li, X.G.: Rapid and effective adsorption of lead ions on fine poly (phenylenediamine) microparticles. Chem. A Eur. J. **12**(16), 4341–4350 (2006)
75. Li, X.G., Liu, R., Huang, M.R.: Facile synthesis and highly reactive silver ion adsorption of novel microparticles of sulfodiphenylamine and diaminonaphthalene copolymers. Chem. Mater. **17**(22), 5411–5419 (2005)
76. Hickner, M.A., Herring, A.M., Coughlin, E.B.: Anion exchange membranes: current status and moving forward. J. Polym. Sci., Part B: Polym. Phys. **51**(24), 1727–1735 (2013)
77. Shao, Q., Niu, Z., Hirtz, M., Jiang, L., Liu, Y., Wang, Z., Chen, X.: High-performance and tailorable pressure sensor based on ultrathin conductive polymer film. Small **10**(8), 1466–1472 (2014)
78. Bredas, J.L., Street, G.B.: Polarons, bipolarons, and solitons in conducting polymers. Accouts Chem. Res. **18**(10), 309–315 (1985)
79. Bredas, J.L., Chance, R.R., Silbey, R.: Comparative theoretical study of the doping of conjugated polymers: polarons in polyacetylene and polyparaphenylene. Phys. Rev. B **26** (10), 5843 (1982)
80. Baughman, R.H., Bredas, J.L., Chance, R.R., Elsenbaumer, R.L., Shacklette, L.W.: Structural basis for semiconducting and metallic polymer dopant systems. Chem. Rev. **82**(2), 209–222 (1982)

81. Liang, Z., Zhang, Y., Souri, M., Luo, X., Boehm, A.M., Li, R., Zhang, Y., Wang, T., Kim, D.Y., Mei, J., Marder, S.R.: Influence of dopant size and electron affinity on the electrical conductivity and thermoelectric properties of a series of conjugated polymers. J. Mater. Chem. A **6**(34), 16495–16505 (2018)
82. Ogle, J., Yehulie, M., Boehme, C., Whittaker-Brooks, L.: Counter-ion and dopant effects on the electronic structure and thermoelectric properties of intrinsically conductive polymers. In: APS Meeting Abstracts (2018)
83. Diaz, A.F., Kanazawa, K.K., Gardini, G.P.: Electrochemical polymerization of pyrrole. J. Chem. Soc., Chem. Commun. **14**, 635–636 (1979)
84. Kaneto, K., Yoshino, K., Inuishi, Y.: Electrical and optical properties of polythiophene prepared by electrochemical polymerization. Solid State Commun. **46**(5), 389–391 (1983)
85. Searson, P.C., Moffat, T.P.: Electrochemical surface modification and materials processing. Critical Rev. Surf. Chem. **3**, 171 (1994)
86. Toshima, N., Hara, S.: Direct synthesis of conducting polymers from simple monomers. Prog. Polym. Sci. **20**(1), 155–183 (1995)
87. Kaur, G., Adhikari, R., Cass, P., Bown, M., Gunatillake, P.: Electrically conductive polymers and composites for biomedical applications. RSC Adv. **5**(47), 37553–37567 (2015)
88. Stejskal, J.: Conducting polymer-silver composites. Chem. Pap. **67**(8), 814–848 (2013)
89. Du, F., Scogna, R.C., Zhou, W., Brand, S., Fischer, J.E., Winey, K.I.: Nanotube networks in polymer nanocomposites: rheology and electrical conductivity. Macromolecules **37**(24), 9048–9055 (2004)
90. Nogales, A., Broza, G., Roslaniec, Z., Schulte, K., Ics, I., Hsiao, B.S., Sanz, A., García-GutiØrrez, M.C., Rueda, D.R., Domingo, C., Ezquerra, T.A.: Macromolecules **37**, 7669 (2004)
91. Winey, K.I., Kashiwagi, T., Mu, M.: Improving electrical conductivity and thermal properties of polymers by the addition of carbon nanotubes as fillers. MRS Bull. **32**(4), 348–353 (2007)
92. Antunes, M., Velasco, J.I.: Multifunctional polymer foams with carbon nanoparticles. Prog. Polym. Sci. **39**(3), 486–509 (2014)
93. Bai, H., Li, C., Shi, G.: Functional composite materials based on chemically converted graphene. Adv. Mater. **23**(9), 1089–1115 (2011)
94. Das, T.K., Prusty, S.: Graphene-based polymer composites and their applications. Polymer-Plast. Technol. Eng. **52**(4), 319–331 (2013)
95. Mittal, V., Nuzzo, R. (eds.): Polymer-Graphene Nanocomposites, vol. 26. Royal Society of Chemistry (2012)
96. Kuilla, T., Bhadra, S., Yao, D., Kim, N.H., Bose, S., Lee, J.H.: Recent advances in graphene based polymer composites. Prog. Polym. Sci. **35**(11), 1350–1375 (2010)
97. Cheng, X., Li, C., Rao, L., Zhou, H., Li, T., Duan, Y.: Platinum wire implants coated with PEDOT/carbon nanotube conducting polymer films in the brain of rats: a histological evaluation. J. Wuhan Univ. Technol. Mater. Sci. Ed. **27**(6), 1053–1057 (2012)
98. Nistal, A., Garcia, E., Pérez-Coll, D., Prieto, C., Belmonte, M., Osendi, M.I., Miranzo, P.: Low percolation threshold in highly conducting graphene nanoplatelets/glass composite coatings. Carbon **139**, 556–563 (2018)
99. Sudduth, R.D.: A percolation threshold model that effectively characterizes the full concentration range for electrical-conducting polymer composites. J. Appl. Polym. Sci. **136** (11), 47184 (2019)
100. Ni, X., Hui, C., Su, N., Cutler, R., Liu, F.: A 3D percolation model for multicomponent nanocarbon composites: the critical role of nematic transition. Nanotechnology **30**(18), 185302 (2019)
101. Gurunathan, T., Rao, C.R., Narayan, R., Raju, K.V.S.N.: Polyurethane conductive blends and composites: synthesis and applications perspective. J. Mater. Sci. **48**(1), 67–80 (2013)
102. dos Santos, K.M., Barbosa, R.D.M., Vargas, F.G.A., de Azevedo, E.P., Lins, A.C.D.S., Camara, C.A., Aragão, C.F., Moura, T.F.D.L.E., Raffin, F.N.: Development of solid

dispersions of β-lapachone in PEG and PVP by solvent evaporation method. Drug Dev. Ind. Pharm. **44**(5), 750–756 (2018)
103. Mutalib, M.A., Aziz, F., Ismail, A.F., Salleh, W.N.W., Yusof, N., Jaafar, J., Soga, T., Sahdan, M.Z., Ludin, N.A.: Towards high performance perovskite solar cells: a review of morphological control and HTM development. Appl. Mater. Today **13**, 69–82 (2018)
104. Das, R., Pattanayak, A.J., Swain, S.K.: Polymer nanocomposites for sensor devices. In: Polymer-based Nanocomposites for Energy and Environmental Applications, pp. 205–218. Woodhead Publishing (2018)
105. Alharbi, A.R., Alarifi, I.M., Khan, W.S., Swindle, A., Asmatulu, R.: Synthesis and characterization of electrospun polyacrylonitrile/graphene nanofibers embedded with $SrTiO_3$/NiO nanoparticles for water splitting. J. Nanosci. Nanotechnol. **17**(8), 5294–5302 (2017)
106. Yun, D.J., Park, S.H., Hwang, J., Ra, H., Kim, J.M., Chung, J., Kim, S.H., Kim, Y.S., Heo, S., Kim, K.H.: In situ photoelectron spectroscopy study on the buffer role of multiwalled carbon nanotubes against thermal degradation in organic conducting composite films with PEDOT: PSS. J. Phys. Chem. C **123**(4), 2238–2247 (2019)
107. Stefanescu, E., Daranga, C., Stefanescu, C.: Insight into the broad field of polymer nanocomposites: from carbon nanotubes to clay nanoplatelets, via metal nanoparticles. Materials **2**(4), 2095–2153 (2009)
108. Avilés, F., Oliva-Avilés, A.I., Cen-Puc, M.: Piezoresistivity, strain, and damage self-sensing of polymer composites filled with carbon nanostructures. Adv. Eng. Mater. **20**(7), 1701159 (2018)
109. Zhang, H.B., Zheng, W.G., Yan, Q., Yang, Y., Wang, J.W., Lu, Z.H., Ji, G.Y., Yu, Z.Z.: Electrically conductive polyethylene terephthalate/graphene nanocomposites prepared by melt compounding. Polymer **51**(5), 1191–1196 (2010)
110. Zagho, M., Hussein, E., Elzatahry, A.: Recent overviews in functional polymer composites for biomedical applications. Polymers **10**(7), 739 (2018)
111. Mohamed, A., Ardyani, T., Bakar, S.A., Sagisaka, M., Umetsu, Y., Hamon, J.J., Rahim, B. A., Esa, S.R., Khalil, H.A., Mamat, M.H., King, S.: Rational design of aromatic surfactants for graphene/natural rubber latex nanocomposites with enhanced electrical conductivity. J. Colloid Interface Sci. **516**, 34–47 (2018)
112. Jurewicz, I., Worajittiphon, P., King, A.A., Sellin, P.J., Keddie, J.L., Dalton, A.B.: Locking carbon nanotubes in confined lattice geometries—a route to low percolation in conducting composites. J. Phys. Chem. B **115**(20), 6395–6400 (2011)
113. Yuan, Y., Zhu, G.: Porous aromatic frameworks as a platform for multifunctional applications. ACS Cent. Sci. (2019)
114. Brigandi, P.J., Cogen, J.M., Pearson, R.A.: Electrically conductive multiphase polymer blend carbon-based composites. Polym. Eng. Sci. **54**(1), 1–16 (2014)
115. Ravindren, R., Mondal, S., Nath, K., Das, N.C.: Synergistic effect of double percolated co-supportive MWCNT-CB conductive network for high-performance EMI shielding application. Polym. Adv. Technol. (2019)
116. Uchiyama, S., Matsumura, Y., de Silva, A.P., Iwai, K.: Fluorescent molecular thermometers based on polymers showing temperature-induced phase transitions and labeled with polarity-responsive benzofurazans. Anal. Chem. **75**(21), 5926–5935 (2003)
117. Cichosz, S., Masek, A., Zaborski, M.: Polymer-based sensors: a review. Polym. Test. **67**, 342–348 (2018)
118. Covington, A.K., Totu, E.: Electrochemical and thermal behaviour of calcium-selective membranes. Analyst **121**(12), 1811–1815 (1996)
119. Aguilar, M.R., San Román, J.: Introduction to smart polymers and their applications. In: Smart Polymers and Their Applications, pp. 1–11. Woodhead Publishing (2019)
120. Ding, Y., Xu, T., Onyilagha, O., Fong, H., Zhu, Z.: Recent advances in flexible and wearable pressure sensors based on piezoresistive 3D monolithic conductive sponges. ACS Appl. Mater. Interfaces (2019)

121. Eom, J., Lee, Y.R., Lee, J.H., Park, S.K., Jeong, Y., Park, J.S., Kim, Y.H.: Highly conductive and stretchable fiber interconnections using dry-spun carbon nanotube fibers modified with ionic liquid/poly (vinylidene fluoride) copolymer composite. Compos. Sci. Technol. **169**, 1–6 (2019)
122. Lindfors, T., Ivaska, A.: Calcium-selective electrode based on polyaniline functionalized with bis [4-(1, 1, 3, 3-tetramethylbutyl) phenyl] phosphate. Anal. Chim. Acta **437**(2), 171–182 (2001)
123. Armstrong, R.D., Horvai, G.: Properties of PVC based membranes used in ion-selective electrodes. Electrochim. Acta **35**(1), 1–7 (1990)
124. Mir, M., Lugo, R., Tahirbegi, I., Samitier, J.: Miniaturizable ion-selective arrays based on highly stable polymer membranes for biomedical applications. Sensors **14**(7), 11844–11854 (2014)
125. Totu, E., Josceanu, A.M., Covington, A.K.: Improved potassium-selective membrane using valinomycin as ionophore for ion-selective microdevices. Mater. Sci. Eng. C **18**(1–2), 87–91 (2001)
126. Korostynska, O., Arshak, K., Gill, E., Arshak, A.: Review on state-of-the-art in polymer based pH sensors. Sensors (Basel, Switzerland) **7**(12), 3027 (2007)
127. Trinh, Q.T., Gerlach, G., Sorber, J., Arndt, K.F.: Hydrogel-based piezoresistive pH sensors: design, simulation and output characteristics. Sens. Actuators B Chem. **117**(1), 17–26 (2006)
128. Wan, Q.J., Zhang, X.J., Zhang, C.G., Zhou, X.Y.: Investigations on carbon fiber pH ultramicrosensor modified by polyaniline film and its applications to the in vivo detection on brassica stigmata. Chem. J. Chin. Univ. Chin. Ed. **18**, 226–228 (1997)
129. Medlock, K., Harmer, H., Worsley, G., Horgan, A., Pritchard, J.: PH-sensitive holograms for continuous monitoring in plasma. Anal. Bioanal. Chem. **389**(5), 1533–1539 (2007)
130. Pistor, P., Chu, V., Prazeres, D.M.F., Conde, J.P.: pH sensitive photoconductor based on poly (para-phenylene-vinylene). Sens. Actuators B Chem. **123**(1), 153–157 (2007)
131. Hegarty, C., McConville, A., McGlynn, R.J., Mariotti, D., Davis, J.: Design of composite microneedle sensor systems for the measurement of transdermal pH. Mater. Chem. Phys. **227**, 340–346 (2019)
132. Huynh, T.P., Haick, H.: Self-healing materials for analyte sensing. In: Nanomaterials Design for Sensing Applications, pp. 325–339. Elsevier (2019)
133. Adhikari, B., Majumdar, S.: Polymers in sensor applications. Prog. Polym. Sci. **29**(7), 699–766 (2004)
134. Leiner, M.J.: Optical sensors for in vitro blood gas analysis. Actuators B Chem. **29**(1–3), 169–173 (1995)
135. Sakai, Y., Sadaoka, Y., Okumura, S., Ikeuchi, K.: A humidity sensor composed of porous polymer film impregnated with a hydrophilic polymer. Kobunshi Ronbunshu **41**(4), 209–214 (1984)
136. Zhang, Q., Wang, X., Fu, J., Liu, R., He, H., Ma, J., Yu, M., Ramakrishna, S., Long, Y.: Electrospinning of ultrafine conducting polymer composite nanofibers with diameter less than 70 nm as high sensitive gas sensor. Materials **11**(9), 1744 (2018)
137. Bai, H., Shi, G.: Gas sensors based on conducting polymers. Sensors **7**(3), 267–307 (2007)
138. Ram, M.K., Yavuz, Ö., Lahsangah, V., Aldissi, M.: CO gas sensing from ultrathin nano-composite conducting polymer film. Sens. Actuators B Chem. **106**(2), 750–757 (2005)
139. Sagara, Y., Kato, T.: Mechanically induced luminescence changes in molecular assemblies. Nat. Chem. **1**(8), 605 (2009)
140. De Baere, I., Van Paepegem, W., Degrieck, J.: On the nonlinear evolution of the Poisson's ratio under quasi-static loading for a carbon fabric-reinforced thermoplastic. Part I: Influence of the transverse strain sensor. Polym. Test. **28**(2), 196–203 (2009)
141. Hu, J., Liu, S.: Responsive polymers for detection and sensing applications: current status and future developments. Macromolecules **43**(20), 8315–8330 (2010)
142. Crenshaw, B.R., Weder, C.: Deformation-induced color changes in melt-processed photoluminescent polymer blends. Chem. Mater. **15**(25), 4717–4724 (2003)

143. Moy, C.L.: Investigations of Disassembling Polymers and Molecular Dynamics Simulations in Molecular Gelation, and Implementation of a Class-Project Centered on Editing Wikipedia (2012)
144. Joy, N., Gopalan, G.P., Eldho, J., Francis, R.: Conducting polymers: biomedical applications. Bio-Based Plast. Mater. Appl. (2017)
145. Zelikin, A.N., Lynn, D.M., Farhadi, J., Martin, I., Shastri, V., Langer, R.: Erodible conducting polymers for potential biomedical applications. Angew. Chem. Int. Ed. **41**(1), 141–144 (2002)
146. Nambiar, S., Yeow, J.T.: Conductive polymer-based sensors for biomedical applications. Biosens. Bioelectron. **26**(5), 1825–1832 (2011)
147. Alarifi, I.M., Alharbi, A., Khan, W.S., Asmatulu, R.: April. Electrospun carbon nanofibers for improved electrical conductivity of fiber reinforced composites. In: Electroactive Polymer Actuators and Devices (EAPAD) 2015, vol. 9430, p. 943032. International Society for Optics and Photonics (2015)
148. Jing, X., Wang, X.Y., Mi, H.Y., Turng, L.S.: Stretchable gelatin/silver nanowires composite hydrogels for detecting human motion. Mater. Lett. **237**, 53–56 (2019)
149. Palza, H., Zapata, P.A., Angulo-Pineda, C.: Electroactive smart polymers for biomedical applications. Materials **12**(2), 277 (2019)
150. Miller, C.L., Stafford, G., Sigmon, N., Gilmore, J.A.: Conductive Nonwoven carbon nanotube-PLA composite nanofibers towards wound sensors via solution blow spinning. IEEE Trans. Nanobiosci. (2019)
151. Rivers, T.J., Hudson, T.W., Schmidt, C.E.: Synthesis of a novel, biodegradable electrically conducting polymer for biomedical applications. Adv. Func. Mater. **12**(1), 33–37 (2002)
152. Wang, J., Hui, N.: Zwitterionic poly (carboxybetaine) functionalized conducting polymer polyaniline nanowires for the electrochemical detection of carcinoembryonic antigen in undiluted blood serum. Bioelectrochemistry **125**, 90–96 (2019)
153. Salesa, B., Martí, M., Frígols, B., Serrano-Aroca, Á.: Carbon nanofibers in pure form and in calcium alginate composites films: new cost-effective antibacterial biomaterials against the life-threatening multidrug-resistant staphylococcus epidermidis. Polymers **11**(3), 453 (2019)
154. Guo, B., Ma, P.X.: Conducting polymers for tissue engineering. Biomacromol **19**(6), 1764–1782 (2018)
155. Ravichandran, R., Sundarrajan, S., Venugopal, J.R., Mukherjee, S., Ramakrishna, S.: Applications of conducting polymers and their issues in biomedical engineering. J. R. Soc. Interface, **7**(suppl 5), S559–S579 (2019)
156. Richardson-Burns, S.M., Hendricks, J.L., Foster, B., Povlich, L.K., Kim, D.H., Martin, D. C.: Polymerization of the conducting polymer poly (3, 4-ethylenedioxythiophene) (PEDOT) around living neural cells. Biomaterials **28**(8), 1539–1552 (2007)
157. Green, R.A., Lim, K.S., Henderson, W.C., Hassarati, R.T., Martens, P.J., Lovell, N.H., Poole-Warren, L.A.: Living electrodes: tissue engineering the neural interface. In: 2013 35th Annual International Conference of the IEEE Engineering in Medicine and Biology Society (EMBC), pp. 6957–6960. IEEE (2013)
158. Ghasemi-Mobarakeh, L., Prabhakaran, M.P., Morshed, M., Nasr-Esfahani, M.H., Baharvand, H., Kiani, S., Al-Deyab, S.S., Ramakrishna, S.: Application of conductive polymers, scaffolds and electrical stimulation for nerve tissue engineering. J. Tissue Eng. Regener. Med. **5**(4), e17–e35 (2011)
159. Xu, H., Holzwarth, J.M., Yan, Y., Xu, P., Zheng, H., Yin, Y., Li, S., Ma, P.X.: Conductive PPY/PDLLA conduit for peripheral nerve regeneration. Biomaterials **35**(1), 225–235 (2014)
160. Runge, M.B., Dadsetan, M., Baltrusaitis, J., Knight, A.M., Ruesink, T., Lazcano, E.A., Lu, L., Windebank, A.J., Yaszemski, M.J.: The development of electrically conductive polycaprolactone fumarate–polypyrrole composite materials for nerve regeneration. Biomaterials **31**(23), 5916–5926 (2010)
161. Martina, M., Hutmacher, D.W.: Biodegradable polymers applied in tissue engineering research: a review. Polym. Int. **56**(2), 145–157 (2007)

162. Sahithi, K., Swetha, M., Ramasamy, K., Srinivasan, N., Selvamurugan, N.: Polymeric composites containing carbon nanotubes for bone tissue engineering. Int. J. Biol. Macromol. **46**(3), 281–283 (2010)
163. Qazi, T.H., Rai, R., Boccaccini, A.R.: Tissue engineering of electrically responsive tissues using polyaniline based polymers: a review. Biomaterials **35**(33), 9068–9086 (2014)
164. Nyholm, L., Nyström, G., Mihranyan, A., Strømme, M.: Toward flexible polymer and paper-based energy storage devices. Adv. Mater. **23**(33), 3751–3769 (2011)
165. Hu, L., Choi, J.W., Yang, Y., Jeong, S., La Mantia, F., Cui, L.F., Cui, Y.: Highly conductive paper for energy-storage devices. Proc. Natl. Acad. Sci. **106**(51), 21490–21494 (2009)
166. Song, Z., Zhou, H.: Towards sustainable and versatile energy storage devices: an overview of organic electrode materials. Energy Environ. Sci. **6**(8), 2280–2301 (2013)
167. Li, G., Zhu, R., Yang, Y.: Polymer solar cells. Nat. Photonics **6**(3), 153 (2012)
168. Wang, X., Zhi, L., Müllen, K.: Transparent, conductive graphene electrodes for dye-sensitized solar cells. Nano Lett. **8**(1), 323–327 (2008)
169. Zhao, Y., Xie, Z., Qin, C., Qu, Y., Geng, Y., Wang, L.: Enhanced charge collection in polymer photovoltaic cells by using an ethanol-soluble conjugated polyfluorene as cathode buffer layer. Sol. Energy Mater. Sol. Cells **93**(5), 604–608 (2009)
170. Sariciftci, N.S., Braun, D., Zhang, C., Srdanov, V.I., Heeger, A.J., Stucky, G., Wudl, F.: Semiconducting polymer-buckminsterfullerene heterojunctions: diodes, photodiodes, and photovoltaic cells. Appl. Phys. Lett. **62**(6), 585–587 (1993)
171. Saito, Y., Kitamura, T., Wada, Y., Yanagida, S.: Application of poly (3, 4-ethylenedioxythiophene) to counter electrode in dye-sensitized solar cells. Chem. Lett. **31**(10), 1060–1061 (2002)
172. Shirakawa, T., Umeda, T., Hashimoto, Y., Fujii, A., Yoshino, K.: Effect of ZnO layer on characteristics of conducting polymer/C60 photovoltaic cell. J. Phys. D Appl. Phys. **37**(6), 847 (2004)
173. Hoppe, H., Sariciftci, N.S.: Organic solar cells: an overview. J. Mater. Res. **19**(7), 1924–1945 (2004)
174. Zhou, Y., Fuentes-Hernandez, C., Khan, T.M., Liu, J.C., Hsu, J., Shim, J.W., Dindar, A., Youngblood, J.P., Moon, R.J., Kippelen, B.: Recyclable organic solar cells on cellulose nanocrystal substrates. Sci. Rep. **3**, 1536 (2013)
175. Park, S.H., Kim, B.K., Lee, W.J.: Electrospun activated carbon nanofibers with hollow core/highly mesoporous shell structure as counter electrodes for dye-sensitized solar cells. J. Power Sources **239**, 122–127 (2013)
176. Pasquier, A.D., Unalan, H.E., Kanwal, A., Miller, S., Chhowalla, M.: Conducting and transparent single-wall carbon nanotube electrodes for polymer-fullerene solar cells. Appl. Phys. Lett. **87**(20), 203511 (2005)
177. Habisreutinger, S.N., Leijtens, T., Eperon, G.E., Stranks, S.D., Nicholas, R.J., Snaith, H.J.: Carbon nanotube/polymer composites as a highly stable hole collection layer in perovskite solar cells. Nano Lett. **14**(10), 5561–5568 (2014)
178. Lee, C.P., Chen, P.Y., Vittal, R., Ho, K.C.: Iodine-free high efficient quasi solid-state dye-sensitized solar cell containing ionic liquid and polyaniline-loaded carbon black. J. Mater. Chem. **20**(12), 2356–2361 (2010)
179. Escorihuela, J., Narducci, R., Compañ, V., Costantino, F.: Proton conductivity of composite polyelectrolyte membranes with metal-organic frameworks for fuel cell applications. Adv. Mater. Interfaces **6**(2), 1801146 (2019)
180. Kreuer, K.D.: On the development of proton conducting polymer membranes for hydrogen and methanol fuel cells. J. Membr. Sci. **185**(1), 29–39 (2001)
181. Dusastre, V.: Materials for sustainable energy: a collection of peer-reviewed research and review articles from Nature Publishing Group. World Scientific (2011)
182. Smitha, B., Sridhar, S., Khan, A.A.: Synthesis and characterization of proton conducting polymer membranes for fuel cells. J. Membr. Sci. **225**(1–2), 63–76 (2003)
183. Kerres, J.A.: Development of ionomer membranes for fuel cells. J. Membr. Sci. **185**(1), 3–27 (2001)

184. Zaidi, S.J., Mikhailenko, S.D., Robertson, G.P., Guiver, M.D., Kaliaguine, S.: Proton conducting composite membranes from polyether ether ketone and heteropolyacids for fuel cell applications. J. Membr. Sci. **173**(1), 17–34 (2000)
185. Meyer, W.H.: Polymer electrolytes for lithium-ion batteries. Adv. Mater. **10**(6), 439–448 (1998)
186. Croce, F., Appetecchi, G.B., Persi, L., Scrosati, B.: Nanocomposite polymer electrolytes for lithium batteries. Nature **394**(6692), 456 (1998)
187. Murugan, A.V., Muraliganth, T., Manthiram, A.: Rapid microwave-solvothermal synthesis of phospho-olivine nanorods and their coating with a mixed conducting polymer for lithium ion batteries. Electrochem. Commun. **10**(6), 903–906 (2008)
188. Snook, G.A., Kao, P., Best, A.S.: Conducting-polymer-based supercapacitor devices and electrodes. J. Power Sources **196**(1), 1–12 (2011)
189. Peng, C., Zhang, S., Jewell, D., Chen, G.Z.: Carbon nanotube and conducting polymer composites for supercapacitors. Prog. Nat. Sci. **18**(7), 777–788 (2008)
190. Frackowiak, E., Khomenko, V., Jurewicz, K., Lota, K., Béguin, F.: Supercapacitors based on conducting polymers/nanotubes composites. J. Power Sources **153**(2), 413–418 (2006)
191. Kim, J.H., Lee, Y.S., Sharma, A.K., Liu, C.G.: Polypyrrole/carbon composite electrode for high-power electrochemical capacitors. Electrochim. Acta **52**(4), 1727–1732 (2006)
192. Wang, K., Wu, H., Meng, Y., Wei, Z.: Conducting polymer nanowire arrays for high performance supercapacitors. Small **10**(1), 14–31 (2014)
193. Lota, K., Khomenko, V., Frackowiak, E.: Capacitance properties of poly (3, 4-ethylenedioxythiophene)/carbon nanotubes composites. J. Phys. Chem. Solids **65**(2–3), 295–301 (2004)
194. Ryu, K.S., Lee, Y., Han, K.S., Park, Y.J., Kang, M.G., Park, N.G., Chang, S.H.: Electrochemical supercapacitor based on polyaniline doped with lithium salt and active carbon electrodes. Solid State Ionics **175**(1–4), 765–768 (2004)
195. Vyas, M.K., Chandra, A.: Synergistic effect of conducting and insulating fillers in polymer nanocomposite films for attenuation of X-band. J. Mater. Sci. **54**(2), 1304–1325 (2019)
196. Chung, D.D.L.: Materials for electromagnetic interference shielding. J. Mater. Eng. Perform. **9**(3), 350–354 (2000)
197. Saini, P., Choudhary, V., Singh, B.P., Mathur, R.B., Dhawan, S.K.: Polyaniline–MWCNT nanocomposites for microwave absorption and EMI shielding. Mater. Chem. Phys. **113**(2–3), 919–926 (2009)
198. Block, H., Kelly, J.P.: U.S. Patent No. 4,687,589. Washington, DC: U.S. Patent and Trademark Office (1987)
199. Kim, J.W., Kim, S.G., Choi, H.J., Jhon, M.S.: Synthesis and electrorheological properties of polyaniline-Na^+-montmorillonite suspensions. Macromol. Rapid Commun. **20**(8), 450–452 (1999)
200. Gast, A.P., Zukoski, C.F.: Electrorheological fluids as colloidal suspensions. Adv. Coll. Interface Sci. **30**, 153–202 (1989)
201. Choi, H.J., Kim, T.W., Cho, M.S., Kim, S.G., Jhon, M.S.: Electrorheological characterization of polyaniline dispersions. Eur. Polymer J. **33**(5), 699–703 (1997)
202. Salaneck, W.P., Lundstörm, I., Ranby, B.: Conjugated Polymers and Related Materials. Oxford University pres, Oxford (1993)
203. Mineo, P.G., Vento, F., Abbadessa, A., Scamporrino, E., Nicosia, A.: An optical sensor of acidity in fuels based on a porphyrin derivative. Dyes Pigm. **161**, 147–154 (2019)
204. Iwata, R., Mizuno, M.: U.S. Patent Application No. 10/196,526 (2019)
205. Margolis, J. (ed.): Conductive Polymers and Plastics. Springer Science & Business Media (2012)
206. Friend, R.H., Gymer, R.W., Holmes, A.B., Burroughes, J.H., Marks, R.N., Taliani, C.D.D. C., Bradley, D.D.C., Dos Santos, D.A., Bredas, J.L., Lögdlund, M., Salaneck, W.R.: Electroluminescence in conjugated polymers. Nature **397**(6715), 121 (1999)
207. Wu, C.Z.: Organic thin-film light-emitting diodes-techniques and application in flat-panel display. Electron. Infor. **4**, 4–12 (1996)

208. Gustafsson, G., Cao, Y., Treacy, G.M., Klavetter, F., Colaneri, N., Heeger, A.J.: Flexible light-emitting diodes made from soluble conducting polymers. Nature **357**(6378), 477 (1992)
209. Pei, Q., Yu, G., Zhang, C., Yang, Y., Heeger, A.J.: Polymer light-emitting electrochemical cells. Science **269**(5227), 1086–1088 (1995)
210. Dudchenko, A.V., Rolf, J., Russell, K., Duan, W., Jassby, D.: Organic fouling inhibition on electrically conducting carbon nanotube–polyvinyl alcohol composite ultrafiltration membranes. J. Membr. Sci. **468**, 1–10 (2014)
211. Duan, W., Ronen, A., de Leon, J.V., Dudchenko, A., Yao, S., Corbala-Delgado, J., Yan, A., Matsumoto, M., Jassby, D.: Treating anaerobic sequencing batch reactor effluent with electrically conducting ultrafiltration and nanofiltration membranes for fouling control. J. Membr. Sci. **504**, 104–112 (2016)
212. de Lannoy, C.F., Jassby, D., Gloe, K., Gordon, A.D., Wiesner, M.R.: Aquatic biofouling prevention by electrically charged nanocomposite polymer thin film membranes. Environ. Sci. Technol. **47**(6), 2760–2768 (2013)
213. Ronen, A., Duan, W., Wheeldon, I., Walker, S., Jassby, D.: Microbial attachment inhibition through low-voltage electrochemical reactions on electrically conducting membranes. Environ. Sci. Technol. **49**(21), 12741–12750 (2015)
214. Duan, W., Dudchenko, A., Mende, E., Flyer, C., Zhu, X., Jassby, D.: Electrochemical mineral scale prevention and removal on electrically conducting carbon nanotube–polyamide reverse osmosis membranes. Environ. Sci. Process. Impacts **16**(6), 1300–1308 (2014)
215. Gao, G., Vecitis, C.D.: Electrocatalysis aqueous phenol with carbon nanotubes networks as anodes: electrodes passivation and regeneration and prevention. Electrochim. Acta **98**, 131–138 (2013)
216. Zaky, A.M., Chaplin, B.P.: Porous substoichiometric TiO_2 anodes as reactive electrochemical membranes for water treatment. Environ. Sci. Technol. **47**(12), 6554–6563 (2013)
217. Uddin, M.K.: A review on the adsorption of heavy metals by clay minerals, with special focus on the past decade. Chem. Eng. J. **308**, 438–462 (2017)
218. Uddin, M.K., Ahmed, S.S., Naushad, M.: A mini update on fluoride adsorption from aqueous medium using clay materials. Desalin. Water Treat. **145**, 232–248 (2019)
219. Uddin, M.K., Rao, R.A.K., Chandra Mouli, K.V.V.: The artificial neural network and Box-Behnken design for Cu^{2+} removal by the pottery sludge from water samples: equilibrium, kinetic and thermodynamic studies. J. Mol. Liq. **266**, 617–627 (2018)
220. Khatoon, A., Uddin, M.K., Rao, R.A.K.: Adsorptive remediation of Pb(II) from aqueous media using Schleichera oleosa bark. Environ. Technol. Innov. **11**, 1–14 (2018)
221. Rao, R.A.K., Kashifuddin, M.: Adsorption studies of Cd(II) on ball clay: comparison with other natural clays. Arab. J. Chem. **9**, S1233–S1241 (2016)
222. Rao, R.A.K., Rehman, F., Kashifuddin, M.: Removal of Cr(VI) from electroplating wastewater using fruit peel of Leechi (Litchi chinensis). Desalin. Water Treat. **49**(1–3), 136–146 (2012)
223. Rao, R.A.K., Kashifuddin, M.: Kinetics and isotherm studies of Cd(II) adsorption from aqueous solution utilizing seeds of bottlebrush plant (Callistemon chisholmii). Appl. Water Sci. **4**(4), 371–383 (2014)
224. Rao, R.A.K., Kashifuddin, M.: Pottery glaze—an excellent adsorbent for the removal of Cu (II) from aqueous solution. Chin. J. Geochem. **31**(2), 136–146 (2012)
225. Pandey, N., Shukla, S.K., Singh, N.B.: Water purification by polymer nanocomposites: an overview. Nanocomposites **3**(2), 47–66 (2017)
226. Weidlich, C., Mangold, K.M., Jüttner, K.: Conducting polymers as ion-exchangers for water purification. Electrochim. Acta **47**(5), 741–745 (2001)
227. Uddin, M.K., Baig, U.: Synthesis of Co_3O_4 nanoparticles and their performance towards methyl orange dye removal: characterisation, adsorption and response surface methodology. J. Clean. Prod. **211**, 1141–1153 (2019)

228. Khan, M.A., Uddin, M.K., Bushra, R., Ahmad, A., Nabi, S.A.: Synthesis and characterization of polyaniline Zr (IV) molybdophosphate for the adsorption of phenol from aqueous solution. React. Kinet. Mech. Catal. **113**(2), 499–517 (2014)
229. Zare, E.N., Motahari, A., Sillanpää, M.: Nanoadsorbents based on conducting polymer nanocomposites with main focus on polyaniline and its derivatives for removal of heavy metal ions/dyes: a review. Environ. Res. **162**, 173–195 (2018)
230. Mallakpour, S., Behranvand, V.: Recycled PET/MWCNT-ZnO quantum dot nanocomposites: adsorption of Cd(II) ion, morphology, thermal and electrical conductivity properties. Chem. Eng. J. **313**, 873–881 (2017)
231. Huang, X., Yang, J., Wang, J., Bi, J., Xie, C., Hao, H.: Design and synthesis of core–shell Fe3O4@ PTMT composite magnetic microspheres for adsorption of heavy metals from high salinity wastewater. Chemosphere **206**, 513–521 (2018)
232. Moja, T.N., Mishra, A.K., Mishra, S.B.: Nano size magnetite particles layered with the blend of conductive polymer and superadsorbent hydrogel: a core-shell based nanocomposite for trivalent arsenide uptake form aqueous solution. J. Inorg. Organomet. Polym Mater. **28**(5), 2131–2142 (2018)
233. Yarandpour, M.R., Rashidi, A., Eslahi, N., Yazdanshenas, M.E.: Mesoporous PAA/dextran-polyaniline core-shell nanofibers: optimization of producing conditions, characterization and heavy metal adsorptions. J. Taiwan Inst. Chem. Eng. **93**, 566–581 (2018)
234. Norouzian, R.S., Lakouraj, M.M.: Preparation and heavy metal ion adsorption behavior of novel supermagnetic nanocomposite of hydrophilic thiacalix [4] arene self-doped polyaniline: conductivity, isotherm, and kinetic study. Adv. Polym. Technol. **36**(1), 107–119 (2017)
235. Chen, J., Wang, N., Liu, Y., Zhu, J., Feng, J., Yan, W.: Synergetic effect in a self-doping polyaniline/TiO_2 composite for selective adsorption of heavy metal ions. Synth. Met. **245**, 32–41 (2018)

Self-supported Electrocatalysts

Paramita Karfa, Kartick Chandra Majhi and Rashmi Madhuri

Abstract Due to depletion of fossil fuels, development of large-scale ground-breaking energy conversion technology like fuel cells, water splitting, air batteries etc. needs pertinent catalyst to ease the process of conversion of chemical energy to electrical energy with greater efficiency in low time consuming. In this chapter, we will discuss the role of self-supported catalyst, which are now trending the era of nanotechnology in electrocatalysis. Self-supported catalyst can be grown on soft substrate, hard substrate or can be free standing. Self-supported electrocatalyst does not needs binder for their attachment on the conductive surface of other electrodes like glassy carbon electrode, platinum electrode, graphite electrode. They have various unique properties like flexible electrode surface, large number of active sites, high electrical conductivity, better catalytic performances, and stability in any pH electrolytic solution. They render much hassle-free electrode synthesis procedure than the powdery electrode material. This chapter mainly focuses on the benefits of using self-supported electrodes in various energy application like water splitting, oxygen reduction reaction (ORR), CO_2 reduction reaction, fuel cells. It has been observed that the self-supported electrocatalyst proves to be the superior electrocatalyst in the immense area of electrocatalysis.

Keywords Self-supported catalysts · Electrocatalysis · Water splitting · Oxygen reduction reaction (ORR) · CO_2 reduction · Fuel cells

1 Introduction

According to the definition, electrocatalysis is a kind of heterogeneous catalysis which involves electrochemical reaction and performed with the help of electrodes and catalysts. Electrocatalysis is an extraordinary field of electrochemistry that has

---

P. Karfa · K. C. Majhi · R. Madhuri (✉)
Department of Applied Chemistry, Indian Institute of Technology (Indian School of Mines), Dhanbad 826 004, Jharkhand, India
e-mail: rashmimadhuriism@gmail.com

Inamuddin et al. (eds.), *Self-standing Substrates*,
Engineering Materials, https://doi.org/10.1007/978-3-030-29522-6_6

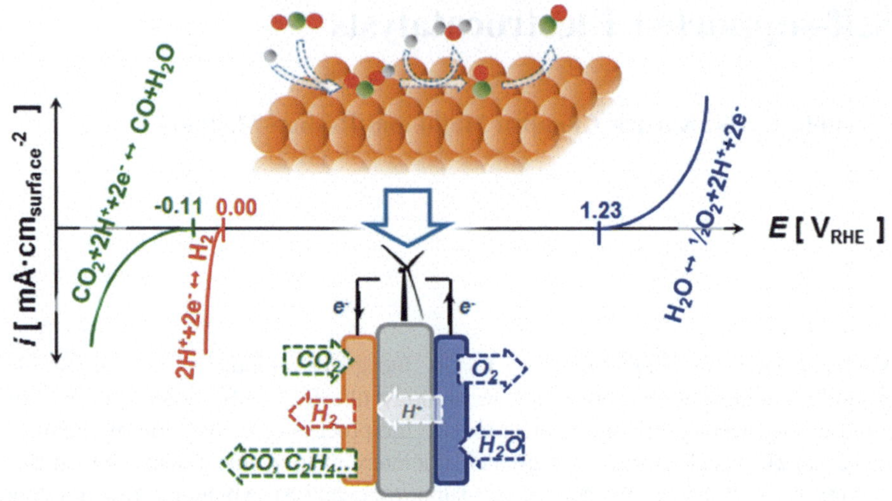

Fig. 1 Illustration showing number of electrochemical reactions that takes place in electrochemical energy conversion and storage (Taken permission from [29])

grown largely in the eighties, due to their major applications in the field of green and renewable energy production as well as conversion [46]. Electrocatalysis have some of the extraordinary features like: (i) they can be performed at normal atmospheric pressure and room temperature; (ii) The reaction rate and selectivity of electrochemical reaction in electrocatalysis depends largely on the electrode potential [74]. Electrocatalysis are of global importance and can be used in large number of research fields including designing of electrochemical sensors, corrosion science, electrochemical reaction based organic synthesis, waste-water treatment and development of energy conversion/storage devices (like batteries, fuel cells and electrochemical water splitting). However, from the year 1980s, electrocatalysis has been evolved as an extraordinary field of electrochemistry for the production of clean energy. A number of electrochemical reactions are reported in literatures that are involved in electrochemical energy conversion/storage or production process shown in Fig. 1 like oxygen evolution reaction (OER), hydrogen evolution reaction (HER), oxygen reduction reaction (ORR), and carbon dioxide reduction reaction (CER) [29].

Prior to the use of electrochemical energy conversion/storage or production, fossil fuels are more popular in terms of green and clean energy source. However, increase in the use of fossil fuels caused detrimental effect to the environment via increase in the CO_2 concentration in the air causing global warming, which roots in increase in sea level, melting of glaciers, ice bergs [28]. As fossil fuels contains large amount of sulfur, consuming the mentioned source of energy also leads to rise in sulfur accumulated fog, increase in SO_2 concentration, acid rains, which damaged the archeological monuments and our health, main impact on respiratory organs [26].

To avoid these problems and stop the depletion of fossil fuel sources, in recent time, researchers and scientists have focused on clean and renewable energy source like electrochemical energy conversion/storage electrolyzers, fuel cells, metal air batteries other than the conventional sustainable energy sources like solar, wind and geothermal energy [24]. Conventional sustainable energy sources depend on the weather, time of the days and other high capacity energy storage devices to store the energy. But, electrocatalysis based new clean energy sources are economically favorable, stores energy in the form of chemical bonds, have high specific energy, and causes no emission of hazardous by products [59].

1.1 Electrocatalysts

In general, most electrochemical reactions involved in energy conversion requires transfer of several electrons, exhibit slow kinetics and therefore needs the material, which can speed up the reactions called as catalyst. The catalyst involved in the electrochemical reactions, commonly known as 'electrocatalyst' usually requires for lowering the overpotential needed to complete the reaction [13]. Electrocatalyst is one of the crucial and key elements of electrocatalysis and therefore must possess some of the fundamental properties like:

(a) It must possess large number of active sites for each unit area;
(b) Exposing high index surface facets of the catalysts, which increase the number of active sites;
(c) Increase in ratio of edges to that of the basal planes;
(d) Large amount of electron transfer on the surface of electrode for better electrolysis;
(e) Material of the catalyst must be earth abundant and less expensive;
(f) Elements by which the catalyst is prepared must be devoid of any oxidation and reduction after certain electrochemical activities that is the catalyst must be stable in all conditions.

In most of the cases, these catalysts are high cost noble metals. For example, Pt/C, Ru, Ir, RuO_2, IrO_2 owns extraordinary electrocatalytic property with great stability due to their inert behavior and famous as benchmark electrocatalyst. Among them, Pt/C is well known for its activity towards ORR and HER in cathodic half-cell reaction, whereas, Ir and Ru based compounds are famous for anodic half reaction for OER [42, 72]. Conversely, there are some demerits associated with these catalysts, like they are very much expensive, which inhibits their large scale industrial and commercial utilization. Similarly, noble metals are scarcely available in the earth, which limits their large consumption, low selectivity and less rigidity. Noble metal-based catalyst needs tough synthesis procedures, which requires valuable instrumental methods [12]. Therefore, current research in several areas has devoted their focus on the development of low-cost catalytic materials that can sort out the above-mentioned problems and promote the electrochemical reactions at low overpotential value.

A promising cost-effective and competent replacement of these conventional noble metal catalysts can be some earth abundant transition metal catalysts (Fe, Co, Ni, Mo, W) and/or non-metallic carbon, boron-based catalysts, which have similar properties to that of the benchmark catalysts [98]. Among the recent edge catalyst, transition metal catalysts are found to be superior in electronic properties, have low cost, corrosion resistance, high thermal and electrical conductivity. They also have synergistic effect with other materials and possess high durability, which make them suitable candidate for electrocatalysis. Other than these, non-metals like carbon-based catalysts have high surface area to volume ratio, variety of morphology, vast dimensionality, unique hybridization of carbon atoms, large amount of π electrons, which makes them an admirable catalyst towards HER, OER, ORR [101].

Both metallic and non-metallic catalysts were synthesized by different methods but one problem, which is associated with each of them is their coating on the electrode surface. After synthesis of any kind of catalytic material, they have to be coated on the electrode surface, after that the catalyst modified electrode surface will be used for electrocatalysis. In other words, to fully utilize the catalytic materials, they have to be dispersed or coated on a conducting support, which is generally an electrode. Coating of the catalysts onto the electrode surface can be done through three simple process: (1) via spin-coating, (2) via drop-coating, or (3) via dip-coating. While coating the catalysts onto the electrode surface, we use some kind of binder, which make the coating stronger and cause their adherence to the electrode surface [55]. If we do not use binders, the catalyst coating may fell off from the electrode or can lead to a non-homogeneous coating. It is true that binders are needed for stable and homogenous coating of catalyst over the electrode surface but it is not true that binder is the solution for perfect coating of catalyst on the electrode surface [64]. Catalysts modified electrodes still have several problems like:

1. Uncontrollable gas evolution in HER, OER, ORR may cause shedding off the electrode materials from the electrodes.
2. The formation of products in different phase to that of the reactants, which causes difficulty in mass transport.
3. Formation of large voids and undesirable dead spaces, which causes impairment in the electronic conductance.
4. Formation of unwanted microstructure, which can be seen after all the electrochemical studies through various scanning microscope that reduces the stability of the electrode.
5. Limited surface area with limited number of active sites for generation of hydrogen and oxygen.

Therefore, we needed a more stable and perfect solution or alternative to solve this coating problem.

1.2 What Is 'Self-supported Catalyst'?

In recent time, a term 'self-supported catalysts' came in focus. Self-supported catalysts are class of catalysts, which does not require any support to work as electrode. They are not coated on the electrode. They are 3-dimensional structure, which is stable enough to stand free and work as electrode [82]. Other than these, self-supported electrocatalyst have several other advantages also, over the powder catalyst:

1. They have flexible electrode structure,
2. Controlled chemical composition,
3. Porous structure,
4. Large surface area,
5. Free from binder,
6. Can be conveniently synthesized,
7. Possess easy flow of electrolytes and therefore have excellent catalytic properties.

Self-supported catalysts can be molded or designed to any flexible morphology like nanowire, foams, nanopillar, nanofilms having chemical composition of metal oxides, selenides, sulfides, phosphides and their respective hybrid materials [17]. Beside their powder catalysts modified electrode counterpart, this self-supported electrode has grown outstanding renown in the field of electrocatalysis, mainly in fuel cell reactions, OER, HER, and ORR. They do not require organic binder, which enables the full interaction of the electrolytic species with that of the conductive surfaces of the self-supported electrode, which retains the stability of the electrode through various cycles for several months [47]. Self-supported electrocatalyst with complex, multifaceted, better exposed nanostructure provides a greater number of active sites than that of the planar powdery coating of catalyst. In addition, 3D, porous, self-supported nanostructured electrode provides better pathways for electrolyte (H^+, OH^-) transport and fast formation of reaction products (H_2, O_2), which results in fast kinetics of the sluggish reactions. Other than electrocatalysis, due to high stability, convenient and time-consuming synthesis procedure, light weight, outstanding potential they are not just limited in sustainable energy devices but also have other applications in fabrication of lithium ion batteries, supercapacitors, sensors, dye degradation, environmental remediation, memory devices etc. [73].

There are various methods reported for the synthesis of self-supported electrode like hydrothermal synthesis process, chemical reduction, vacuum filtration, electrochemical deposition, epitaxial growth, growth on flexible substrate, ball milling, chemical vapour deposition (CVD) etc. Self-supported electrodes synthesis processes come in two categories i.e. template assisted synthesis and free-standing synthesis approach without the template. In the template free synthesis free standing electrode process, 3D porous electrode materials are grown by its own without the help of substrate. In template assisted synthesis process templates are classified in two category i.e. (1) soft substrate and (2) hard substrate. For the soft-substrate, carbon fibre paper or carbon cloth were used, while, metal foams, metal foils, metal meshes and films are used as hard template. In comparison between substrate free and substrate assisted synthesis, substrate free synthesis process is easier and more convenient [44].

Graphene based self-supported electrodes are the most flexible, well known and extensively studied materials than the other self-supported electrode materials, because of its high mechanical strength, better conductivity and high surface area. For example, Dong et al. have synthesized 3D flexible graphene-cobalt oxide electrode through CVD method and used it in supercapacitor application [14]. Xu et al. have synthesized $Zn_xCo_{3-x}O_4$ hierarchical needle like structures using Titanium foil of rhombus shaped pillar as the substrate through simple hydrothermal process [88]. Lee et al. have used the technique of template free synthesis procedure for the fabrication of Co_3O_4 nanowires using cobalt nitrate as the precursor via hydrothermal synthesis procedure. They fabricated nanowire was directly used as bifunctional electrode for ORR/OER and showed better performances than powdery catalyst [41]. Zhu and his coworkers have synthesized flexible, stable reduced graphene oxide films through template free foam film method, which is found very cheap, less time consuming and good application as supercapacitor [103].

1.3 Factors Affecting the Electrocatalytic Activity of Self-supported Catalyst

In the earlier section, we have discussed about fundamental properties of self-supported catalyst. In addition to the fundamental properties of materials used to fabricate self-supported catalysts, there are several other factors also, which can directly or in-directly affect the electrocatalytic activity of catalyst, like:

1. Size
2. Shape
3. Crystalline faces
4. Composition of the catalyst
5. Integration of catalyst and electrode

1.3.1 Size

Electrocatalysis is a surface phenomenon and depends greatly on the size of catalyst, at which the reaction takes place. In general, for electrocatalysis, we mainly prefer nanoparticulate or nano-sized materials over their bulk counterpart because of their high surface to volume ratio, exposed crystal faces, easy wet synthesis process, and quick assembly. It can be very easily concluded that size is a key parameter in determining electrocatalytic behavior of any catalyst. Smaller the size of the particle larger is the number of atoms gathering with more edges and a greater number of corners. Atoms present in the nanomaterials are not fully coordinated, which make them more prone towards molecular adsorption of electrolytes and better interaction with them, require for good electrocatalyst. Therefore, it can be easily concluded that smaller size of particles is better for electrocatalysis [68].

1.3.2 Shape

Tailoring of shape of the nanoparticle largely influence the electrocatalytic properties. For the first time, effect of shape specific nanoparticle was done by El-Sayed and his partners in 1996, where they have performed the synthesis of different shaped Pt nanoparticle from K_2PtCl_4 as the precursor [1]. Different shape of nanoparticles like sphere, cube, rod, triangular, tetrahedron, hexagonal has different atomic arrangement, possess different coordination number of atoms, and their crystal facets have different selectivity and different catalytic activity. For example, the cubic shaped nanoparticle has 6 facets (100) with 8 coordination number and 4-fold symmetry of surface atoms, while rhombohedron have (110) facets and octahedron have (111) facets. Similar to the shape and size of nanoparticles, their hollow or compact nature also affect their electrocatalytic activity. Like, hollow nanoparticles have more surface to area ratio compared to the compact nanoparticle and also have low density with large voids, which help in electrocatalysis [67].

1.3.3 Crystalline Facets

According to the literature, surface energy will be different for different crystalline facets, which affect the selectivity and catalytic property of a material [80]. For face centered cubic nanoparticle the highest indexed facet (hkl) is 111, which represents high density of edges, steps and kinks, which shows higher electrocatalytic activity compared to low indexed facets. Because high indexed facets have atoms of low coordination number at different position making faster electrocatalysis. Molecular adsorption of electrolytic materials depends largely on the crystalline facets, so by tailoring it, we can adjust the catalytic selectivity. It was observed in an experiment that different facets of Pd nanoparticle were synthesized through seed mediated growth process, where the facets can be interchangeable [35]. In different facet nanoparticles, the cubic nanostructure of Pd nanoparticle with (100) facet is found more susceptible to 4 electron ORR pathways than the (111) facet of octahedron nanoparticle.

1.3.4 Composition of the Catalyst

Composition of the catalyst is also very important parameter, which can affect their catalytic property as well as other properties like cost, stability etc. Researcher have tried different compositions like increasing the use of different non precious metals or non-metals by decreasing the concentration of costly metal to reduce the cost of overall material. It was also found that stability of the catalyst may also increase by increasing the use of small amount of noble metal or through noble metal doping. Similarly, use of multi-metal was also tried to change the surface atom composition and design different facets. Till date, for electrocatalysis, mainly, transition metal like Ni, Co, Mo and their different combinations with noble metals are very popular in recent time, owing to their outstanding durability. It is seen that formation of trimetallic or bimetallic nanoparticle using one of the noble metal decreases certain limitation of using noble metals in bulk amount and also causes increase in the catalytic property. Synthesis of catalyst having non-metal like carbon are also popular nowadays and have several advantages like they are abundant in the earth crust, less expensive, easily available, possess tunable molecular structure, stable in different pH electrolyte, and have high electrical conductivity [91].

Therefore, it can be easily understood that we need an electrode material or catalyst, which not only have great electrocatalytic activity, but also found in large abundance, with low cost and high stability.

1.3.5 Integration of Catalyst and Electrode

Integration of catalyst and electrode is another difficult but important task. To be a best catalyst modified electrode, it must possess high surface area, good integration of nanoparticle to the electrode surface with stable coating, which must not be stripped off. These problems can be sorted out, if the electrode is made up of self-supported catalysts. Self-supported catalysts are of flexible in nature, have light weight and porous structure, which provides large surface are and therefore results in great electrocatalytic activity.

2 Role of Electrocatalysts in Energy Applications

Increase in the use of fossil fuels due to worldwide energy claim causes detrimental effect to the environmental through increase in the CO_2 concentration in the air causing global warming which roots in increase in sea level, melting of glaciers, ice bergs. As fossil fuels contains large amount of sulfur and consuming the mentioned source of energy causes rise in sulfur accumulated fog, increase in SO_2 concentration, acid rains, which results in damage in archeological monuments, human health mainly respiratory organs. Similarly, increase in the utilization of fossil fuels also causes energy crisis in the growing industrial civilization, so to meet up this demand, clean and renewable energy sources is required like electrolyzers, fuel cells, metal air batteries other than the conventional sustainable energy sources like solar, wind and geothermal energy [83, 94]. Conventional sustainable energy sources depend on the weather, time of the days and other high capacity energy storage devices to store the energy but new clean energy sources are economically favorable, stores energy in the form of chemical bonds, have high specific energy, and causes no emission of hazardous by products. Chemical fuels are the new form of storage of energy which is now a days a revolutionary field in science and technology to overcome the burgeoning of the energy demand. Therefore, H_2 and O_2 has become the recent alternate chemical fuel sources to generate energy with green environmental approach without any depletion and crisis.

Electrocatalysis is very popular in various fields like fuel cell preparation [direct methanol fuel cells (DMFC), solid oxide fuel cells (SOFC)], organic synthesis, energy conversion and energy storage. However, they are very popular in the field of energy conversion and storage. The most common electrochemical reactions involved in energy conversion are oxygen evolution reaction (OER), hydrogen evolution reaction (HER), oxygen reduction reaction (ORR), carbon dioxide reduction reaction (CER). Applying electrocatalysis, different electrochemical reactions are able to convert electrical energy to chemical energy, like HER and OER occurs in the cathode and anode of the water splitting electrolyzer, OER and ORR occurs in the metal air batteries, methanol oxidation reaction (MOR) and ORR occurs in the two half reactions of DMFCs [30]. In this book chapter, we have focused on the application of self-supported catalysts in the above-mentioned reactions. But, prior to that, discussion on the parameters, which are required to evaluate the performance of a catalysts is also important. In the next section, we have discussed the important parameters, on the basis of which performance of electrocatalyst can be rated.

2.1 Evaluating Parameters of Electrocatalysis

Catalyst is the key material for electrolysis taking part in electrochemical reactions, our first and foremost aim is to increase the electrocatalytic properties of the catalyst by optimizing the various parameters like [3]:

1. Onset and overpotential at particular current density
2. Tafel slope
3. Faradaic efficiency (FE)
4. Electrochemically active surface area (ECSA)
5. Turnover frequency (TOF)
6. Exchange current density (j_0).

In the below section we will discuss about these parameters and will examine their role in judging the catalyst having superior or awful electrocatalytic activity.

2.1.1 Onset Potential and Overpotential

Onset potential is the most reliable activity parameter and is considered as the potential from where the electrolysis starts, but it is difficult to obtain the value of onset potential at zero current density, so it is generally obtained at current density of 10 mA/cm^2 and denoted as E_{onset}.

Overpotential is the potential needed to drive the electrolysis over its reversible potential. Theoretically, reversible potential needed for hydrogen evolution at cathode is 0 V and that for oxygen evolution reaction is +1.23 V. Hence the equation for overpotential can be represented as $\eta_{HER} = E_{onset} - 0$, for HER and for OER is $\eta_{OER} = E_{onset} - 1.23$. To obtain the value of overpotential current density 10 mA/cm^2 is considered as the benchmark for low catalyst loaded electrode and for high catalyst loaded electrode current density of 50 and 100 mA/cm^{2}will be considered [4]. Lower the overpotential value, higher is the catalytic activity of the catalysts for all the cell reactions.

2.1.2 Tafel Slope

Tafel slope mainly explain the kinetics of the electrochemical reaction, pathway or mechanism through which the reaction is proceeding, intermediates of the reaction and to compare the catalytic behavior of different electrocatalysts. Tafel slope is derived from Butler-Volmer equation [18]:

$$I = I_0[\exp(\alpha_A nF/RT \times \eta) - \exp(-\alpha_C nF/RT \times \eta)]$$

where, I represent the current, I_0 designates the exchange current, anodic charge transfer coefficient is α_A and the cathodic charge transfer coefficient is α_C, F is the Faraday constant, n denotes the number of electron and R is real gas constant and T represents the absolute temperature in Kelvin and the overpotential is designated as η. At equilibrium the anodic and cathodic terms are same and I is I_0. At high overpotential, the Tafel equation at anodic and cathodic area can be represented as:

$$\text{Anodic polarization: } \ln \mathrm{I} = \ln \mathrm{I}_0 + (\alpha_A nF/RT \times \eta)$$

$$\text{Cathodic polarization: } \ln \mathrm{I} = \ln \mathrm{I}_0 + (\alpha_C nF/RT \times \eta)$$

This equation just looks like y = mx + c, which gives a linear line when log I is plotted against η.

So, the Tafel slope from the above equation can be represented as:

$$\delta \log \ j/\delta \eta = 2.303\mathrm{RT}/\alpha nF$$

Tafel slope in OER describes the number of electrons involved in the electrocatalytic pathway for particular reaction step. If the Tafel slope for HER is very low that is 30 mV/dec or 40 mV/dec, then it is considered that the kinetics of the reaction is very fast helping in good electrolysis [39]. Unlike HER, the prediction of reaction pathway in OER is difficult because of large number of intermediates involved in the half cell reaction of anode.

2.1.3 Faradaic Efficiency (FE)

To become good electrocatalyst, selectivity is the main criteria for screening catalyst to be suitable for HER, ORR and OER. The better selectivity of the catalyst can be examined through Faradaic efficiency. Faradaic efficiency is free from the influence of catalyst morphology, shape, size so they have same value for particular electrocatalyst obtained through any synthesis procedure or having shape or size. Faradaic efficiency can be calculated through measuring the amount of gas evolved through experimental method to that of the theoretical method support by Faraday law of electrolysis. There are two methods to obtain experimentally the FE, one is by water displacement method which is a very reliable and conventional method, the amount of gas evolved is converted in moles through Avogadro theory for the experimental section. For the theoretical section the moles of gas are calculated through the amount of charge passed following second law of Faradays, if the amount of gas obtained through experiment and theoretical is correspondent to each other then we can have the value of 100 percent Faradaic efficiency that means the electrode is very much selective towards that particular electrolysis reaction [53].

Another method for calculation of FE is through gas chromatography (GC), which is in recent days the most widely accepted method because of its preciseness, and constant overpotential is obtained to evolve gas without any fluctuation, and also time to time measurement of evolved gas is very much easy and accurate. The equation used for calculation of Faradaic efficiency in both the methods are as follows: Mole of H_2/O_2 = I(A).s/2F, I is the current obtained through controlled potential electrolysis, s is the time required, F is the Faradaic constant which has the value of 96,485 A s mol^{-1}. Faradaic efficiency = measured amount of hydrogen × 100/theoretical amount of hydrogen.

2.1.4 Electrochemically Active Surface Area (ECSA)

Electrochemical surface area differs from one electrolyte to other, with different electrochemical technique and with different material, so, it is not a crucial parameter to screen the efficiency of the electrode rather it is used to examine or compare the electrocatalytic activity between different electrode material of comparable composition and catalytic activity. There are two different techniques to calculate ECSA, one is through turnover frequency technique (TOF), other is through double layer capacitance technique.

For the calculation of ECSA, through double layer capacitance technique, cyclic voltammetry (CV) is used. It is scan rate reliant technique, which is used to compare the catalyst having analogous physical and chemical properties with same electrochemical method that is CV. For the measurement, CV runs is taken at different scan rate and their non-Faradaic region will be considered for calculation. From the non-Faradaic region, the anodic and cathodic current density is calculated to obtain the value of $\Delta j = (ja - jc)$ and this is plotted against scan rate, a straight line is obtained, which has a slope and designated as the double layer capacitance (C_{dl}). ECSA can be calculated through dividing the value of double layer capacitance to that of specific capacitance (C_s) using the equation given below: $ECSA = \frac{C_{dl}}{C_S}$, where, C_s is the specific capacitance [63].

In addition to Tafel slope, TOF is also an important activity parameter, which gives idea about the kinetics of the reaction. TOF can be calculated using the formula through CV runs:

$$Q = 1/V \int_{E_1}^{E_2} IdE \qquad (1)$$

$$n = Q/2F \qquad (2)$$

where, E represents the potential, Q is the charge, V denotes the scan rate and '*n*' denotes the per-site turnover frequencies (TOF in s^{-1}), F is the Faraday constant. TOF is used to examine the performance of the electrocatalyst towards different electrolysis. It represents the number of active sites in the catalyst, more the number of active sites more is the catalytic property of the catalyst.

2.1.5 Exchange Current Density (J_o)

It is obtained by extrapolating Tafel plot through linear fitting to the particular logarithm of current density on the reversible potential at equilibrium condition, for example at zero volt for HER and ORR and for OER it is +1.23 V. Exchange current density, like Tafel slope gives an idea of the electrochemical process occurring inside the cell by taking the inverse of logarithm of current density. It also gives an idea about the number of electrons transferred, during the reaction giving an intrinsic idea about the catalytic property of the catalyst. So, higher is the exchange current density higher is the superiority of the catalyst towards electrocatalysis [58].

From large number of articles, experimented by many researchers, we have found (to the best of our knowledge) that now a day's self-supported electrocatalyst have great catalytic properties with considerable electrochemical activity parameters like low onset potential, low overpotential value, low Tafel slope, high electrochemical active area, high exchange current density, nearly 100% Faradaic efficiency, as well as large number of active sites obtained through turn over frequencies (TOF).

3 Hydrogen Evolution Reaction (HER)

In recent time, hydrogen has become the raw material for fuels to drive various chemical reactions with efficient energy carrier property. There is other vast application of hydrogen like metal refining, production of ammonia, refining of petroleum etc. The production of hydrogen through methane reforming (reaction of water and methane to produce hydrogen and carbon dioxide), coal gasification is not environment friendly; because it emits fatal greenhouse gases as well as yield of hydrogen is very low. Among the various methods used to generate H_2, the best and the most clean and green method to produce hydrogen is through electrolysis of water driven by renewable electric power sources [15]. Hydrogen produced through water electrolysis is the purest form of hydrogen with zero carbon emission. In addition, H_2 produced through electrolysis of water have a high calorific value. In water electrolysis, there are two reaction happened at anode and cathode shown in Fig. 2, one is hydrogen evolution reaction (HER), which takes place at the cathode with the evolution of hydrogen gas, and at the anode evolution of oxygen gas takes place, which is called oxygen evolution reaction (OER) [75]. The overall reaction for water splitting can be represented as:

$$H_2O \leftrightharpoons H_2 + {}^1\!/_2 O_2$$

Sometime, if we only want hydrogen as one of the products without any contamination, we set up an electrochemical cell, which must have two chambers, in one chamber hydrogen gas is produced, which is our main requirement and is collected through various method and in other chamber, oxygen is produced. HER largely depends upon the pH of the electrolyte in which the reaction is taking place, means when the pH of the electrolyte in the cell is acidic H^+ is the active ions and when the pH of the electrolyte is basic, OH^- is the active ions. In addition, pH of the electrolyte largely influences the reaction pathway and the intermediate forming in the reaction to produce the hydrogen gas [40]. In the acidic medium the hydrogen evolution goes through the overall reaction given below [16]:

$$2H_2O + \ = O_2 + 4H^+ + 4e^-$$

$$4H^+ + 4e^- = 2H_2$$

In the alkaline reaction the hydrogen evolution reaction goes through the reaction given below:

$$2H_2O + 2e^- = H_2 + 2OH^-$$

The pathway of hydrogen evolution can be explained by considering the value of Tafel slope. Hydrogen gas evolution occurs when the hydrogen ion (H^+) gets adsorbed on the surface of the electrode, which is known as Volmer reaction

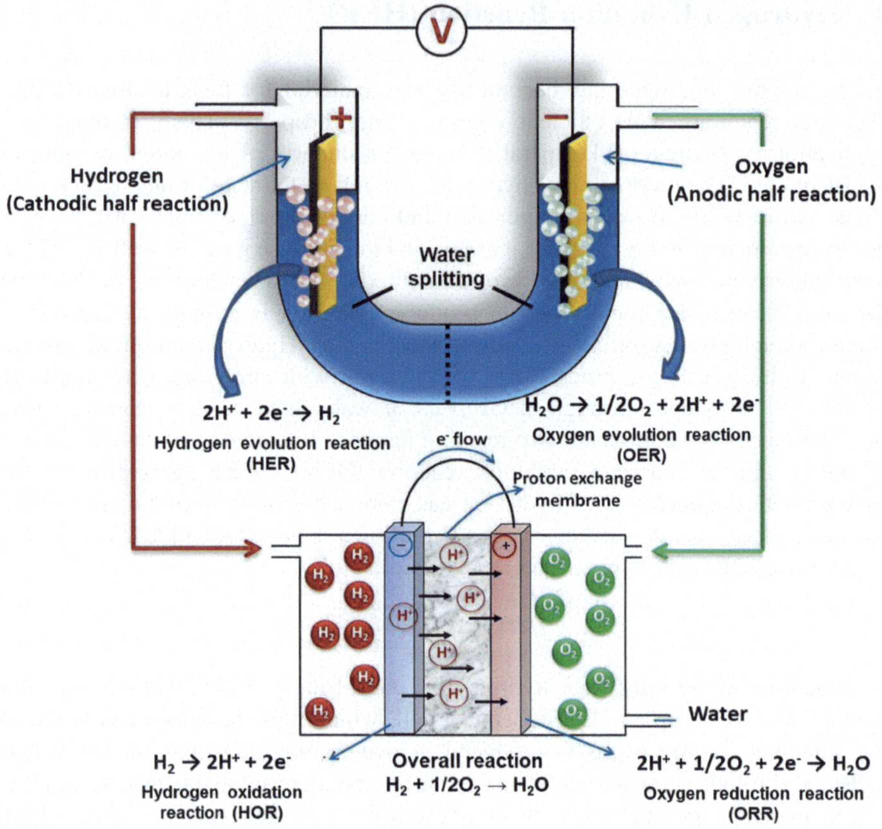

Fig. 2 Anodic and cathodic reaction taking place in water splitting through an efficient electrocatalyst (Taken permission from [75])

step. After Volmer reaction step. the reaction can proceed through two different steps, one is Tafel reaction step and another one is Heyrovsky reaction step for the formation of hydrogen gas [57]. The reaction step in acidic medium as proposed through Tafel slope is as follows:

$$\text{Volmer step: } H^+ + e^- = H_{ads}$$

$$\text{Heyrovsky step: } H^+ + H_{ads} + e^- = H_2$$

$$\text{Tafel step: } H_{ads} + H_{ads} = H_2$$

The reaction pathway for the production of hydrogen in alkaline medium is as follows:

$$\text{Volmer step: } H_2O + e^- = H_{ads} + OH^-$$

$$\text{Heyrovsky step: } H_2O + H_{ads} + e^- = H_2 + OH^-$$

$$\text{Tafel step: } H_{ads} + H_{ads} = H_2$$

H_{ads} is the adsorbed hydrogen atom on the surface of the catalyst, the adsorption of the hydrogen atom on the catalyst is the rate limiting step, which depends largely on the number of active sites and the inherent conductivity of the catalyst. Tafel steps occurs when the desorption of the hydrogen atom from the catalyst surface is very fast having Tafel slope of 30 mV/dec, otherwise it follows Heyrovsky step (40 mV/dec). Hydrogen atom adsorption is the decisive step for choosing the quality of the electrocatalyst, because, easier is the process of adsorption superior will be the catalyst. Till date the benchmark electrocatalyst towards HER is catalyst comprising of Pt, because it has nearly zero onset potential, 30 mv/dec Tafel slope and the free energy of adsorption for hydrogen atom is also equal to zero ($\Delta G_{H+} \sim 0$). 'd' electrons of Pt, contributes a major part towards their good electrocatalytic activity, but due to scarcity, high cost, easy poisoning of the metal limits their large-scale industrial application. For becoming a good catalyst like that of Pt, an electrocatalyst for HER, must have high electrical conductivity to ease charge transfers, must avoid Volmer step as the rate determining step, increase mass transfer to ease electrolysis, have large number of H atom adsorption active sites, and possess stability in different pH electrolytic solution [95].

Non noble metal-based catalyst, which are used now a days are transition metal compounds, like alloys, carbonaceous material, or single active metals. As our chapter is based on self-supported electrocatalyst, we have focused on the self-supported electrocatalysts only, reported for hydrogen evolution reaction. Self-supported catalyst can be grown on hard rigid surface like metal oxide, metal films or can be grown on flexible soft surface like carbon cloth, carbon fiber paper or can be free standing without any template.

Ren et al. has developed a self-supported electrode using amorphous nanosheet array of $CoMoS_4$ grown on flexible carbon cloth ($CoMoS_4$ NS/CC), using simple hydrothermal technique taking Co(OH)F nanosheet array grown in carbon cloth and solution of $(NH_4)_2MoS_4$ as precursors. This 3D electrode shows efficient HER in phosphate buffer saline with very low overpotential value of 183 mV with 100% Faradaic efficiency [61]. Guo et al. have used flexible carbon fibre paper (CFP) as the flexible support for the growth of $NiSe_2$ ultrathin nanowires without using any binder. The electrocatalyst is prepared through hydrothermal technique forming first $Ni_2(CO_3)(OH)_2$ NSs/CFP followed by selenization to form $NiSe_2$ NWs/CFP. The prepared self-supported catalyst shows outstanding HER activity shown in Fig. 3 with small overpotential value of 172 mV having Tafel slope of 32.44 mV/dec and considerable stability [23]. Bandal et al. self-supported bimetallic electrode of iron

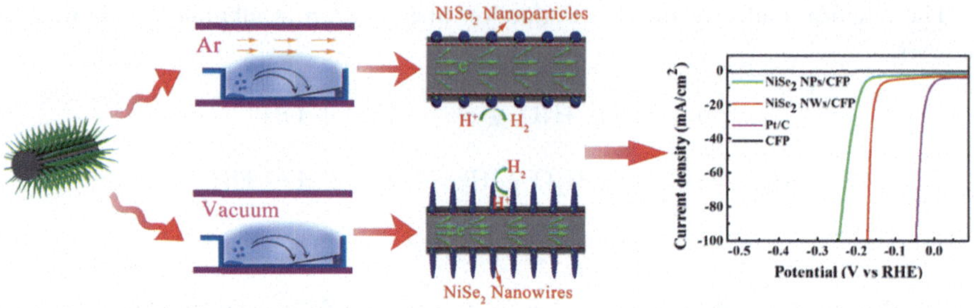

Fig. 3 Graphical abstract showing application of $Ni_2(CO_3)(OH)_2$ NSs/CFP towards hydrogen evolution reaction (Taken permission from [23])

cobalt oxide grown on nickel foam [6]. The spinel like iron cobalt oxide and nickel foam hybrid was synthesized through chemical bath deposition technique, followed by calcinations. The prepared FeCoO-NF electrode shows good water splitting behavior with low overpotential of 205 mV for HER. Zhang and his group members have synthesized transition metal phosphide quantum dots of Co_2P and entrenched them in N and S dual doped carbon and then used flexible carbon cloth to prepare the electrode Co_2P@NPC/CC. The prepared electrode is synthesized through carbonization of phytic acid and polyaniline induced by cobalt ion. The self-supported electrode shows very low overpotential of 116 mV at 10 mA/cm^2 current density and Tafel slope of 63 mV/dec [96]. Liu et al. have synthesized self-supported electrocatalyst using nickel foamand nanosheets of $NiCo_2S_4/Ni_3S_2$. The prepared electrocatalyst is used for HER in alkaline medium because of large number of active sites, good electrical conductivity with large energy transfer ability and high porosity. The catalyst $NiCo_2S_4/Ni_3S_2$/NF, shows low onset potential of 15 mV, small overpotential of 119 mV and Tafel slope of 105.2 mV/dec [48]. Wang et al. have studied the activity of molybdenum phosphide (MoP) grown on N doped porous carbon nanofibre to form a self-supported electrocatalyst for HER. The prepared electrocatalyst MoPNPCNF have large number of phosphorus in the atomic layer for high competence of the catalyst. The catalyst shows small overpotential value of 64 mV, Tafel slope of 65.64 mV/dec and double layer capacitance of 43.46 mF/cm^2 with unaltered current density and onset potential, after 5000 CV cycles and 32-h long term bulk electrolysis [79]. Xiao et al. synthesized biocarbon-fiber cloth decorated with molybdenum carbide (BCF/Mo_2C) nanomaterial as a self-supported electrode for HER shown in Fig. 4. The electrode material was synthesized through simple annealing technique which shows small overpotential value of 88 mV in alkaline and 115 mV in acidic medium [86]. Table 1 showing some of the self-supported binder free electrode for hydrogen evolution reaction (HER).

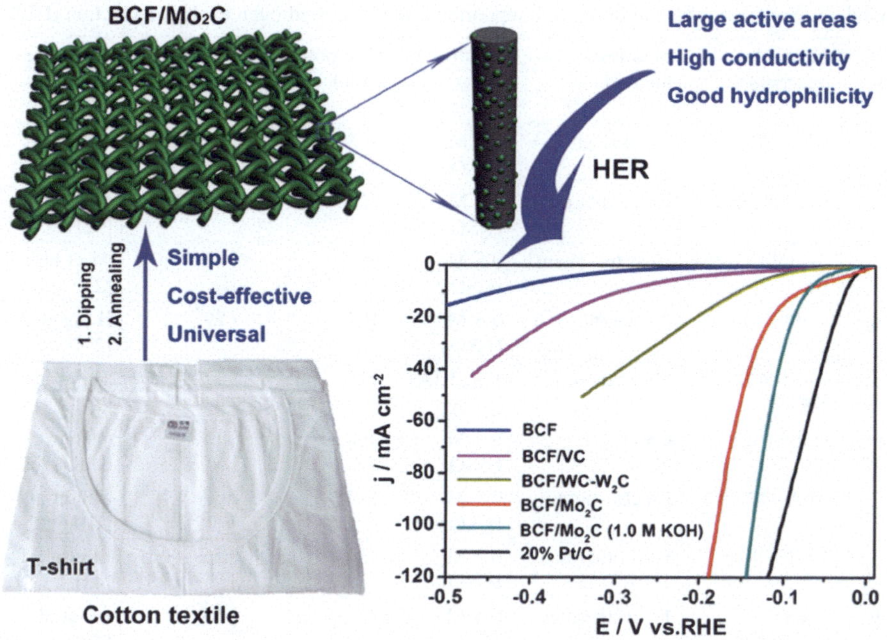

Fig. 4 Illustration showing HER activity of self supported electrode synthesized through fiber cloth decorated with molybdenum carbide (BCF/Mo_2C) (Taken permission from [86])

4 Oxygen Evolution Reaction (OER)

Energy consumption with good environmental balance is the major concern of the current period. Increase in the energy demand and depletion in fossil fuels has turn out to be the major headache for the scientist for development of fossil fuels substitute, which must be environmentally friendly and have infinite reserves [8]. The several energy systems found till date as an alternative of fossil fuels like metal air batteries, water splitting, and fuel cells. In the mentioned energy system, one thing is common that is water oxidation reaction or oxygen evolution reaction. OER is one of the half-cell reactions among the two reaction in water splitting, which occurs at the anode. OER can be coupled with both oxygen reduction reaction (ORR) and that of hydrogen evolution reaction (HER), which occurs at the cathode. OER is very much sluggish reaction for the production of oxygen gas with large number of proton and electron coupled pathways [66]. In the acidic and neutral medium, the reaction process proceeds through oxidation of two water molecules to proton and oxygen following the pathway:

$$\text{Acidic medium: } 2H_2O \leftrightarrow 4H^+ + O_2 + 4e^-$$

Table 1 Summary of some of the self-supported catalyst for hydrogen evolution reaction (HER)

S. no.	Electrode materials	Synthesis method	Electrolyte	Overpotential (mV)	Tafel slope (mV/dec)	References
1.	WC_x@CC	Carburization	0.5 M H_2SO_4	126	79	Wang et al. [77]
2.	MoP NTs/Mo	Electroplating	0.5 M H_2SO_4	224	106.3	Yu et al. [92]
3.	3D OM-FeP/CC	Heat treatment	0.5 M H_2SO_4	68	42	Ai et al. [2]
4.	CoP/S	Annealing	0.5 M H_2SO_4	107	29	Hong et al. [27]
5.	ZnNi-P NSs/NF	Hydrothermal	1 M KOH	175	129	Thi et al. [71]
6.	CuNPs/CNFs	CVD	0.5 M H_2SO_4	200	152	Wang et al. [76]
7.	MoS_2@NHCF	Electrospinning	0.5 M H_2SO_4	112	67	Zhu et al. [104]
8.	Co-Ni-P/NF	Hydrothermal	0.5 M H_2SO_4	134	56	Li et al. [43]
9.	CoP/NF nanoprickles	Hydrothermal	0.5 M H_2SO_4	71	63	Ge et al. [20]
10.	CuS nanowire	Liquid-solid reactions	0.2 M PBS	97	120	Ji et al. [32]
11.	Ni_3S_2@NPC	Galvanic replacement reaction	0.5 M H_2SO_4	91.6	63.5	Yang et al. [89]
12.	$NiCo_2S_4$/Ni_3S_2/Ni Foam	Thermal sulfurization	1 M KOH	119	105.2	Liu et al. [47]

WC_x@CC = WCx nanoparticles onto carbon cloth, 3D-OM-FeP/CC = Three-dimensionally ordered macroporous iron phosphide on carbon cloth, CoP/S = Cobalt phosphosulfate, ZnNi-P NSs/NF = Zinc-nickel phosphide nanosheets on nickel foam, CVD = Chemical vapour deposition, CuNPs/CNFs = Cu nanoparticles on carbon nanofiber, MoS_2@NHCF = MoS_2 on nitrogen doped hollow carbon fiber, Co-Ni-P/NF = Co-Ni-P on a nickel foam, CoP/NF = Cobalt phosphide on Ni foam, Ni_3S_2@NPC = Ni_3S_2 films on a nanoporous copper substrate

whereas, in the alkaline medium four hydroxyl ion are converted into water and oxygen gas followed by the pathway:

$$\text{Alkaline medium: } 4OH^- \leftrightarrow 2H_2O + O_2 + 4e^-$$

A potential difference of 1.23 V is required, which is created by the external current sources taking reversible hydrogen electrode as the reference electrode. OER reaction occurs through multistep reaction process with single electron transfer at each step, which causes accumulation of energy resulting in high overpotential value. The OER reaction occurs through formation of *OH from the adsorbed hydroxyl radical or water molecule in the active sites of the electrode, conversion of *OH to *O and followed by formation of *OOH. The *OOH radical at last leads to the formation of O_2 molecule [70]. The pathway of OER with the reaction intermediates and the active sites of the catalyst is designated as * are as follows:

Acidic medium:

$$H_2O + * \rightarrow H^+ + e^- + {*OH}$$

$$*OH \rightarrow {*O} + H^+ + e^-$$

$$*O + H_2O \rightarrow {*OOH} + H^+ + e^-$$

$$*OOH \rightarrow O_2 + H^+ + e^- + *$$

Basic medium:

$$OH^- + * \rightarrow {*OH} + e^-$$

$$*OH + OH^- \rightarrow H_2O + {*O} + e^-$$

$$OH^- + {*O} \rightarrow {*OOH} + e^-$$

$$*OOH + OH^- \rightarrow * + O_2 + e^-$$

As OER is a very much slow reaction with lots of intermediates like *O, *OOH, *OH, a good catalyst is needed with large number of active sites, large surface area, fast charge transfer ratio and with low overpotential value. Iridium (Ir) and ruthenium (Ru) is considered as the best OER electrocatalyst in acidic medium in the half-cell anodic reaction of water splitting. But there are several disappointing factors regarding this noble metal electrocatalyst like their low abundancy and high cost of synthesis which rule out their commercial use [69].

In recent years, efficient substitution of these catalysts has been explored and transition metal catalyst or some metal free catalyst, which can be used as flat surface electrode bonded with indium doped tin oxide (ITO) substrate, Cu/Ti foil and glassy carbon (GC)or free-standing catalysts on 3D substrate like carbon cloth, carbon paper, metal foam, graphene foam came up as suitable substitute. Flat surface electrodes have certain demerits like current loss, use of binders, single way diffusion pathway, catalyst powder agglomeration, while, 3D self-supported electrode have several advantages like they have hassle free, binder free, direct fabrication method, numerous pathways of electrolyte diffusion, and no powder agglomeration.

For example, Xia et al. have synthesized porous core shell hierarchical structure of cobalt oxide nanowires, through electrochemical deposition technique [84]. The prepared nanoparticle showed small overpotential value of 300 mV at 100 mA/cm^2

and a Tafel slope of 40.3 mV dec^{-1}. Guo et al. have synthesized nanowire structure of CoMnFe layered triple hydroxide (LTH) grown on Ni foam as the substrate through hydrothermal technique with controlled incorporation of iron. The prepared self-supported electrode shows overpotential value of 226 mV at 100 mA/cm^2 and a Tafel slope of 55 mV dec^{-1} having superior durability of 100 h [22]. Yu et al. have synthesized hierarchical nanoprism of Ni-Fe layered double hydroxide, which is hollow in nature and it was synthesized through self-sacrificing template method. The prepared electrocatalyst showed overpotential value of 280 mV and a Tafel slope of 49.4 mV dec^{-1} [93]. Choi et al. have prepared nanosheet arrays of $Cu_{0.81}Co_{2.19}O_4$, which is supported on Ni foam. The honeycomb nanocatalyst was prepared through electrochemical deposition of copper hydroxide and Cobalt on Ni foam via thermal oxidation. The nanosheet array shows an extremely low overpotential value of 290 mV in 1 M KOH and a Tafel slope of 69 mV dec^{-1} for OER [11]. Lu et al. have synthesized nickel-iron composite of mesoporous nanosheet like structure, which is grown in nickel foam with macro-porous pores. This three-dimensional self-supported (3D) amorphous mesoporous electrode showed best OER activity and are synthesized through electrodeposition technique using nickel and iron hydroxide as the precursors. The electrode shows an overpotential value of 270 mV in 80 mA/cm^2 and Tafel slope of 28 mV dec^{-1}, even lower than standard RuO_2 and IrO_2 electrocatalyst [51]. Xiong et al. have also synthesized a 3D oxygen evolution with superior activity having nanowire array like structure with chemical composition MnO_2–CoP_3 and is grown on Ti mesh as hard substrate. This MnO_2–CoP_3/Ti nanowire array electrocatalyst shows good OER stability in basic medium with overpotential value of 288 mV and Tafel slope of 28 mV dec^{-1} [87]. Jia et al. synthesized core-shell NiO@$NiMoO_4$ nanosheets which is supported with nickel foam and the nanomaterial hybrid was synthesized through simple hydrothermal technique. The prepared electrode shows better OER performances which is shown in Fig. 5 with small overpotential value of 280 mV and Tafel slope of 32 mV dec^{-1} [33]. Godínez-Salomón et al. developed iridium-nickel oxide 2D nanoframe through thermal treatment followed by chemical etching [21]. The film

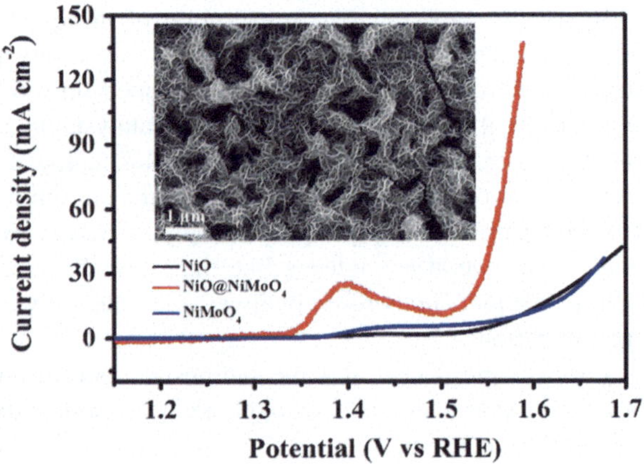

Fig. 5 Diagrammatic illustration showing oxygen evolution activity of nickel foam supported NiO@$NiMoO_4$ nanosheets (Taken permission from [33])

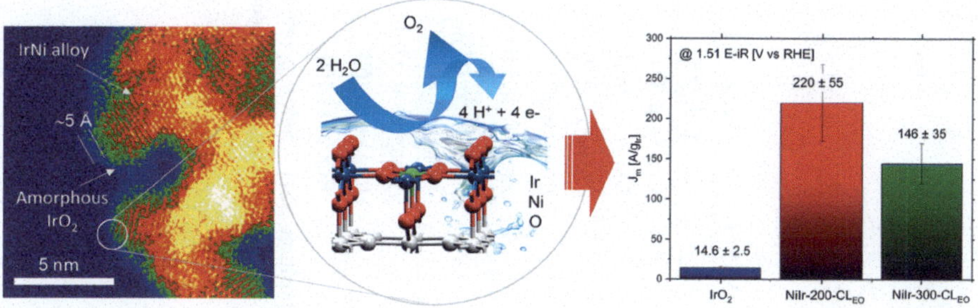

Fig. 6 Diagrammatic arrangement showing the OER activity of iridium-nickel oxide 2D nanoframe (Taken permission from [21])

produced through low thermal reduction temperature of 200 °C showed higher performances towards oxygen evolution reaction which is shown in Fig. 6 having higher current density and high surface area. Table 2 showing some of the self-supported binder free electrode for oxygen evolution reaction (OER).

5 Oxygen Reduction Reaction (ORR)

Fuel cell just like that of batteries can transform chemical energy of the fuels into electrical energies but like batteries they do not require recharging until the raw material or oxidant get finished. Among various fuel cells proton exchange membrane fuel cell (PEMFC) shown in Fig. 7 is actively used in portable electronic devices, automobiles etc. because of its high-power density, low temperature of working, simplicity and quick start up [90]. The main reaction taking place at the cathode is the oxygen reduction reaction, where reduction of oxygen takes place with the help of electrons and protons, which come from oxidation of hydrogen taking place at the anode shown in Fig. 8. The oxygen reduction can follow two pathways one is 2 electron pathways and another one is 4 electron pathways. Because of the presence of large number of intermediates with series of steps make ORR a sluggish kinetic reaction [62].

When the oxygen reduction reaction proceeds through 4 electron pathways, water produced as product and the mechanism is as follows:

$$\text{Acidic medium: } O_2 + 4e^- + 4H^+ \rightarrow 2H_2O$$

$$\text{Basic medium: } O_2 + 4e^- + 2H_2O \rightarrow 4OH^-$$

When ORR proceeds through 2 electron pathways, H_2O_2 is produced as the intermediate and byproduct. Reaction can be described as follows:

Table 2 Summary of some of the self-supported catalyst for oxygen evolution reaction (OER)

S. no.	Electrode materials	Synthesis method	Electrolyte	Overpotential (mV)	Tafel slope (mV/dec)	References
1	$CoSe_2$@VG/CC	PCVD	1 M KOH	418	82	Xia et al. [85]
2	Mo-Ni-O@CC	Thermal reaction	1 M KOH	320	63	Cao et al. [7]
3	$Fe_{4.4}Ni_{17.6}Se_{16}$/NiSe	Thermal selenization	1 M KOH	242 (at 100 mA/cm^2)	61.3	Ma et al. [54]
4	$Co(OH)_2$/Co	Painting-alloying-dealloying	1 M KOH	332	68.3	Zhang et al. [97]
5	$Co(OH)_2$/SSG-FGS	Microwave-polyol process	0.1 M KOH	140	147	Chen et al. [9]
6	FeCoNi-P/NFs	Hydrothermal	1 M KOH	266 (at 50 mA/cm^2)	61.2	Zhou et al. [100]
7	CoOSeP@Co foil	Phosphorization and selenylation	1 M KOH	347	74	Jiang et al. [34]
8	$MnCo_2S_4$/SS	Electrodeposition	1 M KOH	290	67	Jadhav et al. [31]
9	$Ni(OH)_2$/graphene film	Solvothermal	1 M KOH	250	66	Wang et al. [78]
10	Ni_3S_2/NF	Solvothermal	1 M KOH	222	60	Ren et al. [60]
11	NiSe/NF	Solvothermal	1 M KOH	271	80	Ren et al. [60]

$CoSe_2$@VG/CC = $CoSe_2$@vertically-oriented graphene (VG) on carbon cloth, PCVD = Plasma-enhanced chemical vapor deposition, SSG = Self supported graphene, FGS = Flexible graphite sheet, FeCoNi-P/NFs = Iron substituted cobalt-nickel phosphides grown on Ni foam, SS = Stainless steel, NF = Nickel foam

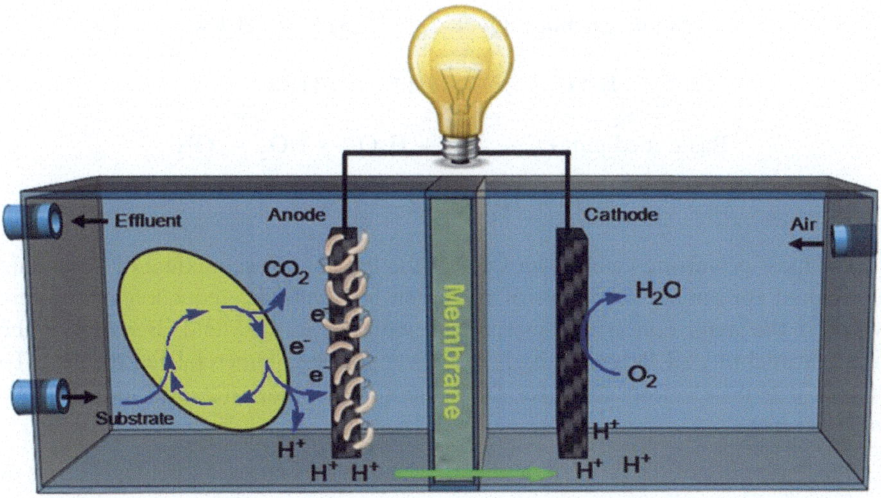

Fig. 7 Illustration showing the arrangement of proton exchange membrane fuel cell (PEMFC) for ORR activity (Taken permission from [62])

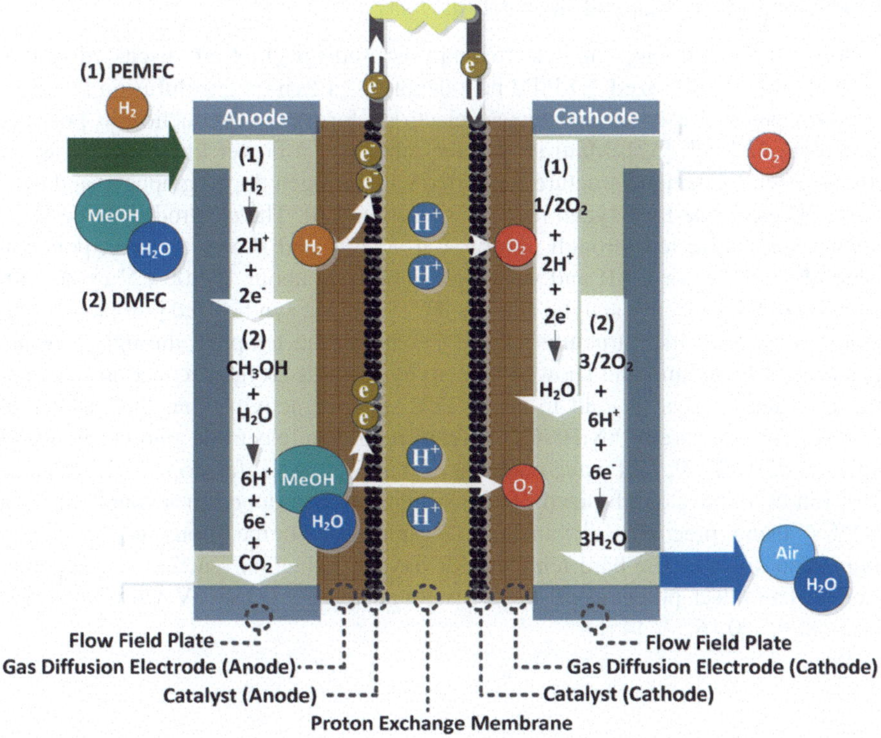

Fig. 8 Electrochemical cell showing the reaction pathways for oxygen reduction reaction (ORR) (Taken permission from [90])

$$\text{Acidic medium: } O_2 + 2e^- + 2H^+ \rightarrow H_2O_2$$

$$H_2O_2 + 2e^- + 2H^+ \rightarrow 2H_2O$$

$$\text{Basic medium: } O_2 + 2e^- + H_2O \rightarrow HO_2^- + OH^-$$

$$HO_2^- + H_2O + 2e^- \rightarrow 3OH^-$$

The most promising catalyst for ORR is the Pt/C but large amount of Pt loading is required for anodic oxidation of hydrogen and cathodic reduction of oxygen, which is the main concern of researchers working in this field. Because of the scarce availability of Pt and its high cost it cannot be commercialized till date. To prepare a good electrocatalyst for ORR there are certain strategies that must be followed like:

1. Weak binding of the oxygen on the metal catalyst is more favorable, because of fast reaction rate;
2. Increase in the pore sizes, which mainly act as the active sites with increased sites activity observed mainly in the self-supported 3D electrodes;
3. Doping of different heterogeneous compound to increase the catalytic activity;
4. Tailoring of the shape of the catalyst.

Some of the self-supported electrocatalysts used for ORR are discussed below. Lu et al. have synthesized 3D PtPd nanodendrites grown on gas diffusion layers of carbon paper and used as an efficient electrode for oxygen reduction in polymer electrolyte fuel cell [52]. Maiti et al. have fabricated a flower like morphology of iron tungsten oxide nanostructure supported with nitrogen doped graphene and used it as a 3D electrode for oxygen reduction reaction [56]. The electrode, 3D Fe-WO_3 NF/NG was fabricated through pyrolysis at 900 °C and shows an onset potential value of −0.98 V for ORR and very high stability to about 15,000 CV cycle with retention of 95% electrocatalytic activity. Fu et al. have synthesized porous 3D PtNi hollow nanochain like structure using Ni foam as the template through galvanic replacement technique and showed its activity towards oxygen reduction reaction. The prepared catalyst is said to show 77% superior activity than Pt/C in 0.1 M $HClO_4$. The prepared PtNiHNCs oxygen reduction electrode shows an onset potential of 0.933 V, half wave potential of 0.856 V and Tafel slope of 27 mV/dec [19]. Lin et al. have synthesized 3D consistent nitrogen doped mesoporous carbon (NMCs) using precursor dicyanamide for nitrogen source, flour as the carbon source and silica as the hard template for oxygen reduction reaction and supercapacitor. The onset potential of NMCs is found to be −0.1886 V with half wave potential of −0.162 V [45].

6 CO_2 Reduction Reaction

The main source of CO_2 is the emission from fossil fuels and as the byproduct of many chemical reactions. Due to increase in CO_2 concentration in the environment, global warming as well as environmental pollution is increasing day by day. The main solution is decreasing consumption of fossil fuels through use of renewable energy sources, nuclear powers and consumption and utilization of CO_2 through its conversion to different chemicals, which can be used as fuels. CO_2 can be converted into formic acid, ethylene, carbon monoxide, methanol, ethanol through various techniques like electrochemical, photochemical, biochemical, radiochemical or thermochemical etc. [36]. Electrochemical conversion of CO_2 is done at ambient temperature and pressure with water as the co-electrolyte using renewable energy sources shown in Fig. 9. Various products obtained by the reduction and are as follows [50]:

$$CO_2 + 2e^- + 2H^+ \rightarrow HCOOH$$

$$CO_2 + 2e^- + 2H^+ \rightarrow CO + H_2O$$

$$CO_2 + 4e^- + 4H^+ \rightarrow HCHO + H_2O$$

$$CO_2 + 6e^- + 6H^+ \rightarrow CH_3OH + H_2O$$

$$CO_2 + 8e^- + 8H^+ \rightarrow CH_4 + 2H_2O$$

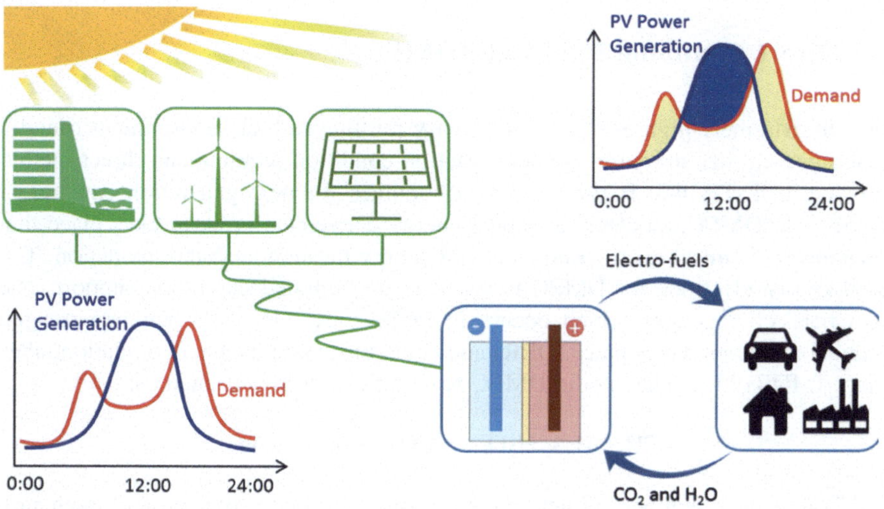

Fig. 9 Illustration showing the Electrochemical conversion of CO_2 and generation of electrofuels (Taken permission from [52])

CO_2 reduction have certain key challenges like high selectivity of the electrode, high overpotential value, multistep reduction process, life time of the catalyst, and high percentage of Faradaic efficiency. Due to these challenges, efficient electrocatalyst is needed for CO_2 reduction.

Some of the self-supported electrocatalyst for CO_2 reduction are reported in the literature and discussed as follows. Choi et al. have synthesized iron porphyrin and made hybrid with graphene hydrogel to prepare a 3D electrode for CO_2 reduction to CO [10]. The prepared FePGH electrode is highly porous in nature, which helps in superior CO_2 reduction with 96.2% Faradaic efficiency. The electrode shows a small overpotential value of 280 mV and 20-h constant electrolysis with great stability. Kas et al. have synthesized 3D compact hollow porous copper electrode, which have three phase boundaries for various gaseous and liquid reaction. The prepared electrode shows CO_2 reduction to CO with 85% faradaic efficiency and small overpotential value of 200 mV [37]. Wu et al. have fabricated 3D graphene foam electrode, which is incorporated with large number of nitrogen defects for reduction of CO_2 to CO. The prepared electrode shows good selectivity with 85% faradaic efficiency and considerable onset overpotential of −0.19 V and having stability of 5 h [81]. He et al. have synthesized palladium incorporated 3D graphene electrode (Pd/3D-RGO) through simple hydrothermal technique for reduction of CO_2 to formate. The synthesized CO_2 reduction electrode have interconnected pores with excellent surface performances showing good faradaic efficiency of 85.3% and onset potential of −0.70 V [25]. Khatavkar and his group members have synthesized self-supported solder alloy electrodes for conversion of CO_2 to formate. The prepared alloy electrode shows 88% faradaic efficiency and great stability for about 19 h [38].

7 Direct Methanol Fuel Cell (DMFC)

Due to extremely high energy density, easy refilling of fuel, environment friendly emission, simpler structure, ambient reaction condition requirement, direct methanol fuel cell (DMFC) is used as energy sources for next generation convenient devices. In DMFC, an electrochemical energy conversion process takes place that transforms chemical energy into electrical energy through methanol oxidation. The well-known electrode for DMFC is Pt electrode saturated on carbon support. The two half-cell reactions which occurs in DMFC are: (1) at the cathode oxygen reduction reaction takes place (ORR) and (2) at the anode methanol oxidation takes place (MOR) [49]. The anodic DMFC reaction can be represented as:

$$CH_3OH + H_2O \rightarrow CO_2 + 6H^+ + 6e^-$$

There is large number of intermediates formed during the oxidation of methanol, as one of the well-known is intermediate is carbon monoxide. The intermediates get adsorbed on the surface of the electrocatalyst and causes poisoning of the electrocatalyst, so to choose electrocatalyst for DMFC great care and knowledge is required to avoid such type of circumstances [5]. Some of the 3D self-supported electrocatalyst for DMFC are reported in the literature and are included here also. For example, Zhao et al. have synthesized 3D nitrogen doped graphene aerogel, which is uniformly scattered with Pt-Ru nanoparticles to improve the activity of DMFC [99]. The prepared N-GA/PtRu showed good stability towards methanol oxidation reaction with peak current density of about 426 mA cm^{-2} and power density of about 93 mW cm^{-2}. Su et al. have synthesized macroporous 3D graphitic structure supported with Pt for a high-performance methanol oxidation on DMFC. The electrode was synthesized through chemical vapour deposition using silica oval as the template [65]. Zhou and his coworkers have synthesized 3D self-supported graphene/carbon nanotube aerogels (GR-CNTs) with the help of freeze-drying method using poly vinyl alcohol (PVA) as the binder. The electrode has multiple application both in supercapacitor as well as direct methanol fuel cell (DMFC). For DMFC the electrode showed maximum current density of 550 mA mg^{-1} [102].

8 Future Aspects and Conclusion

This chapter is mainly based on the discussion of related to self-supported electrocatalysts, prepared by using different types of templates like carbon fibre paper, metal foam, metal and metal films. These types of electrodes are used in various types of electrocatalytic reaction as an electrocatalyst like ORR, OER, HER, MOR, CO_2 reduction etc. Self-supported catalysts are gaining popularity in the field of energy conversion and storage but still needs some major modification or upgradation based on some important criteria, before choosing the self-supported catalyst for any application like:

1. Choose reliable materials, which have high performances electrocatalytic activity that must be studied through both theoretically and electrochemically before incorporation;
2. Development of more hierarchical structure with better composition, complex geometry, large number of active sites, tailoring nano morphology;
3. Mechanical and structural stability study of the electrode for long term and cyclic stability of the electrode;
4. Better information about the interaction of the nanoparticle with the solid substrate through controlled growth with new production methods like 3D printing, inject printing etc.;
5. Spacious electrochemical device for better performances of this self-supported electrode with power source protection with full integrity.

To increase more functional ability of the self-supported electrode, additional features like optical transparency, stretching ability must be increased with optimum knowledge. Due to the unique properties and high performances of the self-supported electrode as the electrocatalyst they are wide possibility to design or fabricate the new generation electrodes, which must be venerated in the future also.

Author Declaration Ms. Karfa has given the major contribution in writing this book chapter along with drawing the figures and tables, taking the copyright permission etc.

References

1. Ahmadi, T.S., Wang, Z.L., Green, T.C., Henglein, A., El-Sayed, M.A.: Shape-controlled synthesis of colloidal platinum nanoparticles. Science **272**(5270), 1924–1925 (1996)
2. Ai, J., Jin, R., Liu, Z., Jiang, J., Yuan, S., Huang, G., Li, N., Li, X.: Three-dimensionally ordered macroporous FeP self-supported structure for high-efficiency hydrogen evolution reaction. Int. J. Hydrogen Energy **44**(12), 5854–5862 (2019)
3. Anantharaj, S., Ede, S.R., Karthick, K., Sankar, S.S., Sangeetha, K., Karthik, P.E., Kundu, S.: Precision and correctness in the evaluation of electrocatalytic water splitting: revisiting activity parameters with a critical assessment. Energy Environ. Sci. **11**(4), 744–771 (2018)
4. Appel, A.M., Helm, M.L.: Determining the overpotential for a molecular electrocatalyst. 630–633 (2014)
5. Arico, A.S., Srinivasan, S., Antonucci, V.: DMFCs: from fundamental aspects to technology development. Fuel cells **1**(2), 133–161 (2001)
6. Bandal, H.A., Jadhav, A.R., Tamboli, A.H., Kim, H.: Bimetallic iron cobalt oxide self-supported on Ni-Foam: an efficient bifunctional electrocatalyst for oxygen and hydrogen evolution reaction. Electrochim. Acta **249**, 253–262 (2017)
7. Cao, S., Wu, Z., Fu, B., Yu, H., Piao, L.: Polymerization pyrolysis derived self-supported Mo-Ni-*O* electrocatalyst for oxygen evolution. Catal. Today **330**, 246–251 (2019)
8. Capellán-Pérez, I., Mediavilla, M., de Castro, C., Carpintero, Ó., Miguel, L.J.: Fossil fuel depletion and socio-economic scenarios: an integrated approach. Energy **77**, 641–666 (2014)
9. Chen, L., Zang, J., Liu, X., Zhang, Y., Jia, S., Tian, P., Wang, Y.: A self-supporting graphene supported cobalt hydroxide for enhanced oxygen evolution catalysis. Electrochim. Acta **281**, 386–393 (2018)
10. Choi, J., Kim, J., Wagner, P., Gambhir, S., Jalili, R., Byun, S., Sayyar, S., Lee, Y.M., MacFarlane, D.R., Wallace, G.G., Officer, D.L.: Energy efficient electrochemical reduction of CO_2 to CO using a three-dimensional porphyrin/graphene hydrogel. Energy Environ. Sci. **12**(2), 747–755 (2019)
11. Choi, W.S., Jang, M.J., Park, Y.S., Lee, K.H., Lee, J.Y., Seo, M.H., Choi, S.M.: Three-dimensional honeycomb-like $Cu_{0.81}Co_{2.19}O_4$ nanosheet arrays supported by Ni foam and their high efficiency as oxygen evolution electrodes. ACS Appl. Mater. Interfaces **10**(45), 38663–38668 (2018)
12. Debe, M.K.: Electrocatalyst approaches and challenges for automotive fuel cells. Nature **486**(7401), 43 (2012)
13. Dincer, I.: Renewable energy and sustainable development: a crucial review. Renew. Sustain. Energy Rev. **4**(2), 157–175 (2000)
14. Dong, X.C., Xu, H., Wang, X.W., Huang, Y.X., Chan-Park, M.B., Zhang, H., Wang, L.H., Huang, W., Chen, P.: 3D graphene–cobalt oxide electrode for high-performance supercapacitor and enzymeless glucose detection. ACS Nano **6**(4), 3206–3213 (2012)
15. Edwards, P.P., Kuznetsov, V.L., David, W.I., Brandon, N.P.: Hydrogen and fuel cells: towards a sustainable energy future. Energy policy **36**(12), 4356–4362 (2008)
16. Eftekhari, A.: Electrocatalysts for hydrogen evolution reaction. Int. J. Hydrogen Energy **42**(16), 11053–11077 (2017)

17. Ellis, B.L., Knauth, P., Djenizian, T.: Three-dimensional self-supported metal oxides for advanced energy storage. Adv. Mater. **26**(21), 3368–3397 (2014)
18. Fletcher, S.: Tafel slopes from first principles. J. Solid State Electrochem. **13**(4), 537–549 (2009)
19. Fu, S., Zhu, C., Song, J., Engelhard, M.H., He, Y., Du, D., Wang, C., Lin, Y.: Three-dimensional PtNi hollow nanochains as an enhanced electrocatalyst for the oxygen reduction reaction. J. Mater. Chem. A **4**(22), 8755–8761 (2016)
20. Ge, K., Zeng, Y., Dong, G., Zhao, L., Wang, Z., Huang, M.: 3D self-standing grass-like cobalt phosphide vesicles-decorated nanocones grown on Ni-foam as an efficient electrocatalyst for hydrogen evolution reaction. Int. J. Hydrogen Energy (2019)
21. Godínez-Salomón, F., Albiter, L., Alia, S.M., Pivovar, B.S., Camacho-Forero, L.E., Balbuena, P.B., Mendoza-Cruz, R., Arellano-Jimenez, M.J., Rhodes, C.P.: Self-supported hydrous iridium–nickel oxide two-dimensional nanoframes for high activity oxygen evolution electrocatalysts. ACS Catal **8**(11), 10498–10520 (2018)
22. Guo, M., Zhou, L., Li, Y., Zheng, Q., Xie, F., Lin, D.: Unique nanosheet-nanowire structured CoMnFe layered triple hydroxide arrays as self-supporting electrodes for high-efficiency oxygen evolution reaction. J. Mater. Chem. A (2019)
23. Guo, Y., Guo, D., Ye, F., Wang, K., Shi, Z., Chen, X., Zhao, C.: Self-supported NiSe2 nanowire arrays on carbon fiber paper as efficient and stable electrode for hydrogen evolution reaction. ACS Sustain. Chem. Eng. **6**(9), 11884–11891 (2018)
24. Hansen, J., Sato, M., Ruedy, R., Lacis, A., Oinas, V.: Global warming in the twenty-first century: an alternative scenario. Proc. Natl. Acad. Sci. **97**(18), 9875–9880 (2000)
25. He, G., Tang, H., Wang, H., Bian, Z.: Highly selective and active Pd-In/three-dimensional graphene with special structure for electroreduction CO_2 to formate. Electroanalysis **30**(1), 84–93 (2018)
26. Hoel, M., Kverndokk, S.: Depletion of fossil fuels and the impacts of global warming. Resour. Energy Econ. **18**(2), 115–136 (1996)
27. Hong, W., Jian, C., Wang, G., He, X., Li, J., Cai, Q., Wen, Z., Liu, W.: Self-supported nanoporous cobalt phosphosulfate electrodes for efficient hydrogen evolution reaction. Appl. Catal. B Environ. (2019)
28. Höök, M., Tang, X.: Depletion of fossil fuels and anthropogenic climate change. A review. Energy Policy **52**, 797–809 (2013)
29. Herranz, J., Durst, J., Fabbri, E., Patru, A., Cheng, X., Permyakova, A.A., Schmidt, T.J.: Interfacial effects on the catalysis of the hydrogen evolution, oxygen evolution and CO2-reduction reactions for (co-) electrolyzer development. Nano Energy **29**, 4–28 (2016)
30. Iwasita, T.: Electrocatalysis of methanol oxidation. Electrochim. Acta **47**(22–23), 3663–3674 (2002)
31. Jadhav, H.S., Roy, A., Thorat, G.M., Chung, W.J., Seo, J.G.: Hierarchical free-standing networks of $MnCo_2S_4$ as efficient electrocatalyst for oxygen evolution reaction. J. Ind. Eng. Chem. **71**, 452–459 (2019)
32. Ji, L., Zhu, L., Wang, J., Chen, Z.: Self-supported CuS nanowire array: an efficient hydrogen-evolving electrode in neutral media. Electrochim. Acta **252**, 516–522 (2017)
33. Jia, D., Gao, H., Xing, L., Chen, X., Dong, W., Huang, X., Wang, G.: 3D self-supported porous NiO@$NiMoO_4$ core-shell nanosheets for highly efficient oxygen evolution reaction. Inorg. Chem. (2019). https://doi.org/10.1021/acs.inorgchem.9b00162
34. Jiang, Y.F., Yuan, C.Z., Zhou, X., Liu, Y.N., Zhao, Z.W., Zhao, S.J., Xu, A.W.: Selenium phosphorus co-doped cobalt oxide nanosheets anchored on Co foil: a self-supported and stable bifunctional electrode for efficient electrochemical water splitting. Electrochim. Acta **292**, 247–255 (2018)
35. Jin, M., Zhang, H., Xie, Z., Xia, Y.: Palladium nanocrystals enclosed by 100 and 111 facets in controlled proportions and their catalytic activities for formic acid oxidation. Energy Environ. Sci. **5**(4), 6352–6357 (2012)
36. Jones, J.P., Prakash, G.S., Olah, G.A.: Electrochemical CO_2 reduction: recent advances and current trends. Isr. J. Chem. **54**(10), 1451–1466 (2014)

37. Kas, R., Hummadi, K.K., Kortlever, R., De Wit, P., Milbrat, A., Luiten-Olieman, M.W., Benes, N.E., Koper, M.T., Mul, G.: Three-dimensional porous hollow fibre copper electrodes for efficient and high-rate electrochemical carbon dioxide reduction. Nat. Commun. **7**, 10748 (2016)
38. Khatavkar, S.N., Ukale, D.U., Haram, S.K.: Development of self-supported 3D microporous solder alloy electrodes for scalable CO_2 electroreduction to formate. New J. Chem. **43**, 6587–6596 (2019)
39. Kong, D., Cha, J.J., Wang, H., Lee, H.R., Cui, Y.: First-row transition metal dichalcogenide catalysts for hydrogen evolution reaction. Energy Environ. Sci. **6**(12), 3553–3558 (2013)
40. Ledezma-Yanez, I., Wallace, W.D.Z., Sebastián-Pascual, P., Climent, V., Feliu, J.M., Koper, M.T.: Interfacial water reorganization as a pH-dependent descriptor of the hydrogen evolution rate on platinum electrodes. Nat. Energy **2**(4), 17031 (2017)
41. Lee, H., Kim, Y.J., Lee, D.J., Song, J., Lee, Y.M., Kim, H.T., Park, J.K.: Directly grown Co_3O_4 nanowire arrays on Ni-foam: structural effects of carbon-free and binder-free cathodes for lithium–oxygen batteries. J. Mater. Chem. A **2**(30), 11891–11898 (2014)
42. Li, C., Tan, H., Lin, J., Luo, X., Wang, S., You, J., Kang, Y.M., Bando, Y., Yamauchi, Y., Kim, J.: Emerging Pt-based electrocatalysts with highly open nanoarchitectures for boosting oxygen reduction reaction. Nano Today (2018)
43. Li, W., Xiong, D., Gao, X., Song, W.G., Xia, F., Liu, L.: Self-supported Co-Ni-P ternary nanowire electrodes for highly efficient and stable electrocatalytic hydrogen evolution in acidic solution. Catal. Today **287**, 122–129 (2017)
44. Liang, H.P., Zhang, H.M., Hu, J.S., Guo, Y.G., Wan, L.J., Bai, C.L.: Pt hollow nanospheres: facile synthesis and enhanced electrocatalysts. Angew. Chem. Int. Ed. **43**(12), 1540–1543 (2004)
45. Lin, G., Ma, R., Zhou, Y., Hu, C., Yang, M., Liu, Q., Kaskel, S., Wang, J.: Three-dimensional interconnected nitrogen-doped mesoporous carbons as active electrode materials for application in electrocatalytic oxygen reduction and supercapacitors. J. Colloid Interface Sci. **527**, 230–240 (2018)
46. Lipkowski, J., Ross, P.N. (eds.): Electrocatalysis (vol. 3). Wiley (1998)
47. Liu, H., Ma, X., Rao, Y., Liu, Y., Liu, J., Wang, L., Wu, M.: Heteromorphic $NiCo_2S_4/Ni_3S_2$/Ni foam as a self-standing electrode for hydrogen evolution reaction in alkaline solution. ACS Appl. Mater. Interfaces **10**(13), 10890–10897 (2018)
48. Liu, J., Zhu, D., Zheng, Y., Vasileff, A., Qiao, S.Z.: Self-supported earth-abundant nanoarrays as efficient and robust electrocatalysts for energy-related reactions. ACS Catal. **8** (7), 6707–6732 (2018)
49. Liu, Z., Ling, X.Y., Su, X., Lee, J.Y.: Carbon-supported Pt and PtRu nanoparticles as catalysts for a direct methanol fuel cell. J. Phys. Chem. B **108**(24), 8234–8240 (2004)
50. Lu, Q., Jiao, F.: (b). Electrochemical CO_2 reduction: electrocatalyst, reaction mechanism, and process engineering. Nano Energy **29**, 439–456 (2016)
51. Lu, X., Zhao, C.: Electrodeposition of hierarchically structured three-dimensional nickel–iron electrodes for efficient oxygen evolution at high current densities. Nat. Commun. **6**, 6616 (2015)
52. Lu, Y., Du, S., Steinberger-Wilckens, R.: Three-dimensional catalyst electrodes based on PtPd nanodendrites for oxygen reduction reaction in PEFC applications. Appl. Catal. B **187**, 108–114 (2016)
53. Lv, W., Zhang, R., Gao, P., Lei, L.: Studies on the faradaic efficiency for electrochemical reduction of carbon dioxide to formate on tin electrode. J. Power Sources **253**, 276–281 (2014)
54. Ma, S., Yuan, H., Cai, L., Wang, X., Long, H., Chai, Y., Tsang, Y.H.: One step synthesis of Fe4. 4Ni17. 6Se16 coupled NiSe foam as self-supported, highly efficient and durable oxygen evolution electrode. Mater. Today Chem. **9**, 133–139 (2018)
55. Ma, T.Y., Dai, S., Qiao, S.Z.: Self-supported electrocatalysts for advanced energy conversion processes. Mater. Today **19**(5), 265–273 (2016)

56. Maiti, K., Balamurugan, J., Gautam, J., Kim, N.H., Lee, J.H.: Hierarchical flowerlike highly synergistic three-dimensional iron tungsten oxide nanostructure-anchored nitrogen-doped graphene as an efficient and durable electrocatalyst for oxygen reduction reaction. ACS Appl. Mater. Interfaces **10**(38), 32220–32232 (2018)
57. Murthy, A.P., Madhavan, J., Murugan, K.: Recent advances in hydrogen evolution reaction catalysts on carbon/carbon-based supports in acid media. J. Power Sources **398**, 9–26 (2018)
58. Neyerlin, K.C., Gu, W., Jorne, J., Gasteiger, H.A.: Study of the exchange current density for the hydrogen oxidation and evolution reactions. J. Electrochem. Soc. **154**(7), B631–B635 (2007)
59. Panwar, N.L., Kaushik, S.C., Kothari, S.: Role of renewable energy sources in environmental protection: a review. Renew. Sustain. Energy Rev. **15**(3), 1513–1524 (2011)
60. Ren, H., Huang, Z.H., Yang, Z., Tang, S., Kang, F., Lv, R.: Facile synthesis of free-standing nickel chalcogenide electrodes for overall water splitting. J. Energy Chem. **26**(6), 1217–1222 (2017)
61. Ren, X., Wu, D., Ge, R., Sun, X., Ma, H., Yan, T., Zhang, Y., Du, B., Wei, Q., Chen, L.: Self-supported CoMoS4 nanosheet array as an efficient catalyst for hydrogen evolution reaction at neutral pH. Nano Res. **11**(4), 2024–2033 (2018)
62. Sawant, S., Han, T., Cho, M.: Metal-free carbon-based materials: promising electrocatalysts for oxygen reduction reaction in microbial fuel cells. Int. J. Mol. Sci. **18**(1), 25 (2016)
63. Shao, M., Odell, J.H., Choi, S.I., Xia, Y.: Electrochemical surface area measurements of platinum-and palladium-based nanoparticles. Electrochem. Commun. **31**, 46–48 (2013)
64. Strmcnik, D., Lopes, P.P., Genorio, B., Stamenkovic, V.R., Markovic, N.M.: Design principles for hydrogen evolution reaction catalyst materials. Nano Energy **29**, 29–36 (2016)
65. Su, F., Zhao, X.S., Wang, Y., Zeng, J., Zhou, Z., Lee, J.Y.: Synthesis of graphitic ordered macroporous carbon with a three-dimensional interconnected pore structure for electrochemical applications. J. Phys. Chem. B **109**(43), 20200–20206 (2005)
66. Suen, N.T., Hung, S.F., Quan, Q., Zhang, N., Xu, Y.J., Chen, H.M.: Electrocatalysis for the oxygen evolution reaction: recent development and future perspectives. Chem. Soc. Rev. **46**(2), 337–365 (2017)
67. Sun, Y., Xia, Y.: Shape-controlled synthesis of gold and silver nanoparticles. Science **298**(5601), 2176–2179 (2002)
68. Tang, Y., Cheng, W.: Key parameters governing metallic nanoparticle electrocatalysis. Nanoscale **7**(39), 16151–16164 (2015)
69. Tao, H., Sun, Z.: Oxygen electrochemistry on two-dimensional nanosheets. Nanotechnol. Res. J. **9**(3), 361–388 (2016)
70. Tao, H., Gao, Y., Talreja, N., Guo, F., Texter, J., Yan, C., Sun, Z.: Two-dimensional nanosheets for electrocatalysis in energy generation and conversion. J. Mater. Chem. A **5** (16), 7257–7284 (2017)
71. Thi, M.L.N., Tran, T.H., Anh, P.H., Nhac-Vu, H.T., Bui, Q.B.: Hierarchical zinc-nickel phosphides nanosheets on 3D nickel foam as self-support electrocatalysts for hydrogen evolution reaction. Polyhedron (2019)
72. Tolmachev, Y.V., Petrii, O.A.: Pt–Ru electrocatalysts for fuel cells: developments in the last decade. J. Solid State Electrochem. **21**(3), 613–639 (2017)
73. Tong, S., Xu, Y., Zhang, Z., Song, W.: Dendritic bimetallic nanostructures supported on self-assembled titanate films for sensor application. J. Phys. Chem. C **114**(49), 20925–20931 (2010)
74. Trasatti, S.: Electrocatalysis: understanding the success of DSA®. Electrochim. Acta **45**(15–16), 2377–2385 (2000)
75. Vij, V., Sultan, S., Harzandi, A.M., Meena, A., Tiwari, J.N., Lee, W.G., Yoon, T., Kim, K. S.: Nickel-based electrocatalysts for energy-related applications: oxygen reduction, oxygen evolution, and hydrogen evolution reactions. ACS Catal. **7**(10), 7196–7225 (2017)
76. Wang, J., Zhu, H., Chen, J., Zhang, B., Zhang, M., Wang, L., Du, M.: Small and well-dispersed Cu nanoparticles on carbon nanofibers: self-supported electrode materials for efficient hydrogen evolution reaction. Int. J. Hydrogen Energy **41**(40), 18044–18049 (2016)

77. Wang, K., Chen, H., Hua, Y., Tong, Y., Wang, Y., Song, S.: Layer-stacking porous WCx nanoparticles on carbon cloth as self-supported integrated electrode for hydrogen evolution reaction. Mater. Today Energy **10**, 343–351 (2018)
78. Wang, L., Zhang, J., Jiang, W., Zhao, H., Liu, H.: Free-standing, flexible β-Ni (OH) 2/ electrochemically-exfoliated graphene film electrode for efficient oxygen evolution. Appl. Surf. Sci. **433**, 88–93 (2018)
79. Wang, M., Ye, C., Xu, M., Bao, S.: MoP nanoparticles with a P-rich outermost atomic layer embedded in N-doped porous carbon nanofibers: self-supported electrodes for efficient hydrogen generation. Nano Res. **11**(9), 4728–4734 (2018)
80. Wu, B., Zheng, N.: Surface and interface control of noble metal nanocrystals for catalytic and electrocatalytic applications. Nano Today **8**(2), 168–197 (2013)
81. Wu, J., Liu, M., Sharma, P.P., Yadav, R.M., Ma, L., Yang, Y., Zou, X., Zhou, X.D., Vajtai, R., Yakobson, B.I., Lou, J.: Incorporation of nitrogen defects for efficient reduction of CO_2 via two-electron pathway on three-dimensional graphene foam. Nano Lett. **16**(1), 466–470 (2015)
82. Xia, B.Y., Ng, W.T., Wu, H.B., Wang, X., Lou, X.W.: Self-supported interconnected Pt nanoassemblies as highly stable electrocatalysts for low-temperature fuel cells. Angew. Chem. Int. Ed. **51**(29), 7213–7216 (2012)
83. Xia, C., Jiang, Q., Zhao, C., Hedhili, M.N., Alshareef, H.N.: Selenide-based electrocatalysts and scaffolds for water oxidation applications. Adv. Mater. **28**(1), 77–85 (2016)
84. Xia, H., Huang, Z., Lv, C., Zhang, C.: A self-supported porous hierarchical core-shell nanostructure of cobalt oxide for efficient oxygen evolution reaction. ACS Catal. **7**(12), 8205–8213 (2017)
85. Xia, Z., Sun, H., He, X., Sun, Z., Lu, C., Li, J., Peng, Y., Kostecki, R., Dou, S., Sun, J., Liu, Z.: In situ construction of CoSe2@ vertical-oriented graphene arrays as self-supporting electrodes for sodium-ion capacitors and electrocatalytic oxygen evolution. Nano Energy (2019)
86. Xiao, J., Zhang, Y., Zhang, Z., Lv, Q., Jing, F., Chi, K., Wang, S.: Self-supported biocarbon-fiber electrode decorated with molybdenum carbide nanoparticles for highly active hydrogen-evolution reaction. ACS Appl. Mater. Interfaces. **9**(27), 22604–22611 (2017)
87. Xiong, X., Ji, Y., Xie, M., You, C., Yang, L., Liu, Z., Asiri, A.M., Sun, X.: MnO_2-CoP_3 nanowires array: an efficient electrocatalyst for alkaline oxygen evolution reaction with enhanced activity. Electrochem. Commun. **86**, 161–165 (2018)
88. Xu, Z., Liu, Y., Zhao, W., Li, B., Zhou, X., Shen, H.: Assembling mesoporous $ZnxCO_3$-xO_4 fibers with interconnected nanocrystals via a topotactic conversion route for enhanced performance Lithium-ion batteries. Electrochim. Acta **190**, 894–902 (2016)
89. Yang, C., Gao, M.Y., Zhang, Q.B., Zeng, J.R., Li, X.T., Abbott, A.P.: In-situ activation of self-supported 3D hierarchically porous Ni3S2 films grown on nanoporous copper as excellent pH-universal electrocatalysts for hydrogen evolution reaction. Nano Energy **36**, 85–94 (2017)
90. Ye, Y.S., Rick, J., Hwang, B.J.: Water soluble polymers as proton exchange membranes for fuel cells. Polymers **4**(2), 913–963 (2012)
91. You, H., Yang, S., Ding, B., Yang, H.: Synthesis of colloidal metal and metal alloy nanoparticles for electrochemical energy applications. Chem. Soc. Rev. **42**(7), 2880–2904 (2013)
92. Yu, H., Cao, S., Fu, B., Wu, Z., Liu, J., Piao, L.: Self-supported nanotubular MoP electrode for highly efficient hydrogen evolution via water splitting. Catal. Commun. (2019)
93. Yu, L., Yang, J.F., Guan, B.Y., Lu, Y., Lou, X.W.: Hierarchical hollow nanoprisms based on ultrathin Ni-Fe layered double hydroxide nanosheets with enhanced electrocatalytic activity towards oxygen evolution. Angew. Chem. Int. Ed. **57**(1), 172–176 (2018)
94. Yuan, C., Wu, H.B., Xie, Y., Lou, X.W.: Mixed transition-metal oxides: design, synthesis, and energy-related applications. Angew. Chem. Int. Ed. **53**(6), 1488–1504 (2014)

95. Zeng, M., Li, Y.: Recent advances in heterogeneous electrocatalysts for the hydrogen evolution reaction. J. Mater. Chem. A **3**(29), 14942–14962 (2015)
96. Zhang, C., Pu, Z., Amiinu, I.S., Zhao, Y., Zhu, J., Tang, Y., Mu, S.: CO_2P quantum dot embedded N, P dual-doped carbon self-supported electrodes with flexible and binder-free properties for efficient hydrogen evolution reactions. Nanoscale **10**(6), 2902–2907 (2018)
97. Zhang, J., Dong, C., Wang, Z., Zhang, C., Gao, H., Niu, J., Zhang, Z.: Flexible, self-supported hexagonal β-Co $(OH)_2$ nanosheet arrays as integrated electrode catalyzing oxygen evolution reaction. Electrochim. Acta **284**, 495–503 (2018)
98. Zhang, L., Chang, Q., Chen, H., Shao, M.: Recent advances in palladium-based electrocatalysts for fuel cell reactions and hydrogen evolution reaction. Nano Energy **29**, 198–219 (2016)
99. Zhao, S., Yin, H., Du, L., Yin, G., Tang, Z., Liu, S.: Three dimensional N-doped graphene/PtRu nanoparticle hybrids as high performance anode for direct methanol fuel cells. J. Mater. Chem. A **2**(11), 3719–3724 (2014)
100. Zhou, C., Mu, J., Qi, Y.F., Wang, Q., Zhao, X.J., Yang, E.C.: Iron-substituted Co-Ni phosphides immobilized on Ni foam as efficient self-supported 3D hierarchical electrocatalysts for oxygen evolution reaction. Int. J. Hydrogen Energy (2019)
101. Zhou, W., Jia, J., Lu, J., Yang, L., Hou, D., Li, G., Chen, S.: Recent developments of carbon-based electrocatalysts for hydrogen evolution reaction. Nano Energy **28**, 29–43 (2016)
102. Zhou, Y., Hu, X., Guo, S., Yu, C., Zhong, S., Liu, X.: Multi-functional graphene/carbon nanotube aerogels for its applications in supercapacitor and direct methanol fuel cell. Electrochim. Acta **264**, 12–19 (2018)
103. Zhu, W., Zhang, R., Qu, F., Asiri, A.M., Sun, X.: Design and application of foams for electrocatalysis. ChemCatChem **9**(10), 1721–1743 (2017)
104. Zhu, X., Mo, L., Wu, Y., Lai, F., Han, X., Ling, X.Y., Liu, T., Miao, Y.E.: Self-supported MoS_2@ NHCF fiber-in-tube composites with tunable voids for efficient hydrogen evolution reaction. Compos. Commun. **9**, 86–91 (2018)

Conductive Polymer Based Flexible Supercapacitor

S. Wazed Ali and Satyaranjan Bairagi

Abstract In the recent days the demand of portable, thin and flexible electronics such as roll-up display, touch screen, smart electronics and wearable sensors, are drawing a great interest in the daily life of human being because of advancement in materials and technology. In order to simplify this growing electronic demand, super capacitor is fascinating as a favourable energy storage devices. Supercapacitor has more power density, faster charge-discharge cycle and higher energy storage capacity with compare to Li-ion batteries. Recently, flexible super capacitor is the main focus in the flexible electronics due to its higher flexibility, high power density and high capacitance performance. Conducting polymers based supercapacitor is more promising candidate compare to other materials in terms of their flexibility, high redox active specific capacitance and essential elastic nature. In this chapter different conducting polymer (CPs) based super capacitor have been explained.

1 Introduction

The demand for portable and flexible electronic goods in our daily life has been increased tremendously in the recent past. For instance, different portable and flexible electronics are mobile, laptop, iPad, roll-up displays, touch screen, smart electronics, wearable electronics, different sensors and so on. Therefore, the need of a flexible, light weight, low cost and sustainable effective energy storage is one of the prime concerns in the modern life for increasing the working efficiency of the above said different portable and flexible electronic goods [24, 30]. The different energy storage systems such as capacitor, batteries, fuel cell and other energy

---

S. Wazed Ali (✉) · S. Bairagi
Department of Textile and Fibre Engineering, Indian Institute of Technology Delhi, Hauz Khas, New Delhi 110016, India
e-mail: wazed@textile.iitd.ac.in

S. Bairagi
e-mail: ranjansatyabairagi@gmail.com

Inamuddin et al. (eds.), *Self-standing Substrates*,
Engineering Materials, https://doi.org/10.1007/978-3-030-29522-6_7

storage devices can fulfil the need of energy storage for the electronic goods. Among them, capacitor is one of the lucrative solution in this regard. Capacitors are of basically two types, conventional capacitor and super capacitor. In the conventional capacitor, electrostatic charges are stored into the dielectric plates, whereas supercapacitor is comprising with electrodes material in the electrolyte solution separated by an insulating membrane [34, 37].

Among the different energy storage systems, supercapacitor has drawn a great interest to fulfil the required energy for the different portable and flexible electronic goods, which can be clearly observed from Fig. 1. It has both high power density, energy storage, faster charge/discharge capability along with longer life span with contrast to battery and other energy storage devices [16, 17, 21]. Moreover, supercapacitor has 10^5 times higher capacitance value than normal conventional capacitor. The supercapacitor can be classified into two sub groups, electric double layer capacitors (EDLCs) and pseudocapacitors or redox capacitors [39]. In EDLCs, the capacitance is used to be developed from the charge build-up at the interface between the electrolyte and electrodes, whereas for the pseudocapacitors the capacitance is occupied through the charge transformation between the electrolytes and electrodes material by simply redox faradaic reactions, which provides higher specific capacitance and power density [7, 13]. However, the specific capacitance value of the pseudo capacitors are 10–100 times more than that of the EDLCs because of the redox faradaic reaction which takes place in the electroactive materials of the pseudocapacitors.

Most recently, researchers are trying to develop flexible supercapacitors due to higher demand of flexible electronic goods for their easy process ability as compared to the conventional capacitors. Generally, three types of materials are employed for the capacitor's electrodes, namely, carbon-based, metal compounds

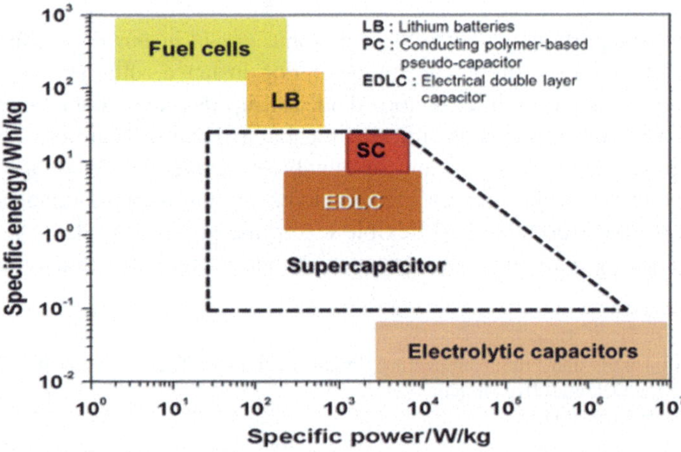

Fig. 1 Ragone plot of energy versus power density comparison of supercapacitor with the other energy storage systems (Adopted from [30])

(like oxide, sulphide, hydroxide and selenide) and conductive polymer-based materials. The materials using carbon such as carbon nanotube (CNT), graphene, carbon black, activated carbon and carbon fibre can be commonly utilised as electro-active devices as non-faradaic EDLCs supercapacitors. Whereas, metal and conductive polymer based electroactive materials are used in pseudocapacitors where redox reaction takes place. The carbon-based materials have limitation in supercapacitor application due to their lower specific capacitance value. However, materials which have carbon in them offers greater power density accompanying stability. The metal-based materials also have some limitations in supercapacitor applications because of their lower conductivity, higher cost and toxic characteristics. These problems with the carbon and metal-based electrode materials in supercapacitor can be solved by the conductive polymer-based materials such as polyaniline, polythiophene and polypyrrole [9, 11, 22, 25, 30, 48]. In this chapter, different conductive polymer based flexible supercapacitors and their fabrication methods have been discussed.

2 Different Components of a Supercapacitor

Generally, a supercapacitor is composed of electrolytes, electrodes, separator and current collector as shown in Fig. 2. Electrodes are the main part of a supercapacitor, which can generate overall cell voltage by providing enough potential difference between them. The function of a separator is mainly to isolate the two electrodes electrically with a transfer of charged ions only. Materials of the separator can be varied, which depends upon the type of electrolytes used in the supercapacitor. For instance, if electrolytes are organic based, polymer (PP) or paper based separator is used and for the aqueous electrolytes glass fibre or ceramic based separator is used. As mentioned above, electrolytes are further sub grouped as solid state electrolyte, organic electrolyte and aqueous electrolyte. Electrolytes help to generate voltage by their dissociation in the supercapacitor. The dissociation

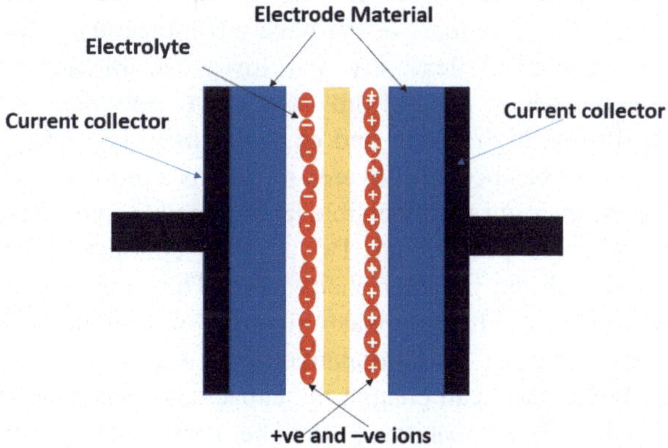

Fig. 2 Components of a supercapacitor (self-created)

voltage is higher for the organic based electrolytes (2.5 V) than aqueous electrolytes (1.23 V). Organic based electrolytes are more effective for developing a highly efficient supercapacitor. As name suggested, the role of current collector is to transport charge carriers from/to active electrode materials during charge-discharge cycle of the supercapacitor [28, 33]. Among the different components of a supercapacitor, electrodes material is one of the most important parts for controlling power density, energy density as well as life cycle of a supercapacitor.

3 Materials for Supercapacitor

As discussed in the earlier section, there are three different materials namely carbon based materials, metal oxide based materials and conductive polymer based materials which have been used in the recent past to develop electrode components in the supercapacitor. Among the different materials, conductive materials have more advantageous properties as an electrode material in the supercapacitor, such as higher conductivity, suitable flexibility, high power and energy density, rapid charge-discharge capability. Mainly carbon based materials like graphene, carbon black, activated carbon have been used in the EDLCs, whereas metal oxide and conductive polymer based materials have been used in the pseudocapacitor. Herein, different conductive polymer based supercapacitor has been discussed in details. The conductive polymers which are generally used in the supercapacitor are polyaniline (PANI), polypyrrole, polythiophene, and derivative of polythiophene [32]. The detailed studies on the above said conductive polymer based supercapacitors have been discussed in the below sub sections.

3.1 Conductive Polymer Based Flexible Supercapacitor

In the recent days researchers have tried to develop supercapacitor based on conductive polymers (CPs), in order to improve the performance of supercapacitor. The conductive polymers (CPs) have advantageous properties like higher charge density, enhanced flexibility and low cost. Because of these lucrative properties supercapacitor made from conductive polymers (CPs) also showed significant flexibility, high power and charge density [20, 26, 27]. Moreover, flexible supercapacitor has been discovered mostly as a promising electrochemical supercapacitor due to their unique reversible faradaic redox landscape, in the last two decades [27]. Conductive polymers (CPs) exhibits conductive nature through the conjugated bond with the polymer chain spine. They are synthesized either by chemical oxidation or electrochemical oxidation of the monomer with parallel insertion of dopant (e.g. Cl^-). The dopant content is less than 1 per polymer unit. The presence of this dopant can change the capacitance value of the conductive polymer based flexible supercapacitor [19]. The most usable conductive polymers (CPs) are

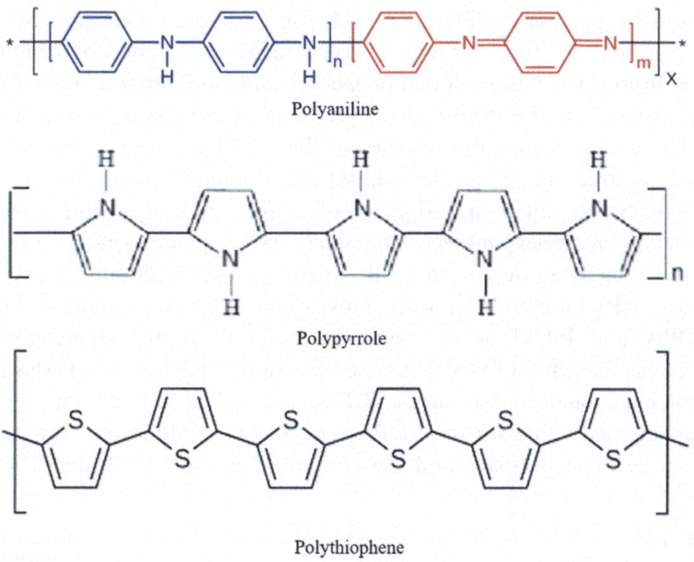

Fig. 3 Chemical structure of the polyaniline, polypyrrole and polythiophene conductive polymers (CPs) (self-created)

polyaniline (PANI), polypyrrole and polythiophene, which are already mentioned. The chemical structure of the conductive polymer is shown in Fig. 3. The specific capacitance and conductivity of the different conductive polymers are higher as compared to conventional polymers. The specific capacitance and conductivity of the conductive polymers (CPs) are shown in Table 1 [31].

3.1.1 Polyaniline Based Flexible Supercapacitor

Among the different conductive polymers, polyaniline (PANI) is one of the most promising candidates for the flexible supercapacitor applications. It has reasonable inherent flexibility, conductivity (shows in Table 1), variable oxidation states,

Table 1 The specific capacitance and conductivity of the different conductive polymers (CPs) (Data from [31])

Conductive polymer	Molecular weight (MW) (g/mol)	Potential change (V)	Theoretical specific capacitance (Fg^{-1})	Experimental specific capacitance (Fg^{-1})	Conductivity (S/cm^{-1})
Polyaniline	93	0.7	750	240	0.1–5
Polypyrrole	67	0.8	620	530	10–50
Polythiophene	84	0.8	485	–	300–400

higher specific capacitance value (400–500 F/g in acidic condition). Moreover, it has good conductivity (0.1–5 S/cm^{-1}, given in Table 1) and environmental stability. In addition, it has high doped capability and easy process ability [5].

The polyaniline can be synthesized by chemical and electrochemical oxidation methods. The electrochemical properties of the PANI is strongly depend upon the morphological, structural properties and type of dopants. Ghenaatian et al. [7] has studied on electrochemical investigation of self-doped polyaniline nanofibers as new electroactive material for electrochemical redox supercapacitor. The self-doped PANI (SDPA) has been deposited on the platinum (Pt) electrode by reverse pulse voltammetric (RPV) method. They have investigated the electrochemical properties of the SDPA and PANI as an electrode material in the supercapacitor. The capacitance was maximum for SDPA based electrode material than PANI due to the higher surface area and porosity in the SDPA material as shown in the Fig. 4a. The specific capacitance value for the SDPA was 480 F/g, calculated from the charge discharge cycle. The porosity and surface area for the SDPA and PANI based

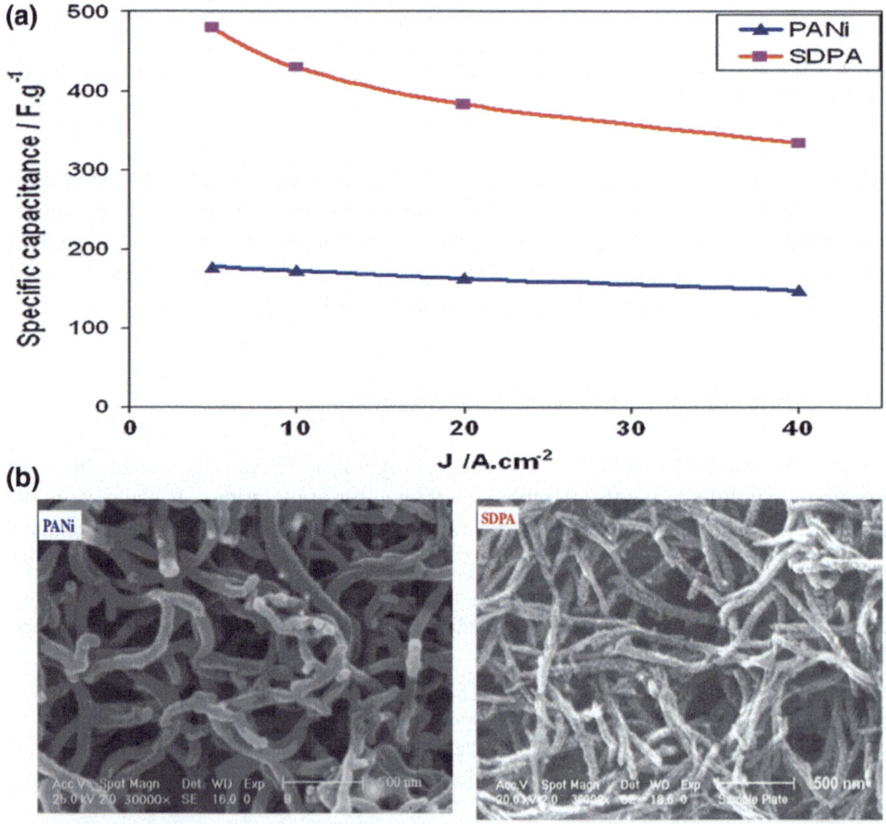

Fig. 4 **a** Specific capacitance of SDPA and PANI, and **b** SEM images of PANI and SDPA (Adopted from [7])

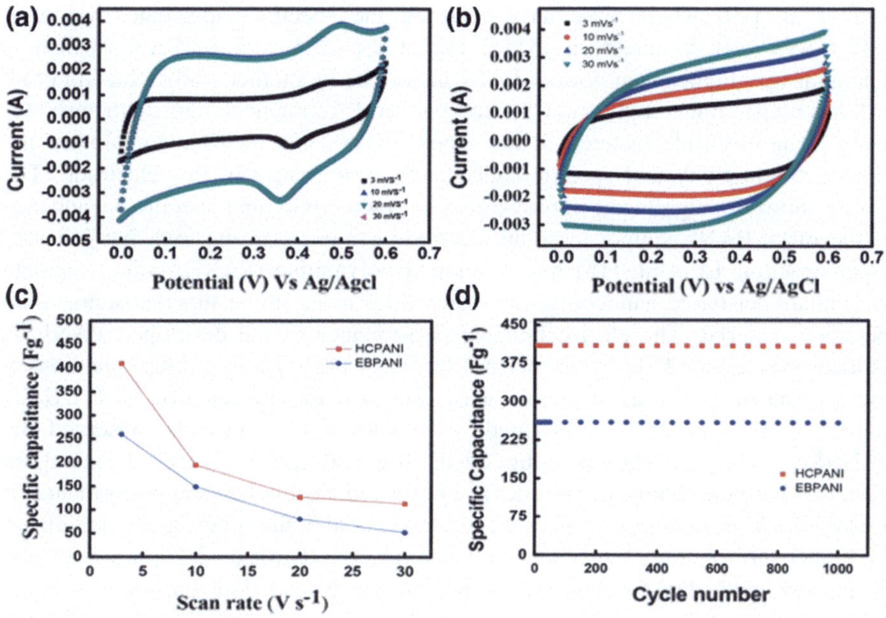

Fig. 5 **a** and **b** Cyclic voltammetry analysis for EBPANI and HCPANI respectively, **c** specific capacitance for EBPANI and HCPANI with different scan rate, and **d** specific capacitance for EBPANI and HCPANI with different cycle number (Adopted from [29])

nanofibers can be seen from the SEM images depicted in Fig. 4b. Zhang et al. [46] has investigated the electrochemical behaviour of conductive polyaniline nanofibers dispersed in an aqueous medium, which is prepared using acidic phosphate ester containing hydrophilic polyethylene glycol segment to dope PANI. They reported that this thin polyaniline nanofibers based electrode material showed a specific capacitance value of 160 F/g at a discharge rate 0.4 A/g with the potential range of −1 to 1 V. They finally concluded that this value for the electrode material has been occurred due to the high surface area of the polyaniline nanofibers.

Sharma et al. [29] has reported the electrochemical performance of a hyper cross-linked nanoporous polyaniline-based electrode material in the supercapacitor. The hyper cross-linked polyaniline (HCPANI) has been developed by treatment of emeraldine base polyaniline (EBPANI) with the help of microwave assisted method. They observed that specific capacitance value for the HCPANI has been increased up to 400 F/g with the scan rate of 3 mV/s and then lowered by incrementing the scan rate from 3 to 10 mV/s, whereas for the EBPANI based material specific capacitance is continuously decreasing with increase in the scan rate. The reason they have explained that the higher surface area (1059 m^2/g) of the HCPANI as compared to EBPANI (40 m^2/g), as depicted in Fig. 5, is mainly reason behind.

The specific capacitance for the sulfonated carbon nanotubes (sMWCNT) modified polyaniline (PANI) nanorods composite materials has been studied by

Zhu et al. [49] where they have reported the specific capacitance value of sMWCNT/PANI composite as 515.2 F/g at the 76.4 wt% of PANI, which is comparatively higher than normal PANI nanorods. In another study, Kazemi et al. [12] has prepared metal (Ni and Cu) incorporated PANI nanocomposite material for using as an electrode material in the supercapacitor and they have evaluated the electric conductivity and specific capacitance performance of this electrode. The results showed a significant improvement in conductivity and specific capacitance of the metal-PANI composite as an electrode candidate in the high performance supercapacitor. Li et al. [15] has developed polyaniline (PANI)/sodium alginate (SA) nanofiber-based nanocomposite material for using in the supercapacitor as an electrode material. The electrochemical performance of the developed PANI/SA nanocomposite-based electrode material has been analysed by cyclic voltammetry and galvanostatic charge-discharge tests. The specific capacitance of PANI/SA nanocomposite was 2093 F/g as stated by the authors. They have also observed that PANI/SA composite showed long cyclic life and fast replicate of oxidation/reduction with the change of current. The improved electrochemical performance of the PANI/SA nanocomposite as an electrode material in the supercapacitor is due to the higher surface area of the nanocomposite material. Girija et al. [8] has evaluated the electrochemical performance of the polyaniline deposited nickel electrode in the supercapacitor application. They found that the specific capacitance value is 2.30×10^3 F/g of the PANI deposited nickel electrode, calculated from cyclic voltammetry and galvanostatic charge analysers. Wang et al. [35] has investigated the electrochemical behaviour of the PANI coated activated carbon film (CA-PANI) as a flexible electrode material in the supercapacitor. They observed that CA-PANI film-based electrode material has a higher specific capacitance i.e. 240 F/g at 0.2 A/g charge- discharge density. This specific capacitance value of the CA-PANI based material is higher than individual component of this composite (activated carbon and PANI). In addition, the electrochemical performance of the different supercapacitor based on polyaniline (PANI) electrode material explained by different researchers can also be seen from the Table 2 [6].

3.1.2 Polypyrrole Based Flexible Supercapacitor

Polypyrrole (PPy) is one of the advantageous polymers for using as an electrode material in the supercapacitor because it has higher flexibility and conductivity than polyaniline and polythiophene polymer. Because of this suitable property of the PPy, flexible supercapacitor can be formed using PPy. Furthermore, PPy has higher charge density, low cost, good thermal stability and suitable charge-discharge capability [1, 10]. The higher amount of flexibility and specific capacitance (400–500 F/cm^3) in PPy polymer mainly comes from the higher density of the PPy polymer. However, it has two shortcomings, one is it can only be used as a cathode electrode in supercapacitor application because it is p-type conductive polymer and other is it has low cycling stability. These drawbacks of the PPy could be solved by modifying synthesis method which gives the better morphology, can make

Table 2 Electrochemical performance of the different supercapacitor based on polyaniline (PANI) electrode material (Data from [6])

Material	Discharge rate	Potential range (V)	Electrolyte	Specific capacitance (Fg^{-1})	Cycling/Capacity retention (Number of Cycles)	Specific energy (Wh kg^{-1})	Specific power (kW kg^{-1})
PANI	0.2 mA cm^{-2}	0 to 0.7	0.5M H_2SO_4	168.0 mF cm^{-2}			
PANI	10 mV s^{-1}	−0.1 to 0.8	1M H_2SO_4	503	85% (10,000)	96	8.88
PANI	2 mA cm^{-2}	0 to 1.0	1M H_2SO_4	480	84% (1000)		
PANI	1.5 A g^{-1}	0 to 0.8	1M H_2SO_4	532			
PANI	0.2 mA cm^{-2}	0 to 0.7	1M HCl	141.4 mF cm^{-2}			
PANI	0.5 A g^{-1}	0 to 0.8	PVA/H_3PO_4	890	89% (1000)		
PANI	0.2 mA cm^{-2}	0 to 0.7	CCl_3COOH	135.4 mF cm^{-2}			
PANI	1 A g^{-1}	0 to 0.8	0.1M Na_2SO_4	330	94% (1000)		
PANI	17 mA cm^{-2}	0 to 1	Gel polymer electrolyte	672		188	1.4
PANI	100 mA cm^{-2}	0 to 1	Gel polymer electrolyte	250		71	7.5
PANI	10 mV s^{-1}	−0.2 to 0.6	Nafion	269	65% (10,000)		
PANI nanowires	23 A g^{-1}	0 to 0.75	1M H_2SO_4	742	92% (1500)	68	16
Nanofibres	0.18 A g^{-1}	0 to 0.8	1M H_2SO_4	548	75% (1000)	36	0.12
Nanofibrous PANI	10 mV s^{-1}	−0.1 to 0.8	1M H_2SO_4	839			
Nanofibrous PANI	10 mV s^{-1}	−0.1 to 0.8	1M H_2SO_4	861	82% (2000)		
Nanowire arrays	1 A g^{-1}	0 to 0.7	1M $HClO_4$	950		130	0.7
PANI nanowires	1 A g^{-1}	−0.2 to 0.8	1M H_2SO_4	470	90% (1000)		
Electroless PANI	10 mV s^{-1}	0 to 1.0	0.1M H_2SO_4	592			
De-doped PANI nanofibres	2.5 A g^{-1}	0.65	1M H_2SO_4	593	70% (5000)		
Hydrogel-assisted PANI microfiber	10 mV s^{-1}	−0.2 to 0.8	1M methane sulfonic acid	703			

(continued)

Table 2 (continued)

Material	Discharge rate	Potential range (V)	Electrolyte	Specific capacitance (Fg^{-1})	Cycling/Capacity retention (Number of Cycles)	Specific energy (Wh kg^{-1})	Specific power (kW kg^{-1})
Self-doped PANI	1 A g^{-1}	−0.6 to 0.7	0.5 Na_2SO_4	408			
Sulfonated PANI	1 A g^{-1}	0 to 1.2	1M H_2SO_4	1107		31.4	14.8
PANI-HCl	1.25 mA cm^{-2}	0 to 1.0	1M Et_4NBF_4	70	57% (1000)		
BF_4-doped	50 mV s^{-1}	0 to 0.75	4.0M HBF	74	5% (20,000)	2.7	1
PANI-$LiPF_6$	1.25 mA cm^{-2}	−0.5 to 0.5	1M Et_4NBF_4	107	78.5% (9000)		
PANI-$LiPF_6$	3.75 A g^{-1}	0 to 0.75	1M Et_4NBF_4	115	78% (5000)		
PANI-$LiPF_6$	2 mA cm^{-2}	0 to 1	Polyolefin separator	100	70% (5000)		
PANI-$LiPF_6$	2 mA cm^{-2}	0 to 1	Polymer electrolyte	80	75% (5000)		
C|PANI	1 mA cm^{-2}	0.8 to 1.6	6M KOH	380	(4000)	18	1.25
PANI/C	2 A g^{-1}	−0.2 to 0.8	0.5M H_2SO_4	1486	82% (500)		
PANI/CNT	0.5 A g^{-1}	−0.2 to 0.8	0.5M H_2SO_4	1065	92% (1000)		
PANI/CNT	5 mV s^{-1}	−0.1 to 0.7	PVA/H_3PO_4	440	93% (1000)	7	5.2
PANI/MWCNT	1 mV s^{-1}	0 to 1.0	0.1M H_2SO_4	560			
PANI/SWCNT	5 mA cm^{-2}	0 to 0.7	1M H_2SO_4	485	94% (1500)	228	2.25
PANI/MWCNT	5 mA cm^{-2}	−0.2 to 0.8	1M $NaNO_3$	328	94% (1000)		
PANI/OMC	6 A g^{-1}	2 to 4 versus Li/Li^+	1M $LiPF_6$	146			
G/PANI/MCM-41	0.8 A g^{-1}	−1.1 to −0.1	6M KOH	405	91.4% (1000)		
	2 mV s^{-1}	−0.1 to 1.0	1M H_2SO_4	470		76.4	

(continued)

Table 2 (continued)

Material	Discharge rate	Potential range (V)	Electrolyte	Specific capacitance (Fg^{-1})	Cycling/Capacity retention (Number of Cycles)	Specific energy (Wh kg^{-1})	Specific power (kW kg^{-1})
Mesoporous C/ PANI							
PANI/OMC	0.1 A g^{-1}	0 to 0.8	1M H_2SO_4	517	91.5% (1000)		
Bamboo carbon/ PANI	0.5 A g^{-1}	−0.2 to 0.7	1M H_2SO_4	244	92% (1000)	47.5	
PANI/C	1 A g^{-1}	−0.4 to 0.6	1M H_2SO_4	439		60.8	0.5
PANI on CNF	5 mV s^{-1}	0 to 0.8	1M H_2SO_4	264			
CNF/PANI	2 A g^{-1}	0 to 0.8	1M H_2SO_4	638	90% (1000)		
PANI nanotube/ layered graphene	1 A g^{-1}	0 to 0.1	1M H_2SO_4	956		12.6	
PANI/G/CNT	1 A g^{-1}	0 to 0.8	1M H_2SO_4	510	91% (5000)	20.5	25
G/PANI	1 A g^{-1}	0 to 1.0	1M H_2SO_4	808		28	0.25
G/PANI	0.2 A g^{-1}	−0.2 to 0.8	2M H_2SO_4	536			
PANI/G	0.4 A g^{-1}	−0.2 to 0.6	1M H_2SO_4	976.5	89% (1000)		
PANI/G	1 A g^{-1}	0.1 to 0.8	1M H_2SO_4	1665			
G/PANI	1 A g^{-1}	0 to 0.8	1M H_2SO_4	790	80% (5000)	17.6	98
G/PANI	0.1 A g^{-1}	−0.2 to 0.8	1M H_2SO_4	676	97% (1000)		
PANI/G	1 A g^{-1}	0 to 0.8	1M H_2SO_4	1255	90% (2000)		
G/PANI	0.5 A g^{-1}	0 to 0.8	1M H_2SO_4	529	85% (1000)		
PANI/G	5 A g^{-1}	−0.2 to 0.8	1M H_2SO_4	413	81% (1000)		
PANI/G	1 A g^{-1}	−0.2 to 0.8	1M H_2SO_4	614	90% (500)		
PANI/G	0.2 A g^{-1}	0 to 0.8	1M H_2SO_4	1045		8.3	

(continued)

Table 2 (continued)

Material	Discharge rate	Potential range (V)	Electrolyte	Specific capacitance (Fg^{-1})	Cycling/Capacity retention (Number of Cycles)	Specific energy (Wh kg^{-1})	Specific power (kW kg^{-1})
PANI/G	1 A g^{-1}	0 to 0.8	6M H_2SO_4	2136	91% (1000)	24	6
PANI/G	1 mV s^{-1}	0 to 0.7	1M $HClO_4$	751	93% (1000)		
PANI/G	1 A g^{-1}	0 to 0.8	1M H_2SO_4 +HQ	553	90% (50,000)		
PANI/GO	0.5 A g^{-1}	−0.2 to 0.8	1M H_2SO_4	375	91% (500) at 3 A g^{-1}		
G/PANI/C	1 A g^{-1}	0 to 0.8	1M H_2SO_4	1145	94% (5000)	25.4	92.2
GO/PANI	1 mV s^{-1}	−0.1 to 0.9	1M H_2SO_4	1136	89% (1000)		
PANI/GO	0.025 mA cm^{-2}	0 to 0.7	1M H_2SO_4	662		64.5	1.16
PANI/GO	0.1 A g^{-1}	−0.2 to 0.8	2M H_2SO_4	480			
GO/PANI	1 A g^{-1}	0 to 0.8	PVA/H_3PO_4	1095		24.3	28.1
PANI-MnO_2	5 A g^{-1}	0 to 1	1M Na_2SO_4	715	96.5% (5000)	200	
PANI/MnO_2	2 A g^{-1}	0 to 0.7	1M H_2SO_4	626		17.8	0.6
PANI/MnO_2	1 A g^{-1}	0 to 0.65	0.5M Na_2SO_4	510			
PANI/MnO_2	0.5 A g^{-1}	−0.1 to 0.9	1M Na_2SO_4	383			
MnO_2/PANI	2 A g^{-1}	0 to 0.8	1M Na_2SO_4	498	95% (5000)		
PANI/MnO_x	1.0 mA cm^{-2}	0 to 0.65	1M NaNO	588	90% (1000)		
G/PANI-MnO_x	1 A g^{-1}	0 to 0.9	1M Na_2SO_4	955	89% (1000) at 20 A g^{-1}	61	4.5
MnO_2/PANI/RGO	0.5 A g^{-1}	−0.4 to 0.4	1 KOH	1090	82% (5000)		
PANI/MnO_2/TiN	1 A g^{-1}	0 to 0.7	0.5M Na_2SO_4	674			
Ag/MnO_2/PANI	1 mA g^{-1}	0 to 1	0.5M $LiClO_4$(in PC)	800			
PANI/SnO_2	1 A g^{-1}	−0.2 to 0.8	1M H_2SO_4	305	95.5% (500)	42	

(continued)

Table 2 (continued)

Material	Discharge rate	Potential range (V)	Electrolyte	Specific capacitance (Fg^{-1})	Cycling/Capacity retention (Number of Cycles)	Specific energy (Wh kg^{-1})	Specific power (kW kg^{-1})
$G/Fe_3O_4/PANI$	1 mV s^{-1}	−1.0 to 0.1	1M KOH	638	92% (5000)		
$PANI/Fe_3O_4$	1 mA cm^{-2}	0 to 0.75	PH 7.4 buffer	213			
$PANI/TiO_2$	2.5 A g^{-1}	−0.2 to 0.7	1M H_2SO_4	495	49.5% (3000)		
$RuO_2/PANI/C$	1 mA cm^{-2}	0 to 1	1M H_2SO_4	531			
$PANI/TiO_2/GO$	1 A g^{-1}	−0.2 to 0.8	1M H_2SO_4	430			
ITO/PANI	4 A g^{-1}		1M H_2SO_4	738			
PANI/C/Ni	0.5 A g^{-1}	0 to 0.8	1M Na_2SO_4	725			
H/Ni co-doped PANI/CNT	0.5 A g^{-1}	−0.2 to 0.8	1M H_2SO_4	781	92% (700)	22	2
$PANI/V_2O_5$	0.5 mA cm^{-2}	−0.9 to 0.7	5M LiCl	443	92% (5000)		
PANI/C.TiN	1 A g^{-1}	0 to 0.6	1M H_2SO_4	1093	98% (2000)	69	0.72
$PANI/MoS_2$	1 A g^{-1}	−0.4 to 0.6	1M H_2SO_4	575	98% (500)		
$MoS_2/PANI/G$	1 A g^{-1}	−0.8 to −0.2	1M H_2SO_4	618	78% (2000)	265	18
$CoMoO_4/PANI$	1 A g^{-1}	−0.2 to 0.7	1M H_2SO_4	380	90% (1000)		
$PANI/NiCo_2O_4$	1 A g^{-1}	0 to 0.8	1M H_2SO_4	901	91% (3000)		

composite combining with carbon or metal-based materials. The PPy can be synthesized by chemical or electrochemical polymerization process.

Xu et al. [41] has synthesized graphene/polypyrrole nanocomposite sheets and evaluated their electrochemical characteristics in the supercapacitor applications. The electrode component in a supercapacitor made from graphene/PPy nanocomposite showed higher specific capacitance value i.e. 318.6 F/g at the scan rate of 2 mV/s. They have also observed that this electrode material has been retained 95% capacitance even after 1000 cycles at the scan rate of 20 mV/s. In another study, Chang et al. [2] has prepared graphene/PPy based nanocomposite by electrochemical synthesis method for using as a promising electrode material in the supercapacitor. They have found the graphene/PPy based nanocomposite material could be a promising candidate for the supercapacitor, where graphene provides much higher area which can be used for segregation of charge for electrolytes and electrode interface and PPy has provided both conductivity and flexibility. Finally, they concluded that this graphene/PPy based electrode material showed 424 F/g specific capacitance value which is comparatively higher than the capacitance value of the graphene and PPy individually.

Yuan et al. [44] has developed a polypyrrole coated cellulosic paper-based composite by soak and polymerization method, for using as an electrode component in the flexible supercapacitor. The soak and polymerization method were helping to develop a uniform coating of the polypyrrole polymer throughout the paper with maintaining desirable porous structure, which can be seen from Fig. 6. After that the electrochemical characteristics have been evaluated by using cyclic voltammetry and galvanostatic charge analyser. The polypyrrole coated paper material showed the areal capacitance value 0.42 F/cm^2 and 1 mWh/cm^3 energy density with the power density of 0.27 W/cm^3 as shown in Fig. 7. Finally, authors have concluded that polypyrrole coated paper can be used as the most promising energy storage material in the flexible supercapacitor applications.

Wu and Lian [40] has reported the cyclic stability and specific capacitance performance of an electrode material in the supercapacitor application composed by flexible graphene/polypyrrole hydrogel (PHG) based material. They found that the graphene/PPy hydrogel-based material could show a specific capacitance value of 363 F/cm^3 at a current density of 1.0 mA/cm^3 and maintain enough mechanical stability. In addition, by using this electrode material 98.6% specific capacitance has been retained after 12000 charge/discharge cycles under bent, twisted and folded conditions. Finally, authors have concluded that the graphene/PPy hydrogel-based material can be used in the supercapacitor as an electrode material to maintain the highest specific capacitance, higher stability and enough flexibility of the supercapacitor. Yun et al. [45] has investigated the electrochemical characteristics of the polypyrrole coated MnO_2/carbon nanotube (CNT) based cotton fabric as a flexible electrode material in the supercapacitor. This electrode material exhibited higher specific capacitance value i.e. 461 F/g with the power and current density of 31.1 Wh/kg and 22.1 kW/kg, respectively. Furthermore, this electrode material can retain 93.8% specific capacitance after 10000 charge/discharge cycles.

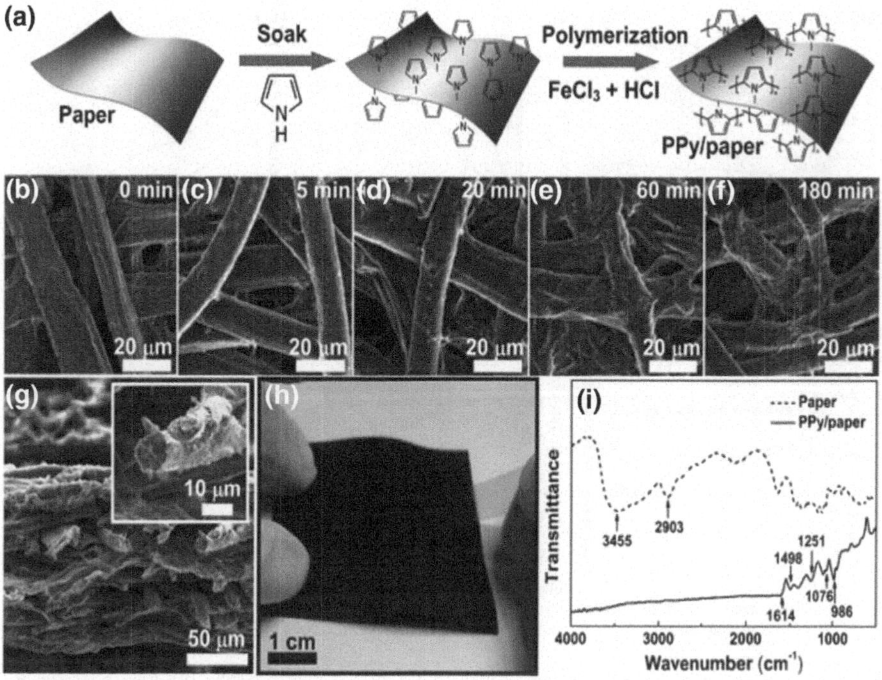

Fig. 6 **a** Preparation of PPy/Paper composite film, **b–f** SEM images of the developed PPy/Paper composite, **g** SEM image of the cross-section of the PPy/Paper composite, **h** Tape test of the developed PPy/Paper composite film to evaluate mechanical stability, and **i** FTIR spectrum of the developed PPy/Paper composite (Adopted from [44])

The electrochemical behaviour of the polypyrrole conductive polymer can be controlled by modification of the PPy structure such as nanoparticles, nanorods, etc. Meanwhile, this topic gained a great interest to modify the electrochemical characteristics of PPy throughout so that it can be used as an electrode material in the supercapacitor application. In this context, Xu et al. [43] has developed a graphene/activated carbon/PPy based electrode material for using as an electrode material in the flexible supercapacitor and have evaluated its electrochemical performance through cyclic voltammetry along with galvanostatic charge analysers. They observed, this electrode material displayed the higher specific capacitance value (906 mF/cm^2) of a current density (0.5 mA/cm^2). This material has also presented the higher specific capacitance value (64.4%) even for 5000 charge/discharge repitations. In another study, Xu et al. [42] has developed a polypyrrole coated cotton fabric for using as an electrode material in the supercapacitor at different preparation conditions (pure PPy, PPy-MO deposition with various time). The morphology of the different engineered samples is shown in Fig. 8. It was inferred that those SEM images showed that, when a deposition time of 4 h was allowed, then some grooves, wrinkle and rough surface was developed which were not

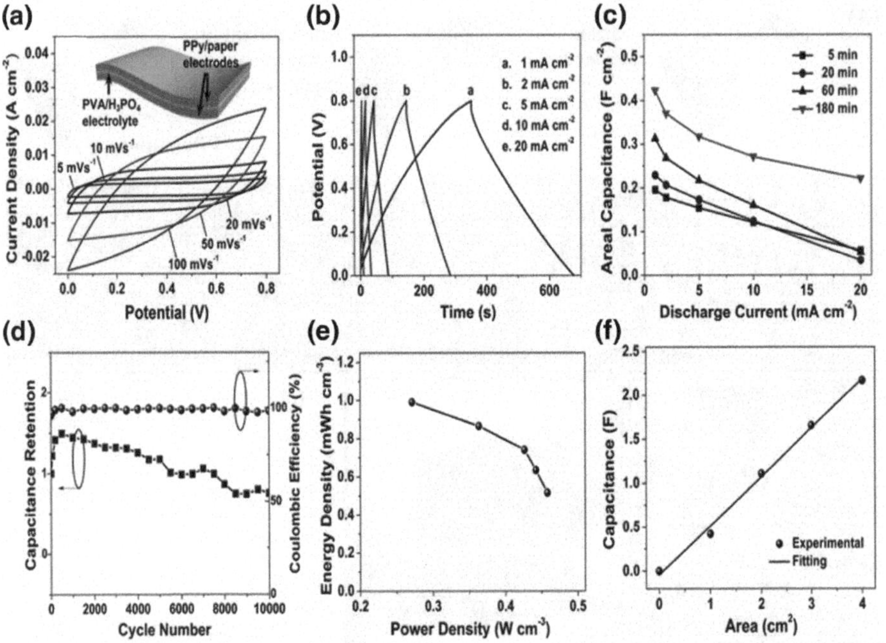

Fig. 7 **a** Cyclic voltammetry results, **b** galvanostatic charge results of the PPy/Paper composite, **c** areal capacitance versus discharge current, **d** capacitance retention versus cycle number, **e** energy density versus power density, and **f** capacitance versus area of the PPy/Paper composite (Adopted from [44])

desirable for electrode material for the supercapacitor. They also observed that PPy-MO-2hr based electrode material showed better specific capacitance value (325 F/g) of current density (0.6 mA/cm^2) when observed with pure PPy and PPy-MO-4hr electrode devices as seen in Fig. 9. Because in case of PPy-MO-2hr based material, the PPy nanorod structure and lower deposition of the PPy on the cotton fabric have provided higher surface area for which resultant interface area between the electrolyte and electrode material has been increased. This electrode material can also retain 200 F/g specific capacitance value after 500 charge/discharge cycles.

3.1.3 Polythiophene and Its Derivatives Based Flexible Supercapacitor

Such materials along with its derivatives have also been used as a conductive polymer-based electrode material in the flexible supercapacitor applications. They have higher potential (0.8–1.2 V) as mentioned in conductive materials for supercapacitor section. However, their conductivity and specific capacitance values

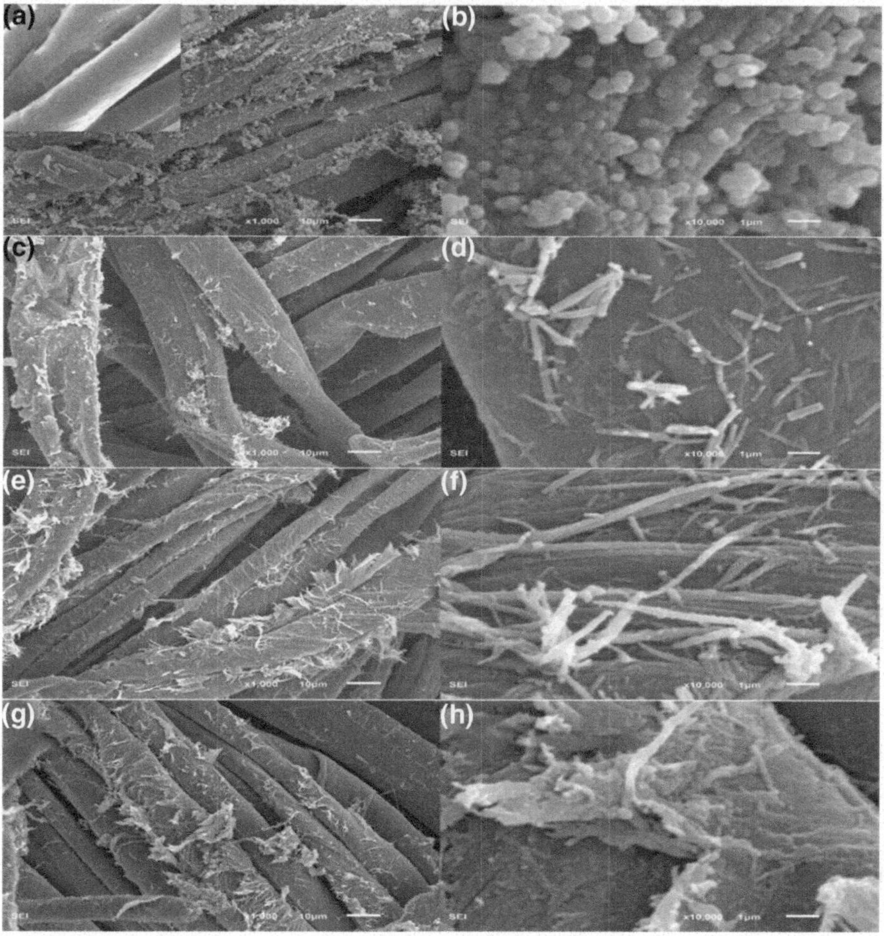

Fig. 8 SEM images of the PPy-coated fabrics prepared under different condition. PPy-2h: (**a**) and (**b**); PPy/MO-1h: (**c**) and (**d**); PPy/MO-2h: (**e**) and (**f**); PPy/MO-4h: (**g**) and (**h**). Inset in (**a**): raw fabric (Adopted from [42])

are comparatively lower than PANI and PPy. They can be p type and n type both, but p-doped polymer is highly stable at the air and moist environments.

Liu et al. [18] has studied on a PEDOT thin film in the form of an electrode device in the supercapacitor application. They have observed that this thin film could generate 130 F/g specific capacitance, which is comparatively higher than another conductive polymer.

Wang et al. [36] has demonstrated the electrochemical potential of a ternary electrode material based on graphene/SnO_2/poly(3,4-ethylene-dioxythiophene (PEDOT) composite through evaluation of data taken from the cyclic voltammetry and galvanostatic charge analyser. The ternary electrode material showed higher

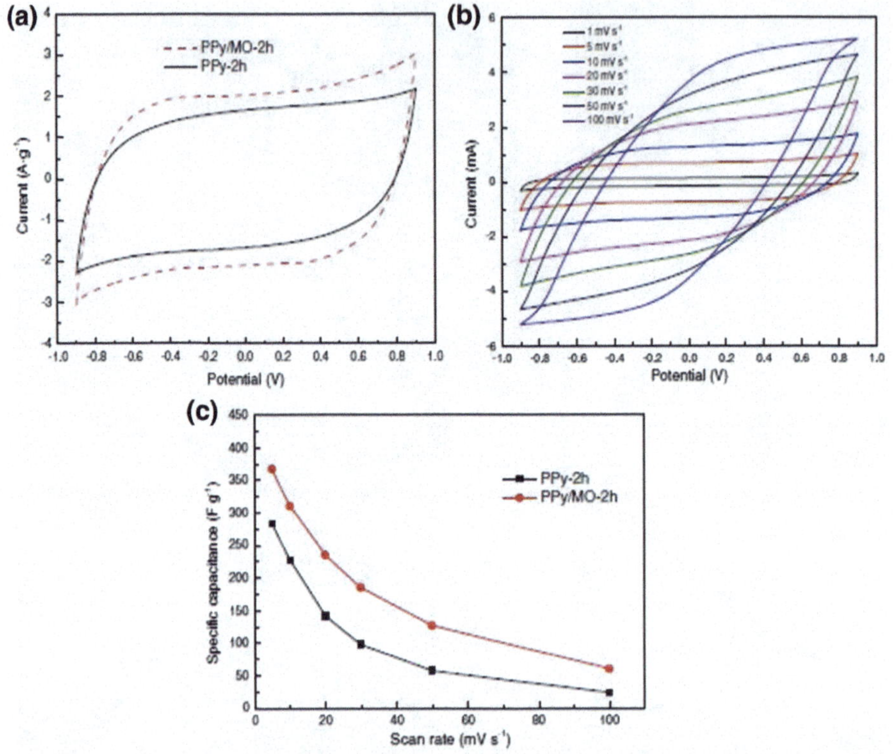

Fig. 9 **a** Cyclic voltammograms of the PPy-coated fabric supercapacitors at a scan rate of 5 mV s^{-1}; **b** cyclic voltammograms of the PPy/MO-2h fabric supercapacitors at different scan rates; **c** specific capacitance from cyclic voltammograms for the PPy/MO-2h and PPy-2h fabric supercapacitors at different scan rates (Adopted from [42])

specific capacitance values, 184 F/g and 180 F/g for the 1M H_2SO_4 and 1M Na_2SO_4 medium, respectively with maintaining desirable power and energy density (238.3 W/kg, 5803.3 W/kg and 22 Wh/kg, 17.1 Wh/kg). In addition, Moussa et al. [23] has also reported a ternary composite based on PEDOT/graphene/MnO_2 for the supercapacitor application as an electrode material. They found that this electrode material not only showed the higher specific capacitance but also higher stretch ability up to 400%. The specific capacitance value was 802.99 F/g at an energy density of 55.76 Wh/kg. Moreover, this ternary composite material also displayed enough ability to retain in specific capacitance that is almost 99% after the stretching cycles of 1000. Furthermore, for supercapacitor applications, Weng et al. [38] has reported a ternary composite-based electrode material based on poly (3,4-ethylene-dioxythiophene): polystyrenesulfonate (PEDOT-PSS)/graphene oxide (GO)/carbon nanotube. The electrochemical properties have been analysed by CV and galvanostatic charge results. The ternary composite-based electrode material showed 365 F/g specific capacitance at an energy density of 100 Wh/kg. In

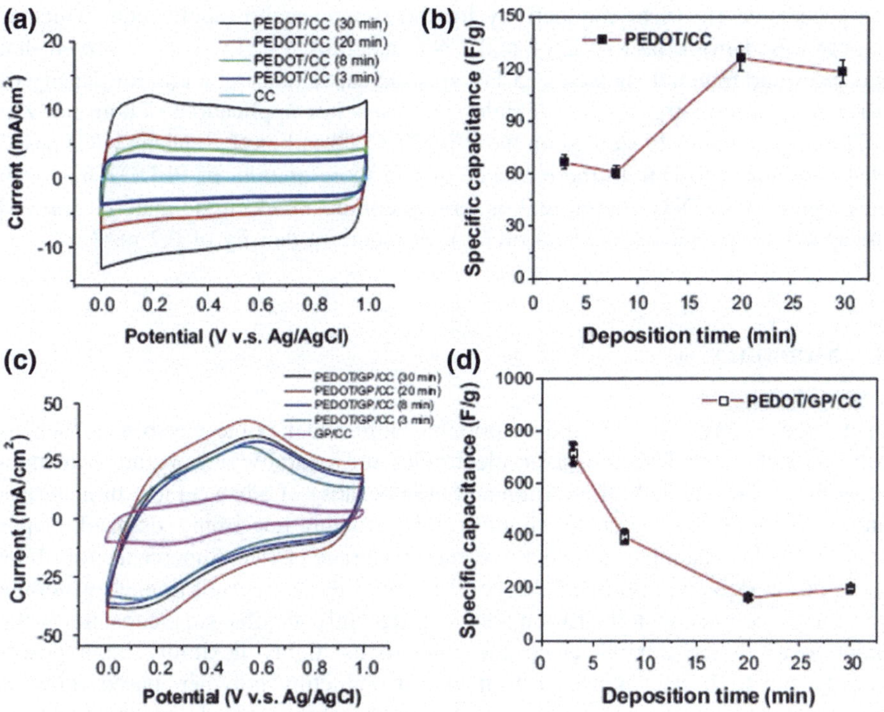

Fig. 10 **a** Cyclic voltammograms of PEDOT/carbon cloth (CC) composite materials with different PEDOT electrodeposition times. Electrolyte: 0.5M H2SO4. Scan rate: 50 mV/s. **b** Electrodeposition time-dependent specific capacitance of PEDOT/CC composite materials derived from (**a**). (**c**) Cyclic voltammograms of PEDOT/graphene paper-modified carbon cloth (GP/CC) with the different PEDOT electrodeposition times. Electrolyte: 0.5M H2SO4. Scan rate: 50 mV/s. **d** Electrodeposition time dependent specific capacitance of PEDOT/GP/CC derived from (**c**) (Adopted from [4])

another study, Lee et al. [14] has also experienced on a PEDOT-PSS/CNT/MnO_2 composite based electrode material for supercapacitor applications. The higher flexibility and better electrochemical characteristics have been observed from this PEDOT-PSS/CNT/MnO_2 composite based materials. The specific capacitance value given by this electrode material was 428.2 F/g with the energy density of 63.8 Wh/kg.

Chu et al. [4] has developed two different composite-based electrode materials for flexible supercapacitor applications, one is composed by PEDOT/carbon cloth (CC) and other is PEDOT/graphene paper (GP)/CC. They found that PEDOT/graphene paper (GP)/CC based electrode showed better performance than PEDOT/carbon cloth (CC), which can be seen from Fig. 10.

Chen et al. [3] has reported a hybrid nanocomposite-based electrode material for flexible supercapacitor, which has been developed by using PEDOT/multi walled carbon nanotube (MWCNT)/graphene materials. The electrode material has showed the acceptable electrical conductivity for the supercapacitor application, which is mainly raised from the MWCNT and CNT materials. Finally, they observed that this electrode material showed 225 F/g specific capacitance at a current density of 0.1 A/g. Additionally, for the flexible supercapacitor applications, Zhou and Zhi [47] has systematically studied on the PEDOT/CNT/carboxyl graphene (CG) based electrode material. Due to the presence of CG more amount of PEDOT has been adsorbed on the CNT. Therefore, this ternary composite electrode material showed the specific capacitance as 150.6 mF/cm^2 at a current density of 0.2 mA/cm^2.

4 Summary

In the recent days, the demand of portable, thin and flexible electronics such as roll-up display, touch screen, smart electronics and wearable sensors, are drawing a great interest in the daily life of human being because of advancement in materials and technology. In order to simplify this growing electronic demand, super capacitor is fascinating as a favourable energy storage device. Supercapacitor offers more power density, improved charge-discharge lifespan as well as energy storage capacity in contrast with the Li-ion batteries. Recently, flexible supercapacitor is the main focus in the flexible electronics due to its higher flexibility, high power density and high capacitance performance. Conducting polymers-based supercapacitor is more promising candidate compare to other materials in terms of their flexibility, high redox active specific capacitance and essential elastic nature. The chapter critically analyses various conducting polymer (CPs) based materials used for the flexible supercapacitor.

References

1. An, H., Wang, Y., Wang, X., Zheng, L., Wang, X., Yi, L., Bai, L., Zhang, X.: Polypyrrole/carbon aerogel composite materials for supercapacitor. J. Power Sources **195**(19), 6964–6969 (2010). https://doi.org/10.1016/j.jpowsour.2010.04.074 (Elsevier B.V.)
2. Chang, H.H., Chang, C.K., Tsai, Y.C., Liao, C.S.: Electrochemically synthesized graphene/polypyrrole composites and their use in supercapacitor. Carbon **50**(6), 2331–2336 (2012). https://doi.org/10.1016/j.carbon.2012.01.056 (Elsevier Ltd.)
3. Chen, J., Jia, C., Wan, Z.: Novel hybrid nanocomposite based on poly (3,4-ethylenedioxythiophene)/multiwalled carbon nanotubes/graphene as electrode material for supercapacitor. Synth. Met. **189**, 69–76 (2014). https://doi.org/10.1016/j.synthmet.2014.01.001 (Elsevier B.V.)
4. Chu, C.Y., Tsai, J.T., Sun, C.L.: Synthesis of PEDOT-modified graphene composite materials as flexible electrodes for energy storage and conversion applications. Int. J. Hydrog. Energy **37**(18), 13880–13886 (2012). https://doi.org/10.1016/j.ijhydene.2012.05.017 (Elsevier Ltd.)
5. Dao, L.H., Talbi, H., Just, P.E.: Electropolymerization of aniline on carbonized polyacrylonitrile aerogel electrodes: applications for supercapacitors. J. Appl. Electrochem. **33**(6), 465–473 (2003)

6. Eftekhari, A., Li, L., Yang, Y.: Polyaniline supercapacitors. J. Power Source **347**, 86–107 (2017). https://doi.org/10.1016/j.jpowsour.2017.02.054 (Elsevier B.V.)
7. Ghenaatian, H.R., Mousavi, M.F., Kazemi, S.H., Shamsipur, M.: Electrochemical investigations of self-doped polyaniline nanofibers as a new electroactive material for high performance redox supercapacitor. Synth. Met. **159**(17–18), 1717–1722 (2009). https://doi.org/10.1016/j.synthmet.2009.05.014
8. Girija, T.C., Sangaranarayanan, M.V.: Polyaniline-based nickel electrodes for electrochemical supercapacitors-influence of Triton X-100. J. Power Sources **159**(2), 1519–1526 (2006). https://doi.org/10.1016/j.jpowsour.2005.11.078
9. Gupta, V., Miura, N.: High performance electrochemical supercapacitor from electrochemically synthesized nanostructured polyaniline. Mater. Lett. **60**(12), 1466–1469 (2006). https://doi.org/10.1016/j.matlet.2005.11.047
10. Hughes, M., Chen, G.Z., Shaffer, M.S.P., Fray, D.J., Windle, A.H.: Controlling the nanostructure of electrochemically grown nanoporous composites of carbon nanotubes and conducting polymers. Compos. Sci. Technol. **64**(15 SPEC. ISS.), 2325–2331 (2004). https://doi.org/10.1016/j.compscitech.2004.01.026
11. Kalaji, M., Murphy, P.J., Williams, G.O.: Study of conducting polymers for use as redox supercapacitors. Synth. Met. **102**(1–3), 1360–1361 (1999). https://doi.org/10.1016/S0379-6779(98)01334-4
12. Kazemi, S.H., Kiani, M.A., Mohamadi, R., Eskandarian, L.: Metal-polyaniline nanofibre composite for supercapacitor applications. Bull. Mater. Sci. **37**(5), 1001–1006 (2014). https://doi.org/10.1007/s12034-014-0037-y
13. Ke, Q., Wang, J.: Graphene-based materials for supercapacitor electrodes—a review. J. Materiomics **2**(1), 37–54 (2016). https://doi.org/10.1016/j.jmat.2016.01.001 (Elsevier Ltd.)
14. Lee, H.Uk., Yin, J.L., Park, S.W., Park, J.Y.: Preparation and characterization of PEDOT:PSS wrapped carbon nanotubes/MnO_2 composite electrodes for flexible supercapacitors. Synth. Met. **228**, 84–90 (2017). https://doi.org/10.1016/j.synthmet.2017.03.016
15. Li, Y., Zhao, X., Qian, X., Zhang, Q., Chen, D.: Facile preparation and enhanced capacitance of the polyaniline/sodium alginate nanofiber network for supercapacitors. Langmuir **27**(10), 6458–6463 (2011). https://doi.org/10.1021/la2003063
16. Lin, Z., Goikolea, E., Balducci, A., Naoi, K., Taberna, P.L., Salanne, M., Yushin, G., Simon, P.: Materials for supercapacitors: when Li-ion battery power is not enough. Mater. Today **21** (4), 419–436 (2018). https://doi.org/10.1016/j.mattod.2018.01.035 (Elsevier Ltd.)
17. Liu, C., Li, F., Lai-Peng, Ma., Cheng, H.M.: Advanced materials for energy storage. Adv. Mater. **22**(8), 28–62 (2010). https://doi.org/10.1002/adma.200903328
18. Liu, K., Zhenglong, H., Xue, R., Zhang, J., Zhu, J.: Electropolymerization of high stable poly (3,4-ethylenedioxythiophene) in ionic liquids and its potential applications in electrochemical capacitor. J. Power Sources **179**(2), 858–862 (2008). https://doi.org/10.1016/j.jpowsour.2008.01.024
19. Lota, K., Khomenko, V., Frackowiak, E.: Capacitance properties of poly (3,4-ethylenedioxythiophene)/carbon nanotubes composites. J. Phys. Chem. Solids **65**(2–3), 295–301 (2004). https://doi.org/10.1016/j.jpcs.2003.10.051
20. Mastragostino, M., Arbizzani, C., Soavi, F.: X Gynecologic Cancer **148**, 1–12 (2007). https://doi.org/10.1016/s0167-2738(02)00093-0
21. Miller, J.R., Simon, P.: Materials science: electrochemical capacitors for energy management. Science **321**(5889), 651–652 (2008). https://doi.org/10.1126/science.1158736
22. Mondal, S.K., Rajendra Prasad, K., Munichandraiah, N.: Analysis of electrochemical impedance of polyaniline films prepared by galvanostatic, potentiostatic and potentiodynamic methods. Synth. Met. **148**(3), 275–286 (2005). https://doi.org/10.1016/j.synthmet.2004.10.010
23. Moussa, M., Shi, G., Hao, W., Zhao, Z., Voelcker, N.H., Losic, D., Ma, J.: Development of flexible supercapacitors using an inexpensive graphene/PEDOT/MnO_2 sponge composite. Mater. Des. **125**, 1–10 (2017). https://doi.org/10.1016/j.matdes.2017.03.075

24. Muzaffar, A., Basheer Ahamed, M., Deshmukh, K., Thirumalai, J.: A review on recent advances in hybrid supercapacitors: design, fabrication and applications. Renew. Sustain. Energy Rev. **101**, 123–145 (2019). https://doi.org/10.1016/j.rser.2018.10.026
25. Prasad, K.R., Koga, K., Miura, N.: Electrochemical deposition of nanostructured indium oxide: high-performance electrode material for redox supercapacitors. Chem. Mater. **16**(10), 1845–1847 (2004). https://doi.org/10.1021/cm0497576
26. Rudge, A., Raistnck, I.A.N.: A study of the electrochemical properties of conducting polymers for application in capacitors **39**(2), 273–87 (1994)
27. Ryu, K.S., Kim, K.M., Park, N.G., Park, Y.J., Chang, S.H.: Symmetric redox supercapacitor with conducting polyaniline electrodes. J. Power Sources **103**(2), 305–309 (2002). https://doi.org/10.1016/S0378-7753(01)00862-X
28. Schneuwly, A., Gallay, R.: [Detlef_Stolten,_Bernd_Emonts]_Hydrogen_Science_an, 1–10 (2000)
29. Sharma, V., Sahoo, A., Sharma, Y., Mohanty, P.: Synthesis of nanoporous hypercrosslinked polyaniline (HCPANI) for gas sorption and electrochemical supercapacitor applications. RSC Adv. **5**(57), 45749–45754 (2015). https://doi.org/10.1039/c5ra03016a (Royal Society of Chemistry)
30. Shown, I., Ganguly, A., Chen, L.C., Chen, K.H.: Conducting polymer-based flexible supercapacitor. Energy Sci. Eng. **3**(1), 1–25 (2015). https://doi.org/10.1002/ese3.50
31. Snook, G.A., Chen, G.Z.: The measurement of specific capacitances of conducting polymers using the quartz crystal microbalance. J. Electroanal. Chem. **612**(1), 140–146 (2008). https://doi.org/10.1016/j.jelechem.2007.08.024
32. Snook, G.A., Kao, P., Best, A.S.: Conducting-polymer-based supercapacitor devices and electrodes. J. Power Sources (2011). https://doi.org/10.1016/j.jpowsour.2010.06.084
33. Taheri, P., Hsieh, S., Bahrami, M.: Investigating electrical contact resistance losses in lithium-ion battery assemblies for hybrid and electric vehicles. J. Power Sources **196**(15), 6525–6533 (2011). https://doi.org/10.1016/j.jpowsour.2011.03.056 (Elsevier B.V.)
34. Wang, G., Zhang, L., Zhang, J.: A review of electrode materials for electrochemical supercapacitors. Chem. Soc. Rev. **41**(2), 797–828 (2012). https://doi.org/10.1039/c1cs15060j
35. Wang, T., Wang, W., Dai, Y., Zhang, H., Shen, Z., Chen, Y., Xiaoping, H.: Electrochemical synthesis of polyaniline films on activated carbon for supercapacitor application. Russ. J. Electrochem. **51**(8), 743–747 (2015). https://doi.org/10.1134/S1023193515080133
36. Wang, W., Lei, W., Yao, T., Xia, X., Huang, W., Hao, Q., Wang, X.: One-pot synthesis of graphene/SnO2/PEDOT ternary electrode material for supercapacitors. Electrochim. Acta **108**, 118–126 (2013). https://doi.org/10.1016/j.electacta.2013.07.012 (Elsevier Ltd.)
37. Wang, X., Han, X., Lim, M., Singh, N., Gan, C.L., Jan, Ma., Lee, P.S.: Nickel cobalt oxide-single wall carbon nanotube composite material for superior cycling stability and high-performance supercapacitor application. J. Phys. Chem. C **116**(23), 12448–12454 (2012). https://doi.org/10.1021/jp3028353
38. Weng, Y.T., Wu, N.L.: High-performance poly(3,4-ethylene-dioxythiophene):polystyrene-sulfonate conducting-polymer supercapacitor containing hetero-dimensional carbon additives. J. Power Sources **238**, 69–73 (2013). https://doi.org/10.1016/j.jpowsour.2013.03.070 (Elsevier B.V.)
39. Winter, M., Brodd, R.J.: What are batteries, fuel cells, and supercapacitors? Chem. Rev. **104** (10), 4245–4270 (2004). https://doi.org/10.1021/cr020730k
40. Wu, X., Lian, M.: Highly flexible solid-state supercapacitor based on graphene/polypyrrole hydrogel. J. Power Sources **362**, 184–191 (2017). https://doi.org/10.1016/j.jpowsour.2017.07.042 (Elsevier B.V.)
41. Xu, C., Sun, J., Gao, L.: Synthesis of novel hierarchical graphene/polypyrrole nanosheet composites and their superior electrochemical performance. J. Mater. Chem. **21**(30), 11253–11258 (2011). https://doi.org/10.1039/c1jm11275a
42. Xu, J., Wang, D., Fan, L., Yuan, Y., Wei, W., Liu, R., Gu, S., Xu, W.: Fabric electrodes coated with polypyrrole nanorods for flexible supercapacitor application prepared via a

reactive self-degraded template. Org. Electron. Phys. Mater. Appl. **26**, 292–299 (2015). https://doi.org/10.1016/j.orgel.2015.07.054 (Elsevier B.V.)

43. Xu, L., Jia, M., Li, Y., Zhang, S., Jin, X.: Design and synthesis of graphene/activated carbon/polypyrrole flexible supercapacitor electrodes. RSC Adv. **7**(50), 31342–31351 (2017). https://doi.org/10.1039/c7ra04566b (Royal Society of Chemistry)
44. Yuan, L., Yao, B., Bin, H., Huo, K., Chen, W., Zhou, J.: Polypyrrole-coated paper for flexible solid-state energy storage. Energy Environ. Sci. **6**(2), 470–476 (2013). https://doi.org/10.1039/c2ee23977a
45. Yun, T.G., Hwang, B.I., Kim, D., Hyun, S., Han, S.M.: Polypyrrole-MnO_2 -coated textile-based flexible-stretchable supercapacitor with high electrochemical and mechanical reliability. ACS Appl. Mater. Interfaces **7**(17), 9228–9234 (2015). https://doi.org/10.1021/acsami.5b01745
46. Zhang, H., Zhao, Q., Zhou, S., Liu, N., Wang, X., Li, J., Wang, F.: Aqueous dispersed conducting polyaniline nanofibers: promising high specific capacity electrode materials for supercapacitor. J. Power Sources **196**(23), 10484–10489 (2011). https://doi.org/10.1016/j.jpowsour.2011.08.066
47. Zhou, H., Zhi, X.: Ternary composite electrodes based on poly(3,4–ethylenedioxythiophene)/carbon nanotubes–carboxyl graphene for improved electrochemical capacitive performances. Synth. Me. **234**, 139–44 (2017). https://doi.org/10.1016/j.synthmet.2017.10.011 (Elsevier)
48. Zhou, Y.K., He, B.L., Zhou, W.J., Huang, J., Li, X.H., Bin, W., Li, H.L.: Electrochemical capacitance of well-coated single-walled carbon nanotube with polyaniline composites. Electrochim. Acta **49**(2), 257–262 (2004). https://doi.org/10.1016/j.electacta.2003.08.007
49. Zhu, Z.Z., Wang, G.C., Sun, M.Q., Li, X.W., Li, C.Z.: Fabrication and electrochemical characterization of polyaniline nanorods modified with sulfonated carbon nanotubes for supercapacitor applications. Electrochim. Acta **56**(3), 1366–1372 (2011). https://doi.org/10.1016/j.electacta.2010.10.070 (Elsevier Ltd.)

Self-healing Substrates: Fabrication, Properties and Applications

Nikiwe Mhlanga and Keletso Mphahlele

Abstract Drawing inspiration from living organisms, inorganic self-healing substrates are the smart materials to revolutionize our world in the next decades. These smart substrates inherit the ability to detect damage and autonomously or non-autonomously heal and restore to its pristine state. The consequence of self-healing offers new route towards sustainable, safer and long lasting materials for multifunctional applications, such as: medicine, energy, construction, food packaging, water treatment and textiles. This chapter explores the preparation of the self-healing substrates from inorganic substrates such as polymers, ceramics and metals; including their healing chemistries and envisioned applications.

Keywords Self-healing · Polymers · Ceramics · Concrete · Metals · Nanocomposites

1 Introduction

Biological systems possess a unique and extraordinary competence to self-heal when exposed to damage, for example, the biggest organ in the human body, skin, can self-heal when subjected to damages such as cuts and burns [1, 2]. On the contrary, synthetic systems lack such capability and as a result have shorter life spans [2]. Inspired by biological systems, the smart self-healing ability has

N. Mhlanga (✉)
NIC/DST Mintek Nanotechnology Innovation Centre, Randburg, South Africa
e-mail: Nikiwem@mintek.co.za

K. Mphahlele
DST/CSIR National Centre for Nanostructured Materials, Council for Scientific and Industrial Research, Pretoria 0001, South Africa
e-mail: KMphahlele@csir.co.za

K. Mphahlele
Department of Chemical, Metallurgical and Material Engineering, Tshwane University of Technology, Pretoria 0001, South Africa

Inamuddin et al. (eds.), *Self-standing Substrates*,
Engineering Materials, https://doi.org/10.1007/978-3-030-29522-6_8

ingressed into real life applications in the last couple of decades. It is explored to fabricate dynamic and tunable systems which mimic natural processes for various applications such as power sustainability and a smarter future. Although it has been decades since their inception, the synthetic self-healing systems still continue to attract the scientific community [2]. This is because self-healing materials which can completely or partially self-heal will open doors to various applications [1].

Ideally, self-healing is the ability of a material to be aware of self-damage and trigger activities which will repair that damage as it is observed on living organisms. The self-healing substrates can be autonomic or non-autonomic, the autonomic substrates internally self-heal without intervention and the latter relies on intervention by stimuli such as heat or light to self-repair [1, 3, 4]. They can be further classified into intrinsic and extrinsic substrates based on how the regeneration process is developed [5]. Figure 1 summarizes the classification of the self-healing substrates based on both the regeneration process development and the healing strategy. The intrinsic substrates internally self-heal using internal chemical and physical interactions whereas extrinsic substrates lack the internal healing ability and therefore are integrated using healing agent monomers [1, 5]. The healing monomers are incorporated via vessels such as fibers, capsules or vascular networks which are compounded with the host or damaged material. Their healing mechanism involves breaking open the vessels to release the healing material when a micro-crack propagates. Then the healing monomer will fill or polymerize to repair the cracks [5]. The extrinsic systems heal autonomically without the use of a stimuli [3]. The healing microcapsules incorporated into substrates are classified into four groups; capsule-catalyst, multi-capsule, latent functionality and phase-separated materials [1]. White et al. [4] pioneered the preparation of self-healing polymer matrix using the capsule/catalyst approach.

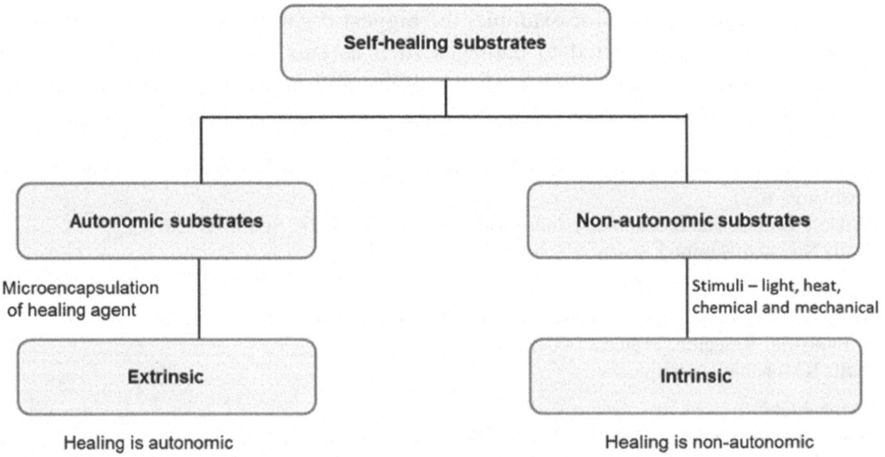

Fig. 1 Classification of self-healing substrates

The intrinsic self-healing substrates orchestrate healing via their physical and chemical interactions at the damaged sites and polymers are good examples of such systems. Polymeric substrates such as thermoplastics undergo self-healing via their physical molecular diffusion and chemical bond interaction. The physical molecular diffusion mechanism relies on the salt concentration of the material; a higher salt concentration will promote the polymer chain's mobility and results in faster chain diffusion and healing. The chemical bond interaction mechanism uses covalent bonding, non-covalent bonding and intermolecular force interactions to self-heal. Intermolecular forces include hydrogen bond, dipole-dipole and van der Waals interactions [1]. The chemical and physical processes are triggered by stimuli, hence intrinsic systems are non-autonomous.

The intrinsic self-healing substrates are reliable and durable compared to the extrinsic substrates. However, they are suitable for small damages whereas extrinsic substrates are used in large damage volumes [1] hence they are suitable for industrial applications. This chapter seeks to expand on the different types of self-healing substrates, their healing chemistries, preparation and applications.

2 Self-healing Substrates and Their Healing Chemistries

Self-healing substrates are prepared from various materials such polymers, ceramics and metals. Polymers are leading the pack and have recorded above 70% restoration of previous functionality [6]. The field of self-healing substrates is growing, as evidenced by a growing number of research articles recorded between year 2000 and 2014 as shown by Fig. 2. The Figure further confirms the dominance of polymeric self-healing substrates compared to other substrates although an increase is also observed for ceramics, concrete, and metals. Researchers are starting to explore ceramics, concrete and metals for self-healing purposes and expand the existing polymeric substrate self-healing chemistries into the latter new substrates. Self-healing is inspired by biological systems and efforts are done to translate it into inorganic systems. The self-healing process in biological systems is complex and occurs via replacement of damaged tissue or repair. Therefore, for synthetic systems to inherit this phenomenon they have to detect/sense damages and be able to respond autonomously by atomic transport to the fractured site and start healing [6]. In synthetic substrates self-healing is governed by two critical activities: physical flow of molecular segments near or at the fractured sites; and reformation of bonds at these sites. These two phenomena can occur continually depending on the kinetics and thermodynamics during the self-healing process although they can be hindered or terminated by environmental factors such as atmospheric oxygen, moisture and other reactive groups. Gibbs energy is influenced by the kinetic energy of the chains and entropic changes created during the movement or diffusion [2].

The self-healing process is stimulated by the mobility of chains, hence rigid chains have to be catalysed by external stimuli to attain mobility [2]. The chain mobility is driven by the creation of voids or free spaces [2]. The governance of

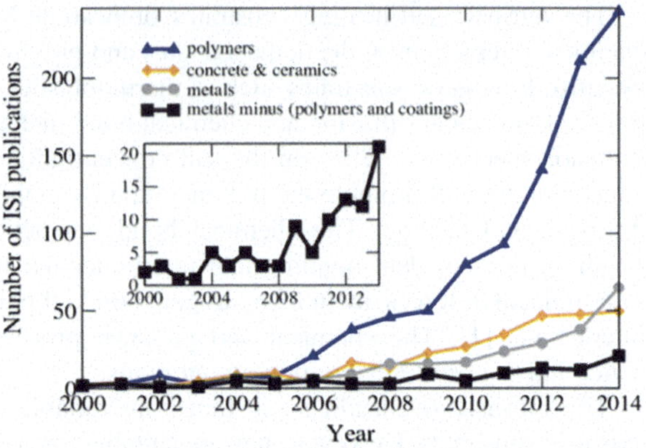

Fig. 2 ISI web of knowledge research article numbers of self-healing substrates between 2000 and 2014. Reprinted by permission from Springer Nature. Self-healing metals, Blazej Grabowski and C. Cem Tasan, Copyright, 2015 [7]

self-healing by chain mobility elucidates the larger numbers of research articles on polymers as opposed to the emerging substrates of ceramics and metals in the field [7] (Fig. 2).

2.1 *Polymeric Substrates*

Intense research has been done on polymeric substrates for self-healing purposes due to their ability to be easily engineered at low temperature to onset the diffusion of chains which is a vital factor for self-healing [8]. Considering the auto-repairing property of the polymers, they are adapted in real life applications to prolong service life and safety e.g., in shipbuilding, wind turbine blades, active implantable medical devices and aeronautics [9]. Thermoplastics and thermosets polymers can each be modified to auto repair, but thermoset polymers aided by their rigid and thermal stability are suitable for various self-healing applications [6, 10]. Ideally, self-healing polymers should continuously detect damage, heal and restore materials performance without altering the polymers' pristine properties. The restored properties include tensile strength, surface smoothness, barrier properties and molecular weight [10].

2.1.1 Thermoplastic Polymers

Thermoplastic polymers have good chain mobility which is used in their self-healing. Thermoplastic self-healing has been intensely studied since the 1980s and it covers amorphous, semi crystalline, block copolymers and fiber reinforced composites [11]. The subsequent sections details the self-healing chemistries in thermoplastics.

Molecular Inter-diffusion

Thermoplastic polymer fractured sites come into contact at a temperature higher than its glass transition state (T_g) and trigger molecular inter-diffusion which results in imperceptible healing of the interface and restoration of its mechanical strength [10]. The molecular inter-diffusion follows the subsequent occurrences: surface rearrangements, surface approach, wetting and diffusion (Fig. 3) [9]. Surface re-arrangement is the relaxation of the polymer chains and their chemical reactions. Surface approach refers to the contact of the damaged or fractured sites by a 10 nm order of closeness via the flow of molecular segments enabled by thermal energy. The level of closeness experienced by the fractured sites will trigger surface forces, electrostatic forces and hydrogen bonds which will result in cohesive healing and

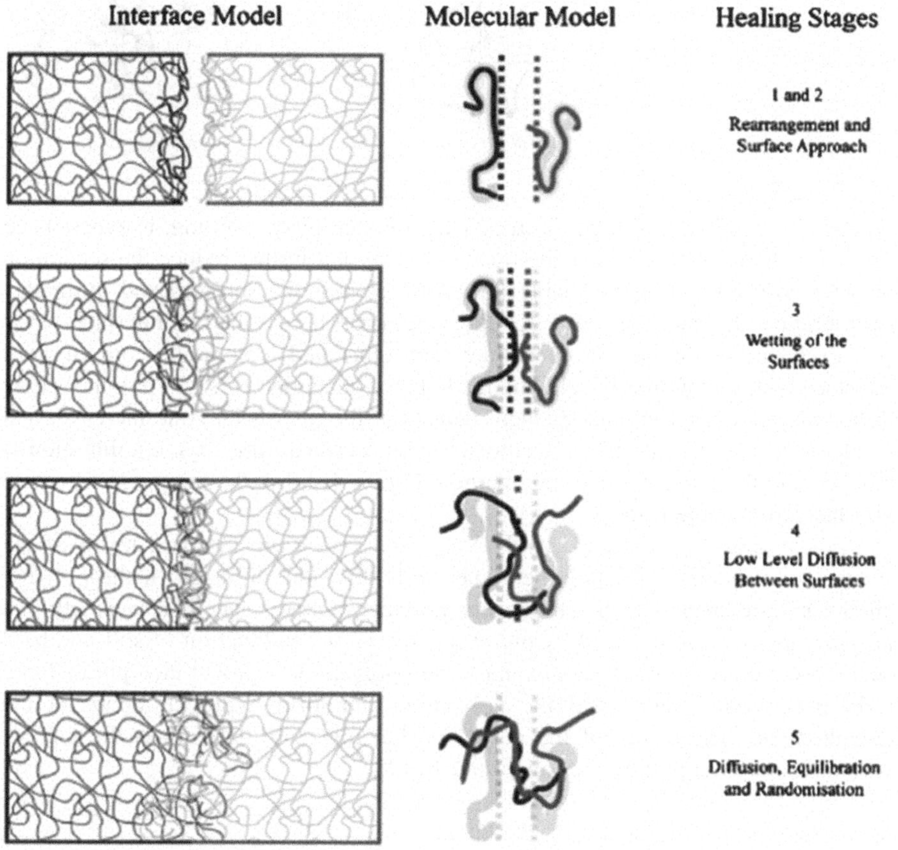

Fig. 3 Self-healing via molecular inter-diffusion in a polymer interface. Reprinted by permission from Springer Nature, Rheologica Acta, Rheological modeling of the diffusion process and the interphase of symmetrical bilayers based on PVDF and PMMA with varying molecular weights, Huagui Zhang, Khalid Lamnawar, Abderrahim Maazou, Copyright, 2012 [72]

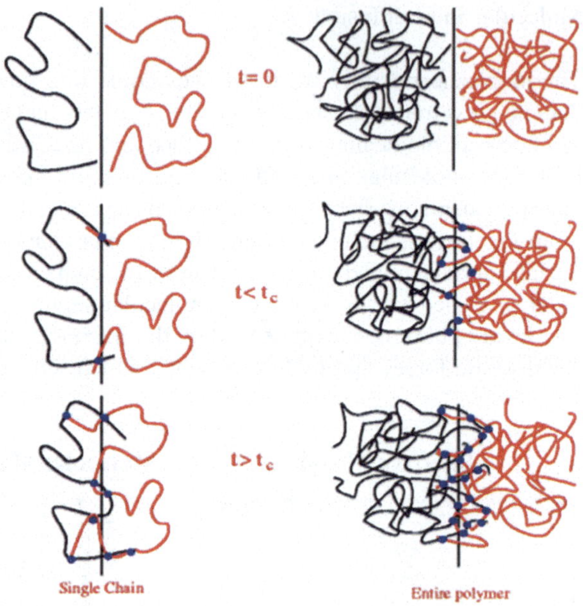

Fig. 4 Schematic diagram of diffusion process at polymer/polymer interface. Reprinted by permission from Springer Nature, Rheologica Acta, Rheological modeling of the diffusion process and the interphase of symmetrical bilayers based on PVDF and PMMA with varying molecular weights, Huagui Zhang, Khalid Lamnawar, Abderrahim Maazou, Copyright, 2012 [72]

gradually reduce surface irregularities. The 3rd occurrence, wetting, is experienced when the chains diffuse across the fractured faces and further reduce the separation distance to 0.3 nm [12] to enable diffusion between the surfaces and ultimately equilibration and randomization. As diffusion occurs across the two interfaces, they will exchange chains as they penetrate each other and replace the compromised interface with a new interfacial zone made up of entanglements. As polymer chain diffusion continues, the entanglement density at the interfacial zone increases and results in a stronger interface. Eventually, after a critical time (t_c), equilibration is reached and the two interfaces are merged. Figure 4 shows the merging of the two interfaces from time 0 to t_c.

Prior to these occurrences, voids would have been created as a result of the fracture or damage in the polymer. The voids drive the diffusion of the polymer chains and are created on heterogeneous polymer matrixes. The heterogeneity in a polymer matrix is introduced by phase separation or combination of soft and hard segments into one network. For example, soft polyamide segment incorporated into rigid polystyrene chains and the polyamide can form hydrogen bonds at the damaged site. The extent of the polymers' heterogeneity determines the type of wound healing magnitude, i.e. micro/macro [2].

Photo-Induced Healing

A photo-responsive thermoplastic fractured site is healed with exposure to the light of an appropriate wavelength. The photo–induced healing is limited to the surface of the polymer where the light can penetrate because light is the catalyst for the healing process [10]. It is a generally fast, simple and environmentally friendly healing approach [13].

Recombination of Chain Ends

This is a recent technique which is limited to polymers prepared via condensation reactions such as polycarbonate, polybutylene. The healing mechanism is driven by a reaction that reverses the chain scission and it entails the following occurrences: degradation of the polymer, which results in chain cleavage; then the cleaved chain recombination by catalytic redox reaction under an oxygen atmosphere catalyzed by copper or amine; and lastly, loss of water as a result of healing [10].

Reversible Bond Formation

Reversible bond formation is attained by self-assembling two mechanisms based on specific chemical and/or physical interactions [8]. These chemical and physical interactions govern self-healing. The physical interactions refer to the flow of molecular segments near or at a damaged site, and chemical interactions are the formation of bonds on the cleaved site [2]. To enhance the physical flow of the chains, polymer matrix can be engineered with reversible bond systems on their backbone [10]. Examples of such systems include organo-siloxane, ionomers (covalent bonds) and supramolecular forces (non-covalent) [8, 10]. The reversible bond systems are suitable for preparation of self-healing polymers because of their high bonding strength and contribution to the mechanical strength of the final product.

Covalent bonds

Covalent bonds used in polymers include ionomers, reversible cycloaddition reaction, exchange reaction, stable free radical mediated reshuffling reaction and heterocyclic carbohydrate facilitated bond reformation. The covalent bonds pose a challenge with their high activation energy requirement. Typically, the polymer's T_g is 100 °C and for self-healing to occur they have to be heated higher than their T_g which surmises that the healing process will occur at higher temperatures. But an ideal self-healing substrate should self-repair at an ambient temperature, hence the temperature issue is mitigated by introducing heterogeneities into the polymer matrix [2, 14].

Ionomers are a class of organic polymers with up to 20 mol% of ionic species incorporated into their structure [15]. Their self-healing is founded on phase separation of ionic groups within non-polar polymeric matrix and is accustomed for semi-crystalline or elastomeric polymers. The development of cracks in polymeric material is attributed to weak bonds and they sacrificially undergo cleavage to be able to reconstruct into improved mechanical strength [8].

Non-covalent interactions

In supramolecular chemistry the polymers are engineered with non-covalent associative groups on their side chain or ends which will bind the polymer and form networks or crosslinking. The non-covalent interaction includes H-bonds, metal coordination, ionic interaction and $\pi - \pi$ stacking host-guest interactions [2]. The supramolecular forces such as H-bond and $\pi - \pi$ stacking are susceptible to breaking under stress, and re-cleave to stronger mechanical strength [8]. Application of the supramolecular forces offers several advantages. Although the H-bond is weaker than most covalent bonds it has increased directionality and affinity producing a wide range of supramolecular polymers [8]. Metal-ligand coordination, i.e., polymers endowed with noncovalent metal centers have distinct properties and stimuli-responsiveness. Their application in self-healing metallic supramolecular networks offers reversibility just like the H-bonds. Their metal core is tunable to attain different properties and shapes such as linear, star-shapes and highly branched dendritic structures. The $\pi - \pi$ stacking thermally-triggered self-healing substrates are customized by a wide temperature range (50–100 °C) attained by changing the spacer and adjusting the composition. They can be combined with H-bonds to gain efficient thermal healing ability [2]. The supramolecular interactions (H-bond, $\pi - \pi$ interactions, van de Waals or hydrophobic interactions) can be integrated to bind two differently structured but dimensionally matching polymers [2].

Living Polymer Approach

Living polymers are normally applied in ionizing radiation and their self-healing ability is stimulated by incorporation of active groups at the chain ends. The active groups resume polymerization upon addition of monomers [10].

Anisotropic Systems Self-repairing

Application of anisotropic materials such as nanoparticles (NPs) in self-healing is emerging and involves their dispersion into a polymer matrix to fill voids or micro-cracks. Continuous healing can be achieved as long as the NPs remain in the matrix [10]. Figure 5 depicts the healing process via incorporation of the NPs. The anisotropic systems such as NPs, nano-rods, nanowires, nanotubes, nanofibers, and graphene sheets can potentially be modified with different functionalities such as amine, carboxylic acid, hydroxyl, thiols, furfuryl, cinnamoyl, and anthracene which enables formation of reversible bonds within the polymer matrix and thus afford self-healing. The polymer of choice should be endowed with matching reactive

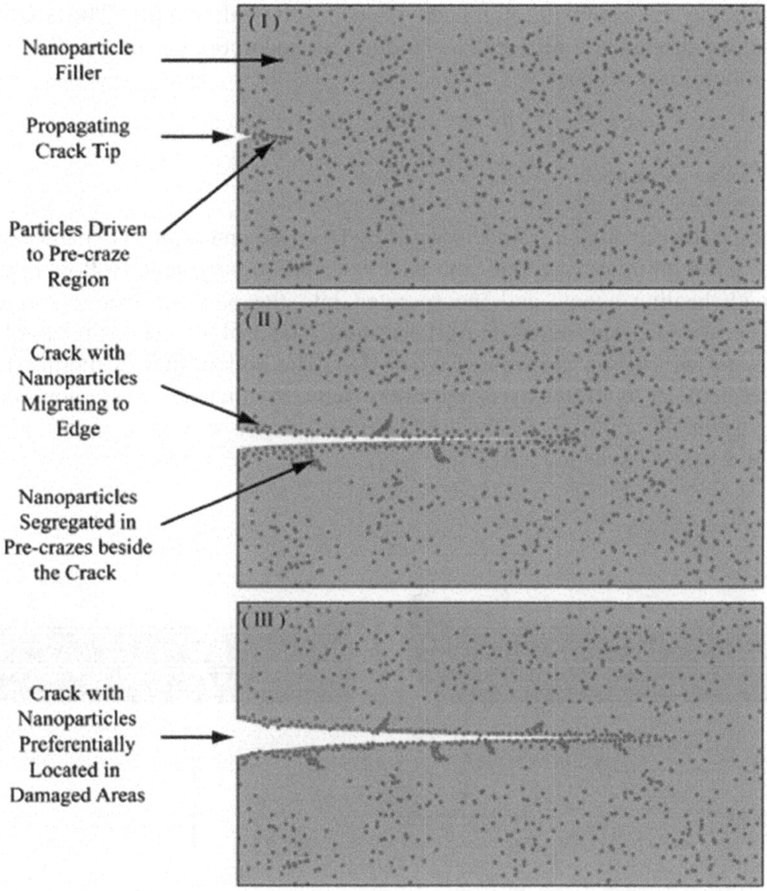

Fig. 5 Schematic representation of NPs dispersed in a thermoplastic. Reprinted with permission from Elsevier. Self-healing polymeric materials: a review of recent development, 33, Dong Yang Wu, Sam Meure and David Solomon, 479–522, Copyright, 2008 [10]

functionalities and, interestingly, stimuli responsive moieties can also be grafted into the anisotropic materials [2].

2.1.2 Thermoset Polymers

Thermosets are hard, rigid and thermally stable polymers. Unlike thermoplastics, thermosets lack chain mobility and their self-healing is achieved by incorporation of self-healing agents. The self-healing agent is loaded into vessels which will fracture and release their contents to fill the micro-cracks. The thermoset's healing mechanism is influenced by the nature and location of the fracture, type of self-healing resin and the environment of the envisioned self-healing [10]. The successive sub-sections discuss the different techniques for incorporating the healing agents into the thermoset matrix.

Hollow Fiber Approach

The hollow fiber approach was developed by Dry [16] and Sottos [17] and was first applied in cementitious materials and then extended to polymers. Hollow fibers are filled with healing agents and incorporated into the polymer matrix and Fig. 6 shows a schematic representation of the system. The healing agent is released upon fracture and will fill up the created voids [18]. The hollow fiber is modified to an accurate diameter, wall thickness and hollowness, to efficiently load enough of the healing agent. In addition, the healing agent of choice should have an ideal viscosity and healing kinetics [18].

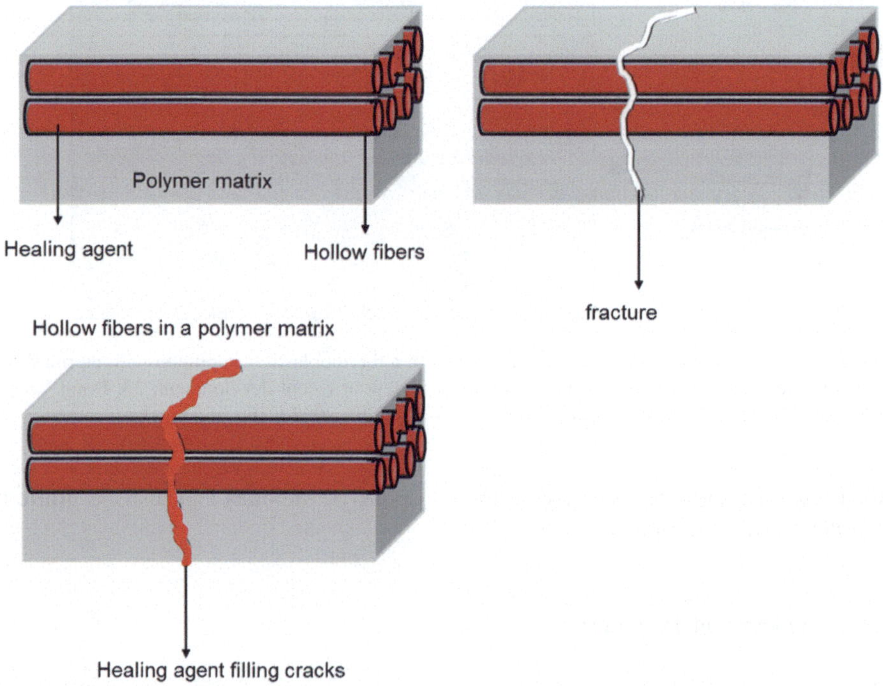

Fig. 6 Schematic representation of the hollow fiber mechanism on polymer matrix

Microencapsulation Approach

The micro-encapsulation approach entails encapsulation of a healing agent into a microcapsule. The microcapsule is then incorporated into a polymer matrix and will rupture and release contents upon damage or fracture [19]. The healing agent is encapsulated into the microcapsule to shield it from environmental influences and ensures its activity upon damage [19]. Zhu et al. [19] gives a comprehensive review of the microencapsulation self-healing approaches and classifies them into 5 systems: single capsule; double capsule; phase-separated droplets/capsule; capsule/dispersed catalyst; and an all-in-one microcapsule. Figure 7 shows a representation of the different micro-encapsulation classes. The single capsule is endowed with one type of healing agent while the double employs multiple healing agent capsules. Reaz et al. [20] reports on a single capsule self-healing approach using urea-formaldehyde microcapsules enclosing room temperature curing epoxy resin system (SC-15) as the healant prepared by in situ polymerization. Microcapsules size range between 30 and 100 μm. Fiber reinforced composite was prepared using three different steps. First, epoxy resin was embedded in urea-formaldehyde shell structure. Secondly, the amine hardener was encapsulated using the vacuum infiltration method. Finally, the different microcapsules were added with epoxy at 10:3 ratio and composite assembly was done with hand layup procedure. Low velocity impact test with 30 and 45 J impact loads established the delamination and micro-crack in composite structures and following healing recovery was observed

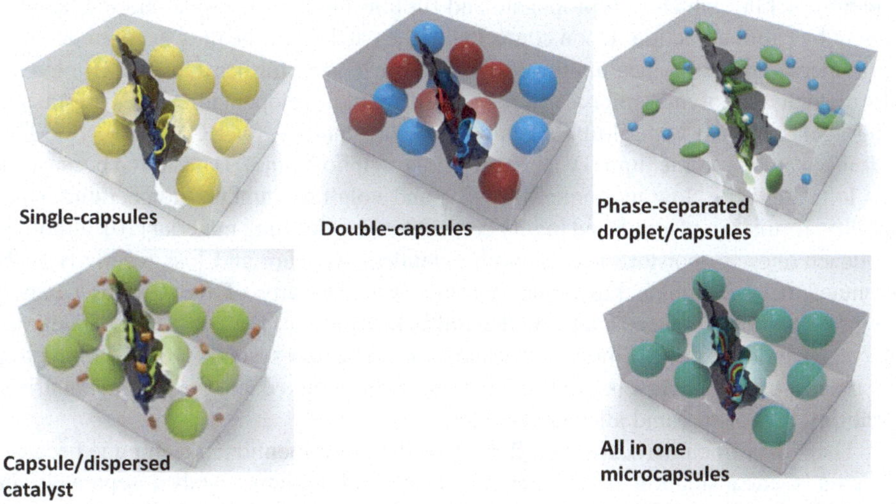

Fig. 7 Representation of the different classes of micro-encapsulation systems. Reprinted with permission from Elsevier. Self-healing polymeric materials based on microencapsulated healing agents: From design to preparation, 49–50, Dong Yu Zhu, Min Zhi Rong and Ming Qiu Zhang, 175–220, Copyright, 2015 [19]

because the damaged area was reduced and the mechanical properties were restored. The filling of damaged samples was observed and the bleeding of healant from the microcapsules was observed using the optical micrograph. Purnomo et al. [21] Investigates a self-healing approach of epoxy-amine thermoset established on the compatible functionalization of the thermoset and encapsulated healant. The thermoset were healed from the reaction of furans in the thermoset and multi-maleimides (MMIs) in the healing agent solution. The healing agent, MMI was dissolved in phenyl acetate and encapsulated by means of a urea-formaldehyde encapsulation procedure. The microcapsules were filled with the healing agent solution and incorporated in a furan-functionalized epoxy-amine thermoset. The resulting self-healing thermoset restored 71% of its original load after fracture.

Jin et al. [22] developed a double–microcapsule using an epoxy-amine chemistry in thermoset epoxy. The capsules were loaded with aliphatic polyamine and an epoxy monomer. Two distinct techniques were used in the preparation of the capsules, vacuum filtration for the amine and situ polymerization for the epoxy microcapsule, respectively. The capsules were amalgamated into an epoxy and tested for fracture recovery. The recovery was measured using a tapered-double-cantilever beam specimens and showed 91% healing ability short term and 68% long term of approximately six months using ambient aging studies [22].

In the phase-separated droplets/capsule, one healing agent goes through phase separation and the remaining component is encapsulated [19]. The Capsule/dispersed catalyst is the most studied technique and is based on ring-opening metathesis polymerization (ROMP). A micro-encapsulated healing agent is incorporated into polymer in the presence of a dispersed catalyst. The polymer will undergo micro-cracking which will propagate and rupture the microcapsule and the healing agent by capillary action. Upon contact with the catalyst the healing agent undergoes polymerization hence healing the cracks and preventing further damage [10]. Urea–formaldehyde polymer microcapsules loaded with dicyclopentadiene (DCPD) healing agent coupled with Grubb's catalyst are the standard for ROMP [10, 18]. The Grubbs catalyst (ruthenium (IV), bis(tricyclohexylphosphine)) is preferred because of its long shell life, low monomer viscosity and volatility, ambient temperature suitability, formation of tough and highly crosslinked crack filling material [10], although it deactivates in moisture and air with extended exposure and loss reactivity with contact to lenetriamine [18]. The stability and reactivity of the Grubbs catalyst improves when it is encapsulated with paraffin but due to its high cost, catalyst options have expanded to poly (dimethyl siloxane) or tin. The new systems proffer elimination of the healing agent vessels (since they can be directly mixed into the polymer matrix), stability in moisture and affordability [18].

The All-in-one microcapsule differs from the afore-mentioned technique because it entails encapsulation of the healing agent and initiator catalyst separately in multi-layered capsules or encapsulated in two capsules of different size which will be incorporated together i.e., capsule-in-capsule. The capsule–in-capsule/ multi-layered capsule is then incorporated into the polymer matrix and healing mechanism is the same as the former capsule [19].

Although ROMP is the standard for micro-encapsulation other chemistries have been explored to counteract the challenges associated with it. Polycondensation is a microencapsulation approach which uses a different catalyst and healing chemistry, and was first attempted in the healing of vinyl ester matrix using siloxane [19]. The approach involves phase separated droplets with hydroxyl ends grafted poly-dimethylsiloxane (HOPDMS), a cross linker polydiethoxysiloxane (PDES) and polyurethane microcapsule loaded with a catalyst (di-n-butyltin dilaurate). Self-healing is attained upon crack formation which will propagate and rupture the microcapsules to release the catalyst. The catalyst will then react with the phase-separated HOPDMS and PDES and seal the crack [13, 23]. Polycondensation offers stability in water and air conditions, elevated temperature stability (facilitating healing in higher-temperature thermoset applications) and uses widely accessible and affordable components [19, 23]. However, the healing agent is in direct contact with the host material matrix and subjected to pre-reacting [19] and a low healing efficiency of about 24% has been reported [13].

Another approach of interest is the anionic ring-opening polymerization. It is based on the epoxy's epoxide group ring-opening reaction and this reaction favors different mechanism hence it can be applied for different healing mechanisms [18]. The ring opening reaction is catalyzed by a latent hardener ($CuCl_2$-imidazole). The catalyst dissociates into $CuCl_2$ and imidazole when heated and prompts a reaction between the imidazole's secondary amine group first hydrogen atom and the epoxide group to form an ionic complex [18]. This healing chemistry is suitable for elevated temperatures, applicable to advanced components characterized by high T_g[18] and offers about 70% healing efficacy [12]. The imidazole catalyst can be replaced with boron trifluoride diethyl etherate $(C_2H_5)_2OBF_3$ to favor ambient to lower temperatures self-healing via cationic polymerization [18]. The cationic polymerization does not require a stoichiometric epoxy/hardener ratio and even distribution of both compounds because the reaction requires maintenance of contact with the latent hardener [12].

Traditional radical polymerization chemistry is also used for self-healing and it uses peroxide initiators such as benzoyl peroxide in Poly(methyl methacrylate) bone cement to induce radical polymerization self-healing [19]. It can be assisted with external stimuli such as oxygen and light [19]. Epoxy is also an exciting option for the healing agent and offers good adhesion and recovering of the pristine polymers properties [17]. The epoxides healing agents can be coupled with hardeners such as amines and thiols because they can react with epoxides via addition polymerization, and self-heal the damage [12].

Microvascular Networks

The hollow fiber and microcapsules suffer from non-continuous healing, i.e., healing can only take place once in a damaged site. To counter this issue microvascular networks have been proposed. The remarkable age of technology, drawing inspiration from living systems integrated with networks serving in sensing, healing and protection from danger, transpired the microvascular networks [18]. The microvascular networks approach is suitable for suppling large volumes of healing monomers to fractured sites via a series of interconnected channels. These channels can even be linked to an external supplier to ensure constant running of the self-healing monomers. They have been tested on the ROMP mechanism using DCPD and a Grubbs catalyst and showed a 70% restoration [24].

Thermally Reversible Cross-Linked Polymers

These are an emerging class of cross-linked polymers that self-heal via thermo-reversible covalent bonds and eliminate the need to incorporate a healing agent but are triggered by heat stimulus e.g., Diels-alder (DA) reaction [25].

Thermoplastic Additives

The thermally reversible crosslinks discussed above can also be replaced by a thermoplastic additive which will not alter the pristine matrix and provides a solidified void filler capable of re-binding the fractured surface [10].

Chain Rearrangement

Arrangement of thermoset polymer chains at ambient to elevated temperature results in self-healing. Similar to thermoplastic inter-diffusion, the micro-cracks are healed via inter-diffusion of dangling chains/chain slippage in the polymer matrix, and Fig. 8 summarizes the process [10].

Metal-Ion Mediated Healing

Metal-ion mediated healing was developed to repair lightly crosslinked hydrophobic polymer gels or hydrogels characterized by flexible hydrophobic side chains endowed with a terminal carboxyl group. It is similar to chain rearrangement because it involves rearrangement of crosslinked polymer networks assisted by metal-ions incorporated in the hydrogels. However, it changes the initial structure and properties of the polymer [10].

Shape Memory Polymers (SMPs)

SMPs can be manipulated to recall their pristine shape by using the right external stimuli such as heat, light, solvent, electric current and magnetic field [26]. The

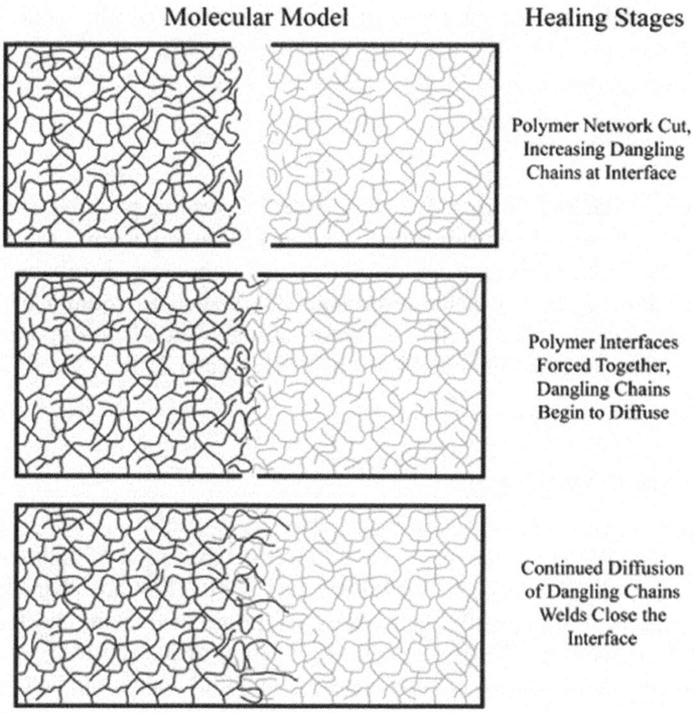

Fig. 8 Self-healing via dangling polymer chain ends. Reprinted with permission from Elsevier. Self-healing polymeric materials: a review of recent development, 33, Dong Yang Wu, Sam Meure and David Solomon, 479–522, Copyright, 2008 [10]

SMPs' healing mechanism can be summarized into the following occurrences: The SMP deforms above its T_g and creates disorientated chains (new shape); The SMP undergoes fixing at lowered temperatures to freeze the temporary shape; then an external stimuli is introduced, i.e., the SMP is heated above its T_g to relax its chains to a pristine shape [26]. The SMPs are also used in thermoplastic although they suffer from limited repeated use because their shape-recovery ability reduces with increasing shape memory cycles. This is caused by the incomplete relaxation of the SMPs chains to the pristine state after cycles of remembering the pristine shape. However, the limitation can be solved by coupling the thermoplastic SMPs with reversible bonds such as H-bonds or self-healing agents [26]. On the other hand, thermoset SMPs, especially those used for recycling applications characterized by high T_g, still suffer from non-reusability and flammability challenges [27]. Feng et al. [27] prepared a novel SMP thermoset epoxy with excellent fire retardancy, recyclability, mechanical performance and 100% shape recovery ratio. The new SMP overcomes the flammability and recyclability challenges because it is triggered by high-temperature flame and is recyclable via solid-state recycling methods. It find use as reinforcing filler for polyester [27].

Swelling of Polymer Matrix

In this case a material is coated with a water absorbing polymer and when it fractures the polymer will absorb water, swell and fill in any fractures or voids, and hence heal the damage. Both water-absorbable and non-water absorbable polymers can be functionalized with water-absorbable fillers such as sodium bentonite or polyethylene oxide to enhance their ability to absorb water. This approach is only suitable for applications which are not sensitive to water [10].

2.2 Ceramic Substrates

Ceramics are one of the prime materials used in the fabrication of high temperature apparatus e.g., gas turbines and fusion reactors [28]. They are characterized by brittle nature and inner porosity, hence they are prone to fracture and their cracks propagates very fast [8, 28]. As a result, introduction of the self-healing capability into their matrix can be beneficial in increasing reliability, cutting maintenance cost, decreasing machining cost, expanding service lifetime, and increasing strength and durability [28]. Extension of the established self-healing approaches using chain mobility and encapsulated healing agents discussed in the polymer section require high activation temperature (above 1000 °C) in ceramics. This is due to the ceramics ionic and covalent nature which favors solid-state diffusion to induce chain mobility [8]. Their high temperature self-healing mechanism is summarized into 4 occurrences:

Crack closure—controlled sintering at the elevated temperatures;
Crack filling—enhanced by viscous flow of the glass phase;
Void filling—cracks filled by oxidation products;
Particle rearrangement by phase transition [8].

These phenomena are grafted into the subsequent self-healing approaches used in ceramics.

2.2.1 Microencapsulation

Microencapsulation used in polymers is also used in ceramics and glass [10]. As previously discussed, the microcapsule technique has been used to encapsulate different healant comprising of chemicals, polymers, and bacteria spores.

2.2.2 Dynamic Oxidation/Diffusion

Ceramics such as silicon composites oxidize and produce products that are capable of filling voids when heated to high temperatures [8]. Osada et al. [29] demonstrated the heat treatment or diffusion self-healing phenomena ceramics by addition of an activator which behaves like the catalyst in the microencapsulation technique discussed in previous sections. The activator is introduced to the fracture pathway where it catalyzes the self-healing process. The activator used was doped with MnO and it rapidly filled the created voids by producing mobile super-cooled melts on the ceramic and enhanced its oxygen uptake. It also aided the recrystallization of the melts to form a mechanically strong healing oxide. The mechanism is initiated by oxidation followed by inflammation, repair and remodeling stages. It can be applied in the fabrication of lightweight aircraft turbine blades [29]. In addition, passivation of stainless steel metal using ceramics such as native aluminium oxide, self-repairs through dynamic oxidation. Interestingly, alumina and silica self-healing is not limited to high temperatures as they can melt into viscous state at ambient temperatures and self-heal by oxidation [14].

Oxidation of an A Element in a MAX Phase System

A MAX phase ceramic system is made up of $M_n = AX_n$ where M = early transition metals, A group = Al, P, S, Ga, Ge, As, Si and X element = C and/or N. They have distinct properties which include: good electrical and thermal conductivity, low hardness, high damage tolerance, and oxidation and corrosion resistance. They also self-heal via oxidation of the A group in the presence of oxygen at elevated temperature which will result in increased volume for filling cracks [8].

2.2.3 Self-healing Using NPs Additives

Rajak et al. [30] simulated self-healing on alumina (Al_2O_3) by embedding SiC/SiO_2 NPs into the alumina core. When damage is induced into the ceramic matrix, it will propagate to the vicinity of the NPs and the NPs will distort and forge channels in the matrix for silica to flow to the damaged site and halt the cracking. At the alumina and NPs interface, facets will be formed that facilitate growth of multiple secondary amorphous grains in the Al_2O_3 to fill the voids or cracks. In addition, diffusion of silica from the NPs will be another mode of healing [30].

2.3 *Metallic Substrates*

The development of metallic self-healing systems is a herculean task because, unlike polymers, diffusion is restrained as a result of their stronger cohesive bindings [6, 31]. Adopting the incorporation of a healing agent into metals requires transportation of the healing monomer via atomic diffusion and it requires higher temperatures [6, 31]. Self-healing metallic substrates can be prepared by adopting the 4 techniques discussed in the following sections.

2.3.1 Precipitation of Solute Atoms to Fill Voids and Micro-cracks

Precipitation reaction is observed in metallic systems characterized by a decreasing solubility for solute atoms with decrease in temperature or vice versa [6]. The metal is heated to high temperatures and the temperature is rapidly reduced and this creates a metastable state with the solute atom. After some time the solute locked in the metal will diffuse and cause nucleation of precipitate on the fractured site and hence restore the strength of the metal [32]. The precipitation of solute atoms is suitable for filling nano-sized voids at both low and high temperatures [31].

2.3.2 Low Melting Point Alloys as Reinforcement

Low melting point alloys are used as reinforcement in metallic matrixes [6]. Hollow fibers or capsules are used to encapsulate the low melting point alloy (solder) and the ceramic tubes are reinforced metal based composites [31, 32]. The healing mechanism is technically similar to the polymeric and ceramics self-healing. The solder is released from the capsule at elevated temperatures when the host material undergoes damage and it wets the damaged surface, fills and closes it up by capillary pressure, surface tension and solidification [31, 32]. Microencapsulation is not challenging in polymers yet it is demanding for metals and proffers challenges in design of the microcapsule and ceramic capsule holder endowed with holes [31].

2.3.3 Additives

Self-healing aiding additives, such as shape memory alloy wires and, NPs, are introduced to reinforce metallic substrates or to fill up voids via swelling, respectively [31].

2.3.4 Electro Healing

Electro healing is an electrochemistry based approach which entails immersion of a damaged metal into an electrolyte solution with a voltage applied to trigger an electrochemical reaction which produces a product used to fill the cracks [31]. Although the material is not altered and produces strong adhesion between the old and new materials, it requires a through-thickness crack to self-heal [31].

2.4 *Concrete Substrates*

Concretes are subjected to harsh preparation and working conditions, prompting ingress with the formation of cracks [8]. Although prone to cracks, they possess a self-healing phenomenon as observed in old concrete structures lined with white crystalline material, suggesting the ability to self-seal the cracks with chemical by-products when exposed to hydration and carbon dioxide in air [33]. This natural healing process can be enhanced by addition of a catalyst and the subsequent techniques details concrete auto-repairing.

2.4.1 Hollow Fiber and Microcapsule Encapsulation

Self-healing agents are loaded into hollow fibers and/or microcapsules and incorporated into the concrete matrix [8]. The mechanism is the same as in the polymers. Pelletier et al. [34] reported on the encapsulation of sodium silicate solution in polyurethane (PU) microcapsules with a diameter range of 40–800 μm were introduced into cement mortar. Recovery of flexural strength was approximately 20–26%, while for control specimens only an approximately of 9–10% of strength was recovered.

2.4.2 Bacteria

Biological microorganisms, bacteria, can also be introduced into the concrete for self-healing purposes. This approach is prominently used in construction and entails encapsulation of bacteria into vessels to onset healing upon damage. The bacteria initiate a reaction which produces calcium carbonate precipitates, and the precipitates fill the available voids [8]. Expanded clay aggregates can be used as vessels for the bacteria and calcium lactate, and porous clay aggregates such as lightweight aggregates (LWA) offer conducive environment for precipitation of the bacteria [35, 36]. However, the effectiveness of the LWA system depend on several factors such as the aggregate water content, aggregate spacing, and aggregate pore arrangement [37, 38]. Alghamri et al. [39], utilised LWA of about 4–8 nm diameter as vessels for sodium silicate healing agent for cracks in concrete. The healing strength of the LWA loaded with sodium silicate were tested by pre-cracking them up to 300 μm crack width at 7 days and immersed in water. They showed about 80% recovery of the pre-cracking after 28 days.

Wang et al. [40] investigated the possibility of using silica gel or polyurethane as the transporter for protecting the bacteria. Their experimental results showed that silica gel immobilized bacteria exhibited a greater activity than polyurethane immobilized bacteria, therefore, the thermogravimetric results showed more $CaCO_3$ precipitated in silica gel (25% by mass) than in polyurethane (11% by mass). However, cracked mortar samples cured by polyurethane immobilized bacteria had a higher strength regain (60%) and lower water permeability coefficient (10^{-10}–10^{-11} m/s), compared with samples cured by silica gel immobilized bacteria which exhibited a strength regain of 5% and a water permeability coefficient of 10^{-7}–10^{-9} m/s. These results showed that polyurethane has more potential to be utilized as a bacterial carrier for self-healing of concrete cracks.

2.4.3 Shape-Memory Polymers (SMPs)

SMPs discussed in the polymer section can be incorporated into the concrete matrix. The technique is based on the assumption that early concrete cracks result from shrinkage and thermal effect during the curing period [8]. The SMPs are temperature dependent hence fill the voids by expansion [8].

2.4.4 Super-Absorbent Polymers

Super absorbent polymers/hydrogels can be incorporated into concrete and fill voids by swelling with hydration [8].

2.4.5 Additives to Aid Hydration of Concrete

Cement admixtures ensure continuous hydration of the cement prior to the initial curing period and this allows intrinsic self-healing [8]. The possible additive range include fly ash, blend of SiO_2 and Al_2O_3, lime and carbonates [8].

2.5 Asphalt Substrates

Asphalt is a self-healing material that is mainly used in the repairing of cracks in pavement constructions in highways and airports [12]. The asphalt self-healing of micro cracks offers reduced pavement repairing cost because it is durable and fully capable of repairing. Its self-healing mechanism is still not well understood but it is hypothesized to be due to intrinsic responses which include, molecular diffusion and inter-diffusion at crack interface [12].

3 Preparation of Self-healing Substrates

Several chemistry approaches are used in the preparation of self-healing substrates and the subsequent sections give an overview of the different techniques.

3.1 *In Situ Emulsification Polymerization*

In situ polymerization application dates back to the 1980s and progressed to industrial applications in the 1990s [19]. It entails vigorous sonication of a biphasic liquid to produce oil emulsions in water or water emulsions in oil and hence monomers undergo polymerization to form a shell encapsulating a suspended healing agent liquid. Appropriate monomers and initiators are required for formation of the capsule wall and they are dissolved in either the dispersed (core material droplet) or continuous phase. The microcapsules will be generated via polymerization at the surface of the core material [19]. It is often used in the micro-encapsulation approach to prepare microencapsulated healing agents and it offers the following advantages: controlled microcapsule size and shell thickness, affordability, simplicity and it can be extended to industrial applications although it requires long reaction durations [19]. Jialan et al. [41] used the in situ polymerization method to prepare self-healing microcapsules with urea formaldehyde resin and epoxy resin E-5 l as wall and core materials, respectively.

3.2 *Interfacial Polymerization*

Interfacial polymerization also produces micro-capsules in oil-in-water emulsions [19]. It uses the two phases (oil and water phase) to form an emulsion. It differs from in situ polymerization by the solvation of the reactive monomers individually into both the phases. The oil-phase is emulsified in the water phase by vigorous sonication. This reaction produces a polymer membrane at the oil and water phase interface, and this membrane surrounds the active compound (healing agent) and thus forms micro/nano-capsules [42]. Liu et al. [43] fabricated two kinds of self-healing microcapsules embedding waterborne epoxy resin and curing agent using the interfacial polymerization method for application to repair micro-damage and micro-crack in asphalt materials. Interfacial polymerization is advantageous because it is a fast, mild and flexible reaction which offers high encapsulation efficiency and its downfall, traces of unreacted shell microcapsules, does not negates the technique [19].

3.3 *Pickering Emulsion Templating*

Pickering emulsion involves two steps: (i) preparation of a stable pickering emulsion (solid-particle-stabilized emulsion) via mechanical actions used to self-assemble the colloidal particles at the interface; (ii) grafting of the colloidal particles adsorbed on the core material droplets using physical and chemical cross linking to form a stable shell [19]. This encapsulation technique prepares highly

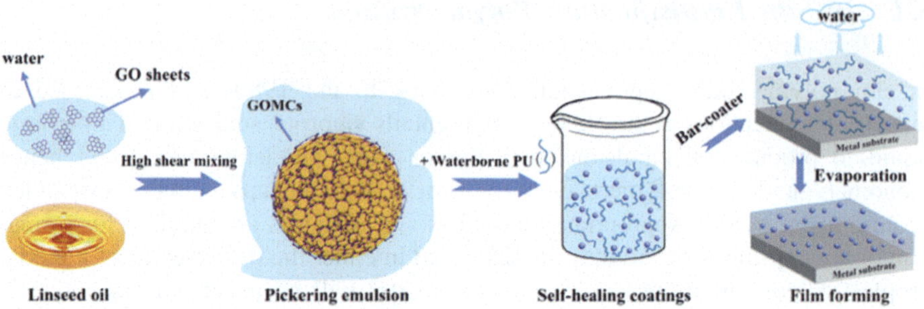

Fig. 9 Schematic representation of the preparation of the GOMCs microcapsules using the pickering emulsion approach. Reprinted with permission from Elsevier. Self-assembled graphene oxide microcapsules in Pickering emulsions for self-healing waterborne polyurethane coatings, 151, Jing Li, Qingkang Feng, Jincan Cui, Qianqian Yuan, Hanxun Qiu, Shanglin Gao, Junhe Yang, 282–290, Copyright, 2017 [44]

stable emulsions endowed with lower toxicity, recyclability, affordability and suitability for chemicals difficult to encapsulate using other techniques [19]. Li et al. [44] reported on the preparation of graphene oxide microcapsules (GOMCs) encapsulating linseed oil healing agent. The nanocapsule shells of GOMCs were prepared via liquid crystalline assembling of graphene oxide (GO) sheets, forming at liquid–liquid interface in pickering emulsions. Ultimately, the GOMCs were compounded into a host material, polyurethane matrix to enable facile fabrication of self-healing composite coatings on hot-dip galvanized steel surfaces. Figure 9 shows the GOMCs preparation.

3.4 *Mini-emulsion Polymerization*

Mini-emulsion uses surfactants to stabilize the emulsions and high-energy homogenization to interface the oil phase to the water phase and results in production of stable droplets with nano-sizes ranging from 50 to 500 nm [19].

3.5 *Solvent Evaporation/Solvent Extraction*

Solvent evaporation entails dissolution of a polymer shell and core material in a water-immiscible volatile organic solvent like dichloromethane and chloroform. This solution is then added dropwise into an aqueous solution containing a surfactant (Poly vinyl alcohol or Polyvinylpyrrolidone) under vigorous stirring and polymer droplets are formed containing the core material. As the stirring continues the droplets will harden to microcapsules, and a third liquid (used to dissolve a healing agent) is used to extract the solvent from the microcapsules. The solvent inside the microcapsules is replaced by the healing agent dissolved in the 3rd liquid [19]. This approach offers simplicity and mild reaction requirements [19].

3.6 *Sol-Gel*

In sol-gel a molecular precursor forms an oxide network via poly-condensation in a liquid. It is suited for low temperature and mild reaction condition microcapsule preparation. It is a fairly time-consuming technique and stability of the microcapsules is a challenge because they are prone to formation of micro-cracks and pores. However, the challenges do not negate the approach for silica capsules, since silica is a good diffusion barrier and is able to maintain the chemical reactivity of the encapsulated healing agent. The silica micro-capsules are prepared via hydrolysis and condensation of an orthosilicate, e.g., tetraethyl orthosilicate [19].

3.7 *Click Chemistry*

Copper(I) catalysed azide-alkyne cycloaddition (CuAAc) is an ideal preparation method for self-healing substrates [45]. Self-healing substrates require formation of chemical bonds at a faster rate at ambient temperatures. CuAAc is capable of attaining these substrates via use of multivalent liquid azides and alkynes in rapid cross-linking chemistries. The liquid azide and alkyne can potentially glue damages or cracks and hence self-heal the damage. Since autonomous self-healing at ambient temperatures is preferred over nonautonomous, the CuAAc click chemistry method uses ambient to low temperatures and the autonomous ability can be incorporated in the CuAAc method e.g., incorporation of materials with internal and autocatalytic chelation effects [45].

4 Applications of Self-healing Substrates

4.1 *Electronic Devices*

The modern age uses batteries and supercapacitors for power source and flexible, lightweight, portable, and wearable electronic energy storages are new developments intended to keep up with the changing world demands. Supercapacitors are mostly built on metal sheets, plastic films, paper, and textile fabric substrates and they are prevalently replacing batteries as power sources [46]. The surge in the application of supercapacitors as high performance energy storage is ascribed to their intriguing properties such as fast charge rates, high power density and long cycle lifetime [17, 47]. Vangari et al. [48] reviewed materials and fabrication methods for supercapacitors and high cost and complicated fabrication processes were some of the challenges in this field. To mitigate these challenges, self-healing substrates can prolong the lifetime of the supercapacitor and hence lower maintenance and production cost. The self-healing substrates can potentially be applied in the fabrication, and coating of a wider range of electronic devices such as electrochemical devices, solar cells, dielectric actuator and batteries [1]. Flexible supercapacitors which are more advantageous should be able to fold and roll, and their performance should not be influenced by their damage. Yet in reality, most of the flexible supercapacitors are made up of polydimethylsiloxane which is a limited material with non-self-healing capabilities and poor interfacial adhesion. Hence the stresses exerted during its application compromise its functionality. The issue can be alleviated with self-healable supercapacitors that can heal after use-induced damages and maintain functionality [47].

Polymers or gel self-healing substrates are preferred for supercapacitors because of their structure and flexibility. But in their natural conformity they take longer to self-heal which relies on a stimulus and the mechanism is vague [47]. For efficient application of the supercapacitors they should be able to heal faster and not compromise functionality. Maitra et al. [49] developed a tribo-electric driven self-charging self-healing asymmetric supercapacitor (power cell) using magnetic cobalt ferrite grown on a stainless steel (ss) fabric as positive electrode and iron oxide decorated reduced graphene oxide grown on a ss fabric as negative electrode, separated by KOH soaked self-healing polymer hydrogel electrolyte membrane. Hu et al. [47] improved the polymer or gel drawbacks by incorporating H-bonds on the skeleton of the polymer. They used known Poly vinyl alcohol (PVA), widely used in supercapacitors, with a natural product Phytic acid (PA). The P=O/P-O bonds of the PA allowed multiple hydrogen bonding with the PVA and improved its performance. The PVA/PA substrates displayed stretchability, self-healing and recyclability properties. The mechanism was investigated using density functional theory which revealed hydrogen bonding between the gels as accountable for the self-healing capability. Hu et al. [50] fabricated a supercapacitor with modified flour electrolyte and electrode. The modern flexible supercapacitor inherits intrinsic self-healing, stretchability after self-healing, and biodegradability properties. Capacitance and full stretchability was successfully retained after 40 autonomous self-healing cycles [50].

Nakahata et al. [51] integrated supramolecular and polymer chemistries to prepare a cyclodextrin and hydrophobic guest molecule on the polymer chain. The hydrogel inherited high toughness and a self-healing ability independent of wet or dry environmental conditions as a result of the reversible nature of host–guest interaction. The host-gel self-healing substrate was evaluated as a scratch healing coating and can expand to vehicle bodies, outer housing for electronic devices, and so on.

Wu et al. [52] investigated a self-healing elastomer prepared via a one-pot poly-condensation chemical reaction. The reaction involved bis(3-aminopropyl)-

terminated poly-(dimethylsiloxane) and 2,4′-tolylene diisocyanate and coordinated by Al(III) metal ions. The elastomer substrate endowed with quadruple hydrogen bonds resulted in self-healing and stretchability properties and the Al-coordination bonds enhanced the toughness and elasticity of the substrate. The elastomer substrate can be applied in the fabrication of flexible electronic skin.

4.2 Food Packaging

Passive food packaging is advancing to active, smart food packaging to improve the safety and quality of food [53]. In smart packaging, the self-healing and self-cleaning (anti-bacterial) properties are the key players [53]. It is critical for food industries to endorse smart packaging because it keeps the food fresh and blocks unwanted foreign material from infecting or spoiling the food. The smart packaging films should be mechanically reliable, endowed with controllable gas permeability and durability to withstand transportation stresses and maintain the freshness of the food [54]. If the standard of food packaging films is compromised it can adversely affect the consumers, due to deteriorated food quality [54]. Xuan et al. [55] fabricated self-healing, antibacterial and metal-ion sensing coatings for food industries. The coatings were prepared using the layer by layer (LBL) self-assembly via host-guest interaction using NPs: MoS_2 nanosheets; beta-cyclodextrin (β-CD)-modified poly (ethylenimine) (PEI); and, adamantane (AD)-modified poly (acrylic acid) (PAA). The MoS_2/β-CD-PEI/AD-PAA coating offers antibacterial, self-healing and sensing properties. Hu et al. [54] used the LBL technique to prepare self-healing and anti-fogging, transparent substrates of acrylamide-modified chitosan and alginate aldehyde for food packaging. The films showed repetitive self-healing properties and successfully healed scratches that penetrated to the underlying substrates.

4.3 Textiles

The Self-healing substrates represent an important class of modern materials with potential application even in woven and non-woven textiles. Textiles are prone to damage due to environmental and biological reasons and hence lack long term durability [56]. This might not be an issue for some individuals but it is a challenge for demanding careers such as the army, fire fighters and medical practitioners. Chen et al. [57] reported the fabrication of intumescent flame retardant and self-healing super hydrophobic coatings on cotton fabric using a technical and simple solution dipping method. Briefly, the coating is made up of a bilayer of ammonium polyphosphate and branched poly(ethylenimine) prepared via dipping LBL assembly, and then layered with fluorinated-decyl polyhedral oligomeric silsesquioxane. The self-healing capability is ascribed to the silsesquioxane and bilayer of ammonium polyphosphate/branched poly(ethylenimine) coupled with micro and nanoscale roughness of the cotton material. Liu et al. [58] prepared a mechanically induced self-healing superhydrophobic textile coating using polydopamine nanocapsule-encapsulated hydrophobic agents such as octadecylamine and octadecanethiol. Loss of the textile hydrophobicity induced by damage will trigger release of the hydrophobic agents to restore its hydrophobicity in addition to self-healing.

4.4 Medical Applications

Self-healing substrates are used in the fabrication and coating of medical devices or applications, e.g., sensors (electrochemical), soft robot health care monitors, E-skin and wound dressing [1]. Li et al. [59] prepared a chitosan based and poly(ethylene glycol) hydrogel with intriguing self-adapting and self-healing properties. The substrate was tested for biomedical application (wound healing and drug delivery). It was used for wound healing in vivo to treat haemorrhaging livers in rats and it surpassed the conventional systems. In addition, the self-adapting and self-healing substrate as a drug carrier can adapt to the ever-changing tissue environments and hence reach areas which are impossible for traditional systems [59]. Qiao et al. [60] designed a ternary hybrid ionic conductor that integrates high sensitivity, stretchability, excellent elasticity, self-healing and self-adhesiveness properties. Their ionic hydrogel consisted of sodium alginate, tannic acid and cross linked polyacrylamide and it showed excellent sensitivity at small strains, low power usage, wide strain window, strong self-adhesiveness and healing properties. It was applied in real time monitoring of limb motions and subtle muscle movements of the human body with high sensitivity and reproducibility.

Self-healing hydrogel sensors are a special niche in science and technology because of their flexibility, biocompatibility, enhanced sensitivity, lightweight, robust stretchability and adhesiveness [61]. The good quality loaded hydrogel sensors are envisioned for wearable sensors to monitor human activities and support diagnosis. The hydrogel sensors can translate external stimuli such as heat, pressure, voice, strain into electrical signals, which is a key property for human motion sensing. The conductivity of the hydrogel is a concern which can be improved by modifying with materials capable of conducting such as carbon nanotubes, graphitic materials, polyaniline and polythiophene. Modifying the hydrogel with graphitic material results in the formation pf a stimulus or photo responsive sensor which is essential in human motion sensing [61].

Wound dressings are also envisioned to be self-healing. An ideal wound dressing should be endowed with unique properties such as self-healing, tissue adhesion, cell affinity, and mechanical toughness, to assist in the repair of skin and muscle damage [62]. However, integration of all these properties into one product to attain the idealized wound dressing is a challenge. Chen et al. [62] fabricated a wound dressing hydrogel with ultra-tough and self-healing cell affinity and tissue adhesiveness properties. The hydrogel was prepared via dopamine oxidized sodium alginate and polyacrylamide chains. The hydrogels were preferred for wound dressing because of their properties: elasticity, flexibility, biocompatibility, high water volume and responsiveness to physiological changes.

Tooth decay is a major concern across ages, culture and race and causes financial burden especially for seniors. Typically, resin composites are used to restore the cavities developed by tooth decay but the resin composite has a shorter life span due to the teeth and mouth environment which promotes its micro cracking. The micro-cracking of resins is still an issue even with major advancements such as whiskers, fibers and alterations in the chemistry of the composites. The ultimate solution is development of resin composite capable of autonomously detecting cracks and repairing them via polymerization [63]. The advantage of prolonged resin service life and self-repairing will lower maintenance and re-installation cost and enacts the importance of self-healing substrates even in dentistry [63].

4.5 Construction/Coatings

Corrosion is an industrial concern globally, as it continues to deteriorate infrastructure, effect economic losses and compromise safety. Polymers have been used as a standard coating due to their properties such as light weight, facile processibility, and chemical stability for coating various materials e.g., vehicles, infrastructure, and sporting equipment. The polymers offer passive protection and when subjected to harsh conditions they lose their protective ability, and this triggers corrosion [5, 64]. To mitigate the corrosion and micro-cracking issues, materials which offer both passive and active protection are a necessity. Passive and active coatings which can protect and heal micro-cracks are envisaged to solve the corrosion issue [64].

The issue of corroding infrastructure is a global concern, and the US infrastructure is deteriorating as a result of corrosion. The US infrastructure industry adopted steel reinforced concrete in the 1800s and that is widely used in infrastructure e.g. in bridges. A deterioration in the bridges decks was observed in the mid-1970s and is known to be catalyzed by salty water. The steel degradation follows the subsequent pattern of occurrences: firstly oxygen or H_2O penetrates the concrete to the steel surface and triggers electrochemical reactions at the surface to produce rust; then the rust will accumulate, resulting in a stressed surface which will produce cracks; detrimentally, the rust does not adhere on the surface long-term but will flake off degrading the steel surface. Epoxy coatings have been used to delay rusting although they are brittle hence susceptible to damage and peeling off. As a better alternative, a self-healing coating could remedy the situation by providing extended protection of the steel and the ability to self-heal in case of damage [65]. Li et al. [66] prepared self-healing rebar coatings for steel-reinforced concrete using microencapsulation. A healing monomer, tung oil was loaded into microcapsules and compounded into epoxy. The self-healing coating showed successful prevention of corrosion after 150 days of accelerated corrosion testing.

Golim et al. [67] reports on prompting self-Healing abilities with intermetallic materials $MoSi_2$ by improving the durability of Thermal Barrier Coating (TBC). Due to high temperature operations self-healing materials become oxidized and lose its healing capabilities. The encapsulation procedure was synthesized through a precipitation process with colloidal aluminum hydroxide as precursor and variations on calcination method. Calcination at 900 °C under low pressure formed the optimum alumina encapsulation and reduction in premature oxidation of the self-healing material was improved.

Jamshidnejad et al. [68] Investigates the manufacturing and characterization of epoxy and urethane based self-healing coatings by encapsulation of olive leaf extract as oxidization inhibitor. HOPDMS and PDES were utilized as healants in urethane capsules. Furthermore, olive leaf extract was added to the capsules which contain the healing agents to increase the corrosion defence of the coating. Polarization and electrochemical impedance spectroscopy tests, were used to evaluate the effectiveness of the olive leaf extract as a corrosion inhibitor for steel in chloride solutions, an inhibition efficiency of approximately 80% was recorded for ethanol extracted inhibitor at 300 ppm. The self-healing capacity of the coatings containing 3,4, and 5% of capsules were studied, and it was observed that with an increasing of capsules content of the coating, the corrosion protection improves, therefore a coating with 5 wt% of capsules showed no corrosion in the adjacent area of a damage.

4.6 Water Treatment

Filtration membranes are widely used in water treatment to filter unwanted or foreign materials from our water [69]. However the membrane technology suffers from compromised functionality due to installation and operation. During installation and operation, the membrane can be damaged which can lower the quality of the water. Membrane integrity monitoring systems currently used may fail to locate the breach in the membranes and cannot repair the damage. Self-healing substrates can revolutionize the membrane technology [70]. Hydrogels are widely used in membrane fabrication owing to their antifouling and antibacterial properties. These hydrogels interestingly possess self-healing abilities via hydrogen bonding and molecular inter-diffusion [70]. Bezawit et al. [70] prepared a hydrogel pore-filled membrane via in situ graft polymerization of 2-acrylamide-2-methyl-1-propanesulfonic acid onto microporous poly(ethersulfone) substrates. Kim et al. [71] used microvascular networks to fabricate an autonomic self-healing membrane by incorporating a healing agent methylene diphenyl diisocyanate housed by an ethylene propylene capillary into the membrane. The mechanism of self-healing is triggered by the damage of the membrane which will rupture the capillary and release the healing monomer. The healing agent will undergo water-induced phase separation to increase its viscosity and it will react with water and swell to fill the created void or cracks. An 82% membrane recovery was recorded by this healing mechanism.

5 Conclusion and Outlook

The remarkable age of technology has birthed self-healing substrates which mimic natural systems to maintain or replace their integrity. Biological systems are uniquely endowed with damage sensing systems which trigger repairing while inorganic systems lack such a niche and hence they are prone to less durability, high maintenance cost and even compromised safety as in the case of industrial construction. Inspiration from these biological systems has transpired self-healing inorganic systems. Polymeric substrates, advantaged by their properties are the most studied self-healing substrates, whereas bottleneck regions such as self-healing metallic and ceramic substrates are still in early stages. Substrates can self-heal autonomously (chemical nature) or non-autonomously via addition of healing promoting monomers as healing agents. This explicates the reason that polymers are the most studied substrates. Polymers naturally possess the chemical and physical interactions to enable self-healing i.e., thermoplastic polymers only require heating above their glass transition state for onset of molecular inter-diffusion resulting in restoration of the damaged structure. The other class of polymers, thermosets, can be modified with external healing agents in various holders such as capsule and micro-vascular networks to attain self-healing which can be extended to industrial applications. These techniques are also for ceramics, concrete, and metallic substrates. Integration of the self-healing substrates in our day to day life, in applications such as construction, medicine, energy, and food packaging will advantageously afford durable and trustworthy products, lower maintenance cost and fewer accidents. The world needs self-healing substrates but the challenge is translating the proof of concept studies reported in patents and research articles into practical applications in the medicinal, energy storage, construction, coatings and food packaging and water purification fields. Although it is a daunting task it has the potential to revolutionize our world.

Acknowledgements The authors would love to acknowledge the Department of Science and Technology (DST), South Africa, Mintek and National Centre for Nano-Structured Materials (CSIR).

References

1. Tan, Y.J., Wu, J., Li, H., Tee, B.C.: Self-healing electronic materials for a smart and sustainable future. ACS Appl. Mater. Interfaces **10**(18), 15331–15345 (2018)
2. Yang, Y., Ding, X., Urban, M.W.: Chemical and physical aspects of self-healing materials. Prog. Polym. Sci. **49**, 34–59 (2015)

3. Joseph, J.P., Singh, A., Pal, A.: Molecular design approaches to self-healing materials from polymer and its nanocomposites. In: Smart Polymer Nanocomposites, pp. 181–218. Springer (2017)
4. Zhang, F., Ju, P., Pan, M., Zhang, D., Huang, Y., Li, G., Li, X.: Self-healing mechanisms in smart protective coatings: a review. Corros. Sci. (2018)
5. Ataei, S., Khorasani, S.N., Neisiany, R.E.: Biofriendly vegetable oil healing agents used for developing self-healing coatings: a review. Prog. Org. Coat. **129**, 77–95 (2019)
6. Alaneme, K., Bodunrin, M.: Self-healing using metallic material systems–a review. Appl. Mater. Today **6**, 9–15 (2017)
7. Blazej Grabowski, C.C.T.: Self-healing metals. In: Advances in Polymer Science. Springer Nature (2015)
8. Lucas, S.S., Von Tapavicza, M., Schmidt, A.M., Bertling, J., Nellesen, A.: Study of quantification methods in self-healing ceramics, polymers and concrete: a route towards standardization. J. Intell. Mater. Syst. Struct. **27**(19), 2577–2598 (2016)
9. Guadagno, L., Naddeo, C., Raimondo, M., Barra, G., Vertuccio, L., Sorrentino, A., Binder, W.H., Kadlec, M.: Development of self-healing multifunctional materials. Compos. Part B: Eng. **128**, 30–38 (2017)
10. Wu, D.Y., Meure, S., Solomon, D.: Self-healing polymeric materials: a review of recent developments. Prog. Polym. Sci. **33**(5), 479–522 (2008)
11. Aïssa, B., Therriault, D., Haddad, E., Jamroz, W.: Self-healing materials systems: overview of major approaches and recent developed technologies. Adv. Mater. Sci. Eng. **2012** (2012)
12. Sun, D., Sun, G., Zhu, X., Guarin, A., Li, B., Dai, Z., Ling, J.: A comprehensive review on self-healing of asphalt materials: Mechanism, model, characterization and enhancement. Adv. Colloid Interface Sci. (2018)
13. Billiet, S., Hillewaere, X.K., Teixeira, R.F., Du Prez, F.E.: Chemistry of crosslinking processes for self-healing polymers. Macromol. Rapid Commun. **34**(4), 290–309 (2013)
14. Yang, Y., Kushima, A., Han, W., Xin, H., Li, J.: Liquid-like, self-healing aluminum oxide during deformation at room temperature. Nano Lett. **18**(4), 2492–2497 (2018)
15. García, S., Fischer, H., Van Der Zwaag, S.: A critical appraisal of the potential of self healing polymeric coatings. Prog. Org. Coat. **72**(3), 211–221 (2011)
16. Dry, C.: Procedures developed for self-repair of polymer matrix composite materials. Compos. Struct. **35**(3), 263–269 (1996)
17. Dry, C.M., Sottos, N.R.: Passive smart self-repair in polymer matrix composite materials. In: Smart Structures and Materials 1993: Smart Materials. International Society for Optics and Photonics (1993)
18. Murphy, E.B., Wudl, F.: The world of smart healable materials. Prog. Polym. Sci. **35**(1–2), 223–251 (2010)
19. Zhu, D.Y., Rong, M.Z., Zhang, M.Q.: Self-healing polymeric materials based on microencapsulated healing agents: from design to preparation. Prog. Polym. Sci. **49**, 175–220 (2015)
20. Chowdhury, R.A., Hosur, M.V., Nuruddin, M., Tcherbi-Narteh, A., Kumar, A., Boddu, V., Jeelani, S.: Self-healing epoxy composites: preparation, characterization and healing performance. J. Mater. Res. Technol. **4**(1), 33–43 (2015)
21. Pratama, P.A., Sharifi, M., Peterson, A.M., Palmese, G.R.: Room temperature self-healing thermoset based on the Diels-Alder reaction. ACS Appl. Mater. Interfaces **5**(23), 12425–12431 (2013)
22. Jin, H., Mangun, C.L., Stradley, D.S., Moore, J.S., Sottos, N.R., White, S.R.: Self-healing thermoset using encapsulated epoxy-amine healing chemistry. Polymer **53**(2), 581–587 (2012)
23. Wilson, G.O., Andersson, H.M., White, S.R., Sottos, N.R., Moore, J.S., Braun, P.V.: Self-healing polymers. In: Encyclopedia of Polymer Science and Technology (2002)
24. Mauldin, T.C., Kessler, M.: Self-healing polymers and composites. Int. Mater. Rev. **55**(6), 317–346 (2010)
25. Chen, X., Wudl, F., Mal, A.K., Shen, H., Nutt, S.R.: New thermally remendable highly cross-linked polymeric materials. Macromolecules **36**(6), 1802–1807 (2003)

26. Xiang, Z., Zhang, L., Yuan, T., Li, Y., Sun, J.: Healability demonstrates enhanced shape-recovery of graphene-oxide-reinforced shape-memory polymeric films. ACS Appl. Mater. Interfaces **10**(3), 2897–2906 (2018)
27. Feng, X., Fan, J., Li, A., Li, G.: Multi-reusable thermoset with anomalous flame triggered shape memory effect. ACS Appl. Mater. Interfaces (2019)
28. Madhan, M., Prabhakaran, G.: Self-healing ability of structural ceramics–a review. In: International Conference on Intelligent Robotics, Automation, and Manufacturing. Springer (2012)
29. Osada, T., Kamoda, K., Mitome, M., Hara, T., Abe, T., Tamagawa, Y., Nakao, W., Ohmura, T.: A novel design approach for self-crack-healing structural ceramics with 3D networks of healing activator. Sci. Rep. **7**(1), 17853 (2017)
30. Rajak, P., Kalia, R.K., Nakano, A., Vashishta, P.: Faceting, grain growth, and crack healing in alumina. ACS Nano **12**(9), 9005–9010 (2018)
31. Manuel, M.V.: Principles of self-healing in metals and alloys: an introduction. In: Self-healing Materials: Fundamentals, Design Strategies, and Applications. Wiley-VCH, Chichester (2009)
32. Srivastava, V., Gupta, M.: Approach to self healing in metal matrix composites: a review. Mater. Today: Proc. **5**(9), 19703–19713 (2018)
33. Li, V.C., Yang, E.-H.: Self healing in concrete materials. In: Self Healing Materials, pp. 161–193. Springer (2007)
34. Pelletier, M.M., Brown, R., Shukla, A., Bose, A.: Self-healing concrete with a microencapsulated healing agent. Cem. Concr. Res. (2011)
35. Jonkers, H.M.: Bacteria-based self-healing concrete. Heron **56**(1/2) (2011)
36. Wiktor, V., Jonkers, H.M.: Assessment of the crack healing capacity in bacteria-based self-healing concrete. In: Proceedings of 3rd International Conference on Self Healing Materials, Bath, UK (2011)
37. Reynolds, D., Browning, J., Darwin, D.: Lightweight aggregates as an internal curing agent for low-cracking high-performance concrete. University of Kansas Center for Research, Inc. (2009)
38. Souradeep, G., Kua, H.W.: Encapsulation technology and techniques in self-healing concrete. J. Mater. Civ. Eng. **28**(12), 04016165 (2016)
39. Alghamri, R., Kanellopoulos, A., Al-Tabbaa, A.: Impregnation and encapsulation of lightweight aggregates for self-healing concrete. Constr. Build. Mater. **124**, 910–921 (2016)
40. Wang, J., Van Tittelboom, K., De Belie, N., Verstraete, W.: Use of silica gel or polyurethane immobilized bacteria for self-healing concrete. Constr. Build. Mater. **26**(1), 532–540 (2012)
41. Jialan, Y., Chenpeng, Y., Chengfei, Z., Baoqing, H.: Preparation process of epoxy resin microcapsules for self-healing coatings. Prog. Org. Coat. **132**, 440–444 (2019)
42. Calabrò, V.: Engineering aspects of membrane bioreactors. In: Handbook of Membrane Reactors, pp. 3–53. Elsevier (2013)
43. Hager, M.D., Greil, P., Leyens, C., van der Zwaag, S., Schubert, U.S.: Self-healing materials. Adv. Mater. **22**(47), 5424–5430 (2010)
44. Li, J., Feng, Q., Cui, J., Yuan, Q., Qiu, H., Gao, S., Yang, J.: Self-assembled graphene oxide microcapsules in Pickering emulsions for self-healing waterborne polyurethane coatings. Compos. Sci. Technol. **151**, 282–290 (2017)
45. Döhler, D., Michael, P., Binder, W.H.: CuAAC-based click chemistry in self-healing polymers. Acc. Chem. Res. **50**(10), 2610–2620 (2017)
46. Zhang, D., Miao, M., Niu, H., Wei, Z.: Core-spun carbon nanotube yarn supercapacitors for wearable electronic textiles. ACS Nano **8**(5), 4571–4579 (2014)
47. Hu, R., Zhao, J., Wang, Y., Li, Z., Zheng, J.: A highly stretchable, self-healing, recyclable and interfacial adhesion gel: preparation, characterization and applications. Chem. Eng. J. **360**, 334–341 (2019)
48. Vangari, M., Pryor, T., Jiang, L.: Supercapacitors: review of materials and fabrication methods. J. Energy Eng. **139**(2), 72–79 (2012)

49. Maitra, A., Paria, S., Karan, S.K.,Bera, R., Bera, A., Das, A.K., Si, S.K., Halder, L., De, A., Khatua, B.B.: Triboelectric nanogenerator driven self-charging and self-healing flexible asymmetric supercapacitor power cell for direct power generation. ACS Appl. Mater. Interfaces (2019)
50. Hu, M., Wang, J., Liu, J., Wang, P., Feng, Y., Wang, H., Nie, N., Wang, Y., Huang, Y.: A flour-based one-stop supercapacitor with intrinsic self-healability and stretchability after self-healing and biodegradability. Energy Storage Mater. (2018)
51. Nakahata, M., Takashima, Y., Harada, A.: Highly flexible, tough, and self-healing supramolecular polymeric materials using host–guest interaction. Macromol. Rapid Commun. **37**(1), 86–92 (2016)
52. Wu, X., Wang, J., Huang, J., Yang, S.: Robust, stretchable and self-healable supramolecular elastomers synergistically crosslinked by hydrogen bonds and coordination bonds. ACS Appl. Mater. Interfaces (2019)
53. Mlalila, N., Kadam, D.M., Swai, H., Hilonga, A.: Transformation of food packaging from passive to innovative via nanotechnology: concepts and critiques. J. Food Sci. Technol. **53**(9), 3395–3407 (2016)
54. Hu, B., Chen, L., Lan, S., Ren, P., Wu, S., Liu, X., Shi, X., Li, H., Du, Y., Ding, F.: Layer-by-layer assembly of polysaccharide films with self-healing and antifogging properties for food packaging applications. ACS Appl. Nano Mater. **1**(7), 3733–3740 (2018)
55. Xuan, H., Dai, W., Zhu, Y., Ren, J., Zhang, J., Ge, L.: Self-Healing, antibacterial and sensing nanoparticle coating and its excellent optical applications. Sens. Actuators B: Chem. **257**, 1110–1117 (2018)
56. Gaddes, D., Jung, H., Pena-Francesch, A., Dion, G., Tadigadapa, S., Dressick, W.J., Demirel, M.C.: Self-healing textile: enzyme encapsulated layer-by-layer structural proteins. ACS Appl. Mater. Interfaces **8**(31), 20371–20378 (2016)
57. Chen, S., Li, X., Li, Y., Sun, J.: Intumescent flame-retardant and self-healing superhydrophobic coatings on cotton fabric. ACS Nano **9**(4), 4070–4076 (2015)
58. Liu, Y., Liu, Y., Hu, H., Liu, Z., Pei, X., Yu, B., Yan, P., Zhou, F.: Mechanically induced self-healing superhydrophobicity. J. Phys. Chem. C **119**(13), 7109–7114 (2015)
59. Li, Y., Wang, X., Fu, Y.-N., Wei, Y., Zhao, L., Tao, L.: Self-adapting hydrogel to improve the therapeutic effect in wound-healing. ACS Appl. Mater. Interfaces **10**(31), 26046–26055 (2018)
60. Qiao, H., Qi, P., Zhang, X., Wang, L., Tan, Y., Luan, Z., Xia, Y., Li, Y.-H., Sui, K.: Multiple weak H-bonds lead to highly sensitive, stretchable, self-adhesive and self-healing ionic sensors. ACS Appl. Mater. Interfaces (2019)
61. Deng, Z., Hu, T., Lei, Q., He, J., Ma, P.X., Guo, B.: Stimuli-responsive conductive nanocomposite hydrogels with high stretchability, self-healing, adhesiveness and 3D printability for human motion sensing. ACS Appl. Mater. Interfaces (2019)
62. Chen, T., Chen, Y., Rehman, H.U., Chen, Z., Yang, Z., Wang, M., Li, H., Liu, H.: Ultratough, self-healing, and tissue-adhesive hydrogel for wound dressing. ACS Appl. Mater. Interfaces **10**(39), 33523–33531 (2018)
63. Wu, J., Xie, X., Zhou, H., Tay, F.R., Weir, M.D., Melo, M.A.S., Oates, T.W., Zhang, N., Zhang, Q., Xu, H.H.: Development of a new class of self-healing and therapeutic dental resins. Polym. Degrad. Stab. (2019)
64. Liu, C., Zhao, H., Hou, P., Qian, B., Wang, X., Guo, C., Wang, L.: Efficient graphene/cyclodextrin-based nanocontainer: synthesis and host-guest inclusion for self-healing anticorrosion application. ACS Appl. Mater. Interfaces **10**(42), 36229–36239 (2018)
65. Weishaar, A., Carpenter, M., Loucks, R., Sakulich, A., Peterson, A.M.: Evaluation of self-healing epoxy coatings for steel reinforcement. Constr. Build. Mater. **191**, 125–135 (2018)
66. Chen, Y., Xia, C., Shepard, Z., Smith, N., Rice, N., Peterson, A.M., Sakulich, A.: Self-healing coatings for steel-reinforced concrete. ACS Sustain. Chem. Eng. **5**(5), 3955–3962 (2017)

67. Golim, O., Prastomo, N., Izzudin, H., Hastuty, S., Sundawa, R., Sugiarti, E., Thosin, K.: Synthesis of alumina ceramic encapsulation for self-healing materials on thermal barrier coating. J. Phys. Conf. Ser. (2018) (IOP Publishing)
68. Jamshidnejad, Z., Afshar, A., RazmjooKhollari, M.A.: Synthesis of self-healing smart epoxy and polyurethane coating by encapsulation of olive leaf extract as corrosion inhibitor. Int. J. Electrochem. Sci. **13**, 12278–12293 (2018)
69. Nqombolo, A., Mpupa, A., Moutloali, R.M., Nomngongo, P.N.: Wastewater treatment using membrane technology. In: Wastewater and Water Quality. IntechOpen (2018)
70. Getachew, B.A., Kim, S.-R., Kim, J.-H.: Self-healing hydrogel pore-filled water filtration membranes. Environ. Sci. Technol. **51**(2), 905–913 (2017)
71. Kim, S.-R., Getachew, B.A., Kim, J.-H.: Toward microvascular network-embedded self-healing membranes. J. Membr. Sci. **531**, 94–102 (2017)
72. Zhang, H., Lamnawar, K., Maazouz, A.: Rheological modeling of the diffusion process and the interphase of symmetrical bilayers based on PVDF and PMMA with varying molecular weights. Rheol. Acta **51**(8), 691–711 (2012)

Self-supported Materials for Flexible/Stretchable Sensors

Zhangpeng Li, Jingxia Huang and Jinqing Wang

Abstract There is an increasing demand on the fabrication of robust, flexible, cost-effective, and eco-friendly self-supporting materials, such as yarns, fibers, papers-like, films, and monoliths, etc., due to their promising applications in flexible sensing fields. With the unique structure and outstanding properties, various functional materials with zero-dimensional, one-dimensional, and two-dimensional structures have been employed as promising building blocks for the assembly of self-supporting materials. In this chapter, the type, key parameter, and working principles of sensors are presented. Meanwhile, we summarize the sensing properties of different self-supporting materials through detailed cases. In addition, the challenges and opportunities of current sensors based on self-supporting materials are briefly discussed.

Keywords Self-supported materials · Sensors · Flexible · Devices

1 Introduction

Flexible electronics are flourishing toward complex platforms consisting of multi-functional components due to the eagerness of portable consumer electronics to be lightweight, flexible, and even wearable, and their potential usefulness in intelligent systems and wearable electronics [1]. As an important component of flexible electronics, flexible/stretchable sensors have attracted widespread attention in recent

Z. Li (✉) · J. Huang · J. Wang
State Key Laboratory of Solid Lubrication, Lanzhou Institute of Chemical Physics, Chinese Academy of Sciences, Lanzhou 730000, China
e-mail: zhangpengli@licp.cas.cn

J. Wang
e-mail: jqwang@licp.cas.cn

J. Huang
Center of Materials Science and Optoelectronics Engineering, University of Chinese Academy of Sciences, Beijing 100049, China

Inamuddin et al. (eds.), *Self-standing Substrates*, Engineering Materials, https://doi.org/10.1007/978-3-030-29522-6_9

years. Compared to traditional metal or inorganic semiconductor-based sensors with limited detection range (5% of strain) and complex preparation processes as well as the high fabrication cost [2], they possess several intriguing advantages, such as high sensitivity, favorable flexibility and stretchability, lightweight, biocompatibility, durability, and low power consumption, and so forth. As a result, they have great application prospects in various fields, including human body disease diagnosis/therapy/health detection [3, 4], artificial muscles [5], soft robotic skin [6], human-machine interfaces [7], human motion detection [8], and virtual reality and entertainment technology [9], etc.

Viewed from the present research status of flexible/stretchable sensors, high sensitivity and resolution, wide sensing range, rapid response, and excellent durability and environmental stability are the prerequisites for sensor application. Generally, the type and geometric morphology of sensing element significantly affects the performance of the sensor. Thus, the construction of a special structure sensing material is an effective approach to regulate the behavior of the sensor. In addition, new sensing mechanisms and fabrication techniques can also promote the evolution of flexible and stretchable sensors. Hence, in order to achieve above-mentioned outstanding sensing characteristics of the flexible/stretchable sensor, the choice of sensing material type and fabrication of unique morphology as well as design of suitable structure of sensor by a special processing technology is the crucial factors to be considered. Note that the published review papers have concentrated on the electrically conductive polymer composites [10, 11], functional nanomaterials [12], and carbon materials [13] for flexible and wearable strain sensors [14, 15], bio-electronics [16], and tactile sensors [17], etc., but flexible/stretchable sensors based on self-supporting materials have not been systematically addressed in any reviews. Self-supporting materials, such as yarns, fabrics, papers, films, and there-dimensional (3D) structure materials such as monoliths or bulks, refer to a kind of materials with basic rigidity and without deformation under natural state or non-stress. They are assembled from building blocks, including zero-dimensional (0D) nanoparticels, one-dimensional (1D) nanowires or nanofibers, and two-dimensional (2D) nanosheets materials, which have become potential for the development of highly efficient flexible/stretchable sensors, as shown in Fig. 1.

Motivated by the above reviews, the recent advances and applicable prospects for the progress of flexible/stretchable sensors are summarized in this chapter. Moreover, the basic sensing mechanism of flexible/stretchable sensors is presented in second part. Meanwhile, a comprehensive summary of sensing materials, including types, structures, properties, and applications, is introduced with detailed examples in the third part. In particular, the self-supporting sensing materials comprising yarns, fibers, like-papers, films, monoliths such as aerogels, and hydrogels, are discussed in detail. The recent achievement in the flexible/stretchable sensors by designing special conductive networks (such as porous structure [18], micro-crack structure [19], segregated structure [20], and wrinkled structure [21])

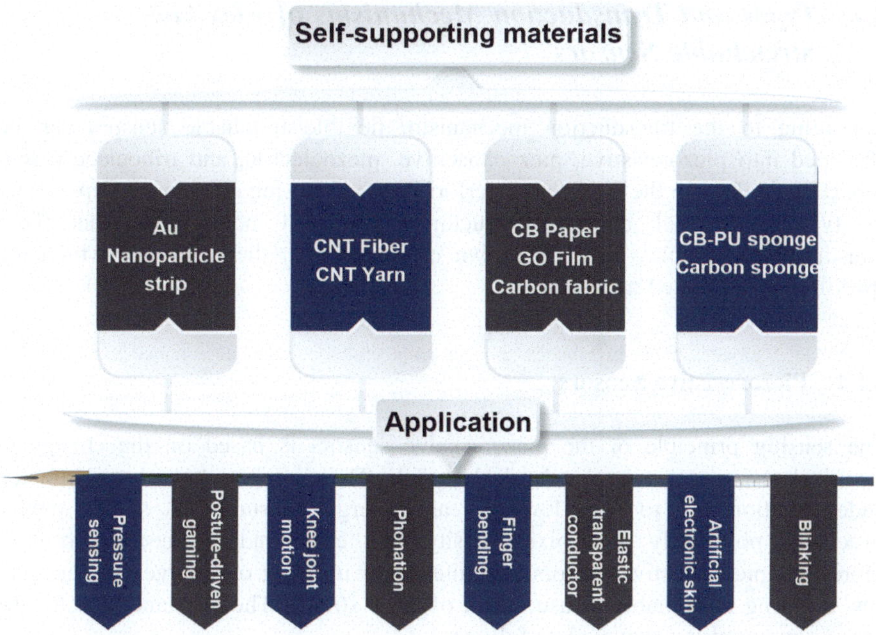

Fig. 1 Diagram of self-supporting materials and corresponding applications (*CNT* carbon nanotube; *CB* carbon black; *GO* graphene oxide; *PU* polyurethane)

and introducing multifunctional features (such as self-healing capability [22], good light transmission [23], and superhydrophobicity [24]) are also presented. In addition, the challenges and future opportunities for employ of self-supporting materials in designing flexible sensors will be discussed in the end of this chapter.

2 Fundamental of Flexible/Stretchable Sensors

A flexible/stretchable sensor represents an electronic device with various components that can bend freely, fold, and adhere to a regular or irregular surface to percept the stimulation from the physical world. Among them, the sensing material directly translates external stimuli into electrical signals. In general, the practical mechanical stimulation is very complex, but basically, there are some common types, such as strain, pressure, vibration, and shear force. Especially, a great deal of research interest is focused on strain and pressure monitoring. A detailed discussion of the types, transduction mechanisms, and key parameters of flexible/stretchable sensors are provided below.

2.1 Types and Transduction Mechanisms of Flexible/Stretchable Sensors

According to the transduction mechanism, flexible/stretchable sensors can be classified into piezoresistive, piezocapacitive, piezoelectric, and triboelectric sensors (Fig. 2). Due to the conductive performance of sensing component depends on the type of material, micro/nanostructure, fabrication method, and thus these transduction principles have their own characteristics, the details of which are specifically elaborated as follows.

2.1.1 Piezoresistive Sensors

The sensing principle of the piezoresistive sensors is based on the change in electrical resistance under mechanical stimuli. These sensors have been attracted wide attention due to their low cost and energy consumption, simple device structures, potentially high pixel density, and easy read-out mechanism [25]. Moreover, piezoresistive sensors are suitable for utilizing over large pressure ranges, enabling the reliable measurement of large strains. The resistance (R) for the sensing material is calculated as follow:

$$R = \frac{\rho L}{S}$$

where ρ and L are resistivity and length, respecitively. S represents cross-sectional area. In the piezoresistive sensors, the resistivity caused by the contact between different components and the change of geometric structure for the sensing materials are the main cause of the resistance change ($\Delta R/R$), which can be denoted as below:

$$\frac{\Delta R}{R} = (1 + 2v)\varepsilon + \frac{\Delta \rho}{\rho}$$

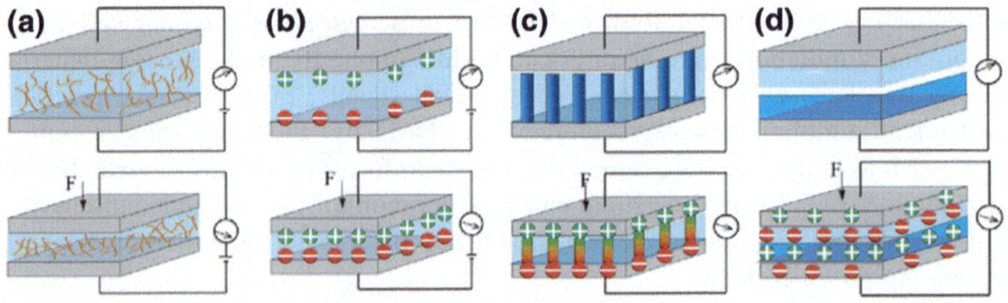

Fig. 2 Schematic diagram of **a** piezoresistive, **b** piezocapacitive, **c** piezoelectric, and **d** triboelectric mechanism. Reproduced with permission from Ref. [17]. Copyright 2017 Elsevier Ltd

where v represent Poisson's ratio and ε is applied strain. The sensitivity is expressed as $(\Delta R/R_0)/P$ for the pressure sensor and $(\Delta R/R_0)/\varepsilon$ for the strain sensor, respectively, where P is the pressure intensity. Generally, gauge factor (GF) represents the sensitivity of the strain sensor, and S refers to the sensitivity of the pressure sensor. The obtained GF or S value is constant when the relationship of resistance to strain or pressure is linear. Therefore, the value of GF or S is correlative with the deformation range strongly.

2.1.2 Piezocapacitive Sensors

The sensing principle of capacitive sensors is based on capacitance changes under the external stimuli, such as strain, normal forces, and shear forces. Capacitance sensors are one of the most economical and viable sensing devices due to the advantages of high strain sensitivity, high energy efficient, excellent dynamic response, good tolerance to temperature change, and simple structure requirements [26, 27], but are easily disturbed by external noise sources [28], such as in the form of electromagnetic waves. The capacitance (C) can be calculated by the following equation:

$$C = \frac{\varepsilon_0 \varepsilon_r A}{d}$$

where ε_0 and ε_r are the free space dielectric constant and relative static dielectric constant of the dielectric layer between the parallel plates, respectively. A is the overlap area of the parallel plates, and d represents the distance between the parallel plates. Generally, three variables (ε_r, A, and d) are sensitive to force changes. Among them, the change in A and d is respectively employed to measure shear forces [29] and normal forces [30], where the external forces usually cause the displacement of material structure with variables d and A, while the change in ε_r is typically used to estimate forces or strains through specially designed materials [31], however, this method has not been studied in depth. The sensitivity is respectively given by $(\Delta C/C_0)/\varepsilon$ for the capacitive strain sensor and $(\Delta C/C_0)/P$ for the capacitive pressure sensor, in which ε and P are the applied strain and the intensity of pressure, respectively. Since the invention of Si-based capacitive sensor in 1970s, these kinds of sensors have developed rapidly [32].

2.1.3 Piezoelectric Sensors

Piezoelectricity is also a common transduction principle for sensing devices. It refers to the electric polarization of the material under the applied mechanical stimulus [33], which is caused by the electric dipole moments observed in some materials with anisotropic crystalline structure [34], as shown in Fig. 2c. Piezoelectric sensors have been widely employed to detect the vibrations associated with slip and sound [35, 36] due to the high sensitivity to dynamic force and fast responsiveness, and the transient sensing ability is similar to those of the rapid adaptors in human skin [37]. However, static sensing is not applicable because of its sensitivity to temperature and the instability of the sensing response over time. The piezoelectric coefficient (d_{33}) is one of the most important physical quantities that reflects the capability of the material to convert normal forces into electrical charges.

2.1.4 Triboelectric Sensors

As a common phenomenon, the triboelectric effect is widely employed to design the triboelectric sensors in which the material is electrically charged after touch with different materials [38]. As exhibited in Fig. 2d, in the original state, a small gap exists between two different materials with opposite electric polarities, when the two materials rub against or contact with each other, different charges on both surfaces are generated. When the two materials are separated, opposite charges are simultaneously separated and corresponding compensation charges are naturally generated at the other side of each material. The charge can be incompletely neutralized and thus a potential difference is formed due to the air gap between the two surfaces. In general, the amount of the generated charge is directly related to the triboelectric polarity difference between the two contact materials [38]. This mechanism allows the triboelectric sensors to give electrical signals under various stimuli, such as physical touching, vibration, motion, mechanical stimuli, and biological movement, and thus makes the device can be employed as self-powered sensors. For instance, Fan et al. reported transparent triboelectric nanogenerators based on micropatterned plastic films, which were employed as self-powered pressure sensors for sensing a falling feather (contact pressure $\sim$0.4 Pa) and giving a low detection limit ($\sim$13 mPa) [39]. As for triboelectric effect-based sensors, numerous researches are still being conducted to improve the energy conversion efficiency.

2.2 *Key Parameters for Flexible/Stretchable Sensors*

Key parameters of flexible/stretchable sensors include sensitivity, response time, limit of detection, stability and repeatability, etc.

Sensitivity, defined as the electrical response magnitude upon an external stimuli, reflects the accuracy of the sensors. As mentioned above, the sensitivity of the piezoresistive sensors and piezocapacitive sensors are expressed as $S = (\Delta R/R_0)/P$ and $(\Delta C/C_0)/P$, or $GF = (\Delta R/R_0)/\varepsilon$ and $(\Delta C/C_0)/\varepsilon$, respectively, where P and ε represent the applied force and strain, respectively, and the unit of S is kPa^{-1} while the unit of GF is dimensionless.

Response time reflects the speed at which the sensor response to the stimulus, and defined as the time consumption from the start of applying stimulus to the generation of a steady output signal. To describe the dynamic response, the response and recovery time are typically used, which represent the time required to rises to the maximum value and recovers to the initial state, respectively.

Limit of detection is a parameter that represents the measurement area including the minimum and maximum values of the external stimulus. For sensors, high limit of detection is crucial for sensitive detection, safe and accurate manipulation of objects.

Stability and repeatability represent the durability of the sensors to the periodic deformation/restoration processes without causing performance failures, such as electrical and mechanical properties.

In addition to above-mentioned parameters, stretchability is a critical parameter that determines the working range of the sensors. Generally, good stretchability and flexibility allow the sensors without incurring damage during the motion of measured objects. Linearity is another important parameter, which reflects the proportional relationship between the sensor output and the input signals, such as the capacitance and applied force in a piezocapacitive sensor. Moreover, good linearity is facility to be calibrated and dealt with data. In addition, the applied voltage should also sometimes be considered because it relates to the power consumption of the sensor.

3 Functional Materials and Applications of Sensors

Flexible/stretchable sensors are usually composed of several essential components: (1) flexible substrate that supports the entire circuit and simultaneously conform to irregular surfaces; (2) Specific flexible components for transducing the external stimuli into detectable electronic signals; (3) Electrodes for electronic signal export.

3.1 Substrate Materials

Substrate materials with the good chemical and physical properties play an important role in the flexible/stretchable sensor construction. Currently, various flexibility substrates are reported, including synthetic polymers (such as polyethylene terephthalate (PET) [40, 41], polyimide (PI) [42, 43], polyethylene naphthalate (PEN) [44], silicone (such as PDMS [45, 46], Ecoflex [47]), natural polymer (silk fiber) [48], and unconventional substrates (such as papers [49, 50], metal foil [51]), self-healing materials [52]. Their physical and chemical properties are summarized in Table 1.

Table 1 The typical flexibility substrates and their physical and chemical properties

	Flexible substrates	Stretchable/ bendable	Transparency	Thermal stability	Chemical resistance
Synthesis polymer	PET	Bendable	High transparency (85%)	Temperature resistance (<100 °C)	Dissolvable in acetone
	PI	Bendable	Low transparency (yellow color)	Temperature resistance (<450 °C)	Resist weak acids and alkali
	PDMS	Stretchable	High transparency (>95%)	Temperature resistance (<100 °C)	Resist ethanol, acetone, moisture, and oxygen
	PEN	Bendable	High transparency (85%)	Temperature resistance (<180 °C)	Easily permeated by oxygen and water
	Ecoflex	Stretchable	High transparency (>91%)	Temperature resistance (<100 °C)	Resist ethanol and acetone
Natural polymer	Silk fiber	Bendable	No	Heat resistance	No
Unconventional substrates	Papers	Bendable	No	Temperature resistance (<100 °C)	No
	Metal foil	Bendable	No	Temperature resistance (~250 °C)	Resist ethanol, acetone, moisture, and oxygen
	Self-healing	Stretchable	Low transparency	Thermal healing	Resist ethanol, acetone, moisture, and oxygen

3.2 *Representative Self-supporting Sensing Materials*

As the key source of information for sensors, flexible active components determine the sensitivity, linearity, reliability, and application areas of the sensor. Therefore, selecting suitable active sensing materials or synthesizing new multifunctional materials utilizing a novel approach can significantly promote the development of flexible/stretchable sensors. To date, various materials with different dimensions or structures, including metal materials (silver nanoparticles (AgNPs) [53], silver nanowires (AgNWs) [54], gold nanowires (AuNWs) [55, 56], Au films [57], and copper nanowires (CuNWs) [58], CuNWs aerogel [59]), organic conductive materials (polyaniline (PANI) nanowires [60], polypyrrole (PPy) film [61], PPy nanowires composite aerogel [62], and PEDOT:PSSfilm [63]), polymer matrixes filled with conductive fillers (such as polyurethane (PU) foam enriched with PANI/ multi-walled carbon nanotube (MWCNT) nanocomposite [64] or PPy composite [65]), as well as inorganic nonmetallic materials (carbon materials such as CNT [31, 66], graphene films [25, 67], porous graphene [68, 69], and carbon fiber [70]), etc., have been employed as the sensing elements. Among them, the self-supporting materials, which assembled from building blocks with 0D, 1D, or 2D structure, have turn out to be potential candidates for developing highly efficient flexible/ stretchable sensors. They possess outstanding properties (such as chemical, thermal, mechanical, and electrical properties) and structural diversity (such as laminar film, porous network, microarray, and other various pattern structures). More recently, tremendous efforts have been devoted to the investigation of new self-supporting materials for flexible/stretchable sensors.

3.2.1 Metallic Materials and Inorganic Semiconductors

Metallic Nanomaterials

With the development of flexible electronics, traditional pressure/strain sensors generally made of inorganic semiconductor have not dissatisfy the imperative requirement for high flexible/stretchable sensing capability due to their limited detection range (5%) [71]. Metallic nanomaterials, including nanoparticles, nanowires, films, and aerogels, etc., have been applied as sensing elements for the flexible sensors. For instance, Qian et al. fabricated a flexible piezoresistive strain sensor by coating Au NPs monolayer film onto elastic substrate (Fig. 3a (i)), which exhibited a fast dynamic response to external acoustic vibrations from 1 to 20,000 Hz [72], as shown in Fig. 3a (ii). Cheng et al. constructed a flexible piezoresistive pressure sensor by dip-coating Au NWs onto a tissue paper and then sandwiched between a pair of PDMS sheets, which possessed a quick response (about 17 ms), high sensitivity of 1.14 kPa^{-1}, a low detection limit of 13 Pa, and good stability (over 50,000 cycles) [55]. Inspired by the spider's sensory system, Choi et al. reported a cut-through crack-based flexible sensor via coating a solid platinum (Pt) thin layer with the thickness of about 20 nm onto viscoelastic polyurethane acrylate substrate, which displayed a ultrasensitivity (*GF* of 2000) within the strain range of 0–2% due to unique crack feature [19]. Notably, the cracked-Pt based strain sensor could also detect accurately subtle vibration at amplitude of about 10 nm. Furthermore, the relationship between the crack depth and sensitivity or signal-to-noise ratio has also been investigated [73]. The results show that the sensitivity (*GF* of 16,000) of crack-based sensor is improved greatly as the crack depth increases. Similarly, Zhang et al. fabricated cuttable piezoresistive strain sensor by integrating microcracked Au nanofilms with commercial abrasive papers, as shown in Fig. 3b, which possessed a relative high sensitivity (*GF* of 75.8) within the range of 0–0.59% strain, rapid response time, and ultrahigh stability and durability during repeated bend-release of 18,000 cycles [57].

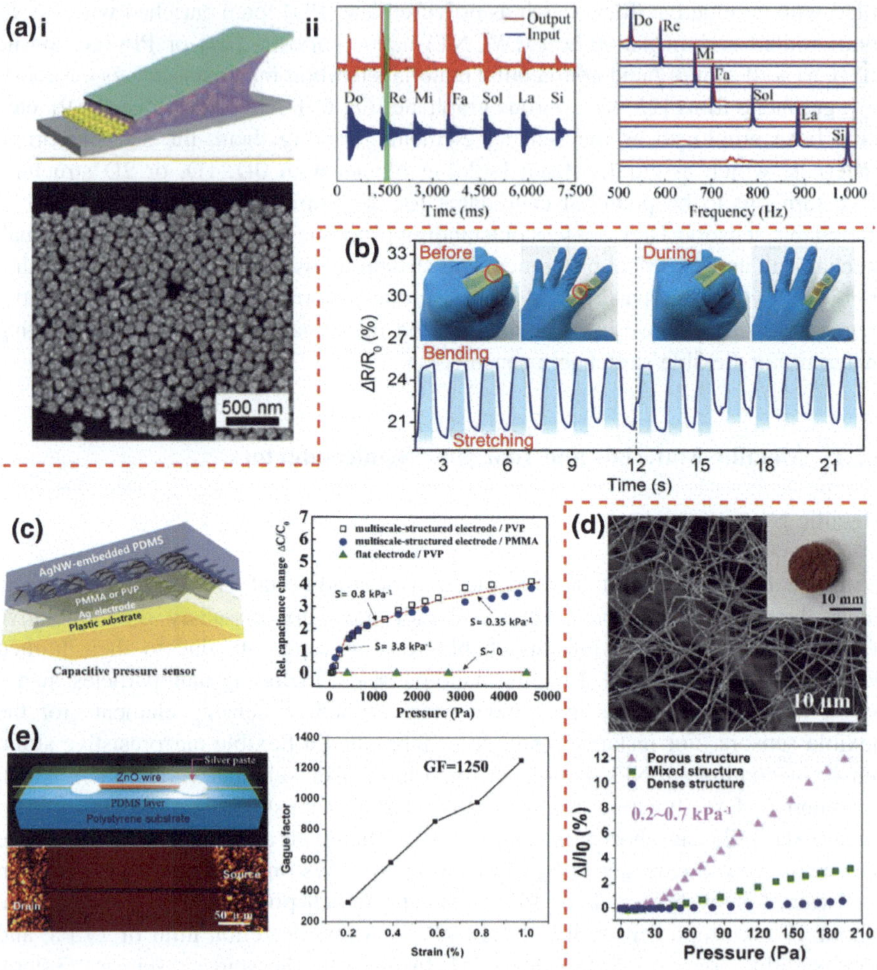

Fig. 3 Metallic materials and inorganic semiconductors based sensors. **a** (i) Schematic illustration for the assembly of one strip of Au NPs and its SEM images, **a** (ii) dynamic response of Au NPs monolayer to musical tones and analysis of their vibratory frequencies. Reproduced with permission from Ref. [72] Copyright 2015, Tsinghua University Press and Springer-Verlag Berlin Heidelberg. **b** Microcracked Au nanofilms@paper based strain sensor for finger bending detection. Reproduced with permission from Ref. [57]. Copyright 2017, the American Chemistry Society. **c** Piezocapacitive sensor composed of Ag NWs embedded patterned matrix and its sensitivity. Reproduced with permission from Ref. [74]. Copyright 2015, the Royal Society of Chemistry. **d** Digital (the inset) and SEM photographs of porous CuNWs aerogel, as well as its sensitivity. Reproduced with permission from Ref. [59]. Copyright 2017, the American Chemistry Society. **e** Schematic and optical image of a single ZnO based strain sensor, as well as its sensitivity. Reproduced with permission from Ref. [75]. Copyright 2008, the American Chemistry Society

In the past decade, numerous reports have focused on the room temperature synthesis and solution-processed metal nanowires (Ag, Cu) because they are more conductive than inorganic semiconductors (Si, Ge, ZnO, etc.). For example, Ag NWs are embedded PDMS matrix with patterned structure to fabricate a high-performance piezocapacitive sensor [74], which showed a high sensitivity of over 3.8 kPa^{-1}, as shown in Fig. 3c. Although the conductive is slightly low, Cu is 100 times cheaper than Ag, making Cu NWs another popular candidate for flexible sensor. Sun et al. fabricated Cu NWs aerogel via a one-pot method for piezoresistive pressure sensor. Thanks to high porosity and low density of the aerogel, the device with a good sensitivity with range of 0.02–0.7 kPa^{-1} and a low detection limit of 14 Pa has been observed [59], as shown in Fig. 3d. Their sensing mechanisms are the change of electrical resistance induced by geometry change of monolith structure under applied pressure.

Inorganic Semiconductors

Recently, inorganic piezoelectric materials (ZnO, GaN, InN, etc.) with wurtzite structure have received extensive attention for flexible sensing, especially ZnO NWs hold great potential for wearable sensors due to their high electron mobility, unique hierarchical structure, and novel piezoelectric properties [75]. The strain sensor based on such ZnO NWs is first prepared by laterally bonding ZnO on a polymer substrate in 2008 [75]. Due to the combination of the piezoelectric effect induced by strain and change in band structure result in the change in height of Schottky barrier, the strain sensor displays a high sensitivity with *GF* of 1250 at a strain of 1% (Fig. 3e), but very limited stretchability. Moreover, this design has also been applied for the formation of flexible energy devices [76] and self-power devices [77]. Additionally, a hybrid structure composed of ZnO NWs grown on elastic fibers (such as polystyrene (PS) [75, 78], PU [79]) exhibits a highly stretchability of 150% and shows an additional features with multifunctional detection for strain, temperature, and UV [79]. It is worth mentioning that Wang et al. launched a pioneering work on ZnO to greatly promote its practical application [80, 81].

Obviously, the above-mentioned semiconductors exhibit outstanding electromechanical properties, but the high cost for the production of nanowires devices and stability issues related with poor control of surface effect limit their practical application.

3.2.2 Carbon Nanomaterials

Carbon materials exhibit excellent performance in flexible/stretchable sensors owing to their fantastic properties, such as remarkable mechanical flexibility, tunable metallic/semiconducting property, and easy assembled into hierarchical macroscopic structures. For example, sensors based on carbon nanofibers and CB show good electromechanical properties. Moreover, macroscopic carbon materials obtained from biological derivatives by carbonization processes are also widely investigated for the advanced flexible/stretchable sensors. In this section, carbon materials, especially representative carbon nanomaterials (CNTs and graphene) with different structures and their sensing performance are reviewed.

CNT

CNTs, a unique carbon allotrope with 1D configuration are favorable candidates for flexible/stretchable sensor application owing to their remarkable charge carrier mobility, high chemical stability, outstanding mechanical strength and elasticity, and excellent processibility [82]. In general, CNTs with suitable chiral angles have considerable sensitivity to nanoscale strains because of the band structure change under applied force [83]. Several studies have concentrated on CNT materials (such as fibers, films, monoliths, etc.) based flexible/stretchable sensors. For example, Ryu et al. reported a CNT fiber-based strain sensor, which exhibited an ultrahigh stretchability (>900% strain) along the axial direction of the CNT fibers [8] (Fig. 4a). Similarly, Baughman et al. fabricated a buckled fiber dielectric layer by coating CNTs onto a stretched rubber fiber for capacitive sensor [84], which showed high sensitivity as well as good linearity in a wide strain range. Moreover, this stretchable and twistable sandwich CNT fiber dielectric layer can be fixed into a glove to modulate strains (Fig. 4b), as reported by Choi et al. [85] Obviously, flexible sensors based on CNT fibers exhibit high conductivity and elasticity, but relative low sensitivity limits their further application in wearable sensors. To optimize CNT fiber-based sensors' performance, CNT microyarns were assembled from good knittability of CNT fibers by Kim et al. [86], and the microyarns-based capacitive sensor exhibited a low detection limit (0.38 Pa) (Fig. 4c (i)) and high sensitivity of 0.5 MPa^{-1} (Fig. 4c (ii)), as well as excellent mechanical compatibility to various deformations, such as stretching, twisting, and folding (Fig. 4c (iii)).

Carbon films developed by several strategies have also been widely studied for the preparation of advanced flexible/stretchable sensors. For instance, a CNT network/polymer composite film prepared by premixing elastomers with CNTs followed by curing, exhibited a *GF* of around 20 at a little strain of 0.5% and a nonlinear piezoresistivity when the loading of CNT is closer to the percolation threshold [87]. Moreover, CNT network film deposited directly on the 3 M VHB 4905 substrate maintained its electric conductivity at strain as large as 700%, however, due to local cracks and irrecoverable deformation, an obvious loss of electric conductivity was observed during repeated stretch/release [88]. To improve the structural stability of CNT network, Zhu et al. prepared a robust CNT-graphene hybrid network-based strain sensor via in situ graphene hybridization (Fig. 5a (i)), which exhibited high sensitivity, rapid response, and good durability, and thus can be applied to detect human motions (Fig. 5a (ii)) owning to effectively resist the bundling and bulking deformation of CNT networks under repeated stretch and release [89]. Notably, Bao et al. fabricated a single-walled CNT (SWCNT)-based

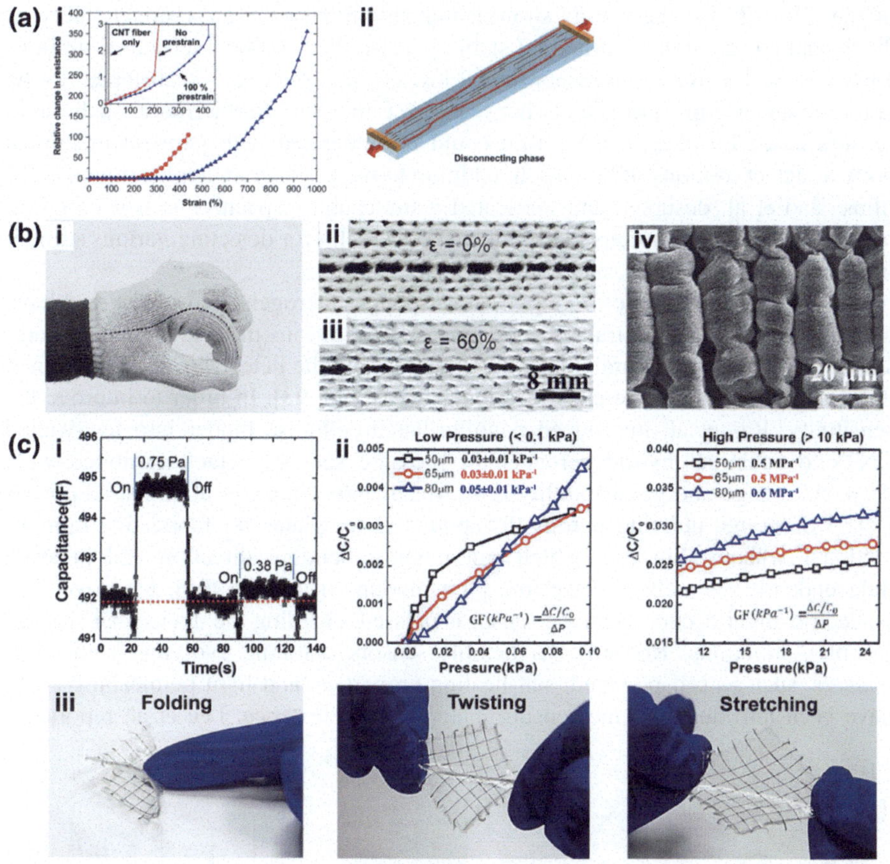

Fig. 4 CNT fibers-based flexible sensors. **a** (i) Relative resistance change vs. strain of CNT fibers-basded senseor and **a** (ii) schematic morphology of CNT fibers under strain. Reproduced with permission from Ref. [8]. Copyright 2015, the American Chemistry Society. **b** (i) Composite fibers composed of CNT and elastic fibers fixed into a grove, **b** (ii) magnified images before and **b** (iii) after 60% tensile strain, and **b** (iv) SEM images of a twisted sandwich CNT fibers. Reproduced with permission from Ref. [85]. Copyright 2016, the American Chemical Society. **c** (i)–(ii) Electromechanical performance of CNT microyarns based capacitive sensor and **d** (iii) its flexible properties. Reproduced with permission from Ref. [86]. Copyright 2015, Wiley-VCH

spring-like structure by exploiting the irrecoverable structure change (Fig. 5b (i), (ii)), which showed a high conductivity of 2200 S cm^{-1} in the strain range of 0–150%. This is attributed to the high ratio of CNT in the film can be deformed and recovered by twisting or rotating individual CNT to release stress, thereby avoiding crack propagation. Meanwhile, this spring-like structure film can be utilized as capacitive sensor to detect the external force [31]. Additionally, aligned SWCNT film was also investigated for piezoresisitve sensors in which the film was attached on a PDMS substrate and stretched along the perpendicular direction to the length of the CNTs [90]. The results showed that this film achieved a stretchability of 280% and an remarkable structural stability even after 10,000 cycles at a strain of 150%, as well as the rapid response and low creep, which can be attributed to the deformation of gaps and islands between CNT bundles. Furthermore, such strain sensors based on aligned CNT film could be integrated with garment or human body to detect motions of human. In addition to the large strain application of CNT films, Bao et al. designed and fabricated a stretchable capacitive sensor based on hybrids composed of porous PDMS and SWCNT film for detecting various stimuli, including strain, pressure, and sound [91].

In addition, CNT monolith (e.g., foams, sponges, aerogels, hydrogels, etc.) have also been employed in strain or pressure sensors due to their favorable piezoresistive effect and good durability to deformation, but the detectable range is limited due to their good intrinsic electrical conductivity [92, 93]. In order to improve the sensitivity, Ras et al. introduced nanofibrillated cellulose binder into few-walled CNTs to construct hybrid aerogels for pressure sensors, which exhibited wide detectable range and good stability [94]. Meanwhile, Zhang et al. prepared porous CNT-PU sponge utilizing porous PU sponge as templates for fabricating omnidirectional sensors (Fig. 5c), which can monitor bending direction and pressure independently, as well as detect the joint motions of human [95]. Moreover, triboelectric effect occurs when the sensor is pressed, enabling the device can work in complex situations. Recently, as flexible sensors continue evolving, some new features, such as self-powered, self-healing properties, and light permeability, etc., have been introduced to the sensing materials. For instance, Lee et al. reported a

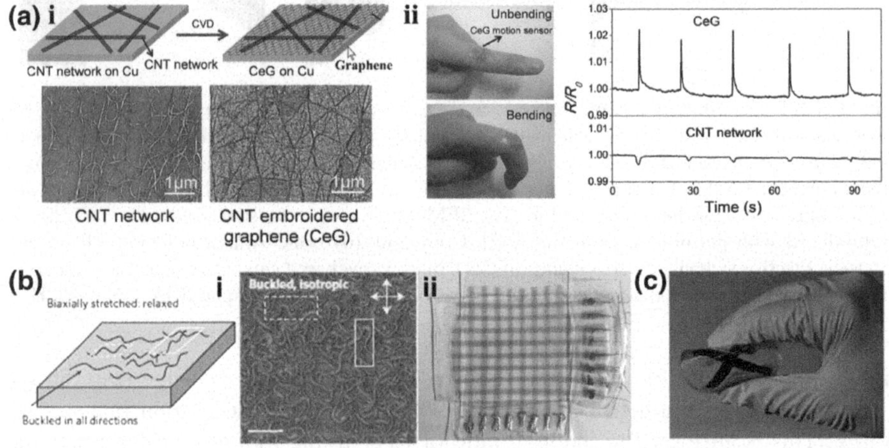

Fig. 5 CNT films based sensors. **a** (i) CNT-graphene network film and **a** (ii) as-prepared strain sensor for detection of finger bending. Reproduced with permission from Ref. [89]. Copyright 2016, Wiley-VCH. **b** (i) SWCNT spring-like structure and SEM image and **b** (ii) CNT based sensor array. Reproduced with permission from Ref. [31]. Copyright 2011. Nature Publishing Group. (**c**) Optical pictures of the porous CNT-PU sponge based omnidirectional sensors. Reproduced with permission from Ref. [95]. Copyright 2017, Wiley-VCH

SWCNT hybrid hydrogel with self-healing properties for stretchable strain sensors [96], which possessed rapid self-healing ability and extremely high stretchability as high as 1000%. Apparently, the self-healing hydrogels are potential candidates for applying in flexible device due to their ability to heal the broken sites and restore the properties of the active materials.

Graphene and Derivative

Similar to CNT materials, graphene as another carbon allotrope with an atomical thickness, has drawn tremendous attention since in 2004. There is growing interest in the investigation of various graphene flexible sensors driven by its unique physco-chemical properties, such as excellent mechanical flexibility, stability, high intrinsic carrier mobility of 200,000 $cm^2\ V^{-1}\ S^{-1}$, superior optical transparency, and large-scale fabrication [97]. Generally, the considerable change in resistance and electronic band structure of graphene are caused by the partial destroy of hexagonal honeycomb structure at the edge of the graphene layer under stretching [98]. Currently, several researches have been conducted to investigate the potential applications of graphene-based materials (including fibers, films, papers, and monoliths, etc.) in flexible sensors. For examples, Hu et al. reported highly conductive and stretchable core-sheath reduced graphene oxide (rGO) nanofibers fabricated from graphene bundles on the Cu wires substrate (Fig. 6a), and a small strain range was observed when serves as a strain sensor due to lack of stretchability [99]. To enhance the electromechanical performance of graphene, Zhu et al. fabricated graphene fabrics by chemical vapor deposition method employing Cu mesh as the catalyst, which was then attached to a PDMS substrate and subsequently assembled into strain sensors for the detecting human motions, such as phonation, hand griping, blink, expression change, pulse, and breath, showing unique advantages of ultrahigh sensitivity with *GF* of $\sim 10^3$ ranging from 2 to 6% strain and superior physical robustness [69]. In another work, a graphene-nanocellulose composite for application in strain sensor is developed, in which 3D macroporous nanopaper from nanocellulose and crumpled graphene is implanted into PDMS matrix. A high stretchability of 100% and a high sensitivity (*GF* of 7.1) were achieved [100] (Fig. 6b (i)), enabling application in fingers motion detection, as shown in Fig. 6b (ii).

Structural design is also an important way to obtain high-performance flexible sensors. For example, Shi et al. prepared a fish-scale-like graphene-sensing layer to fabricate strain sensor through stretching-releasing rGO film on pre-stretched substrate [101] (Fig. 6c (i)). The sensor gave a high sensitivity with *GF* of 16.2–150, good stretchability (strain of 82%), as well as low detection limit (strain of 0.1%), which can be used to detect finger bending, pulses at human wrist, and throat of a person [101] (Fig. 6c (ii)). Ren et al. developed a foam-like graphene-based flexible piezoresistive sensor by a laser-scribing approach [102], which displayed a high sensitivity of 0.96 kPa^{-1} at a low pressure region within 50 kPa. Moreover, an array pressure sensor was placed in a chess board to achieve the electronically

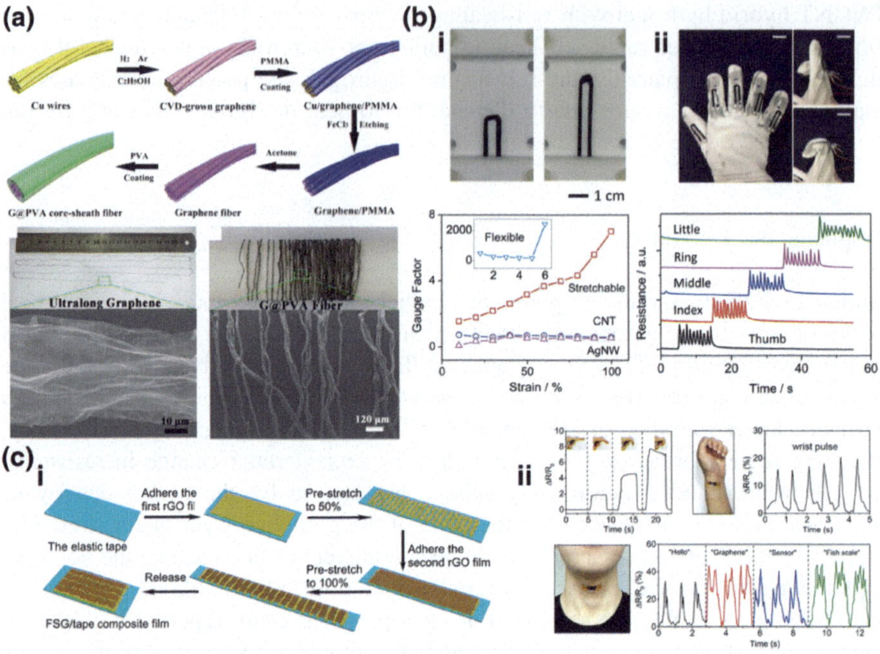

Fig. 6 Graphene fibers based sensors. **a** Schematic illustration of the preparation process, digital and SEM photographs of the core-sheath graphene fibers. Reproduced with permission from Ref. [99]. Copyright 2015, the American Chemical Society. **b** (i) 3D macroporous graphene and nanocellulose nanopaper based strain sensor can be stretched up to 100%, and **b** (ii) its application for finger movement detection. Reproduced with permission from Ref. [100]. Copyright 2014, Wiley-VCH. **c** (i) Schematic illustration of fabricating a fish-scale-like graphene strain sensor and **c** (ii) its application for monitoring the human-motion. Reproduced with permission from Ref. [101]. Copyright 2016, the American Chemical Society

identifiable position of chess pieces. Similarly, Cho et al. constructed hierarchical graphene films by electively etched Cu foils substrate for flexible pressure sensors [103] (Fig. 7a (i)), which exhibited a high linear sensitivity (8.5 kPa^{-1}) cross a large pressure range (Fig. 7a (ii)), low detection limit (1 Pa), and good durability (Fig. 7a (iii)). In addition, flexible pressure sensors based on free-standing honeycomb-like graphene network monolith decorated with bubbles were also developed, which showed an extremely high sensitivity within 0.56 kPa. Moreover, this sensor can be operated at a voltage as low as 10 mV and thus promising its applications in wearable flexible sensors [104]. To further optimize the electromechanical sensing performance, combination of graphene with CNTs or metal nanomaterials has also been fabricated [89, 105]. For instance, a strain sensor made of graphene bridged the adjacent Ag nanoparticles was fabricated, which exhibited a highly stretchable of 1000% and detection limit as low as 0.5% [105].

It is worth noting that graphene monoliths are also employed for fabricating pressure sensors to monitor high-frequency forces, human motions and sound

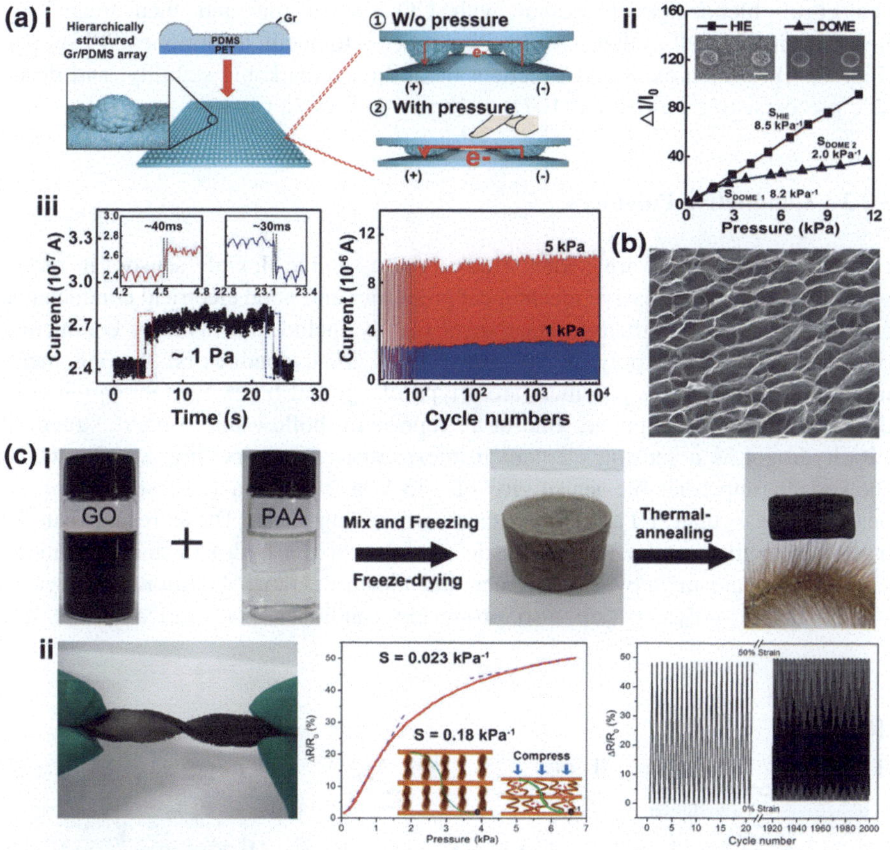

Fig. 7 Structural design of graphene based sensors. **a** (i) Illustration of the prepared pressure sensor composed of hierarchical graphene/PDMS arrays and **a** (ii) its sensitivity, as well as **a** (iii) detection limit and durability. Reproduced with permission from Ref. [103]. Copyright © 2016, Wiley-VCH. **b** SEM image of graphene aerogel. Reproduced with permission from Ref. [106]. Copyright 2015, Wiley-VCH. **c** (i) Schematic illustration of the fabrication process of rGO/PI nanocomposite and **c** (ii) its flexibility as well as electromechanical properties. Reproduced with permission from Ref. [107]. Copyright 2015, the American Chemical Society

pressures due to their high porosity [68, 106] (Fig. 7b), ideal dynamic electromechanical property and excellent piezoresistive behavior [68]. However, pure graphene monolith is an easy collapse structure that reduces its sensing reliability. To enhance the stability, blending graphene with various polymers, such as PI [107], PS [108], PU [109], natural rubber [110, 111], have been employed widely to fabricate high-performance flexible sensors. For example, Wang et al. demonstrated a nitrogen-doped graphene sponge for flexible pressure sensor, from which a high sensitivity of 1.33 kPa^{-1} below 6 kPa was achieved, owing to their well-ordered network structure and excellent compressible recovery of 80% strain [112].

Qin et al. blended PI precursor with GO suspensions and then treated by freeze-drying as well as thermal annealing process to obtain the graphene/PI sponge (Fig. 7c (i)), which exhibited excellent flexibility, remarkable stability, and desirable piezoresistive properties [107], as shown in Fig. 7c (ii).

3.2.3 Conductive Polymers

Conductive elastomers are widely studied for assemble flexible sensors in recent years due to their impressive mechanical properties and good electrical conductance at strained states. Conductive elastomers mainly include intrinsically conducting polymers, conductive polymer composite, and ionic conductors. Among them, intrinsically conducting polymers are a type of optimal fillers for fabrication conductive elastomer composite. Bao et al. reported a hollow microsphere structural PPy hydrogel as a sensing element in piezoresistive sensors (Fig. 8a (i)), which showed an unpredictable sensitivity of 133 kPa^{-1} (Fig. 8a (ii)), small limit of detection (less than 1 Pa), and a short responding time. These results can be attributed to the small effective elastic modulus of the hydrogel and the unique sensing mechanism between polymer and electrode [25]. Additionally, organic semiconductor polymers are also promising candidates for fabricating flexible

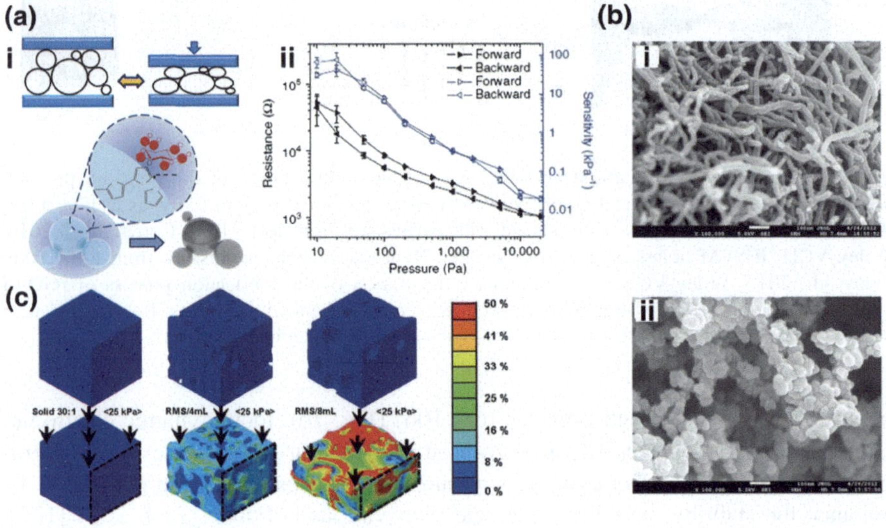

Fig. 8 **a** (i) Pressure sensor fabricated from PPy hydrogel with hollow sphere structure and their synthetic mechanism, and **a** (ii) sensitivity of the sensor. Reproduced with permission from Ref. [25]. Copyright 2014. Nature Publishing Group. SEM image of **b** (i) CNT and **b** (ii) CB. Reproduced with permission from Ref. [114]. Copyright 2013. Elsevier B.V. **c** Strain distribution of porous pressure-sensitive rubber. Reproduced with permission from Ref. [118]. Copyright 2014, Wiley-VCH

devices because their fast carrier transport is comparable to that of Si-based electronic, as well as the flexibility and stretchability are comparable to those of typical plastics. However, a competition between deformability and mobility often exits due to high carrier mobility occurs in the high crystallinity conjugated polymers, which has poor deformability. In fact, Bao et al. discussed the effect of nanoconfinement on the stretchability of semiconductor polymers by fabricating semiconductor polymer nanofibrils into elastomer matrix. Because the nanoconfinement increases disorder of crystallization and polymer chain dynamics but is not sufficient to destroy the charge transport pathways, the stretchability of semiconductor polymer nanofibrils is remarkably enhanced up to 100%, and its charge transport mobility is still higher than that of amorphous silicon [113].

As for conductive elastomer composite, which is made of insulated elastomers enriched with conductive fillers of the different types and dimensions, such as 0D CB, 1D CNT or metal fibers, and 2D graphene, etc. The electromechanical sensing performances of these materials have showed great potential application due to the desired properties. In addition, the conductive elastomer composite has low fabrication cost. General speaking, the type, morphology, dispersion, and concentration of the fillers, geometry and modulus of the matrix elastomer, as well as interaction between the fillers and matrix elastomers could influence the performance of the final composite and the sensors. Among them, conductive polymer filled with CNT usually possesses a very low percolation threshold because of its large aspect ratio and special entangled structure (Fig. 8b (i)), enabling conductive elastomer exhibits a distinct strain sensing behavior [114]. Conductive networks consisted of polyvinylchloride filled with high structure CB (Fig. 8b (ii)) possesses good sensing performance owing to its stability during tension and a greater degree of recoverability [115]. Moreover, filler particles with sharp protrusions appear to be superior to filler particles with a smooth surface due to the protrusions enrich the electric field in the matrix, thereby facilitating transportation of tunneling currents from one site to another [116]. In view of graphene filler, a lower loading, larger strain amplitude, and higher strain rate are beneficial for the increase of sensitivity [117], and the maximum *GF* can be achieved when its concentration near to the percolation threshold, in which the piezoresistive feature of conductive elastic is mainly determined by breaking and reforming of the percolating pathway. Additionally, the introduction of hollow structure into matrix materials, such as hollow spheres (Fig. 8c), displays a relative high sensitivity for monitoring mechanical stimuli [118]. Meanwhile, strong bonding between elastic matrix and filler is also contributing to further increase of sensitivity for conductive elastic composite due to the improved stretchability, hysteresis performance, and cyclic stability [47, 119], where conductive filler is often functionalized with surfactants to enhance the interfacial interaction [120]. In addition, uniform distribution of the conductive fillers in the matrix can lead to further improvements in sensitivity [121].

Ionic conductors, such as ionogels and hydrogels, are also promising candidates for flexible sensors due to their impressive stretchability, good conductivity and transparency [122, 123]. Polymers swelled in brine are known as hydrogels, which are usually biocompatible and softer than tissues, thus enabling application in biometric sensors to detect soft tissue by ions as charge carriers instead of electrons [122]. Suo et al. prepared a highly transparent, stretchable, and biocompatible ionic skin-based sensor [124], which showed a super-low detection limit of 1 Pa and a wide range of strain up to 700%. In addition, conductive fluids (Ga-In alloys) have proven to be unique methods for designing new stretchable sensors with remarkably stretchability with deformation up to 1000% without significant hysteresis [125], but the high cost, thus limiting its practical application.

3.3 *Flexible Electrodes*

High conductivity and stretchability is very important to the flexible electrodes for sensing application to ensure reliable performance under repeated and extended stretching operations. Generally, metallic conductive wires such as Au, Ag, Cu, and aluminum (Al) are often applied as electrode materials mainly due to their high electric conductivities. However, for flexible/stretchable sensors, electrode materials are required to mechanically match sensing materials, and thus excellent mechanical compliance is also an essential feature. In fact, an electrode sometimes also acts as sensing elements for some flexible sensors. For instance, the resistance variation of some piezoresistive sensors under the applied stimuli occurs at the contact between the electrodes, leading to an effective electronic signal output [126].

Typically, metal thin films (lower than 100 nm) deposited onto elastomeric substrates with strong adhesion is a candidate materials for stretchable electrode because they may form cracks under stretching, but resist to damage under bending or folding. As demonstrated by Graz et al. [127], the gold film with 50 nm of thickness deposited on PDMS matrix can form an interconnected network under stretching, and exhibited a reversible change within 20% applied strain without electrical failure over 250,000 stretching cycles. For the purpose of acquiring better electromechanical property on a large scale (over 100% strain), appropriate geometrical design for metal films have been reported, such as surface roughing of substrate [128], and metal encased in an elastomer [129], serpentine structure [129], bulking [130], meshing of metal thin film [131] (Fig. 9a), foaming [132], etc. Metal nanowire networks made from Au, Ag, and Cu is another promising sensing electrode because they are both highly conductive and mechanically compliant compared to the bulk counterparts. For instance, the AgNW network electrode, which is developed by percolating AgNW network onto the elastic substrate using the solution-processing and post low-temperature nano-welding, showing low resistance of $\sim 9\ \Omega\ sq^{-1}$ at the transparency of 90% and high stretchable strain of over 460% [133], as shown in Fig. 9b.

The composite of conductive elastomer can also be employed as advanced electrode materials. For instance, Lee et al. prepared a flexible strain sensor built from the electrode of the conductive composite of PU-PEDOT: PSS [134], which

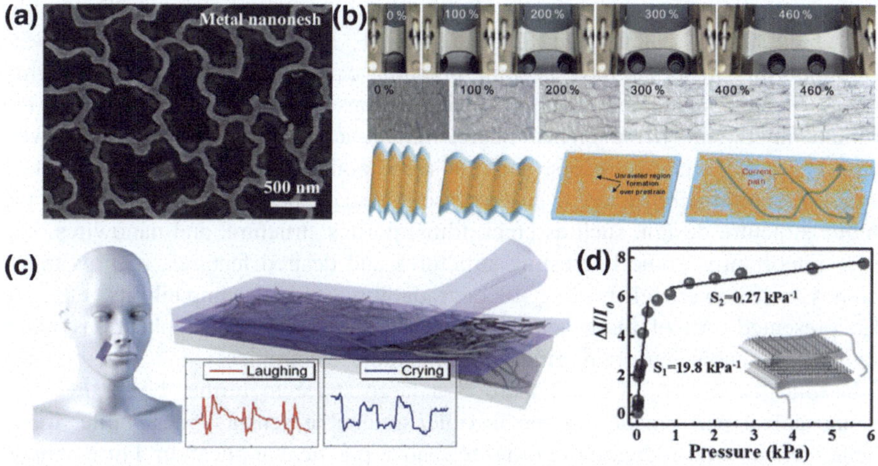

Fig. 9 Flexible electrodes materials and properties. **a** SEM image of metal thin film mesh. Reproduced with permission from Ref. [131]. Copyright 2014, Nature Publishing Group. **b** Percolating AgNW network electrode showing high stretchable strain of over 460%. Reproduced with permission from Ref. [133]. Copyright 2012, Wiley-VCH. **c** Electrode of the conductive composite of PU-PEDOT:PSS can detect small deformation induced by crying and laughing. Reproduced with permission from Ref. [134]. Copyright 2015, the American Chemical Society. **d** Aligned CNTs/graphene electrode show a high sensitivity. Reproduced with permission from Ref. [135]. Copyright 2017, Wiley-VCH

showed a favorable stretchability of 100% and a *GF* of 62, allowing it can detect subtle deformations induced by facial expressions (Fig. 9c). Conductors such as CNTs or graphene are another type of promising alternative as electrode materials. Zhang et al. fabricated a aligned CNTs/graphene electrode-based pressure sensor, which exhibited impressive sensitivity and stability [135], as exhibited in Fig. 9d. In addition, liquid metal embedded in the soft materials has attracted increasing research interest as flexible electrodes in the past few years. Thanks to the flux of liquid metal under external stimuli, it can reconfigure quickly and resist large strains without causing significant failure of electrical conductivity.

4 Conclusions and Outlook

Over the past decade, inspired by the booming of flexible electronics, the fabrication of high-performance flexible/stretchable devices has attracted extensive attention in the scientific community. In this chapter, we systematically review significant progress made in the field of flexible/stretchable sensors in recent years. A variety of self-supporting materials, such as yarns or fibers, like-papers, films, and 3D structure materials such as monoliths or bulks assembled from building blocks, including 0D nanoparticels, 1D nanowires or nanofibers, and 2D nanosheets, have been categorized and discussed in detail. Apparently, these self-supporting materials have become potential candidates for developing highly efficient flexible/stretchable sensors, serving as sensing elements, to monitoring the artificial intelligent, personal healthcare, human-activity, and beyond. Moreover, some key parameter of sensors, such as stretchability, sensitivity, limit of detections, and reproducibility, are even surpassed those of the traditional sensors by the unique structure design, such as crack films, porous structure, and nanowires, etc. More importantly, some interesting structures and desired features, such as transparency, self-power, self-healing, biodegradability, and biocompatibility, etc., are also presented. All of these achievements and their potential applications make flexible/stretchable sensors a critical part of future electronics.

Despite the above progress, there is a large space to further boost the electromechanical performance of the flexible/stretchable sensors; meanwhile, many challenges must be addressed in order to realize practical application. For example, achieving both high mechanical compliance and electromechanical sensitivity remains a significant challenge, especially for personalized healthcare applications where small strain or pressure changes on the human skin surface under stimuli, such as heartbeat, pulse, voice sensing, and blood pressure, etc., need to be monitored. To resolve this problem, the investigation of new functional materials and the design of new structural materials to improve the sensitivity, stretchability, and durability of flexible/stretchable sensors is an interesting research topic. Additionally, exploring novel sensing mechanism and multifunction are also urgent factors in detecting of the multiple stimuli, such as various forces, temperature, and humidity, etc. Moreover, the package of flexible sensors with communication, power supply, signal processing, as well as data management units, is another major challenge. Therefore, integrating components into a completely sensing system/platform is another research focus for applications in the intelligent medical treatment, transportation, and industry.

In summary, flexible/stretchable sensors have tremendous potential for application in innovative sensor products and systems, such as personalized health care and artificial intelligence, etc. The proliferation of flexible/stretchable sensors is inseparable from the development of related flexible, free-standing functional materials, including active sensing materials, substrate and electrode materials. It is foreseeable that with the development of preparation technology of flexible materials and the development and utilization of new flexible materials, flexible/stretchable sensors and corresponding integrated systems will greatly improve the quality of human life and may even change the global technical and social environment.

Acknowledgements This work was supported by the National Natural Science Foundation of China (Nos. 51502306, 51675514).

References

1. William, S., Wong, A.: Flexible Electronics: Materials and Applications, pp. 1–461. Springers Science+Business Media, New York (2009)
2. Zhang, W., et al.: Highly sensitive and flexible strain sensors based on vertical zinc oxide nanowire arrays. Sens. Actuators A Phys. **205**, 164–169 (2014)
3. Son, D., et al.: Multifunctional wearable devices for diagnosis and therapy of movement disorders. Nat. Nanotechnol. **9**(5), 397–404 (2014)
4. Schwartz, G., et al.: Flexible polymer transistors with high pressure sensitivity for application in electronic skin and health monitoring. Nat. Commun. **4**, 1859 (2013)
5. Stoyanov, H., et al.: Soft conductive elastomer materials for stretchable electronics and voltage controlled artificial muscles. Adv. Mater. **25**(4), 578–583 (2013)
6. Lu, N., Kim, D.-H.: Flexible and stretchable electronics paving the way for soft robotics. Soft Robot. **1**(1), 53–62 (2014)
7. Jeong, J.W., et al.: Materials and optimized designs for human-machine interfaces via epidermal electronics. Adv. Mater. **25**(47), 6839–6846 (2013)
8. Ryu, S., et al.: Extremely elastic wearable carbon nanotube fiber strain sensor for monitoring of human motion. ACS Nano **9**(6), 5929–5936 (2015)
9. Lange, B., et al.: Markerless full body tracking- depth-sensing technology within virtual environments. Interserv. Ind. Train. Simul. Educ. Conf. **11363**, 1–8 (2011)
10. Alamusi, et al.: Piezoresistive strain sensors made from carbon nanotubes based polymer nanocomposites. Sensors (Basel) **11**(11), 10691–10723 (2011)
11. Liu, H., et al.: Electrically conductive polymer composites for smart flexible strain sensors: a critical review. J. Mater. Chem. C **6**(45), 12121–12141 (2018)
12. Rim, Y.S., et al.: Recent progress in materials and devices toward printable and flexible sensors. Adv. Mater. **28**(22), 4415–4440 (2016)
13. Baptista, F.R., et al.: Recent developments in carbon nanomaterial sensors. Chem. Soc. Rev. **44**(13), 4433–4453 (2015)
14. Rodgers, M.M., Pai, V.M., Conroy, R.S.: Recent advances in wearable sensors for health monitoring. IEEE Sens. J. **15**(6), 3119–3126 (2015)
15. Amjadi, M., et al.: Stretchable, skin-mountable, and wearable strain sensors and their potential applications: a review. Adv. Funct. Mater. **26**(11), 1678–1698 (2016)
16. Yang, Y., et al.: Recent progress in flexible and wearable bio-electronics based on nanomaterials. Nano Res. **10**(5), 1560–1583 (2017)
17. Wan, Y., Wang, Y., Guo, C.F.: Recent progresses on flexible tactile sensors. Mater. Today Phys. **1**, 61–73 (2017)
18. Liu, H., et al.: Lightweight conductive graphene/thermoplastic polyurethane foams with ultrahigh compressibility for piezoresistive sensing. J. Mater. Chem. C **5**(1), 73–83 (2017)
19. Kang, D., et al.: Ultrasensitive mechanical crack-based sensor inspired by the spider sensory system. Nature **516**(7530), 222–226 (2014)
20. Wang, M., et al.: Enhanced electrical conductivity and piezoresistive sensing in multi-wall carbon nanotubes/polydimethylsiloxane nanocomposites via the construction of a self-segregated structure. Nanoscale **9**(31), 11017–11026 (2017)
21. Park, S.-J., et al.: Highly flexible wrinkled carbon nanotube thin film strain sensor to monitor human movement. Adv. Mater. Technol. **1**(5), 1600053 (2016)
22. Liu, S., et al.: A high performance self-healing strain sensor with synergetic networks of poly(ε-caprolactone) microspheres, graphene and silver nanowires. Compos. Sci. Technol. **146**, 110–118 (2017)
23. Cai, L., et al.: Highly transparent and conductive stretchable conductors based on hierarchical reticulate single-walled carbon nanotube architecture. Adv. Funct. Mater. **22** (24), 5238–5244 (2012)
24. Li, L., et al.: A superhydrophobic smart coating for flexible and wearable sensing electronics. Adv. Mater. **29**(43) (2017)

25. Pan, L., et al.: An ultra-sensitive resistive pressure sensor based on hollow-sphere microstructure induced elasticity in conducting polymer film. Nat. Commun. **5**, 3002 (2014)
26. Pang, L., et al.: Enhanced pressure and proximity sensitivities of a flexible transparent capacitive sensor with PZT nanowires. IOP Conf. Ser. Mater. Sci. Eng. **479**, 012035 (2019)
27. Muhammad, H.B., et al.: A capacitive tactile sensor array for surface texture discrimination. Microelectron. Eng. **88**(8), 1811–1813 (2011)
28. Cheng, M.Y., et al.: A polymer-based capacitive sensing array for normal and shear force measurement. Sensors (Basel) **10**(11), 10211–10225 (2010)
29. Huang, Y., et al.: A flexible three-axial capacitive tactile sensor with multilayered dielectric for artificial skin applications. Microsyst. Technol. **23**(6), 1847–1852 (2016)
30. Mitrakos, V., et al.: Design, manufacture and testing of capacitive pressure sensors for low-pressure measurement ranges. Micromachines **8**(2), 41 (2017)
31. Lipomi, D.J., et al.: Skin-like pressure and strain sensors based on transparent elastic films of carbon nanotubes. Nat. Nanotechnol. **6**(12), 788–792 (2011)
32. Dobrzynska, J.A., Gijs, M.A.M.: Flexible polyimide-based force sensor. Sens. Actuators A Phys. **173**(1), 127–135 (2012)
33. Park, K.I., et al.: Highly-efficient, flexible piezoelectric PZT thin film nanogenerator on plastic substrates. Adv. Mater. **26**(16), 2514–2520 (2014)
34. Wegener, M., Wirges, W., Gerhard-Multhaupt, R.: Piezoelectric polyethylene terephthalate (PETP) foams–specifically designed and prepared ferroelectret films. Adv. Eng. Mater. **7** (12), 1128–1131 (2005)
35. Hammock, M.L., et al.: 25th anniversary article: the evolution of electronic skin (e-skin): a brief history, design considerations, and recent progress. Adv. Mater. **25**(42), 5997–6038 (2013)
36. Cotton, D.P.J., et al.: A novel thick-film piezoelectric slip sensor for a prosthetic hand. IEEE Sens. J. **7**(5), 752–761 (2007)
37. Dargahi, J., Najarian, S.: Human tactile perception as a standard for artificial tactile sensing—a review. Int. J. Med. Robot. Comput. Assist. Surg. **1**(1), 23–35 (2004)
38. Wang, Z.L.: Triboelectric nanogenerators as new energy technology for self-powered systems and as active mechanical and chemical sensors. ACS Nano **7**(11), 9533–9557 (2013)
39. Fan, F.R., et al.: Transparent triboelectric nanogenerators and self-powered pressure sensors based on micropatterned plastic films. Nano Lett. **12**(6), 3109–3114 (2012)
40. Lau, P.H., et al.: Fully printed, high performance carbon nanotube thin-film transistors on flexible substrates. Nano Lett. **13**(8), 3864–3869 (2013)
41. Nomura, K., et al.: Room-temperature fabrication of transparent flexible thin-film transistors using amorphous oxide semiconductors. Nature **432**, 488–492 (2004)
42. Yokota, R., et al.: Molecular design of heat resistant polyimides having excellent processability and high glass transition temperature. High Perform. Polym. **13**(2), S61–S72 (2016)
43. Lanzara, G., et al.: A spider-web-like highly expandable sensor network for multifunctional materials. Adv. Mater. **22**(41), 4643–4648 (2010)
44. Kaltenbrunner, M., et al.: An ultra-lightweight design for imperceptible plastic electronics. Nature **499**(7459), 458–463 (2013)
45. Sekitani, T., et al.: A rubberlike stretchable active matrix using elastic conductors. Science **321**(5895), 1468–1472 (2008)
46. Chun, S., et al.: A graphene force sensor with pressure-amplifying structure. Carbon **78**, 601–608 (2014)
47. Amjadi, M., Yoon, Y.J., Park, I.: Ultra-stretchable and skin-mountable strain sensors using carbon nanotubes-ecoflex nanocomposites. Nanotechnology **26**(37), 375501 (2015)
48. Zhu, B., et al.: Silk fibroin for flexible electronic devices. Adv. Mater. **28**(22), 4250–4265 (2016)
49. Li, R.Z., et al.: Direct writing on paper of foldable capacitive touch pads with silver nanowire inks. ACS Appl. Mater. Interfaces **6**(23), 21721–21729 (2014)

50. Shafiee, H., et al.: Paper and flexible substrates as materials for biosensing platforms to detect multiple biotargets. Sci. Rep. **5**, 8719 (2015)
51. Zou, Q., et al.: Dielectric properties of lead zirconate titanate thin films deposited on metal foils. Appl. Phys. Lett. **77**(7), 1038 (2000)
52. Cao, Y., et al.: A transparent, self-healing, highly stretchable ionic conductor. Adv. Mater. **29**(10) (2017)
53. Su, X., et al.: Highly stretchable and conductive superhydrophobic coating for flexible electronics. ACS Appl. Mater. Interfaces **10**(12), 10587–10597 (2018)
54. Amjadi, M., et al.: Highly stretchable and sensitive strain sensor based on silver nanowire-elastomer nanocomposite. ACS Nano **8**(5), 5154–5163 (2014)
55. Gong, S., et al.: A wearable and highly sensitive pressure sensor with ultrathin gold nanowires. Nat. Commun. **5**, 3132 (2014)
56. Li, Q., et al.: Wide-range strain sensors based on highly transparent and supremely stretchable graphene/ag-nanowires hybrid structures. Small **12**(36), 5058–5065 (2016)
57. Liao, X., et al.: Flexible, cuttable, and self-waterproof bending strain sensors using microcracked gold nanofilms@paper substrate. ACS Appl. Mater. Interfaces **9**(4), 4151–4158 (2017)
58. Tang, Y., et al.: Manufacturable conducting rubber ambers and stretchable conductors from copper nanowire aerogel monoliths. ACS Nano **8**(6), 5707–5714 (2014)
59. Xu, X., et al.: Copper nanowire-based aerogel with tunable pore structure and its application as flexible pressure sensor. ACS Appl. Mater. Interfaces **9**(16), 14273–14280 (2017)
60. Ge, G., et al.: A flexible pressure sensor based on rGO/polyaniline wrapped sponge with tunable sensitivity for human motion detection. Nanoscale **10**(21), 10033–10040 (2018)
61. Shao, Q., et al.: High-performance and tailorable pressure sensor based on ultrathin conductive polymer film. Small **10**(8), 1466–1472 (2014)
62. Xue, J., et al.: Wearable and visual pressure sensors based on Zn2GeO4@polypyrrole nanowire aerogels. J. Mater. Chem. C **5**(42), 11018–11024 (2017)
63. Lipomi, D.J., et al.: Electronic properties of transparent conductive films of PEDOT:PSS on stretchable substrates. Chem. Mater. **24**(2), 373–382 (2012)
64. Hong, S.Y., et al.: Polyurethane foam coated with a multi-walled carbon nanotube/polyaniline nanocomposite for a skin-like stretchable array of multi-functional sensors. NPG Asia Mater. **9**(11), e448–e448 (2017)
65. Li, M., et al.: Stretchable conductive polypyrrole/polyurethane (PPy/PU) strain sensor with netlike microcracks for human breath detection. ACS Appl. Mater. Interfaces **6**(2), 1313–1319 (2014)
66. Chen, S., et al.: Acid-interface engineering of carbon nanotube/elastomers with enhanced sensitivity for stretchable strain sensors. ACS Appl. Mater. Interfaces **10**(43), 37760–37766 (2018)
67. Sun, H., Xu, Z., Gao, C.: Multifunctional, ultra-flyweight, synergistically assembled carbon aerogels. Adv. Mater. **25**(18), 2554–2560 (2013)
68. Jeong, Y.R., et al.: Highly stretchable and sensitive strain sensors using fragmentized graphene foam. Adv. Funct. Mater. **25**(27), 4228–4236 (2015)
69. Wang, Y., et al.: Wearable and highly sensitive graphene strain sensors for human motion monitoring. Adv. Funct. Mater. **24**(29), 4666–4670 (2014)
70. Gallo, G.J., Thostenson, E.T.: Electrical characterization and modeling of carbon nanotube and carbon fiber self-sensing composites for enhanced sensing of microcracks. Mater. Today Commun. **3**, 17–26 (2015)
71. Yang, S., Lu, N.: Gauge factor and stretchability of silicon-on-polymer strain gauges. Sensors (Basel) **13**(7), 8577–8594 (2013)
72. Yi, L., et al.: Nanoparticle monolayer-based flexible strain gauge with ultrafast dynamic response for acoustic vibration detection. Nano Res. **8**(9), 2978–2987 (2015)
73. Park, B., et al.: Dramatically enhanced mechanosensitivity and signal-to-noise ratio of nanoscale crack-based sensors: effect of crack depth. Adv. Mater. **28**(37), 8130–8137 (2016)

74. Joo, Y., et al.: Silver nanowire-embedded PDMS with a multiscale structure for a highly sensitive and robust flexible pressure sensor. Nanoscale **7**(14), 6208–6215 (2015)
75. Zhou, J., et al.: Flexible piezotronic strain sensor. Nano Lett. **8**(9), 3035–3040 (2008)
76. Zhang, X., et al.: Tungsten oxide nanowires grown on carbon cloth as a flexible cold cathode. Adv. Mater. **22**(46), 5292–5296 (2010)
77. Nour, E.S., et al.: A flexible anisotropic self-powered piezoelectric direction sensor based on double sided ZnO nanowires configuration. Nanotechnology **26**(9), 095502 (2015)
78. Xiao, X., et al.: High-strain sensors based on ZnO nanowire/polystyrene hybridized flexible films. Adv. Mater. **23**(45), 5440–5444 (2011)
79. Liao, X., et al.: A highly stretchable zno@fiber-based multifunctional nanosensor for strain/temperature/UV detection. Adv. Funct. Mater. **26**(18), 3074–3081 (2016)
80. Wang, Z.L.: From nanogenerators to piezotronics—a decade-long study of ZnO nanostructures. MRS Bull. **37**(9), 814–827 (2012)
81. Pan, C., et al.: Piezotronic effect on the sensitivity and signal level of Schottky contacted proactive micro/nanowire nanosensors. ACS Nano **7**(2), 1803–1810 (2013)
82. Park, S., Vosguerichian, M., Bao, Z.: A review of fabrication and applications of carbon nanotube film-based flexible electronics. Nanoscale **5**(5), 1727–1752 (2013)
83. Kanoun, O., et al.: Flexible carbon nanotube films for high performance strain sensors. Sensors (Basel) **14**(6), 10042–10071 (2014)
84. Liu, Z.F., et al.: Hierarchically buckled sheath-core fibers for superelastic electronics sensors and muscles. Science **349**(6246), 400–404 (2015)
85. Choi, C., et al.: Twistable and stretchable sandwich structured fiber for wearable sensors and supercapacitors. Nano Lett. **16**(12), 7677–7684 (2016)
86. Kim, S.Y., et al.: Highly sensitive and multimodal all-carbon skin sensors capable of simultaneously detecting tactile and biological stimuli. Adv. Mater. **27**(28), 4178–4185 (2015)
87. Hu, N., et al.: Tunneling effect in a polymer/carbon nanotube nanocomposite strain sensor. Acta Mater. **56**(13), 2929–2936 (2008)
88. Hu, L., et al.: Highly stretchable, conductive, and transparent nanotube thin films. Appl. Phys. Lett. **94**(16), 161108 (2009)
89. Shi, J., et al.: Graphene reinforced carbon nanotube networks for wearable strain sensors. Adv. Funct. Mater. **26**(13), 2078–2084 (2016)
90. Yamada, T., et al.: A stretchable carbon nanotube strain sensor for human-motion detection. Nat. Nanotechnol. **6**(5), 296–301 (2011)
91. Park, S., et al.: Stretchable energy-harvesting tactile electronic skin capable of differentiating multiple mechanical stimuli modes. Adv. Mater. **26**(43), 7324–7332 (2014)
92. Zhao, W., et al.: Highly stable carbon nanotube/polyaniline porous network for multifunctional applications. ACS Appl. Mater. Interfaces **8**(49), 34027–34033 (2016)
93. Bryning, M.B., et al.: Carbon nanotube aerogels. Adv. Mater. **19**(5), 661–664 (2007)
94. Wang, M., et al.: Modifying native nanocellulose aerogels with carbon nanotubes for mechanoresponsive conductivity and pressure sensing. Adv. Mater. **25**(17), 2428–2432 (2013)
95. Chen, H., et al.: Omnidirectional bending and pressure sensor based on stretchable CNT-PU sponge. Adv. Funct. Mater. **27**(3), 1604434 (2017)
96. Cai, G., et al.: Extremely stretchable strain sensors based on conductive self-healing dynamic cross-links hydrogels for human-motion detection. Adv. Sci. (Weinh) **4**(2), 1600190 (2017)
97. Allen, M.J., Tung, V.C., Kaner, R.B.: Honeycomb carbon: a review of graphene. Chem. Rev. **110**(1), 132–145 (2010)
98. Zang, J., et al.: Multifunctionality and control of the crumpling and unfolding of large-area graphene. Nat. Mater. **12**(4), 321–325 (2013)
99. Wang, X., et al.: Highly stretchable and conductive core-sheath chemical vapor deposition graphene fibers and their applications in safe strain sensors. Chem. Mater. **27**(20), 6969–6975 (2015)

100. Yan, C., et al.: Highly stretchable piezoresistive graphene-nanocellulose nanopaper for strain sensors. Adv. Mater. **26**(13), 2022–2027 (2014)
101. Liu, Q., et al.: High-performance strain sensors with fish-scale-like graphene-sensing layers for full-range detection of human motions. ACS Nano **10**(8), 7901–7906 (2016)
102. Tian, H., et al.: A graphene-based resistive pressure sensor with record-high sensitivity in a wide pressure range. Sci. Rep. **5**, 8603 (2015)
103. Bae, G.Y., et al.: Linearly and highly pressure-sensitive electronic skin based on a bioinspired hierarchical structural array. Adv. Mater. **28**, 5300–5306 (2016)
104. Sheng, L., et al.: Bubble-decorated honeycomb-like graphene film as ultrahigh sensitivity pressure sensors. Adv. Funct. Mater. **25**(41), 6545–6551 (2015)
105. Chen, S., et al.: A highly stretchable strain sensor based on a graphene/silver nanoparticle synergic conductive network and a sandwich structure. J. Mater. Chem. C **4**(19), 4304–4311 (2016)
106. Qiu, L., et al.: Ultrafast dynamic piezoresistive response of graphene-based cellular elastomers. Adv. Mater. **28**(1), 194–200 (2016)
107. Qin, Y., et al.: Lightweight, superelastic, and mechanically flexible graphene/polyimide nanocomposite foam for strain sensor application. ACS Nano **9**(9), 8933–8941 (2015)
108. Zhang, P., et al.: Superelastic, macroporous polystyrene-mediated graphene aerogels for active pressure sensing. Chem. Asian J. **11**(7), 1071–1075 (2016)
109. Yao, H.B., et al.: A flexible and highly pressure-sensitive graphene-polyurethane sponge based on fractured microstructure design. Adv. Mater. **25**(46), 6692–6698 (2013)
110. Lin, Y., et al.: Graphene-elastomer composites with segregated nanostructured network for liquid and strain sensing application. ACS Appl. Mater. Interfaces **8**(36), 24143–24151 (2016)
111. Lin, Y., et al.: A highly stretchable and sensitive strain sensor based on graphene–elastomer composites with a novel double-interconnected network. J. Mater. Chem. C **4**(26), 6345–6352 (2016)
112. Huang, J., et al.: Extremely elastic and conductive N-doped graphene sponge for monitoring human motions. Nanoscale **11**(3), 1159–1168 (2019)
113. Xu, J., et al.: Highly stretchable polymer semiconductor films through the nanoconfinement effect. Science **355**, 59–64 (2017)
114. Zhao, J., et al.: A comparison between strain sensing behaviors of carbon black/polypropylene and carbon nanotubes/polypropylene electrically conductive composites. Compos. A Appl. Sci. Manuf. **48**, 129–136 (2013)
115. Yazdani, H., et al.: Strain-sensitive conductivity of carbon black-filled PVC composites subjected to cyclic loading. Carbon **79**, 393–405 (2014)
116. Ma, R., et al.: Extraordinarily high conductivity of stretchable fibers of polyurethane and silver nanoflowers. ACS Nano **9**(11), 10876–10886 (2015)
117. Liu, H., et al.: Electrically conductive thermoplastic elastomer nanocomposites at ultralow graphene loading levels for strain sensor applications. J. Mater. Chem. C **4**(1), 157–166 (2016)
118. Jung, S., et al.: Reverse-micelle-induced porous pressure-sensitive rubber for wearable human-machine interfaces. Adv. Mater. **26**(28), 4825–4830 (2014)
119. Lin, L., et al.: Towards tunable sensitivity of electrical property to strain for conductive polymer composites based on thermoplastic elastomer. ACS Appl. Mater. Interfaces **5**(12), 5815–5824 (2013)
120. Kim, Y., et al.: Stretchable nanoparticle conductors with self-organized conductive pathways. Nature **500**(7460), 59–63 (2013)
121. Jurewicz, I., et al.: Locking carbon nanotubes in confined lattice geometries–a route to low percolation in conducting composites. J. Phys. Chem. B **115**(20), 6395–6400 (2011)
122. Keplinger, C., et al.: Stretchable, transparent, ionic conductors. Science **341**, 984–987 (2013)
123. Bai, Y., et al.: Cyclic performance of viscoelastic dielectric elastomers with solid hydrogel electrodes. Appl. Phys. Lett. **104**(6), 062902 (2014)
124. Sun, J.Y., et al.: Ionic skin. Adv. Mater. **26**(45), 7608–7614 (2014)

125. Zhu, S., et al.: Ultrastretchable fibers with metallic conductivity using a liquid metal alloy core. Adv. Funct. Mater. **23**(18), 2308–2314 (2013)
126. Choong, C.L., et al.: Highly stretchable resistive pressure sensors using a conductive elastomeric composite on a micropyramid array. Adv. Mater. **26**(21), 3451–3458 (2014)
127. Graz, I.M., Cotton, D.P.J., Lacour, S.P.: Extended cyclic uniaxial loading of stretchable gold thin-films on elastomeric substrates. Appl. Phys. Lett. **94**(7), 071902 (2009)
128. Lambricht, N., Pardoen, T., Yunus, S.: Giant stretchability of thin gold films on rough elastomeric substrates. Acta Mater. **61**(2), 540–547 (2013)
129. Gray, D.S., Tien, J., Chen, C.S.: High-conductivity elastomeric electronics. Adv. Mater. **16** (5), 393–397 (2004)
130. Lacour, S.P., et al.: Stretchable gold conductors on elastomeric substrates. Appl. Phys. Lett. **82**(15), 2404–2406 (2003)
131. Guo, C.F., et al.: Highly stretchable and transparent nanomesh electrodes made by grain boundary lithography. Nat. Commun. **5**, 3121 (2014)
132. Vandeparre, H., et al.: Localization of folds and cracks in thin metal films coated on flexible elastomer foams. Adv. Mater. **25**(22), 3117–3121 (2013)
133. Lee, P., et al.: Highly stretchable and highly conductive metal electrode by very long metal nanowire percolation network. Adv. Mater. **24**(25), 3326–3332 (2012)
134. Roh, E., et al.: Stretchable, transparent, ultrasensitive, and patchable strain sensor for human-machine interfaces comprising a nanohybrid of carbon nanotubes and conductive elastomers. ACS Nano **9**(6), 6252–6261 (2015)
135. Jian, M., et al.: Flexible and highly sensitive pressure sensors based on bionic hierarchical structures. Adv. Funct. Mater. **27**(9), 1606066 (2017)

Graphene-Based Materials for Flexible Supercapacitors

Preety Ahuja and Sanjeev Kumar Ujjain

Abstract Rapid growth of flexible/wearable electronic devices has encouraged the swift development in the direction of fabricating high performance flexible supercapacitors with superior electrochemical performance. Recently, graphene based electrodes have received a great deal of attention for flexible supercapacitors on account of their outstanding properties; including excellent mechanical flexibility, enormous surface area and high electrical conductivity. This chapter summarizes the recent research progress on flexible supercapacitors and identifies the existing challenges related to preparation of electrodes and their device fabrication with regulated electrical and electrochemical properties. Furthermore, it draws attention towards the recent flexible prototype supercapacitors, in plane supercapacitors and fiber-type supercapacitors development. Moreover, it aims for insightful understanding on opportunities and challenges, endowing stimulation of further research progress in this fascinating field.

Keywords Graphene · Flexible · Symmetric · Asymmetric · Ionic liquid · Gel polymer electrolyte · Supercapacitor

Abbreviations

SC	Supercapacitor
EDLC	Electrical double layer capacitors
CNT	Carbon Nanotube
MWCNT	Multi-walled carbon nanotube
CMWCNT	Carboxylated multi-walled carbon nanotube
GO	Graphene oxide
rGO	Reduced graphene oxide
CVD	Chemical vapor Deposition
PANI	Polyaniline
PPy	Polypyrrole

P. Ahuja (✉) · S. K. Ujjain
Research Initiative for Supra-Materials, Shinshu University, 4-17-1 Wakasato, Nagano 380-8553, Japan
e-mail: drpreetyahuja@gmail.com

Inamuddin et al. (eds.), *Self-standing Substrates*,
Engineering Materials, https://doi.org/10.1007/978-3-030-29522-6_10

PEDOT	Poly(34-ethylenedioxythiophene)
$HClO_4$	Perchloric acid
BMIM BF_4	1-butyl-3-methylimidazolium tetrafluoroborate
GONR	Graphene oxide nanoribbon
RGONR	Reduced graphene oxide nanoribbon
PVDF	Polyvinylidene fluoride
PTFE	Polytetrafluoroethylene
SEM	Scanning Electron Microscopy
GR	Graphene
CB	Carbon black
PET	Poly(ethylene terephthalate)
PVA	Poly(vinyl alcohol)
CV	Cyclic Voltammetry
PU	Polyurethane
NF	Nanofiber
GS	Graphene sheets
GNS	Graphene nanosheet
PSS	Poly(sodium 4-styrenesulfonate)
CFP	Carbon fiber paper
IL	Ionic liquid
CMG	Chemically modified graphene
CAN	Acetonitrile
PC	Propylene carbonate
PEO	Polyethyleneoxide
PMMA	Poly(methylmethacrylate)
ILGPE	Ionic liquid based gel polymer electrolyte
BMIMI	1-butyl-3-methylimidazolium iodide
GCD	Galvanostatic charge-discharge
MPG	Methane plasma assisted graphene

1 Introduction

The rapid strides of social change increased demands for efficient energy production and storage in eco-friendly and sustainable manner. In order to meet these demands, flexible and portable electronics represents an important trend with wide variety applications in flexible circuits, displays, touch screens, flexible solar cells and supercapacitors (SCs) [1–3]. Among these, SCs have shown pronounced potential as a new type of energy storage device because to their excellent rate performance, elevated power density and large cycling stability along with extra advantage of

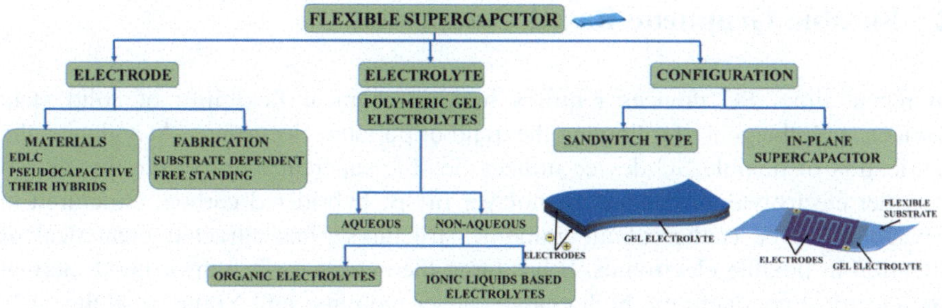

Fig. 1 Schematic representation of flexible supercapacitor with their important components

mechanical flexibility. As with conventional SCs, additional optimization of flexible SCs is critically oriented towards improving the device performance through the new electrode materials, electrolytes and device architectures.

The potential performance of supercapacitor is determined by its charge storage mechanism, i.e. electrical double layer capacitors (EDLC), where energy storage occurs through adsorption-desorption of ions and pseudocapacitance involves redox reactions for energy storage. EDLCs store charge more like ideal capacitors whereas pseudo-capacitors work more like batteries; thereby holding a unique position that bridges the slit between conventional capacitors and batteries with respect to power and energy densities [4]. Carbonaceous materials dominantly utilize EDLC storage, which depends on the available surface sites of the carbon based materials. However, high surface area carbon materials are still limited with the electrochemical accessibility of whole surface area during their communication with electrolyte; thereby resulting in lesser energy density. To augment their energy density, EDLCs are often coupled with pseudo-capacitive materials to get benefitted from fast Faradic reactions or redox reactions of electroactive materials. Therefore, maintaining a balance in energy density without decline in the power capability is the most critical aspect for ultimate performance of the SCs [5, 6].

Recent research on SC devices has broadly focused on carbonaceous materials. Among these, outstanding electrical and mechanical properties of graphene and carbon nanotubes (CNTs) bestow them as potential candidate for flexible electrodes. Also, the mechanical flexibility of SC device with lightweight, portable and high electrochemical performance brings new challenges for energy storage [7–9]. Since last decade, numerous attempts have been reported to fabricate variety of flexible SC electrodes, electrolytes and devices [10, 11] (Fig. 1). In this chapter, we will present latest progress towards graphene based flexible electrodes, electrolytes and design of flexible symmetric and asymmetric SCs along with the advances and future scenario in this stimulating field.

2 Flexible Graphene Based Composites

In recent time, SC devices requires high mechanical flexibility of solid state devices, with the goal of following the trend of portable electronics. At a glance, the fabrication of flexible SC device utilizes flexible supercapacitor electrode and gel polymer electrolyte. Graphene, monolayer of sp^2 hybridized carbon, structured in hexagonal lattice (father of all graphitic structures), has attracted great deal of attention in flexible electronics. It has large theoretical surface area, good thermal and electrical conductivity, high charge carrier mobility, and Young modulus [12–18]. On the other hand, gel polymer electrolyte because of its semisolid framework, delivers maximum adhesion and strong interaction among electrode-electrolyte interfaces. This consequently adds an extra advantage to the device in preventing the leakage from the solution and offers easy mechanical bending or twisting of the fabricated flexible device.

2.1 Electrode Material

Among different materials utilized for SCs, graphene has emerged as most promising candidate for EDLC electrode. Pristine graphene due to electrostatic interactions is prone to agglomerate in solvents, rendering it to be unapt to take advantage of outstanding properties of graphene monolayer. Consequently, solution based processes are not applicable for its fabrication [19–23]. They are produced by mechanical cleavage method or high temperature treatment, yet associated with limitations of high cost, elevated operating temperature and low yield [24–32]. Noteworthy, its peer colleagues, graphene oxide (GO) and reduced graphene oxide are easier to be processed into usable counterparts.

2.1.1 Synthesis Approach

Chemical Vapor Deposition (CVD)

This process is primarily used to synthesize carbonaceous materials involving CNTs, graphene and other nanostructures. As the name described, it takes place under vapor phase and produce defect free and even structured graphene. The starting material for graphene synthesis in vapor phase is subjected to high temperature (800–1200 °C), along with the targeted substrate, which provides surface for deposition of graphene growth. Generally, graphene growth occurs in two ways on the metal substrate, either by precipitation or by surface-mediated reaction, which depends on the properties of substrates. The metal carbide stability determines the further precipitation of carbon from particular carbide to form graphene. For example, the metastable Ni_2C phase with Ni substrate readily promotes the graphene formation [33–38]. On the contrary, other metals (Ti, Zr, Hf, V, Nb, Ta, Mo, and W) have been reported as substrates to grow monolayer graphene [39].

Electrodeposition

This method involves application of current density at room temperature in electrolytic solution to grow the film on particular substrate. The thickness of the resulting film can be easily adjusted by controlling the current density. Not only thickness of the film, but also the dimensions of the resulting morphology can be easily controlled. In addition, it is convenient to polymerize the material along one dimension rather than forming new nuclei by optimizing electrodeposition parameters. It has the benefits of mass production and low capital investment costs [40]. This is broadly utilized in synthesis of conducting polymers, polyaniline (PANI), Polypyrrole (PPy) and poly(3,4-ethylenedioxythiophene) (PEDOT) etc.; however they have poor mechanical strength resulting in limited cycling life during electrochemical evaluation [41, 42]. Numerous attempts have been reported to improve their performance by introducing EDLC materials such as graphene, CNTs to overcome their limitations.

Recently, polyaniline nanocone arrays were grown on three-dimensional graphene by electrodeposition involving 1M perchloric acid ($HClO_4$) and 0.1M aniline aqueous solutions as electrolyte. In this process, 3D graphene was used to grow PANI nanocones uniformly distributed over whole graphene surface [43]. Also, electrochemically polymerized PEDOT and GO films in ionic liquid, 1-butyl-3-methylimidazolium tetrafluoroborate ($BMIMBF_4$) were used as flexible electrode substrates [44, 45]. Not only polymers, metal oxides can also be easily hybridized with EDLC material following this approach. The nanosheets of rGO-$Co_{(1-x)}Ni_x(OH)_2$ have been successfully fabricated by following similar electro-chemical deposition on flexible polyimide substrate [46].

On the contrary, electroless deposition (autocatalytic variant of electrochemical deposition) involves the electrons from reducing agent in the solution unlike external power supply as source of electrons in electrochemical deposition. It is suitable for deposition of material on non-conducting surfaces. This has been used for ruthenium oxide-carbon composite for SC [47].

In-Situ Polymerization

In general, in-situ polymerization involves solution processable reaction in which monomer usually disperse in a solvent with ultrasonication for fixed time interval. Further, an oxidizing agent is added to initiate the polymerization and the precipitated product can be used to fabricate electrode by spray coating or direct filtration technique. This method is applied to make EDLC and pseudo-capacitive composites without the addition of binder as addition of binder compromised the conductivity and complicate the process of electrode fabrication. It is quite a simple method, and during the fabrication process, the graphene material and PANI were mixed under stirring or ultrasonication [48–50]. However, there processes are time consuming and stability is a concern but with certain modifications, it becomes advantageous with short time and operational simplicity [51–53].

Chemical Precipitation Method

It involves reaction of precursors in solution phase, which resulted in the formation of solid product. Li et al. discussed synthesis of manganese oxide on multiwalled carbon nanotube impregnated porous nickel current collector by chemical precipitation [54, 55]. Also, nickel oxide-carbon composite has been synthesized by chemical precipitation of nickel hydroxide on active carbon followed by heat treatment at 300 °C in air and investigated as SC electrode to achieve a highest capacitance of 107 F/g [56]. Similarly, cobalt oxide Co_3O_4/reduced graphene oxide nanoribbon (RGONR) hybrid electrodes derived from solvothermal reaction of cobalt acetate $Co(CH_3COO)_2 \cdot 4H_2O$ in the presence of graphene oxide nanoribbon (GONR) have shown high capacitance 525 F/g [57, 58].

Direct Coating

This involves the application of active material directly on the substrate itself, so, it would require efficient interfacial contact with good adhesion of the electrode material on the surface. Particularly some additives have been reported to be used for this purpose say carbon black (CB), polyvinylidine fluoride (PVDF), polytetrafluoroethylene (PTFE) as they can offer maximum adhesion with minimal disturbance to the conductivity of active material-substrate interface. This was prepared in slurry form by mixing with the active material in particular solvent followed by spray-coated directly on the substrate [57, 59, 60]. However, when the composite is mixed with binders or additives to produce electrodes, the structural organization and electrical conductivity are considerably altered [61, 62]. This can also be done by vacuum filtration method to obtain nanocomposite from physical mixing of different materials. The active materials are premixed in a particular solvent followed by vacuum filtration process. After drying, the nanocomposite film can be detached from the membrane and its composition can be easily monitored by optimizing the percentage of the constituents during the initial physical mixing process.

Based on these approaches, the flexible electrodes have been prepared in binary ways which involve a substrate on which active material can be adhered via binder or additive and other without any substrate or free standing film. The mechanical properties of substrates determine the mechanical strength or stability of the resulted flexible SC. So far, different substrate materials have been exploited including nickel foam, aluminum foil or foam, graphite sheets and carbon cloths, owing to their high flexibility mechanical strength and porosity [63–67]. In addition, they offer a conducting framework for the active material and fulfill the purpose of current collector. Recently, various articles also reported carbon nanofibers in the fabrication of flexible SC electrodes [48, 68, 69].

2.1.2 Fabrication of Electrode for Symmetric Supercapacitor

Electrodes with Additives (Carbonaceous or Pseudo-Capacitive Materials)

Among 2-dimensional (2D) layered carbonaceous materials, reduced graphene oxide (few layered graphene) offers porous networks resulting in large electrochemically active sites for ion diffusion [70]. For successful utilization of the excellent properties of graphene for electrochemical applications, their re-stacking should be minimized [71, 72]. This has been achieved by introducing electroactive substances within the graphene framework to act as spacers and thereby hindering the close contact between the graphene sheets. The CNTs intercalated reduced graphene oxide (rGO) film (CNTs as spacer between rGO sheets) electrodes assembling symmetrically to form both positive and negative electrodes demonstrated elevated specific capacitance, excellent rate capability and enhanced cyclic performance as compared to rGO in 1M sodium sulphate (Na_2SO_4) aqueous electrolyte [73]. Scanning electron micrograph revealed the layered stacking of rGO nanosheets with approx. thickness of 3–4 μm. CNTs act as spacers between rGO sheets to generate a sandwich structured rGO/CNT hybrid film. On the contrary, pristine CNTs due to electrostatic interactions form agglomerated bundles which hamper their electrochemical and mechanical properties, consequently opening a challenge to form their stable dispersion [74–77]. Graphene oxide sheets complimentarily enable CNTs to solubilize in solvent through π-stacking interaction.

Also, Deng et al. reported synthesis of graphene/multiwall carbon nanotubes (GR/MWCNT) composite and analyzed as SC electrode with a maximum capacitance of 91 F/g in aqueous Na_2SO_4 solution, which was much higher than the pristine GR or MWCNT [78]. This ascertains the synergistic interaction between them which consequently enhanced the capacitive performance. Further, reduced graphene oxide/carbon black (rGO/CB) films with varying CB contents have been obtained by vacuum filtration method [79]. These CB particles prevented the restacking of rGO by distributing uniformly between the graphene layers and also provided electrical contact between the layered rGO sheets. Further, their solid state flexible SC fabricated on poly(ethylene terephthalate) (PET) substrate with polymeric gel electrolyte (poly(vinyl alcohol)/Sulphuric acid, PVA/H_2SO_4) showed rectangular CV curve with a maximum capacitance of 79.6 F/g at 1 V/s. Remarkably, the SC device exhibited similar areas under curves in normal and bent state indicating perfect structural integrity of the fabricated device. Moreover, cycling stability was also investigated with first 3000 cycles in normal state followed by 2000 cycles in bent state. Notably, it exhibited high capacitive retention (94%), even after 500 cycles.

Other electroactive materials like metal oxides and polymers can also be used as active spacers and augment the overall electrochemical performance of the device by increasing the energy density. For instance, Ma et al. fabricated hierarchically structured manganese oxide (MnO_2) nanowires/graphene hybrid fibers which possessed with excellent capacitive performance [80]. MnO_2 nanowires played dual role of inhibiting the restacking of graphene nanosheets as well as providing pseudo-capacitance towards overall performance improvement. Meanwhile, the conductive graphene wrapping over the MnO_2 nanowires not only inhibited the MnO_2 nanowires aggregation but also assisted as e^- conducting networks. Additionally, the graphene nanosheets also enhance the structural integrity of the composite and proposed flexibility. Fabricated flexible solid-state device exhibited a maximum volumetric capacitance of 66.1 F/cm^3 with high energy as well as power density and capacitive retention of 96% over 10,000 charge/discharge cycles. These devices also preserved their electrochemical performance under mechanical distortion generated by bending the device to 180°, specifying good flexibility of the devices.

Electrodes with Binders

Primarily, to fabricate thin film of graphene based SC electrodes, low surface area carbon black additives have been utilized to facilitate electrical charge transfer among the active electrode materials. Its conductivity approached to that of graphite [81]. This consequently resulted in an upgraded electrode material to collector ratio, thereby augmenting the energy density of the device. In order to produce a flexible and mechanically stable graphene film, it requires binders which are chosen based on some important properties. They should be inert to electrolytes as well as should not degrade the internal configuration of the electrode, and must be electrochemically stable in large potential window [82]. Generally, the binders used are some electrically insulating polymers, so their concentration should be kept around 5–10%. Being insulating, their excess amounts escalate the internal resistance resulting in pore blockade, thereby diminishing ionic transport and lowering the charge storage of the device [83, 84]. Among existing binders, polytetrafluoroethylene PTFE and PVDF have been frequently used [85]. Besides being electrically insulating, they also require to use toxic organic solvents for their processing. Consequently, environmentally friendly, water soluble binders of conductive nature are required to generate high performance electrodes. In recent times, various binders like polyvinyl alcohol (PVA), polyethylenedioxythiophene (PEDOT:PSS) have been developed to fulfill the requirements of an apt binder [86, 87]. Song et al. explored the outcome of different binders on the charge storage of rGO based electrode in SCs [82]. Besides PTFE and PVDF, their group investigated the electrochemical performance in presence of other water dispersible binder (electrodag). PTFE/rGO showed better capacitance retention with increasing scan rate whereas PVDF/rGO displayed tilted CVs and large capacitance decay with increase in scan rate. This is attributed to hydrophobic nature of PVDF which hinders ionic diffusion in aqueous electrolytes. On the contrary, a maximum specific capacitance of 103 F/g, achieved for rGO electrodag electrode, showed about 28.7 and 14.4% improvement over the PVDF and PTFE binders, respectively. It also maintains excellent rate capability owing to the conductive module of the electrodag which resulted in faster charge transfer in the active electrode material. In addition, they also optimize the ratio of rGO to binder as large amount of binder can considerably lessen the overall performance whereas insufficient amount results in weak adhesion for flexible film fabrication.

In contrast, graphene sheets prepared by conventional methods tends to restack with one another, consequently decreasing the surface accessibility by impeding the diffusion of electrolyte ions into the layered structured rGO; thereby ensuing significantly lower capacitance.

Binderless Electrodes

Introducing binder or additives in the construction of electrode, resulted in sluggish ionic diffusion; not only due to high internal resistance of the binder material but also high interfacial resistances [88]. Recently, binderless electrodes offering advantageous platform for flexible devices have been reported. The reduced cost (eliminating the cost of binder), and diminished electrical interference at the interfacial region of the components have been main advantages. This concept has been utilized during direct growth of CNTs on carbon cloth (flexible substrate) with improved electron transport and therefore the device performance [89]. Such binder free materials could achieve high capacitance and low interfacial resistance, facilitating development of most promising flexible SCs.

Pure Graphene Electrode

Ervin et al. formulated aqueous graphene oxide dispersion as an ink for inkjet printing onto Kapton flexible substrate. Here, Kapton because of its excellent thermal stability to 400 °C, good dielectric properties and low outgassing properties was utilized as flexible substrate. Also it has been coated with thermoplastic fluorinated ethylene propylene (FEP) to permit heat sealing, as Kapton decomposed on heating rather than melting. The printed graphene oxide electrode was thermally reduced at 200 °C to produce conductive graphene flexible electrodes [90]. These electrodes were then utilized for fabricating SC device with Celgard separator injected with H_2SO_4 electrolyte or ionic liquid 1-Butyl-3-methylimidazolium tetrafluoroborate (BMIM BF_4). With the aim of enhancing wettability of GO electrodes, the BMIM BF_4 was deposited on them and then annealed at 250 °C for 30 min. After that, the fabricated devices were sealed diminishing trapped air in the device. The inkjet printed device displayed a maximum capacitance of 192 F/g (20 mV/s) in 1M H_2SO_4 electrolyte, while for ionic liquid BMIM BF_4, 73 F/g (20 mV/s) was observed. Remarkably, graphene films obtained by direct laser reduction of graphite oxide have also been reported, where GO film casted on flexible substrate was treated with laser by LightScribe CD/DVD optical drive [91]. The resulted graphene offered excellent conductivity (1738 S/cm) with 1% change in the resistivity over 1000 bending cycles. Hence, it can be easily employed as SC electrode without binders. Moreover, it exfoliates the graphene sheet simultaneously during the process, preventing the agglomeration of graphene and facilitates electrolyte diffusion into the open network of graphene. Subsequently, thin SC device fabricated using gel polymer electrolyte showed ideal rectangular cyclic voltammetry (CV) curve shape with high capacitance retention as 95.6% after 10,000 cycles.

Bilayered composites of rGO bonded to polyurethane (PU) film have also been reported. Benefitted from shape holding capacity of PU, it can be changed into various shapes to maintain a satisfactory electrochemical performance [92]. Macroporous bubble rGO film by uniting vacuum filtration and polymethyl methacrylate latex spheres has been developed by hard template directed method [93]. Device fabricated from this offered low resistance ionic diffusion providing high rate capability of the SC electrode.

Likewise, graphene hydrogels were also utilized for flexible solid state SC [94]. They were synthesized by revised hydrothermal treatment followed by immersion overnight in 1M H_2SO_4 aqueous solution. For fabricating SC device, graphene hydrogels were cut and pressed over gold coated polyimide substrate with pressure $\sim$1 MPa and sandwiched PVA/H_2SO_4 polymer electrolyte in between (Fig. 2a and

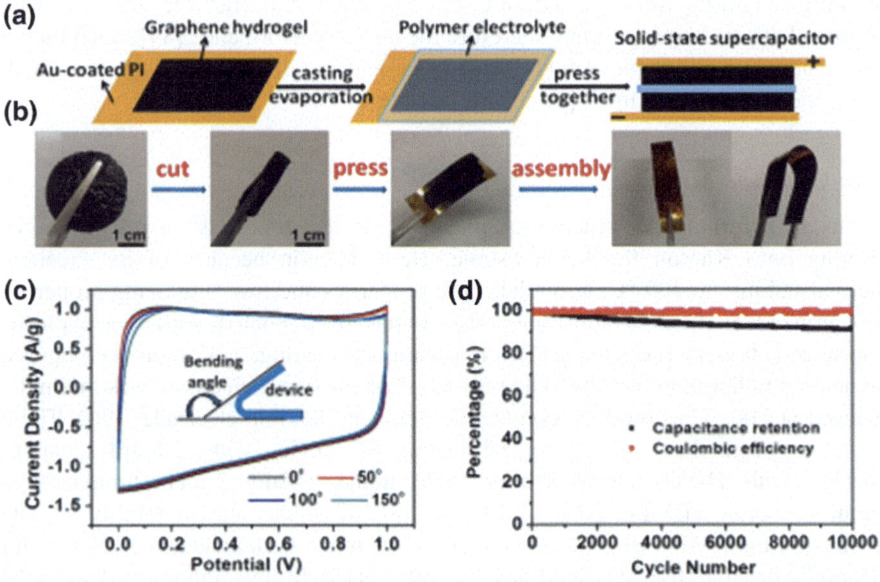

Fig. 2 **a** Schematic diagram and **b** photographs of the fabrication process of flexible solid-state supercapacitors based on graphene hydrogel films, **c** CV curves of the flexible solid-state device at 10 mV/s for different bending angles and **d** cycling stability of the device at a current density of 10 A/g. Reprinted with permission from [83]. Copyright (2013) American Chemical Society

b). The morphology of the graphene structure was maintained even after mechanical pressing, resulted in decrease in electrode thickness from 3 mm to 120 μm with large difference in electrical conductivity (7.8–192 S/cm) before and after pressing. A symmetrical triangular shaped charge-discharge curve was reported during electrochemical evaluation with high rate capability (specific capacitance of 186 F/g at 1 A/g to 130 F/g at 20 A/g). Also, remarkable flexibility was achieved at various bending angles (Fig. 2c) with 8.4% decline in specific capacitance over 10,000 cycles under 150° bending angle (Fig. 2d), attributed to mechanically stable and flexible framework.

Another strategy of enhancing capacitive storage inside the 2D graphene lattice involves substitutional heteroatom doping of the graphene framework or by developing organic functional groups which significantly alter their physical and chemical properties. This may augment the charge carrier density effectively and subsequently increases surface energy and related wettability. In addition, synergetic contributions from both double layer capacitance and pseudocapacitance further enhance device performance. In this direction, Jeong et al. prepared nitrogen doped graphene by sequential reduction of GO by hydrogen and nitrogen plasma [95]. Flexible electrode was fabricated by coating the prepared slurry on flexible conductive carbon textile electrodes along with organic electrolyte tetraethyl ammonium tetra fluoro-borate. The performance was improved about four times that of undoped graphene based devices. The device exhibited high energy density of 48 Wh/kg at a power density of 8×10^5 W/kg with excellent stability over 2,30,000 consecutive cycles. Later, boron and nitrogen co-doped graphene was reported to possess superior performance as compare to single heteroatom doped graphene [96]. Solid state SC of this co-doped monolithic graphene aerogel with PVA/H_2SO_4 as electrolyte displayed high energy and power densities with excellent rate capability.

Likewise, graphene hybrid with various organic molecules also showed similar superior performance. Hydroquinone functionalized 3D graphene hydrogel flexible SCs were fabricated by single step process involving in-situ reduction and functionalized GO by hydroquinone [97]. All solid state flexible SCs with semi-solid (PVA/H_2SO_4) electrolyte demonstrated maximum capacitance of 421 F/g at 1 A/g with a capacitive retention of 86% over 10,000 cycle at 10 A/g. Further, nafion has also been reported to functionalize reduced graphene oxide (Nf-rGO) by supramolecular assembly method [98]. Here, nafion not only used for the functionalization but also acted as an electrolyte during fabrication of flexible SC. The Nf-rGO SC device exhibited a two-fold enhanced capacitance and improved rate capability than simple graphene based device.

Conducting Polymer/Graphene Composites

Different pure graphene structures have been utilized successfully for SC electrodes with moderate energy density. This energy density of the SC device can be improved by introducing electroactive conducting polymers in graphene based electrodes [99–103]. rGO/PANI nanofiber (NF) films were prepared by vacuum filtration method with very high conductivity (5.5×10^2 S/m) as compared to pure PANI-NF. These films were directly used for SC without any insulating binder owing to their flexibility and mechanical strength [48]. Electrochemical evaluation displayed maximum capacitance of 210 F/g, which was higher as compared to the pristine PANI-NF and rGO. Moreover, a capacitive retention of 94% with increased current density from 0.3 to 3 A/g was mainly attributed to the rGO framework which sustains PANI-NF from severe swelling and shrinking process. Further, layer by layer assembly was used to fabricate multilayered rGO/PANI flexible films by alternate deposition of negatively charged GO nanosheets and positively charged PANI onto PET substrate, further chemically reduced to form rGO [104]. Resulted intimate contact between EDLC of rGO nanosheets with pseudocapacitance of PANI augmented electrochemical stability and electronic conductivity throughout the charge/discharge cycling. Not only films, PANI-NF/graphene paper has also been prepared by electrostatic adsorption of positively charged PANI-NFs over negatively charged graphene sheets (PSS GS) generated by poly(sodium 4-styrenesulfonate) (PSS). The resulting PSS GS/PANI NF suspension was vacuum filtered and dried to fabricate electrode [105]. Specific capacitance of 301 F/g was attained from the fabricated electrodes due to synergistic interaction between graphene and PANI-NF.

Another promising polymer, polypyrrole (PPy) with large pseudocapacitance has been developed by pulsed electro-polymerization to form flexible graphene/PPy composite and utilized for fabricating flexible SC [106]. The high energy and power density of the fabricated SC, owe to favorable nucleation of PPy chains at graphene sheets defective sites. Likewise, flexible membrane of rGO-PPy was synthesized via self-assembly method involving in-situ reduction of GO in presence of PPy rendering the formation of rGO-PPy foam [107]. The SC fabricated using rGO-PPy foam electrodes accomplished enhanced areal capacitance (175 mF/cm^2) and outstanding cycling stability.

In addition, stretchable SC electrode of graphene/ PANI has also been reported via series of chemical methods involving two templates of nickel foam and graphene, out of which nickel foam template was completely removed leaving behind porous graphene [64]. The symmetrical SC device was fabricated by utilizing these working electrodes with polymeric electrolyte from PVA and phosphoric acid (PVA/H_3PO_4) enclosed in Ecoflex. The highest specific capacitance, 261.24 F/g was reported at 0.38 A/g. Nyquist plot demonstrated nearly $-90°$ slope along with lesser equivalent series resistance (6.4 Ω). This improved performance with high stability of the fabricated device was credited to the highly stable and conducting structure of dense PANI film supported on porous graphene. Remarkably, this device illustrated negligible differences on bending and stretching indicating their susceptibility to large bending stress and mechanical deformations.

Metal Oxides or Hydroxides/Graphene Composite

Metal oxides are promising electroactive candidates for SCs owing to their higher pseudocapacitance via bulk redox reactions [108–110]. However, they have limited electrical conductivity but considerable volumetric variations during cycling which limits their stability, leading to rapid capacitance decay. However, graphene provides the conducting channels and stability with expected significant synergistic effects during their composite formation; resulting in flexible SCs with augmented performance. Direct electrochemical deposition of MnO_2 nanoflower on graphene fiber network has been achieved and flexible supercapacitor was fabricated with polymer gel electrolyte (PVA/H_2SO_4) [111]. Besides high electrochemical performance, there was no interference in the response during deformations as confirmed by no distortion in the CV rectangular shape recorded in deformed state. This device further retained a stable capacitance of 70–73 μF, even after 1000 cycles of straightening-bending-straightening process. Likewise, MnO_2 nanowires/graphene fibers were fabricated by wet-spinning method [80] where hierarchical MnO_2 structures inhibited the restacking of graphene nanosheets by acting as spacer and consequently, improved the electrochemical activity. Complimentarily, graphene nanosheets covered MnO_2 nanowires and provided superhighways for electronic transport in the fiber by inhibiting MnO_2 aggregation. Correspondingly fabricated flexible SC exhibited high volumetric capacitance (66.1 F/cm^3), with excellent power and energy densities (0.51 W/cm^3 and 5.8 mWh/cm^3), respectively. Their structure and electrochemical performance were well maintained during mechanical deformation, indicating their high susceptibility and excellent mechanical strength.

Apart from MnO_2, other metal oxides also played important role in enhancing the performance of graphene based flexible SCs [112]. A bendable freestanding paper was developed from rGO-wrapped flowery nickel hydroxide $Ni(OH)_2$ by hydrothermal method [113]. The flexible electrode exhibited areal capacitance, 10.44 F/cm (877.1 F/g) and excellent cycling stability along with 93.6% capacitive retention after 15,000 cycles. Additionally, it showed excellent mechanical properties including bending to arbitrary angles (even 180°) and good tensile strength of 48.8 MPa. The reduced graphene oxide with molybdenum oxide composite (rGO/MoO_3) was prepared by simply mixing and annealing of molybdenum-based metal organic frameworks (Mo-MOFs) and GO sheets [114]. Resulted porous rGO/MoO_3 was used for flexible symmetric SC which exhibited excellent electrochemical and mechanical stabilities.

Some active materials require flexible substrates as framework for working electrode fabrication. All solid state flexible SC of ß-$Ni(OH)_2$/graphene nanocomposite with gold coated PET as substrate with PVA/KOH electrolyte was fabricated [115]. The device showed specific capacitance of 3304 μF/cm calculated from discharge slope at 0.1 A/m^2. Moreover, remarkable flexibility was confirmed by consistency in galvanostatic charge-discharge curves under normal and bend conditions. Another attempt to fabricate MnO_2-rGO on flexible carbon fiber paper (CFP) has been reported [68]. rGO was prepared by chemical reduction method and MnO_2 by ethylene glycol assisted synthesis followed by premixing of MnO_2 and rGO and spray coated on CFP. To improve the electrochemical stability and capacitance retention of ruthenium oxide (RuO_2) prepared on flexible Cu foil current collector by cathodic electrodeposition, graphene was inserted between RuO_2 and the Cu-foil current collector [116]. Moreover, the performance and stability of the flexible RuO_2/Gr/Cu electrode under bent condition remained unchanged owing to improved mechanical adhesion between the RuO_2 and the current collector.

Recently, mixed metal oxides have been considered as prospective active materials for SCs after realizing high equivalent series resistance with restricted energy and power densities of pristine transition metal oxides [117]. With excellent electrical conduction, they offer richer sites for redox reactions, thereby augmenting electrochemical activity. Among mixed metal oxides, cobalt and zinc manganite (spinel structures) were hybridized with graphene nanoribbon for all solid state SC [118, 119]. Cobalt manganite graphene nanoribbon based SC showed a large potential window, 3 V with high energy density, 44.6 Wh/kg [119]. Moreover, it retained the electrochemical performance under different bending conditions, offering a high performance graphene based flexible devices.

Additionally, nickel cobaltite $NiCo_2O_4$/graphene oxide, flexible and porous sheets have been fabricated by electrodeposition method followed by heat treatment [120]. $NiCo_2O_4$-GO displayed specific capacitance, 1078 F/g and remarkable cyclic stability, attributed to their high porosity. The porous structure provided large accessible surface area and significantly enhanced the active sites for the Faradic redox reaction of $NiCo_2O_4$. The applicability for flexible charge storage device was investigated by measuring the change in electrical resistivity as a function of bending angles. Apparently, no change in the electrical resistivity of the $NiCo_2O_4$-GO electrode was observed on bending from 0 to 170°. Moreover, $NiCo_2O_4$-GO electrode at various angles of bending demonstrated identical CV shape signifying electrochemical stability of the electrode on bending. Almost no change in electrical and charge storage properties of $NiCo_2O_4$-GO under bending conditions suggested their successful flexibility and mechanical strength. Additionally, mixed cobalt and nickel hydroxide-rGO composite (rGO-$Co_{(1-x)}Ni_x(OH)_2$ was prepared by one-step electrochemical deposition approach [46]. The distinct $Ni(OH)_2$, integrated with cobalt hydroxide resulted in broad redox bands; thereby modifying the charge storage leading to large rate capability and Columbic efficiency of the SC device. Also, capacitive behaviors of manganese ferrite $MnFe_2O_4$/graphene hybrids towards high performance and flexible electrodes were analyzed [121] by fabricating flexible graphite sheets as substrate for $MnFe_2O_4$/graphene inks. The flexible SC device displayed excellent electrochemical performance, specific capacitance of 120 F/g at 0.1 A/g along with outstanding cycling performance. Notably, it also exhibited outstanding mechanical strength during bending and twisting, as evidenced by no significant changes in electrochemical properties under different bending angles or in twisted condition.

Graphene Based Ternary Composite

For the realm hybrid SCs, various EDLCs and pseudo-capacitive materials have been explored as electrodes. However, each of them has limitations like lesser energy density (EDLCs), restricted cyclic stability (pseudo-capacitive materials) and large resistivity (pseudo-capacitive materials). Consequently, to improvise them, binary composites constituting metal oxides and conducting polymers with carbon have been reported [122]. Recently, ternary composites have become the latest choice of the researchers constituting metal oxides, polymers and carbon based materials [123, 124]. Fascinatingly, they provided large energy density without compensation in power density, low internal resistance, excellent specific capacitance, reasonable cycle life, and large diffusion efficiency [125–130]. Composite of Polypyrrole, graphene oxide and zinc oxide (PPy/GO/ZnO) was developed on flexible nickel foam by potentiostatic electrochemical polymerization [131]. The SC device was fabricated with PPy/GO/ZnO and sodium sulfate solution saturated filter paper as working electrode and electrolyte. Their pseudo-capacitive behavior synergistically resulted in high energy density of 10.65 Wh/kg and power density, 258.26 W/kg along with good capacitive retention.

Assembling MnO_2/PANI on rGO paper for flexible SCs has been recently reported in which MnO_2 nanoflakes were developed on rGO by electrodeposition method followed by oxidative polymerization of aniline between intersecting MnO_2 [132]. This unique nanoarchitecture ensured fast electron transfer and more electrochemical active sites for reversible Faradaic reactions, thereby easing the ionic diffusion of the electrolyte. Based on the fascinating synergistic effects, it displayed significantly enhanced electrochemical properties, high specific capacity (636.5 F/g at 1.0 A/g) and outstanding cycling stability. MnO_2/PANI hybrid strongly bind to the surface of rGO paper, negating the use of binder in electrode fabrication. Furthermore, the incorporation of PANI enhances the conductivity and charge transfer process; favoring higher capacitance. Likewise, nano nickel oxide (NiO) coated with graphene nanosheet/polyaniline (NiO-GNS/PANI) was utilized for flexible SC [133]. Firstly, nano NiO was grown on the GNS surface and subsequently, followed by in-situ polymerization of PANI on their surface. For testing their electrochemical performance, the NiO-GNS/PANI film was assembled on gold-coated polyimide substrate. Further, CV test was recorded at 2 mV/s by bending the device at different angles from 0° to 90°. The response showed no changes in CV curves under different bending angle (0°, 30°, 60° and 90°). Capacity retention (86%) was maintained after 1000 bending cycles; attributed to the graphene nanosheets offering high electrical conductivity and mechanical strength. Further, reduced graphene oxide/polypyrrole/Cu_2O–$Cu(OH)_2$ (RGO/PPy/Cu_2O–$Cu(OH)_2$) was synthesized by electrochemical deposition of graphene oxide–polypyrrole (GO/PPy) nanocomposite on Ni foam followed by its conversion to RGO/PPy. Then, chronoamperometry was used to develop a thin layer of Cu_2O–$Cu(OH)_2$ on RGO/PPy film [134]. The performance of the composite was analyzed using three-electrode system which indicated the excellent specific capacitance of 997 F/g. On the contrary, in two electrode assembly, it displayed specific capacitance of 225 F/g with good energy and power density of 20 Wh/kg and 8000 W/kg. It is also capable of retaining 90% specific capacitance, after 2000 cycles. Many other flexible substrates like cellulose papers and textiles have been frequently explored in the graphene based ternary flexible SC electrodes; offering lightweight and robust properties promoting flexibility of the electrode with improved mechanical strength. More importantly, the sustainable and recyclable flexible substrates are favorable for new generation wearable and environmental friendly devices.

2.1.3 Asymmetric Supercapacitor

In addition to the investigation on symmetrical devices, asymmetric solid state flexible SCs have also been extensively explored. As the name suggests, they are asymmetric with respect of the electrodes. Generally, one electrode is EDLC type as negative electrode and other pseudo-capacitive constitutes positive electrode. This choice of using two different electrodes resulted in increased operating voltage, leading to noticeable enhancement in the energy density of fabricated flexible SCs [135, 136]. Hybrid film of reduced graphene oxide and carboxylated multiwalled carbon nanotube (rGO/CMWCNT) and carbon fiber paper (CFP) supported with PPy composite (CFP/PPy) were prepared [137]. They were assembled in asymmetric solid state SC constituting polymeric gel electrolyte along with rGO/CMWCNT and CFP/PPy as negative and positive electrodes. The utilization of asymmetric electrodes enabled the large operating potential window (i.e. 1.8 V) with stable and rectangular shaped CV curve with large retention (93%) of initial capacity. Further, the flexibility was revealed by the similar CV curves under various curvatures of the device.

Meanwhile another flexible asymmetric device was constructed by employing self-supporting films of modified graphene functionalized with ionic liquid (IL-CMG) as negative electrode and ruthenium oxide modified graphene, RuO_2-IL-CMG as positive electrode [138]. Graphene was modified by mixing ionic liquid to graphene oxide (sonication for 60 min.) followed by thermal hydrazine reduction at 95 °C for 6 h. While positive electrode was synthesized by sol gel process and additional thermal treatment resulting in uniform distribution of RuO_2 nanoparticles along IL-CMG surface. Their asymmetric assembly provided 1.8 V potential window with stable galvanostatic charge-discharge cycles with 97% capacitive retention. Also, under twisting condition, the device was able to retain 95% capacitance after 200 cycles.

2.2 *Electrolyte*

Additionally, to boost the overall performance of flexible supercapacitors, besides electrode materials, apt choice of electrolytes also plays a crucial role. It is well established that large potential range is the competent approach to augment the energy density of SC device. The stability of active potential window are related to electrolytes. Replacing aqueous counterparts with limited potential range, non-aqueous particularly organic electrolytes are attracting more attention. The most frequently used organic solvents are acetonitrile (ACN) and propylene carbonate (PC), which can dissolve salts like tetraethylammonium tetra fluoroborate, tetraethylphosphonium tetra fluoroborate and triethylmethylammonium tetra fluoroborate [138]. ACN can easily dissolve larger amount of salt but being toxic in nature, be replaced by PC. On the contrary, PC based electrolytes are environmental benign and offer lower but relatively good conductivity [139].

Recently, solid-state SCs have attracted great interest for polymeric gel electrolytes. They can be aqueous and non-aqueous gel electrolytes. Usually aqueous gel electrolytes have aqueous electrolytes (e.g., H_2SO4, KOH, and Na_2SO_4) with polymeric matrix, say poly(vinylidene fluoride) (PVDF), poly(ethylene oxide) (PEO) [140]. Also, many other types of gel electrolytes have been explored like PVA/H_2SO_4, PVA/KOH, polyethyleneoxide (PEO)/$LiClO_4$ and poly(methylmethacrylate) (PMMA)/$LiClO_4$ [96, 141–144].

Non-aqueous polymeric gels comprised electrolytic salts and organic electrolytes with polymer complexes [145]. However, aqueous electrolytes have limited working potential (1.2 V) owing to water decomposition [146, 147], resulting in limited energy density. As far as organic electrolytes are concerned, they possess serious health and safety issues. Recently, ionic liquid based gel electrolytes (ILGPE) have been considered as a boon for solid state SCs [148–150]. Ionic liquids are pure organic salts with no solvent and melting point below 100 °C. They are composed of dissociated ions constituting a delocalized charged ion (aromatic) and organic part preventing the stable crystal formation as components. Various substitutions in them controlled the resultant physico-chemical properties [151]. By incorporating these ILs into polymeric matrices, highly conductive, electrochemically and thermally stable IL gel polymer electrolytes can be produced [152].

Modern era need for flexible energy devices has triggered the development of gel polymeric electrolytes [153]. Zhong et al. fabricated ILGPE by cross-linking between IL and PEO, along with a separator (FPC3018) [154]. Reported large ionic conductivity up to 6.7 mS/cm was achieved, at room temperature. Also, electrochemical evaluation was explored to optimize the ratio of IL/PEO in the resultant gel polymeric electrolytes. Their flexibility and stability were explored by analyzing the electrochemical performance under different bending angles by using in SC device with activated carbon electrodes. The rectangular shaped CV curves and symmetric GCD curves with no significant change at different bending angles were observed. Also, slight decrement in capacitance following bending angles order ($0° > 50° > 100° > 150° > 180°$) of the device was observed.

Another redox reactive gel polymeric electrolyte was reported by introducing a neutral electrolyte Li_2SO_4 with an ionic liquid 1-butyl-3-methylimidazolium iodide (BMIMI) in PVA and utilized in fabricating flexible SC with activated carbon as electrodes [149]. Fabricated GPE also showed high tensile strength 2.0 MPa, evaluated from stress-strain curve. Besides, capacitances of the device calculated from GCD curves were 384.1, 384.0, 382.0 and 374.7 F g^{-1} at corresponding bending angles of 0°, 45°, 90°, 135° respectively. The capacitance showed negligible change (only 2.4% capacitance loss at 135°) with bending angles, suggesting device flexibility under bending conditions.

Noteworthy, the electrochemical activity of the device was strongly determined by the ionic conductivity [155–157], which can be enhanced by introducing some additives in GPE. Previous reports proposed the incorporation of boron modified segments into the polymer matrix [158, 159], and also to the electrolytic salts [160, 161] of GPE. Likewise, modified cross-linkers functionalized with boron were synthesized by transesterification reaction and their inclusion in polymer network of GPE augmented the ionic conductivity and mechanical stability [148]. Moreover, all-solid-state symmetric SC with reduced graphene oxide (rGO) as the electrode was fabricated, which offered high energy and power delivery along with good tolerance to high temperature.

2.3 *Architectural Variations*

The era of flexible electronics such as photovoltaic cells, rolling displays, and other wearable devices require the fabrication of micro-scale flexible energy storage devices. Micro-SCs generally, follow one of the two device architecture; sandwich or in-plane. As a matter of fact, most of the articles discussed in this chapter follow the sandwich device fabrication with symmetrical or asymmetrical electrodes. However, in-plane devices have proved to be more promising as the micro-SC fabrication process involved interdigital electrodes on 2D flexible plane, and electrolyte ions in the narrow interspaces between electrode fingers; resulting in rapid ionic diffusion leading to higher charge storage capacity of the device [162].

Niu et al. fabricated flexible micro-SCs by utilizing photolithography and electrophoretic deposition to develop graphene microelectrodes [163]. These graphene microelectrodes facilitated the electrolyte ions transportation, allowing more accessible surface area. Also, the reported specific capacitance of the fabricated graphene based micro-SC was 285 F/g, nearly three times more as compared to conventional graphene based SC (86 F/g).

In conventional SCs, introducing different capacitive insertions such as CNTs, metal oxides, or conductive polymers provide a prospective strategy to electrochemically improve the device [164–166]. Likewise, Beidaghi et al. reported rGO–CNTs interdigitated microelectrodes, fabricated by electrostatic spray deposition and photolithography for micro-SC [167]. The introduction of CNTs as spacer between graphene with planar interdigital microelectrodes fabrication enhances electrolyte ions diffusivity, thereby augmented performance of the fabricated device. Also, graphene with MnO_2 as pseudo-capacitive spacer was developed by Peng et al, introducing more redox active sites for efficient ionic diffusion of electrolyte [168]. The solid state micro-SC was fabricated which exhibited specific capacitance of 267 F/g at low current density and retained capacitance of 208 F/g at 10 A/g indicating good rate capability of the resulting device. Klaus Mullen group developed planar interdigital micro-SC with electrodes of plasma assisted reduced graphene (MPG) and PVA-H_2SO_4 as electrolyte [162]. Areal capacitance, 80.7 $\mu F/cm^2$ with stack capacitance, 17.9 F/cm^3 were realized, higher than classical sandwich SC of MPG based electrodes.

In recent times, a novel fiber shaped design has emerged to fabricate flexible SCs with 1D fibers as electrodes, which can provide maximum flexibility, as compared to conventional 2D planar SCs [169–175]. Fiber SCs are 1D wires (dia. micrometers to millimeters), they can be easily woven into fabrics and offer unique wearability. These unique architectures also deliver adaptability to versatile designs, leading to their fabrication into different desired shapes [176]. Remarkably, their omni-directional flexibility encouraged the rapid development of portable and flexible electronics by providing restriction free operational space. Till now, they can be assembled in three different types of configurations say, parallel, twisted and coaxial [169]. They evolved from the conventional sandwiched structure of 2D SC (electrode/separator/electrode). The parallel-fiber are scaled by integrating fiber electrodes on planar substrate, limiting the applicability of modern era flexible devices. On the contrary, with the other two, twisted and coaxial configurations, planar substrate are not required; instead they utilize stand-alone fibers. The twisted-fiber configuration of SC involves twisting of two stand-alone fiber electrodes along with solid state electrolytes between them. Whereas, the coaxial-fiber device can be constituted with layering assembly of core fiber electrode, a electrolyte, and an outer electrode layer.

Fascinatingly, fiber SC has been reported by Li et al. with rGO on gold wires as electrodes, produced by electrochemical reduction of GO [177]. The capacitance of 11.4 mF/cm was achieved corresponding to an areal capacitance of 0.73 mF/cm^2 with unit length of the fiber type flexible SC. Moreover, their remarkable performance was demonstrated by negligible capacitance decrease ($\sim$1%) during CV, when the SC was bent to 120° or S-shaped structure.

In particular, graphene fibers (GFs) have high strength and electrical conductivities [178, 179] which can be directly employed to build the fiber SCs; two fiber electrodes were twisted together with an indispensable separator, which could be the porous membrane or solid electrolyte [180]. Recently, an all-in-one graphene SC has been reported where no additional bonding agent and assembling process are required and the device is fabricated on graphene fiber only [181]. This was achieved by region-specific reduction of GO fiber under laser irradiation, where rGO layers acted as electrodes of the resultant device, rGO/GO/rGO and GO, the separator and the reservoir for holding electrolyte. In addition, the fiber capacitor showed no degradation after 1000 cycles of charge-discharge process evidencing their high stability.

3 Conclusions and Future Perspectives

Here, we have discussed the recent developments in graphene based flexible SC devices. Innovative strategies have been utilized to modify graphene based electrodes for improving the conductivity and effective surface accessibility in order to boost the overall electrochemical activity of the fabricated SC devices. Despite the aforementioned notable progress, it is still a large modern era requirement to develop highly efficient flexible SC for wearable practical applications. In our view, the barricades or futuristic research prospects lie in the following aspects:

1. Graphene based new materials with better pseudo-capacitive contribution would be the next focusing point for flexible SCs. In particular, pseudo-capacitive materials, have already been integrated into graphene electrodes to enhance the capacitances. Despite the extensive developments regarding hybrid composites, major research efforts should be directed towards other aspects with the capacitance increment i.e. without sacrificing their cycling stability, and retaining their mechanical strength as well.
2. Beside electrode materials, device arrangement or configuration should be properly covered for the development of next generation flexible SCs. Mostly, in-plane and fiber based SCs are preferred over sandwitch structure conventional SCs. In planar SCs, collector, electrode and electrolyte are in the same horizon to provide improved ion migration networks which consequently, enhanced the performances of energy storage devices.
   With regard to fiber based SCs, most of these have metal wires current collectors. They contribute significantly to the total weight and volume of the device but very little to the total capacitance, hence compromise with the gravimetric and volumetric performance of the device. Thus, utilization of metal wires or metal coated other counterparts should be minimized as current collectors and other alternatives must be explored for efficient performances of the device. In addition, most of the fiber based SCs by far are symmetrical devices, limiting their voltage range to 0.6–1 V. For practical applications, high-voltages SCs required a large number of these individual SC to be connected in series to meet the requirements. A probable solution is to fabricate asymmetric SC, comprising two different electrodes; capacitive and battery-type materials for fiber based SC, which is still uncovered.
3. Alongside, mechanical flexibility is the most important parameter to evaluate the flexibility of the fabricated device. Preferentially, standard methods are not available to accurately estimate the flexibility of the device. Hence, it is necessary to understand the charge storage mechanism of flexible SC including interfacial reaction at electrode/electrolyte and active material/current.
4. Noteworthy, there is a lack of standardized methods to evaluate the device performances. Without the complete experimental details and relevant data, i.e. active mass loading, capacitance, equivalent series resistance per length or area of active materials along with the physical dimensions, it has no meaning to report the comparative performances of the devices. A probable solution is to establish or clarify on some standardized guidelines for reporting and monitoring the performance in terms of capacitance, energy or power density of flexible SC device. This can be actually realized by the combined contribution from journal reviewers and as well as authors.

It is anticipated that the directions provided in this chapter will guide the researchers to pursue further work in this exciting field to find solutions to the existing challenges in fabricating graphene based flexible devices. Besides, innovative solutions to the existing challenges and design of the device will bring a break-through in energy storage technology.

References

1. Miller, J.R., Simon, P.: Electrochemical capacitors for energy management. Science **321**, 651–652 (2008)
2. Lu, X., Shen, C., Zhang, Z., Barrios, E., Zhai, L.: Core-shell composite fibers for high-performance flexible supercapacitor electrodes. ACS Appl. Mater. Interfaces. **10**, 4041–4049 (2018)
3. Hou, Y., Cheng, Y., Hobson, T., Liu, J.: Design and synthesis of hierarchical MnO_2 nanospheres/carbon nanotubes/conducting polymer ternary composite for high performance electrochemical electrodes. Nano Lett. **10**, 2727–2733 (2010)
4. Ghosh, S., Inganas, O.: Conducting polymer hydrogels as 3D electrodes: applications for supercapacitor. Adv. Mater. **11**, 1214–1218 (1999)
5. Eftekhari, A., Li, L., Yang, Y.: Polyaniline supercapacitors. J. Power Sources **347**, 86–107 (2017)
6. Wang, X., Yang, C., Jin, J., Li, X., Cheng, Q., Wang, G.: High-performance stretchable supercapacitors based on intrinsically stretchable acrylate rubber/MWCNTs@conductive polymer composite electrodes. J. Mater. Chem. A **6**, 4432–4442 (2018)
7. Liu, L., Ye, D., Yu, Y., Liu, L., Wu, Y.: Carbon-based flexible micro-supercapacitor fabrication via mask-free ambient micro-plasma-jet etching. Carbon **111**, 121–127 (2017)
8. Ghosh, K., Yue, C.Y., Sk, M., Jena, R.K.: Development of 3D urchin-shaped coaxial manganese dioxide@polyaniline (MnO_2@PANI) composite and self-assembled 3D pillared graphene foam for asymmetric all-solid-state flexible supercapacitor application. ACS Appl. Mater. Interfaces **9**, 15350–15363 (2017)
9. Wang, L., Zhang, G., Zhang, X., Shi, H., Zeng, W., Zhang, H., Liu, Q., Li, C., Liu, Q., Duan, H.: Porous ultrathin carbon nanobubbles formed carbon nanofiber webs for high-performance flexible supercapacitors. J. Mater. Chem. A **5**, 14801–14810 (2017)
10. Ujjain, S.K., Sahu, V., Sharma, R.K., Singh, G.: High performance, all solid state, flexible supercapacitor based on ionic liquid functionalized graphene. Electrochim. Acta **157**, 245–251 (2015)
11. Ujjain, S.K., Ahuja, P., Bhatia, R., Attri, P.: Printable multi-walled carbon nanotubes thin film for high performance all solid state flexible supercapacitors. Mater. Res. Bull. **83**, 167–171 (2016)
12. Morozov, S.V., Novoselov, K.S., Katsnelson, M.I., Schedin, F., Elias, D.C., Jaszczak, J.A., Geim, A.K.: Giant intrinsic carrier mobilities in graphene and its bilayer. Phys. Rev. Lett. **100**, 016602(1–4) (2008)
13. Getty, S.A., Cobas, E., Fuhrer, M.S.: Extraordinary mobility in semiconducting carbon nanotubes. Nano Lett. **4**, 35–39 (2004)
14. Stoller, M.D., Park, S., Zhu, Y., An, J., Ruoff, R.S.: Graphene-based ultracapacitors. Nano Lett. **8**, 3498–3502 (2008)
15. Lee, C., Wei, X., Kysar, J.W., Hone, J.: Measurement of the elastic properties and intrinsic strength of monolayer graphene. Science **321**, 385–389 (2008)
16. Balandin, A.A., Ghosh, S., Bao, W., Calizo, I., Teweldebrhan, D., Miao, F., Lau, C.N.: Superior thermal conductivity of single-layer graphene. Nano Lett. **8**, 902–907 (2008)
17. Zhu, B.Y., Murali, S., Cai, W., Li, X., Suk, J.W., Potts, J.R., Ruoff, R.S.: Graphene and graphene oxide: synthesis, properties and applications. Adv. Mater. **22**, 3906–3924 (2010)

18. Putri, A.D., Chotimah, N., Ujjain, S.K., Wang, S., Futamura, R., Vallejos-burgos, F., Khoerunnisa, F., Morimoto, M., Wang, Z., Hattori, Y., Sakai, T., Kaneko, K.: Charge-transfer mediated nanopore-controlled pyrene derivatives/graphene colloids. Carbon **139**, 512–521 (2018)
19. Dikin, D., Stankovich, S., Zimney, E.J., Piner, R.D., Dommett, G.H.B., Evmenenko, G., Nguyen, S.T., Ruoff, R.S.: Preparation and characterization of graphene oxide nanosheets. Nature **448**, 457–460 (2012)
20. Chen, D., Tang, L., Li, J.: Graphene-based materials in electrochemistry. Chem. Soc. Rev. **39**, 3157–3180 (2010)
21. Dreyer, D.R., Ruoff, R.S., Bielawski, C.W.: From conception to realization: An historial account of graphene and some perspectives for its future. Angew. Chem. Int. Ed. **49**, 9336–9344 (2010)
22. Cui, X., Zhang, C., Hao, R., Hou, Y.: Liquid-phase exfoliation, functionalization and applications of graphene. Nanoscale **3**, 2118–2126 (2011)
23. Li, C., Shi, G.: Three-dimensional graphene architectures. Nanoscale **4**, 5549–5563 (2012)
24. Novoselov, K.S., Geim, A.K., Morozov, S.V., Jiang, D., Zhang, Y., Dubonos, S.V., Grigorieva, I.V., Firsov, A.A.: Electric field effect in atomically thin carbon films. Science **306**, 666–669 (2004)
25. Li, X., Cai, W., An, J., Kim, S., Nah, J., Yang, D., Piner, R., Velamakanni, A., Jung, I., Tutuc, E., Banerjee, S.K., Colombo, L., Ruoff, R.S.: Large-area synthesis of high-quality and uniform graphene films on copper foils. Science **324**, 1312–1314 (2009)
26. Zhang, B., Lee, W.H., Piner, R., Kholmanov, I., Wu, Y., Li, H., Ji, H., Ruoff, R.S.: Low-temperature chemical vapor deposition growth of graphene from toluene on electropolished copper foils. ACS Nano **6**, 2471–2476 (2012)
27. Wu, W., Jauregui, L.A., Su, Z., Liu, Z., Bao, J., Chen, Y.P.: Growth of single crystal graphene arrays by locally controlling nucleation on polycrystalline Cu using chemical vapor deposition. Adv. Mater. **23**, 4898–4903 (2011)
28. Wassei, J.K., Mecklenburg, M., Torres, J.A., Fowler, J.D., Regan, B.C., Kaner, R.B., Weiller, B.H.: Chemical vapor deposition of graphene on copper from methane, ethane and propane: evidence for bilayer selectivity. Small **8**, 1415–1422 (2012)
29. Wang, Y., Zheng, Y., Xu, X., Dubuisson, E., Bao, Q., Lu, J., Loh, K.P.: Electrochemical delamination of CVD-grown graphene film: toward the recyclable use of copper catalyst. ACS Nano **5**, 9927–9933 (2011)
30. Films, C., Tao, L., Lee, J., Chou, H., Holt, M., Ruoff, R.S., Akinwande, D.: Synthesis of high quality monolayer graphene at reduced temperature on hydrogen-enriched evaporated. ACS Nano **6**, 2319–2325 (2012)
31. Robertson, A.W., Warner, J.H.: Hexagonal single crystal domains of few-layer graphene on copper. Nano Lett. **11**, 1182–1189 (2011)
32. Rasool, H.I., Song, E.B., Mecklenburg, M., Regan, B.C., Wang, K.L., Weiller, B.H., Gimzewski, J.K.: Atomic-scale characterization of graphene grown on copper (100) single crystals. J. Am. Chem. Soc. **133**, 12536–12543 (2011)
33. Bae, S., Kim, H., Lee, Y., Xu, X., Park, J.S., Zheng, Y., Balakrishnan, J., Lei, T., Ri Kim, H., Il Song, Y., Kim, Y.J., Kim, K.S., Özyilmaz, B., Ahn, J.H., Hong, B.H., Iijima, S.: Roll-to-roll production of 30-inch graphene films for transparent electrodes. Nat. Nanotechnol. **5**, 574–578 (2010)
34. Karu, A.E., Beer, M.: Pyrolytic formation of highly crystalline graphite films. J. Appl. Phys. **37**, 2179–2181 (1966)
35. Reina, A., Jia, X., John, H., Daniel, N., Hyungbin, S., Vladimir, B., Dresselhaus, M.S., Kong, J.: Layer area, few-layer graphene films on arbitrary substrates by chemical vapor deposition. Nano Lett. **9**, 3087–3088 (2009)
36. Reina, A., Thiele, S., Jia, X., Bhaviripudi, S., Dresselhaus, M.S., Juergen, A.: Growth of large-area single- and bi-layer graphene by controlled carbon precipitation on polycrystalline Ni surfaces. Nano Res. **2**, 509–516 (2009)
37. Nielson, J.R.: Mechanisms of defect formation. J. Catal. **48**, 155–165 (1977)

38. Lahiri, J., Miller, T.S., Ross, A.J., Adamska, L., Oleynik, I.I., Batzill, M.: Graphene growth and stability at nickel surfaces. New J. Phys. **13**, 025001 (2011)
39. Zou, Z., Fu, L., Song, X., Zhang, Y., Liu, Z.: Carbide-forming groups IVB-VIB metals: a new territory in the periodic table for CVD growth of graphene. Nano Lett. **14**, 3832–3839 (2014)
40. Vangari, M., Pryor, T., Jiang, L.: Supercapacitors: review of materials and fabrication methods. J. Energy Eng. **139**, 72–79 (2013)
41. Zhang, Q., Li, Y., Feng, Y., Feng, W.: Electropolymerization of graphene oxide/polyaniline composite for high-performance supercapacitor. Electrochim. Acta **90**, 95–100 (2013)
42. Chang, H.H., Chang, C.K., Tsai, Y.C., Liao, C.S.: Electrochemically synthesized graphene/polypyrrole composites and their use in supercapacitor. Carbon **50**, 2331–2336 (2012)
43. Yu, M., Ma, Y., Liu, J., Li, S.: Polyaniline nanocone arrays synthesized on three-dimensional graphene network by electrodeposition for supercapacitor electrodes. Carbon **87**, 98–105 (2015)
44. Zhang, X., Li, X., Zhu, M., Li, X., Zhen, Z., He, Y., Wang, K., Wei, J., Kang, F., Zhu, H.: Graphene/polyaniline woven fabric composite films as flexible supercapacitor electrodes. Nanoscale **7**, 7318–7322 (2015)
45. Lehtimäki, S., Suominen, M., Damlin, P., Tuukkanen, S., Kvarnström, C., Lupo, D.: Preparation of supercapacitors on flexible substrates with electrodeposited PEDOT/graphene composites. ACS Appl. Mater. Interfaces **7**, 22137–22147 (2015)
46. Grote, F., Yu, Z.Y., Wang, J.L., Yu, S.H., Lei, Y.: Self-stacked reduced graphene oxide nanosheets coated with cobalt-nickel hydroxide by one-step electrochemical deposition toward flexible electrochromic supercapacitors. Small **11**, 4666–4672 (2015)
47. Ramani, M., Haran, B.S., White, R.E., Popov, B.N.: Synthesis and characterization of hydrous ruthenium oxide-carbon supercapacitors. J. Electrochem. Soc. **148**, A374 (2001)
48. Wu, Q., Xu, Y., Yao, Z., Liu, A., Shi, G.: Supercapacitors based on flexible graphene/polyaniline nanofiber composite films. ACS Nano **4**, 1963–1970 (2010)
49. Ahuja, P., Ujjain, S.K., Arora, I., Samim, M.: Hierarchically grown NiO-decorated polyaniline-reduced graphene oxide composite for ultrafast sunlight-driven photocatalysis. ACS Omega **3**, 7846–7855 (2018)
50. Ujjain, S.K., Ahuja, P., Sharma, R.K.: Facile preparation of graphene nanoribbon/cobalt coordination polymer nanohybrid for non-enzymatic H_2O_2 sensing by dual transducer: electrochemical as well as fluorescence. J. Mater. Chem. B **38**, 7614–7622 (2015)
51. Zhou, S., Zhang, H., Zhao, Q., Wang, X., Li, J., Wang, F.: Graphene-wrapped polyaniline nanofibers as electrode materials for organic supercapacitors. Carbon **52**, 440–450 (2013)
52. Li, Y., Peng, H., Li, G., Chen, K.: Synthesis and electrochemical performance of sandwich-like polyaniline/graphene composite nanosheets. Eur. Polym. J. **48**, 1406–1412 (2012)
53. Wang, L., Lu, X., Lei, S., Song, Y.: Graphene-based polyaniline nanocomposites: preparation, properties and applications. J. Mater. Chem. A **2**, 4491–4509 (2014)
54. Li, J., Yang, Q.M., Zhitomirsky, I.: Composite electrodes for electrochemical supercapacitors. Nanoscale Res. Lett. **5**, 512–517 (2010)
55. Ahuja, P., Ujjain, S.K., Kanojia, R.: MnOx/C nanocomposite: an insight on high-performance supercapacitor and non-enzymatic hydrogen peroxide detection. Appl. Surf. Sci. **404**, 197–205 (2017)
56. Lota, K., Sierczynska, A., Lota, G.: Supercapacitors based on nickel oxide/carbon materials composites. Int. J. Electrochem. **2010**, 1–6 (2011)
57. Deori, K., Ujjain, S.K., Sharma, R.K., Deka, S.: morphology controlled synthesis of nanoporous Co_3O_4 nanostructures and their charge storage characteristics in supercapacitors. ACS Appl. Mater. Interfaces **5**, 10665–10672 (2013)
58. Ujjain, S.K., Singh, G., Sharma, R.K.: Co_3O_4@Reduced graphene oxide nanoribbon for high performance asymmetric supercapacitor. Electrochim. Acta **169**, 276–282 (2015)

59. Park, S., Kim, S.: Effect of carbon blacks filler addition on electrochemical behaviors of Co_3O_4/graphene nanosheets as a supercapacitor electrodes. Electrochim. Acta **89**, 516–522 (2013)
60. Ujjain, S.K., Bhatia, R., Ahuja, P., Attri, P.: Highly conductive aromatic functionalized multi-walled carbon nanotube for inkjet printable high performance supercapacitor electrodes. PLoS One **10**(1–12), e0131475 (2015)
61. Stoller, M.D., Ruoff, R.S.: Best practice methods for determining an electrode material's performance for ultracapacitors. Energy Environ. Sci. **3**, 1294–1301 (2010)
62. Hassan, M., Raghava, K., Haque, E., Nayeem, S., Ghasemi, S., Minett, A.I., Gomes, V.G.: Hierarchical assembly of graphene/polyaniline nanostructures to synthesize free-standing supercapacitor electrode. Compos. Sci. Technol. **98**, 1–8 (2014)
63. Vidhyadharan, B., Khayyriah, N., Zain, M., Misnon, I.I., Aziz, R.A., Ismail, J., Yusoff, M. M., Jose, R.: High performance supercapacitor electrodes from electrospun nickel oxide nanowires. J. Alloys Compd. **610**, 143–150 (2014)
64. Xie, Y., Liu, Y., Zhao, Y., Tsang, Y.H., Lau, S.P., Huang, H., Chai, Y.: Stretchable all-solid state supercapacitor with wavy shaped polyaniline/graphene electrode. J. Mater. Chem. A **2**, 9142–9149 (2014)
65. Zhang, D., Yan, H., Lu, Y., Qiu, K., Wang, C., Tang, C.: Hierarchical mesoporous nickel cobaltite nanoneedle/carbon cloth arrays as superior flexible electrodes for supercapacitors. Nanoscale Res. Lett. **9**, 139 (2014)
66. He, Y., Chen, W., Li, X., Zhang, Z., Fu, J., Zhao, C., Xie, E.: Freestanding three-dimensional graphene/MnO_2 composite networks as ultralight and flexible supercapacitor electrodes. ACS Nano **7**, 174–182 (2013)
67. Zilong, W., Zhu, Z., Qiu, J., Yang, S.: High performance flexible solid-state asymmetric supercapacitors from MnO_2/ZnO core-shell nanorods/specially reduced graphene oxide. J. Mater. Chem. C **2**, 1331–1336 (2014)
68. Sawangphruk, M., Srimuk, P., Chiochan, P., Krittayavathananon, A., Luanwuthi, S., Limtrakul, J.: High-performance supercapacitor of manganese oxide/reduced graphene oxide nanocomposite coated on flexible carbon fiber paper. Carbon **60**, 109–116 (2013)
69. Sahoo, S., Dhibar, S., Hatui, G., Bhattacharya, P., Das, C.K.: Graphene/polypyrrole nanofiber nanocomposite as electrode material for electrochemical supercapacitor. Polymer **54**, 1033–1042 (2013)
70. Jung, N., Kwon, S., Lee, D., Yoon, D., Park, Y.M., Benayad, A., Choi, J., Park, J.S.: Synthesis of chemically bonded graphene/carbon nanotube composites and their application in large volumetric capacitance supercapacitors. Adv. Mater. **25**, 6854–6858 (2013)
71. Ji, J., Li, Y., Peng, W., Zhang, G., Zhang, F., Fan, X.: Advanced graphene-based binder-free electrodes for high-performance energy storage. Adv. Mater. **27**, 5264–5279 (2015)
72. Mahmood, N., Zhang, C., Yin, H., Hou, Y.: Graphene-based nanocomposites for energy storage and conversion in lithium batteries, supercapacitors and fuel cells. J. Mater. Chem. A **2**, 15–32 (2014)
73. Cui, X., Lv, R., Ur, R., Sagar, R., Liu, C., Zhang, Z.: Reduced graphene oxide/carbon nanotube hybrid film as high performance negative electrode for supercapacitor. Electrochim. Acta **169**, 342–350 (2015)
74. Vaisman, L., Wagner, H.D., Marom, G.: The role of surfactants in dispersion of carbon nanotubes. Adv. Colloid Interface Sci. **128–130**, 37–46 (2006)
75. Ujjain, S.K., Bhatia, R., Ahuja, P.: Aziridine-functionalized graphene: Effect of aromaticity for aryl functional groups on enhanced power conversion efficiency of organic photovoltaic cells. J. Saudi Chem. Soc. (2018). https://doi.org/10.1016/j.jscs.2018.11.006
76. Bhatia, R., Ujjain, S.K.: Soluble single-walled carbon nanotubes for photovoltaics. Mater. Lett. **190**, 165–168 (2017)
77. Salunkhe, R.R., Lin, J., Malgras, V., Xue, S., Ho, J., Yamauchi, Y.: Large-scale synthesis of coaxial carbon nanotube/$Ni(OH)_2$ composites for asymmetric supercapacitor application. Nano Energy **11**, 211–218 (2015)

78. Deng, L., Hao, Z., Wang, J., Zhu, G., Kang, L., Liu, Z.H., Yang, Z., Wang, Z.: Preparation and capacitance of graphene/multiwall carbon nanotubes/MnO_2 hybrid material for high-performance asymmetrical electrochemical capacitor. Electrochim. Acta **89**, 191–198 (2013)
79. Wang, Y., Chen, J., Cao, J., Liu, Y., Zhou, Y., Ouyang, J.H., Jia, D.: Graphene/carbon black hybrid film for flexible and high rate performance supercapacitor. J. Power Sources **271**, 269–277 (2014)
80. Ma, W., Chen, S., Yang, S., Chen, W., Cheng, Y., Guo, Y., Peng, S., Ramakrishna, S., Zhu, M.: Hierarchical MnO_2 nanowire/graphene hybrid fibers with excellent electrochemical performance for flexible solid-state supercapacitors. J. Power Sources **306**, 481–488 (2016)
81. Stankovich, S., Dikin, D.A., Piner, R.D., Kohlhaas, K.A., Kleinhammes, A., Jia, Y., Wu, Y.: Synthesis of graphene-based nanosheets via chemical reduction of exfoliated graphite oxide. Carbon **45**, 1558–1565 (2007)
82. Song, B., Wu, F., Zhu, Y., Hou, Z., sik Moon, K., Wong, C.P.: Effect of polymer binders on graphene-based free-standing electrodes for supercapacitors. Electrochim. Acta **267**, 213–221 (2018)
83. Fang, B., Binder, L.: A novel carbon electrode material for highly improved EDLC performance. J. Phys. Chem. B **110**, 7877–7882 (2006)
84. Dyatkin, B., Presser, V., Heon, M., Lukatskaya, M.R., Beidaghi, M.: Development of a green supercapacitor composed entirely of environmentally friendly materials. Chemsuschem **6**, 2269–2280 (2013)
85. Ruiz, V., Blanco, C., Granda, M., Menéndez, R., Santamaría, R.: Influence of electrode preparation on the electrochemical behaviour of carbon-based supercapacitors. J. Appl. Electrochem. **37**, 717–721 (2007)
86. Higgins, T.M., Park, S.H., King, P.J., Zhang, C., McEvoy, N., Berner, N.C., Daly, D., Shmeliov, A., Khan, U., Duesberg, G., Nicolosi, V., Coleman, J.N.: A commercial conducting polymer as both binder and conductive additive for silicon nanoparticle-based lithium-ion battery negative electrodes. ACS Nano **10**, 3702–3713 (2016)
87. Park, B.H., Choi, J.H.: Improvement in the capacitance of a carbon electrode prepared using water-soluble polymer binder for a capacitive deionization application. Electrochim. Acta **55**, 2888–2893 (2010)
88. Chabi, S., Peng, C., Hu, D., Zhu, Y.: Ideal three-dimensional electrode structures for electrochemical energy storage. Adv. Mater. **26**, 2440–2445 (2014)
89. Wang, C., Du, H., Tsai, Y., Chen, C., Huang, C., Chen, L.C., Chen, K.H., Shih, H.: High performance of low electrocatalysts loading on CNT directly grown on carbon cloth for DMFC. J. Power Sources **171**, 55–62 (2007)
90. Ervin, M.H., Le, L.T., Lee, W.Y.: Inkjet-printed flexible graphene-based supercapacitor. Electrochim. Acta **147**, 610–616 (2014)
91. El-Kady, M.F., Veronica, S., Sergey, D., Richard, B.K.: Laser scribing of high-performance and flexible graphene-based electrochemical capacitors. Science **335**, 1326–1330 (2012)
92. Tai, Z., Yan, X., Xue, Q.: Shape-alterable and -recoverable graphene/polyurethane bi-layered composite film for supercapacitor electrode. J. Power Sources **213**, 350–357 (2012)
93. Chen, C.M., Zhang, Q., Huang, C.H., Zhao, X.C., Sen Zhang, B., Kong, Q.Q., Wang, M.Z., Yang, Y.G., Cai, R., Sheng Su, D.: Macroporous "bubble" graphene film via template-directed ordered-assembly for high rate supercapacitors. Chem. Commun. **48**, 7149–7151 (2012)
94. Xu, Y., Lin, Z., Huang, X., Liu, Y., Huang, Y., Duan, X.: Flexible solid-state supercapacitors based on three-dimensional graphene hydrogel films. ACS Nano **7**, 4042–4049 (2013)
95. Jeong, H.M., Lee, J.W., Shin, W.H., Choi, Y.J., Shin, H.J., Kang, J.K., Choi, J.W.: Nitrogen-doped graphene for high-performance ultracapacitors and the importance of nitrogen-doped sites at basal planes. Nano Lett. **11**, 2472–2477 (2011)

96. Wu, Z., Winter, A., Chen, L., Sun, Y., Turchanin, A., Feng, X., Müllen, K.: Three-dimensional nitrogen and boron co-doped graphene for high-performance all-solid-state supercapacitors. Adv. Mater. **24**, 5130–5135 (2012)
97. Xu, Y., Lin, Z., Huang, X., Wang, Y., Huang, Y., Duan, X.: Functionalized graphene hydrogel-based high-performance supercapacitors. Adv. Mater. **25**, 5779–5784 (2013)
98. Choi, B.G., Hong, J., Hong, W.H., Hammond, P.T., Park, H.: Facilitated ion transport in all-solid-state flexible supercapacitors. ACS Nano **5**, 7205–7213 (2011)
99. Zhang, L.L., Zhao, S., Tian, X.N., Zhao, X.S.: Layered graphene oxide nanostructures with sandwiched conducting polymers as supercapacitor electrodes. Langmuir **26**, 17624–17628 (2010)
100. Xu, J., Wang, K., Zu, S.Z., Han, B.H., Wei, Z.: Hierarchical nanocomposites of polyaniline nanowire arrays on graphene oxide sheets with synergistic effect for energy storage. ACS Nano **4**, 5019–5026 (2010)
101. Wang, D., Li, F., Zhao, J., Ren, W., Chen, Z., Tan, J., Wu, Z., Gentle, I., Lu, G.Q., Cheng, H.: Fabrication of graphene/polyaniline composite paper via in situ anodic electropolymerization for high performance flexible electrode. ACS Nano **3**, 1745–1752 (2009)
102. Zhang, K., Zhang, L.L., Zhao, X.S., Wu, J.: Graphene/polyaniline nanofiber composites as supercapacitor electrodes. Chem. Mater. **22**, 1392–1401 (2010)
103. Kumar, N.A., Choi, H.J., Shin, Y.R., Chang, D.W., Dai, L., Baek, J.B.: Polyaniline-grafted reduced graphene oxide for efficient electrochemical supercapacitors. ACS Nano **6**, 1715–1723 (2012)
104. Lee, T., Yun, T., Park, B., Sharma, B., Song, H.K., Kim, B.S.: Hybrid multilayer thin film supercapacitor of graphene nanosheets with polyaniline: Importance of establishing intimate electronic contact through nanoscale blending. J. Mater. Chem. **22**, 21092–21099 (2012)
105. Liu, S., Liu, X., Li, Z., Yang, S., Wang, J.: Fabrication of free-standing graphene/polyaniline nanofibers composite paper via electrostatic adsorption for electrochemical supercapacitors. New J. Chem. **35**, 369–374 (2011)
106. Davies, A., Audette, P., Farrow, B., Hassan, F., Chen, Z., Choi, J., Yu, A.: Graphene-based flexible supercapacitors: pulse-electropolymerization of polypyrrole on free-standing graphene films. J. Phys. Chem. C **115**, 17612–17620 (2011)
107. Zhang, J., Chen, P., Oh, B.H.L., Chan-Park, M.B.: High capacitive performance of flexible and binder-free graphene-polypyrrole composite membrane based on in situ reduction of graphene oxide and self-assembly. Nanoscale **5**, 9860–9866 (2013)
108. Li, S.M., Wang, Y.S., Yang, S.Y., Liu, C.H., Chang, K.H., Tien, H.W., Wen, N.T., Ma, C.C. M., Hu, C.C.: Electrochemical deposition of nanostructured manganese oxide on hierarchically porous graphene–carbon nanotube structure for ultrahigh-performance electrochemical capacitors. J. Power Sources **225**, 347–355 (2013)
109. Hsu, C.T., Hu, C.C.: Synthesis and characterization of mesoporous spinel $NiCo_2O_4$ using surfactant-assembled dispersion for asymmetric supercapacitors. J. Power Sources **242**, 662–671 (2013)
110. Chen, Y.C., Hsu, Y.K., Lin, Y.G., Lin, Y.K., Horng, Y.Y., Chen, L.C., Chen, K.H.: Highly flexible supercapacitors with manganese oxide nanosheet/carbon cloth electrode. Electrochim. Acta **56**, 7124–7130 (2011)
111. Chen, Q., Meng, Y., Hu, C., Zhao, Y., Shao, H., Chen, N., Qu, L.: MnO_2-modified hierarchical graphene fiber electrochemical supercapacitor. J. Power Sources **247**, 32–39 (2014)
112. Drieschner, S., von Seckendorff, M., del Corro, E., Wohlketzetter, J., Blaschke, B.M., Stutzmann, M., Garrdo, J.A.: Uniformly coated highly porous graphene/MnO_2 foams for flexible asymmetric supercapacitor. Nanotechnology 1–18 (2018)
113. Ma, L., Liu, R., Liu, L., Wang, F., Niu, H., Huang, Y.: Facile synthesis of $Ni(OH)_2$/graphene/bacterial cellulose paper for large areal mass, mechanically tough and flexible supercapacitor electrodes. J. Power Sources **335**, 76–83 (2016)
114. Cao, X., Zheng, B., Shi, W., Yang, J., Fan, Z., Luo, Z., Rui, X., Chen, B., Yan, Q., Zhang, H.: Reduced graphene oxide-wrapped MoO_3 composites prepared by using metal-organic

frameworks as precursor for all-solid-state flexible supercapacitors. Adv. Mater. **27**, 4695–4701 (2015)
115. Xie, J., Sun, X., Zhang, N., Xu, K., Zhou, M., Xie, Y.: Layer-by-layer β-$Ni(OH)_2$/graphene nanohybrids for ultraflexible all-solid-state thin-film supercapacitors with high electrochemical performance. Nano Energy **2**, 65–74 (2013)
116. Cho, S., Kim, J., Jo, Y., Ahmed, A.T.A., Chavan, H.S., Woo, H., Inamdar, A.I., Gunjakar, J. L., Pawar, S.M., Park, Y., Kim, H., Im, H.: Bendable RuO_2/graphene thin film for fully flexible supercapacitor electrodes with superior stability. J. Alloys Compd. **725**, 108–114 (2017)
117. Ahuja, P., Ujjain, S.K., Sharma, R.K., Singh, G.: Enhanced supercapacitor performance by incorporating nickel in manganese oxide. RSC Adv. **4**, 57192–57199 (2014)
118. Ahuja, P., Sharma, R.K., Singh, G.: All solid state, high performance supercapacitor using zinc manganite embedded graphene nanoribbons. J. Mater. Chem. A **3**, 4931–4937 (2015)
119. Ujjain, S.K., Ahuja, P., Sharma, R.K.: Graphene nanoribbon wrapped cobalt manganite nanocubes for high performance all-solid-state flexible supercapacitors. J. Mater. Chem. A **3**, 9925–9931 (2015)
120. Mitchell, E., Jimenez, A., Gupta, R.K., Gupta, B.K., Ramasamy, K., Shahabuddin, M., Mishra, S.R.: Ultrathin porous hierarchically textured $NiCo_2O_4$/graphene oxide flexible nanosheets for high performance supercapacitor. New J. Chem. **39**, 2181–2187 (2015)
121. Cai, W., Lai, T., Dai, W., Ye, J.: A facile approach to fabricate flexible all-solid-state supercapacitors based on $MnFe_2O_4$/graphene hybrids. J. Power Sources **255**, 170–178 (2014)
122. Vijaya Sankar, K., Kalai Selvan, R.: Fabrication of flexible fiber supercapacitor using covalently grafted $CoFe_2O_4$/reduced graphene oxide/polyaniline and its electrochemical performances. Electrochim. Acta **213**, 469–481 (2016)
123. Garcia-Torres, J., Crean, C.: Ternary composite solid-state flexible supercapacitor based on nanocarbons/manganese dioxide/PEDOT:PSS fibres. Mater. Des. **155**, 194–202 (2018)
124. Azman, N.H.N., Lim, H.N., Mamat, M.S., Sulaiman, Y.: Synergistic enhancement of ternary poly(3,4-ethylenedioxythiophene)/graphene oxide/manganese oxide composite as a symmetrical electrode for supercapacitors. Energies **11**, 1510 (2018)
125. Tang, P., Han, L., Zhang, L.: Facile synthesis of graphite/PEDOT/MnO_2 Composites on commercial supercapacitor separator membranes as flexible and high-performance supercapacitor electrodes. ACS Appl. Mater. Interfaces **6**, 10506–10515 (2014)
126. Chen, H., Zhou, S., Wu, L.: Porous nickel hydroxide-manganese dioxide-reduced graphene oxide ternary hybrid spheres as excellent supercapacitor electrode materials. ACS Appl. Mater. Interfaces **6**, 8621–8630 (2014)
127. Li, P., Yang, Y., Shi, E., Shen, Q., Shang, Y., Wu, S., Wei, J., Wang, K., Zhu, H., Yuan, Q., Cao, A., Wu, D.: Core-double-shell, carbon nanotube@polypyrrole@MnO_2 sponge as freestanding, compressible supercapacitor electrode. ACS Appl. Mater. Interfaces **6**, 5228–5234 (2014)
128. Dhibar, S., Bhattacharya, P., Ghosh, D., Hatui, G., Das, C.K.: Graphene-single-walled carbon nanotubes-poly(3-methylthiophene) ternary nanocomposite for supercapacitor electrode materials. Ind. Eng. Chem. Res. **53**, 13030–13045 (2014)
129. Sankar, K.V., Selvan, R.K.: The preparation of $MnFe_2O_4$ decorated flexible graphene wrapped with PANI and its electrochemical performances for hybrid supercapacitors. RSC Adv. **4**, 17555–17566 (2014)
130. Sankar, K.V., Selvan, R.K.: The ternary $MnFe_2O_4$/graphene/polyaniline hybrid composite as negative electrode for supercapacitors. J. Power Sources **275**, 399–407 (2015)
131. Chee, W.K., Lim, H.N., Harrison, I., Chong, K.F., Zainal, Z., Ng, C.H., Huang, N.M.: Performance of flexible and binderless polypyrrole/graphene oxide/zinc oxide supercapacitor electrode in a symmetrical two-electrode configuration **157**, 88–94 (2015)
132. Li, H., He, Y., Pavlinek, V., Cheng, Q., Saha, P., Li, C.: MnO_2 nanoflake/polyaniline nanorod hybrid nanostructures on graphene paper for high-performance flexible supercapacitor electrodes. J. Mater. Chem. A **3**, 17165–17171 (2015)

133. Wu, X., Wang, Q., Zhang, W., Wang, Y., Chen, W.: Nano nickel oxide coated graphene/polyaniline composite film with high electrochemical performance for flexible supercapacitor. Electrochim. Acta **211**, 1066–1075 (2016)
134. Asen, P., Shahrokhian, S.: A high performance supercapacitor based on graphene/polypyrrole/Cu_2O-$Cu(OH)_2$ ternary nanocomposite coated on nickel foam. J. Phys. Chem. C **121**, 6508–6519 (2017)
135. Ahuja, P., Kumar, S., Kanojia, R.: Electrochemical behaviour of manganese and ruthenium mixed oxide@reduced graphene oxide nanoribbon composite in symmetric and asymmetric supercapacitor. Appl. Surf. Sci. **427**, 102–111 (2018)
136. Ahuja, P., Sahu, V., Ujjain, S.K., Sharma, R.K., Singh, G.: Performance evaluation of asymmetric supercapacitor based on cobalt manganite modified graphene nanoribbons. Electrochim. Acta **146**, 429–436 (2014)
137. Yang, C., Shen, J., Wang, C., Fei, H., Bao, H., Wang, G.: All-solid-state asymmetric supercapacitor based on reduced graphene oxide/carbon nanotube and carbon fiber paper/polypyrrole electrodes. J. Mater. Chem. A **2**, 1458–1464 (2014)
138. Choi, B.G., Chang, S.J., Kang, H.W., Park, C.P., Kim, H.J., Hong, W.H., Lee, S., Huh, Y.S.: High performance of a solid-state flexible asymmetric supercapacitor based on graphene films. Nanoscale **4**, 4983–4988 (2012)
139. Zhou, D., Zhou, R., Chen, C., Yee, W.A., Kong, J., Ding, G., Lu, X.: Non-volatile polymer electrolyte based on poly(propylene carbonate), ionic liquid, and lithium perchlorate for electrochromic devices. J. Phys. Chem. B **117**, 7783–7789 (2013)
140. Yu, H., Wu, J., Fan, L., Lin, Y., Xu, K., Tang, Z., Cheng, C., Tang, S., Lin, J., Huang, M., Lan, Z.: A novel redox-mediated gel polymer electrolyte for high-performance supercapacitor. J. Power Sources **198**, 402–407 (2012)
141. Liu, X., Osaka, T.: All-solid-state electric double-layer capacitor with isotropic high-density graphite electrode and polyethylene oxide/$LiClO_4$ polymer electrolyte. J. Electrochem. Soc. **143**, 3982–3986 (1996)
142. Chen, Q., Li, X., Zang, X., Cao, Y., He, Y., Li, P., Wang, K., Wei, J., Wu, D., Zhu, H.: Effect of different gel electrolytes on graphene based solid-state supercapacitors. RSC Adv. **4**, 36253–36256 (2014)
143. Hashmi, S.A., Kumar, A., Tripathi, S.K.: Investigations on electrochemical supercapacitors using polypyrrole redox electrodes and PMMA based gel electrolytes. Eur. Polym. J. **41**, 1373–1379 (2005)
144. Tripathi, S.K., Kumar, A., Hashmi, S.A.: Electrochemical redox supercapacitors using PVdF-HFP based gel electrolytes and polypyrrole as conducting polymer electrode. Solid State Ionics **177**, 2979–2985 (2006)
145. Yang, P., Mai, W.: Flexible solid-state electrochemical supercapacitors. Nano Energy **8**, 274–290 (2014)
146. Environ, E., Fic, K., Lota, G., Meller, M., Frackowiak, E.: Novel insight into neutral medium as electrolyte for high-voltage supercapacitors. Energy Environ. Sci. **5**, 5842–5850 (2012)
147. Béguin, F., Presser, V., Balducci, A., Frackowiak, E.: Carbons and electrolytes for advanced supercapacitors. Adv. Mater. **26**, 2219–2251 (2014)
148. Jin, M., Zhang, Y., Yan, C., Fu, Y., Guo, Y., Ma, X.: A high-performance ionic liquid based gel polymer electrolyte incorporating anion-trapping boron sites for all-solid-state supercapacitors application. ACS Appl. Mater. Interfaces. **10**, 39570–39580 (2018)
149. Tu, Q.M., Fan, L.Q., Pan, F., Huang, J.L., Gu, Y., Lin, J.M., Huang, M.L., Huang, Y.F., Wu, J.H.: Design of a novel redox-active gel polymer electrolyte with a dual-role ionic liquid for flexible supercapacitors. Electrochim. Acta **268**, 562–568 (2018)
150. Feng, L., Wang, K., Zhang, X., Sun, X., Li, C., Ge, X.: Flexible solid-state supercapacitors with enhanced performance from hierarchically graphene nanocomposite electrodes and ionic liquid incorporated gel polymer electrolyte. Adv. Funct. Mater. **28**, 1704463 (1–9) (2018)

151. González, A., Goikolea, E., Andoni, J., Mysyk, R.: Review on supercapacitors: technologies and materials. Renew. Sustain. Energ. Rev. **58**, 1189–1206 (2016)
152. Galiński, M., Lewandowski, A., Stepniak, I.: Ionic liquids as electrolytes. Electrochim. Acta **51**, 5567–5580 (2006)
153. Na, R., Lu, N., Zhang, S., Huo, G., Yang, Y., Zhang, C., Mu, Y., Luo, Y., Wang, G.: Facile synthesis of a high-performance, fire-retardant organic gel polymer electrolyte for flexible solid-state supercapacitors. Electrochim. Acta **290**, 262–272 (2018)
154. Zhong, X., Tang, J., Cao, L., Kong, W., Sun, Z., Cheng, H., Lu, Z., Pan, H., Xu, B.: Cross-linking of polymer and ionic liquid as high-performance gel electrolyte for flexible solid-state supercapacitors. Electrochim. Acta **244**, 112–118 (2017)
155. Muchakayala, R., Song, S., Wang, J., Fan, Y., Bengeppagari, M., Chen, J., Tan, M.: Development and supercapacitor application of ionic liquid-incorporated gel polymer electrolyte films. J. Ind. Eng. Chem. **59**, 79–89 (2018)
156. Wee, G., Larsson, O., Srinivasan, M., Berggren, M., Crispin, X., Mhaisalkar, S.: Effect of the ionic conductivity on the performance of polyelectrolyte-based supercapacitors. Adv. Funct. Mater. **20**, 4344–4350 (2010)
157. Yang, X., Zhang, F., Zhang, L., Zhang, T., Huang, Y., Chen, Y.: A high-performance graphene oxide-doped ion gel as gel polymer electrolyte for all-solid-state supercapacitor applications. Adv. Funct. Mater. **23**, 3353–3360 (2013)
158. Shim, J., Lee, J.S., Lee, J.H., Kim, H.J., Lee, J.: Gel polymer electrolytes containing anion-trapping boron moieties for lithium-ion battery applications. ACS Appl. Mater. Interfaces. **8**, 27740–27752 (2016)
159. Yuan, P., Cai, C., Tang, J., Qin, Y., Jin, M., Fu, Y.: Anion acceptors dioxaborinane contained in solid state polymer electrolyte: preparation, characterization, and DFT calculations. Adv. Funct. Mater. **26**, 5930–5939 (2016)
160. Zhang, Y., Cai, W., Rohan, R., Pan, M., Liu, Y., Liu, X.: Toward ambient temperature operation with all-solid-state lithium metal batteries with a sp^3 boron-based solid single ion conducting polymer electrolyte. J. Power Sources **306**, 152–161 (2016)
161. Zhu, Y., Xiao, S., Shi, Y., Yang, Y., Hou, Y., Wu, Y.: A composite gel polymer electrolyte with high performance based on poly(vinylidene fluoride) and polyborate for lithium ion batteries. Adv. Energy Mater. **4**, 1300647 (2014)
162. Wu, Z.S., Parvez, K., Feng, X., Müllen, K.: Graphene-based in-plane micro-supercapacitors with high power and energy densities. Nat. Commun. **4**, 1–8 (2013)
163. Niu, Z., Zhang, L., Liu, L., Zhu, B., Dong, H., Chen, X.: All-solid-state flexible ultrathin micro-supercapacitors based on graphene. Adv. Mater. **25**, 4035–4042 (2013)
164. Xue, M., Li, F., Zhu, J., Song, H., Zhang, M., Cao, T.: Structure-based enhanced capacitance: in situ growth of highly ordered polyaniline nanorods on reduced graphene oxide patterns. Adv. Funct. Mater. **22**, 1284–1290 (2012)
165. Lin, J., Zhang, C., Yan, Z., Zhu, Y., Peng, Z., Hauge, R.H., Natelson, D., Tour, J.M.: 3-Dimensional graphene carbon nanotube carpet-based microsupercapacitors with high electrochemical performance. Nano Lett. **13**, 72–78 (2013)
166. Online, V.A., Liu, W., Yan, X., Chen, J., Feng, Y., Xue, Q.: Novel and high-performance asymmetric micro-supercapacitors based on graphene quantum dots and polyaniline nano fibers. Nanoscale **5**, 6053–6062 (2013)
167. Beidaghi, M., Wang, C.: Micro-supercapacitors based on interdigital electrodes of reduced graphene oxide and carbon nanotube composites with ultrahigh power handling performance. Adv. Funct. Mater. **22**, 4501–4510 (2012)
168. Peng, L., Peng, X., Liu, B., Wu, C., Xie, Y., Yu, G.: Ultrathin two-dimensional MnO_2/graphene hybrid nanostructures for high-performance, flexible planar supercapacitors. Nano Lett. **13**, 2151–2157 (2013)
169. Cai, X., Peng, M., Yu, X., Fu, Y., Zou, D.: Flexible planar/fiber-architectured supercapacitors for wearable energy storage. J. Mater. Chem. C **2**, 1167–1348 (2014)

170. Lu, X., Zhai, T., Zhang, X., Shen, Y., Yuan, L., Hu, B., Gong, L., Chen, J., Gao, Y., Zhou, J., Tong, Y., Wang, Z.L.: WO_{3-x}@Au@MnO_2core-shell nanowires on carbon fabric for high-performance flexible supercapacitors. Adv. Mater. **24**, 938–944 (2012)
171. Ren, J., Li, L., Chen, C., Chen, X., Cai, Z., Qiu, L., Wang, Y., Zhu, X., Peng, H.: Twisting carbon nanotube fibers for both wire-shaped micro-supercapacitor and micro-battery. Adv. Mater. **25**, 1155–1159 (2013)
172. Fu, Y., Cai, X., Wu, H., Lv, Z., Hou, S., Peng, M., Yu, X.: Fiber supercapacitors utilizing pen ink for flexible/wearable energy storage. Adv. Mater. **24**, 5713–5718 (2012)
173. Huang, G., Hou, C., Shao, Y., Zhu, B., Jia, B., Wang, H., Zhang, Q., Li, Y.: High-performance all-solid-state yarn supercapacitors based on porous graphene ribbons. Nano Energy **12**, 26–32 (2015)
174. Yu, D., Qian, Q., Wei, L., Jiang, W., Goh, K., Wei, J., Zhang, J., Chen, Y.: Emergence of fiber supercapacitors. Chem. Soc. Rev. **44**, 647–662 (2015)
175. Ding, X., Zhao, Y., Hu, C., Hu, Y., Dong, Z., Chen, N., Zhang, Z., Qu, L.: Spinning fabrication of graphene/polypyrrole composite fibers for all-solid-state, flexible fibriform supercapacitors. J. Mater. Chem. A **2**, 12355–12360 (2014)
176. Lu, X., Yu, M., Wang, G., Tong, Y., Li, Y.: Flexible solid-state supercapacitors: design, fabrication and applications. Energy Environ. Sci. **7**, 2160–2181 (2014)
177. Online, V.A., Li, Y., Sheng, K., Yuan, W., Shi, G.: A high-performance flexible fibre-shaped electrochemical capacitor based on electrochemically reduced graphene oxide. Chem. Commun. **49**, 291–293 (2013)
178. Xu, Z., Gao, C.: Graphene chiral liquid crystals and macroscopic assembled fibres. Nat. Commun. **2**, 571–579 (2011)
179. Cong, H.P., Ren, X.C., Wang, P., Yu, S.H.: Wet-spinning assembly of continuous, neat, and macroscopic graphene fibers. Sci. Rep. **2**, 1–6 (2012)
180. Tran, C., Kalra, V.: Fabrication of porous carbon nanofibers with adjustable pore sizes as electrodes for supercapacitors. J. Power Sources **235**, 289–296 (2013)
181. Hu, Y., Cheng, H., Zhao, F., Chen, N., Jiang, L., Feng, Z., Qu, L.: All-in-one graphene fiber supercapacitor. Nanoscale **6**, 6448–6451 (2014)

Free-Standing Graphene Materials for Supercapacitors

Paramita Karfa, Kartick Chandra Majhi and Rashmi Madhuri

Abstract Free standing graphene belongs to the pioneering group of carbonaceous nanomaterial and has gained large appreciation in the field storage and conversion of energy. Free standing or self-supported graphene materials are made up of dense graphene sheets arranged in the form of three dimensional structures like foam, films, monoliths, papers, and aerogels with hierarchical porous structure. In last few decades, speedy growth in the binder free graphene based super-capacitors can be credited to their influential properties like flexibility, large surface to volume ratio, high mechanical durability, electrical and thermal conductivity, and light weight. The free standing graphene also delivers us with short and easy diffusion pathway for ions (generated from electrolytes), channels for electron transport and composites with active materials which provide a synergistic effect. This chapter will provide the information about synthesis of free standing graphenes and their recent advancements as the efficient electrode materials for supercapacitors.

Keywords Supercapacitor · Graphene foam · Graphene film · Aerogel · Free-standing graphene

List of Abbreviations

2D	Two dimensional
3D	Three dimensional
AES	Augar electron spectroscopy
AFM	Atomic force microscopy
CV	Cyclic voltammetry
CVE	Centrifugal vacuum evaporation
CVD	Chemical vapour deposition
DMF	Dimethyl formamide
EDLCs	Electric double layer capacitors
FT-IR	Fourier transform infrared spectroscopy

P. Karfa · K. C. Majhi · R. Madhuri (✉)
Department of Applied Chemistry, Indian Institute of Technology (Indian School of Mines), Dhanbad 826004, Jharkhand, India
e-mail: rashmimadhuriism@gmail.com

Inamuddin et al. (eds.), *Self-standing Substrates*,
Engineering Materials, https://doi.org/10.1007/978-3-030-29522-6_11

FESEM	Field emission scanning electron microscope
HRTEM	High resolution transmission electron microscope
MCVD	Microwave assisted chemical vapour deposition
MS	Mass spectroscopy
NEC	Nippon electric company
NMR	Nuclear magnetic resonance spectroscopy
r-GO	Reduced Graphene oxide
PANI	Polyaniline
PL	Photoluminescence spectroscopy
SC	Supercapacitors
SIMS	Secondary ion mass spectrometry
UCs	Ultracapacitors
XPS	X-Ray photo electron spectroscopy
XRD	X-ray diffraction spectroscopy

1 Introduction

Graphene is the "wonder material" of 21st decennium, which has stimulated a gold rush after its discovery by two eminent scientists Geim and Novoselov [1]. Graphene is transparent sheet of single atom thickness. The reason behind calling graphene as a "wonder material" is that it has interesting chemical and physical properties. It is 100 times harder than steel yet it is more flexible with outstanding mechanical strength. It has incredible electrical conductivity, high value of young's modulus ($\sim$1 TPa), good electron mobility (approx 200,000 $cm^2\ V^{-1}\ s^{-1}$), large double layer capacitance, high thermal conductivity and excellent optical transparency [2]. In addition, possessing high surface area, it has good adsorption capacity, biocompatibility, impervious to gases, and easy to modify or functionalized. Due to these exceptional properties, graphene is a promising material for various applications in different fields like supercapacitors, sensors, environmental pollution remediation, fuel cells, transistors, batteries, solar cells, memory devices, bio applications, food packaging, dyes degradation, electrocatalysts, optical modulators and energy storage devices etc. [3, 4]. In 2004, Geim and—coworkers for the first time synthesized single layered graphene by mechanical exfoliation method using graphite as the precursor. They won the noble prize for single layered graphene separation in the year 2010, which paved the way for the discovery of various graphene analogues. Every year, more than thousands of papers and hundreds of patents are appearing on the commercial interest or research application of graphene based materials/devices in various fields [5].

1.1 Classification of Graphene Based Nanomaterials

Graphene with all the sp^2 hybridization carbon atoms has inimitable two dimensional honey comb like arrangement, which is considered as the elementary unit of different dimensional carbonaceous materials [6]. For example, zero dimensional fullerene—formed from graphene can be rolled up into nanotubes of one dimensional materials. Similarly, staked layer by layer structures of graphene can form 3D graphene material. Graphene or graphene based materials can be classified in different ways, as proposed by Bianco et al. [7]. Graphene itself has single atom thickness with hexagonal arrangement with lateral dimension ranging from nanometers to micrometer. Graphene oxide (GO) is the most popular derivative of graphene family, formed by chemical modification of graphene through oxidation or chemical exfoliation of basal planes. Similarly, reduced graphene oxide (r-GO), is formed by various reductive methods (like thermal, chemical, mechanical, microbials, photo-chemicals) by reducing the number of oxygen contents present on the surface of graphene. Graphene nanosheets which are considered as the breakthrough in the field of nanotechnology are sheets of single atom thickness, where the carbon atoms are hexagonally arranged with dimension of <100 nm. Graphene nanoribbon is a strip of single atom thickness, which may be used as freely suspended or substrate bound. The lateral dimension of length and breadth of these nanoribbons differs largely with length having larger dimension. Graphite nanoplates have dimension or thickness <100 nm, also called as ultra-thin graphite. Multilayers of graphene can be stacked together as layer by layer and used as free-standing three dimensional graphene or can be substrate coated to make film on the solid surface. Free standing graphene can be made from graphene oxide or graphene consists of few atomic single layers without any support system, where all the intrinsic surfaces of the graphene can be utilized with 3 open edges [8]. Free standing graphene materials are hoped to—dominate the research area of supercapacitors and energy devices in next few years due to their large porous surface, useful pore volume, better electrical and thermal conductivities and high mechanical strength [9]. Free standing graphenes can be further classified by different terminologies shown in Fig. 1:

- Free standing graphene film,
- Free standing graphene foam,
- Free standing graphene monolith,
- Free standing graphene sponges,
- Free standing graphene framework, and
- Free standing graphene ribbons.

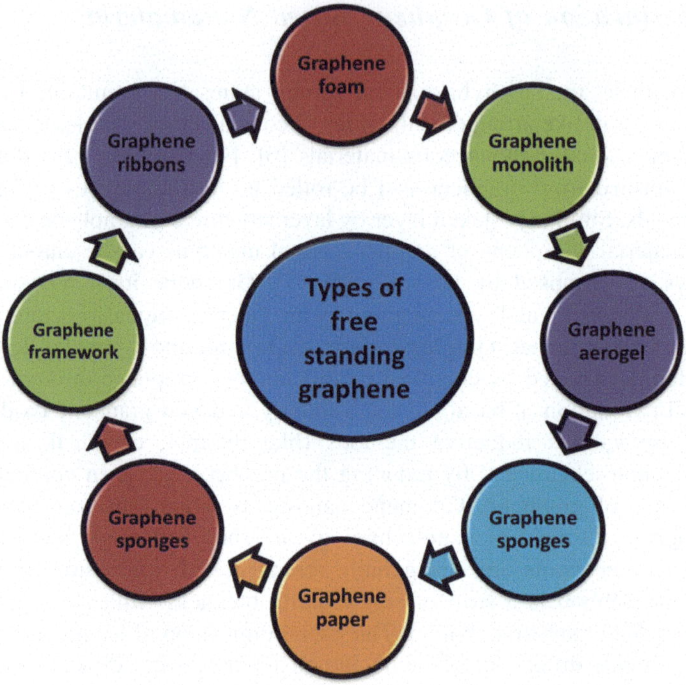

Fig. 1 Diagrammatic presentation showing different types of free standing graphene

1.2 Synthesis of Graphene Based Nanomaterials

Most commonly, fabrication of graphene can be done through two different approaches i.e. top down approach and bottom up approach. Top down approach can be best explained with very common day to day life examples of cutting down a plant or tree or slicing a cheese cube into many pieces or cleavage of any large segment into fragments. The famous top down approach for the synthesis of graphene is Hummers method, which is used to fabricate desirable nano-size, oxidized sheets/layers of graphene [10]. Briefly, in Hummers method the graphite powders or flakes are chemically oxidized to various oxygen containing functional groups like carboxyl, epoxy, alkoxy and finally exfoliated to form layered graphene with the help of ultrasonication, thermal or mechanical method. Other common top down methods for the synthesis of graphene are: (a) intercalation method; (b) solvent and organic ligand based exfoliation; (c) arc discharge method and (d) micromechanical exfoliation method. Chemical exfoliation process is the best method for the synthesis of graphene from graphite oxide, where the precursor is naturally abundant graphite [11]. Sometime, graphene is exfoliated with some solvent and surfactant using graphite as the precursor called as liquid exfoliation method. Surfactants are used to reduce the hydrophobicity of graphite and provide better dispersion.

Splitting of large number of stacked layers of graphite into monolayers by shear forces, ultra-sonication is another popular process of exfoliation. Less defects and high quality graphene with low oxidized surface can be prepared by performing exfoliation of graphite with the help of external electrical field named as electrochemical exfoliation method. In mechanical exfoliation method, Van der Waals forces of graphene sheets are teared apart to form single graphene layers through scotch tape method. Exfoliation method is very popular method for graphene synthesis, however, sometimes the final product resulted from this synthesis method gives agglomerated and damaged graphene sheets, which is the main disadvantage of this method [3, 12].

Other than these, the popular bottom up methods for the synthesis of graphene include carbonization, hydrothermal, solvothermal method and epitaxial growth on substrate methods. A different bottom down method for graphene synthesis is chemical vapour deposition (CVD) method, which synthesizes layered graphene sheets with different sizes grown on diverse substrate. In the CVD method, for the synthesis of graphene, precursor, carrier and substrate are required. Methane gas is among the most popular carbon sources, with hydrogen gas as the carrier gas, and metal substrates like copper or nickel. The mechanism behind the synthesis can be described as; first, the mixture of methane gas and the carrier gas passes through the evacuated heated chamber. The methane gas gets bounded with the metal substrate during the reaction leaving behind carbon atoms on the substrate with the formation of graphene. CVD produces graphene sheets of high quality and less number of defects containing graphene sheets. The major disadvantage associated with CVD is the requirement of high temperature, poisonous gases and fully vacuum environment [13, 14].

1.3 Synthesis of Free Standing Graphene

Free standing graphene other than substrate bound horizontal graphene can be synthesized through various methods as shown in Fig. 2 like vacuum filtration, centrifugal vacuum evaporation (CVE), direct evaporation, sugar blowing synthesis, ball milling, epitaxial growth, hydrothermal reduction or plasma method. Template assisted synthesis is also one of the popular methods for synthesis of free standing graphene, where templates get removed after the synthesis. Here, various kinds of easily available and abundant templates like nickel foam, Pt(111), Ni(111), W(110), SiO_2 substrate can be used for synthesis, which show diversity in morphology of free standing graphene with different porous structures. Templated synthesis method can be performed in combination with different synthetic procedures like chemical vapor deposition (CVD), electrodeposition, sol-gel synthesis, and hydrothermal synthesis [15, 16].

Graphene foam is a most common type of self-supported or free-standing graphene. It can be synthesized through template assisted synthesis using metal foam, polystryrene sphere or bi-metallic foam as the template. For example, Ma et al. have

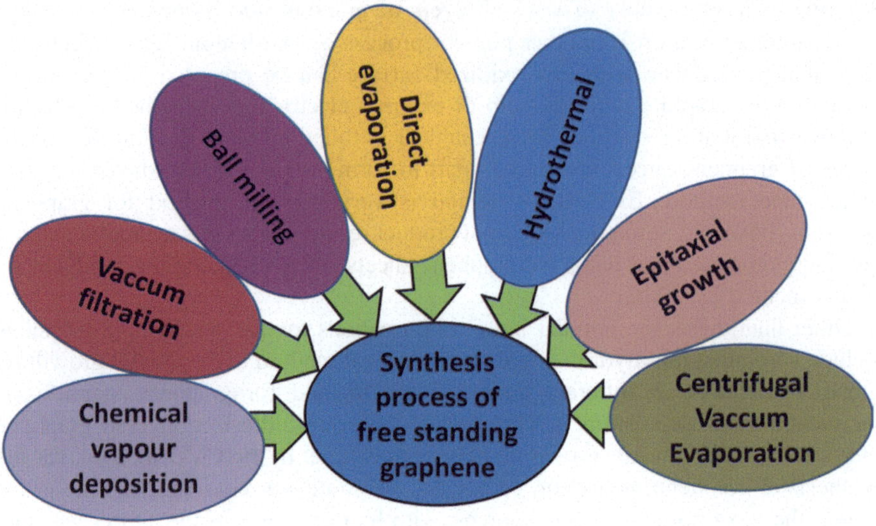

Fig. 2 Schematic diagram showing diverse methods of production of free standing graphene

synthesized free standing graphene through a special form of CVD that is segregation-adsorption chemical vapour deposition. Here, they have synthesized a monolayer of graphene film with grain size ranging from 200 nm to 1 μm using Pt as substrate [17]. After the synthesis the graphene film is carefully removed from the substrate. Other than this, substrate free synthesis of graphene was done by Hu et al. [18]. In this work, the research group coupled spray coating and CVD for the synthesis of CNT-graphene composite. The CVD method was used for the synthesis of carbon nanotube and the graphene oxide solution was sprayed onto the CNT for the synthesis of the composite. Cheng et al. have synthesized self-supported graphene foam using nickel foam as the substrate via CVD [19]. At first, the carbon atoms get attached to the Ni metal foam substrate, followed by diffusion and formation of graphene layers at Ni foam. After the formation of graphene foam, template was etched to make the graphene foam free from substrate. Yin et al. synthesized free standing graphene foam through hydrothermal process using nickel foam as the template, graphene oxide as the precursor, and N_2H_4 as the reducing agent [20]. These are the few examples, where graphene foams were synthesized through template assisted synthesis approach.

Instead of these, template free methods have also been reported in the literature for synthesis of free standing graphene. Some of which are discussed below. Template free synthesis is more expedient and proficient, because it has high yield without any hassle of etching or removing the product from template, which causes lesser loss in product. Hummer's method is the most convenient template free synthesis process, where the graphene oxide and reduced graphene oxide both self-assembled themselves in liquid phase to form self-supported graphene [10]. Vacuum filtration is another template free technique, where a solid product is extracted from liquid reaction mixture [21]. In this process, Buchner funnel is attached with vacuum pump and used to filter out the solid-liquid mixture, where the solid part remains in the filter paper and the liquid part comes out. Vacuum filtration is a very cost effective technique and was firstly explored by Ruoff group for the synthesis of graphene paper using filter paper made up of anodisc membrane [22]. Li et al. have also synthesized porous free standing nanometer scaled graphene paper in a non-volatile electrolytic solution with large number of pores for ion transport [23]. In addition to these, graphene sponge can also be synthesized through centrifugal vacuum evaporation technique, which was discovered by Liu and allies [24]. They have found that evaporating colloidal suspension of graphene oxide in temperature variable vacuum chamber leads to formation of layered graphene membrane. Another method for self-supported graphene synthesis is electrochemical leavening method, where graphite paper is used as the precursor. The intercalation of graphite paper with ions present in the electrolytic solution resulted in the expansion of interlayer spaces to form self-supported graphene layers [25]. Some other techniques (with template and without template) used for the synthesis of free standing graphene listed in Table 1 [26–35].

1.4 Characterization of Free Standing Graphene

Characterization of prepared free-standing graphenes is another very important aspect, which provides information about their properties, morphology and structure. First of all the basic characterization of prepared graphene material can be done by different spectroscopic techniques like UV-Visible spectroscopy, Raman spectroscopy, Nuclear magnetic resonance spectroscopy (NMR), Fourier transform infra-red spectroscopy (FT-IR), and mass spectrometer (MS) etc. For the determination of chemical composition, X-ray diffraction analysis (XRD) is commonly used technique, to estimate the crystalline size of particles using Scherer equation. The morphological features can be visualized by different microscopic techniques like high-resolution transmission electron microscopy (HR-TEM), field emission scanning electron microscope (FE-SEM), and atomic force microscopy (AFM). The functional group presence at the surface of graphene can be identified by FT-IR spectroscopy, secondary ion mass spectrometry (SIMS), auger electron spectroscopy (AES), and NMR spectroscopy. To determine the oxidation state of elements present in the graphene based nanomaterials, X-Ray photo electron spectroscopy (XPS) is commonly used. Photoluminescence (PL) spectroscopy can also be used sometimes for the determination of the optical property of the graphene based quantum dots. Galvanostat/potentiostat is in general used for electrochemical or electroanalytical studies of graphene materials [36]. The role of prepared graphene in energy storage or energy conversion is also explored using galvanostat/potentiostat. In electrochemical studies, cyclic voltammetry (CV), AC impedance or

Table 1 Different methods or techniques used for the synthesis of free standing graphene and their composite with different materials

S. no.	Materials	Template used	Methods of synthesis	Surface area (m^2 g^{-1})	Application	Ref.
1.	Fe_3O_4/3D-graphene	Template free	Sol-gel chemistry	195.4	Lithium/sodium storage and supercapacitors	[26]
2.	Fe_3O_4/GA composite	Template free	Hummers methods and freeze drying	126.4	Lithium ion batteries	[27]
3.	CoO/NGA	Template free	Solvothermal and self-assembly	237.3	Lithium storage	[28]
4.	BC-graphene-TiO_2	Template free	Hummers method	3.8×10^4	Supercapacitor	[29]
5.	3D graphene/Ni_3S_2 nanostructure	Nickel foam	CVD	125.6	Supercapacitor	[30]
6.	3D graphene hydrogels	Template free	Hydrothermal	921	Supercapacitor	[31]
7.	MnO_2/nanoporous3D graphene	NH_2-SiO_2 as sacrificial template	Modified hummers method	–	Supercapacitor	[32]
8.	$NiCoO_2$@rGO/NF	Nickel foam	Hydrothermal	–	Supercapacitor	[33]
9.	Bi_2WO_6/3D graphene	Template free	Hydrothermal	–	Lithium storage	[34]
10.	g-C_3N_4-rGO/LDH-1	Template free	Hydrothermal	240	Supercapacitor	[35]

3D = Three dimensional; BC = Biocarbon, NGA = Nitrogen doped graphene aerogel; (g-C_3N_4) = graphitic carbon nitride; LDH = double layered hydroxide; rGO = reduced graphene oxide; CVD = Chemical vapour density; NF = Nickel foam

electrochemical impedance spectroscopy, liner sweep voltammetry (LSV), and bulk electrolysis are the most common methods to explore the electrochemical ability of prepared graphene.

2 What Is Supercapacitor?

Energy storage devices have plentiful industrial purposes, where they can be used in energy storage, energy conversion or as a transient energy buffer. In current epoch rechargeable batteries are in great demand, due to their good life cycle, high energy density, and enhanced safety features. But they also have certain limitations, like they have high internal resistance, low power density, large area utilization, and low power discharge potential under huge current loading and highly expensive too [37, 38].

To overcome these problems, the researchers have developed and designed supercapacitors (SCs). Huge advancements have been accomplished in the domain of supercapacitor with the use of diverse types of nanoparticles with large charge/discharge rates, high power density, long cycle life, speedy charging system, excellent durability, and high recyclability potential. With the growing enthusiasm, in the coming 5 years, supercapacitors are expected to surpass batteries with their stupendous properties to become a preeminent energy storage device for portable electronics and/or wearable hybrid vehicles [39]. Supercapacitors have certain prominent features like large extent of operating temperature window, low inner resistance, and high efficiency. Sometime in spite of functioning independently, supercapacitors are coupled with high energy devices for power sourcing or power sinking and therefore supercapacitors can be categorized in two main classes [40], as follows:

1. Electric double layer capacitors (EDLCs) or ultracapacitors (UCs): In this category, by altering the specific surface area, morphology, and pore size of components, we can accumulate large amount of charges at the interfaces of electrode and electrolyte.
2. Another one is called as pseudocapacitor, which has higher energy density and superior specific capacitance than EDLCs. This type of capacitor uses conductive polymer and metal oxide as the electrode material. Energy storage occurs through transfer of Faradic charges between electrode and electrolyte, as a result of multi-electron reversible oxidation and reduction Faradaic reactions.

EDLCs were first discovered by Helmholtz in the year 1853 [41]. While, Becker, a general electric company got a patent in 1957 by using electrode of carbon material with porous structure having large specific area for electric double layer formation [42]. First commercialization of supercapacitor was done in 1971 by Nippon electric company (NEC). In general, supercapacitors consist of two electrodes, electrolyte and membrane for insulation of two electrodes as shown in Fig. 3, also illustrating the formation of electrical double-layer capacitance. Electrolyte plays main role in supercapacitor, where the specific capacitance can be improved by means of various kinds of electrolytes like ionic liquid, organic electrolyte, and/or aqueous electrolyte. Electrolyte to be used for supercapacitor must have immense voltage window, soaring electrochemical stability, lesser volatility, and low resistivity. Among all of these, aqueous electrolytes are considered as the excellent electrolyte for preparation of supercapacitor with good power liberation, high capacitance, and higher ionic concentration [43]. Similarly, the membrane separator allows the ions to pass through them without any electrical contact. The main process to fabricate an EDLC supercapacitor is the development of double layer capacitance at the junction of electrode and electrolyte. For this, the negative and positive charges are arranged in opposition to each other at little

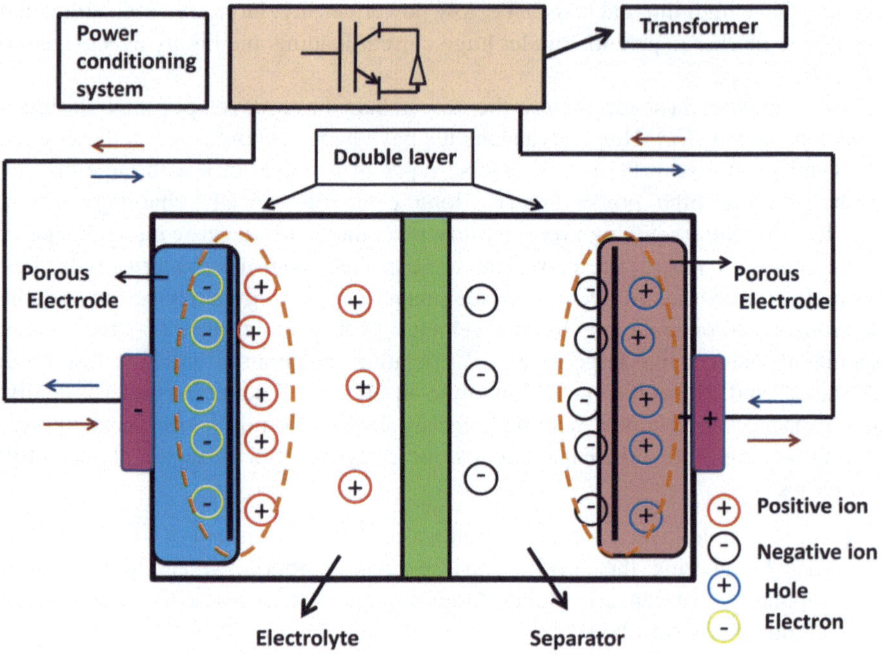

Fig. 3 Schematic diagram showing electrical double-layer capacitance in supercapacitor (Reproduced with permission from Ref. [38])

distance around solid electrode and liquid electrolyte interface. The charge gathered in the interface can be represented as [44]:

$$C = \frac{\varepsilon}{4\pi\delta} \int dS \tag{1}$$

where, ε = dielectric constant, δ = distance amid the interface and center of the electrolytic ions and S = surface area of the electrode edge. We can calculate specific capacitance through area calculation of electrochemical cyclic voltammetry curve, using the equation given below [45]:

$$C_m = k \frac{\int i}{ms} \tag{2}$$

C_m = specific capacitance, $\int i$ = total area of cyclic voltammetry, m = electrode mass, s is the particular scan rate used to record the CV run. As we have said earlier, the high capacitance of the supercapacitors depends on the morphology, specific surface area, porosity and conductivity of the material used to fabricate the

Table 2 Summary of the free standing graphene based material for supercapacitor application

S. no.	Electrode used	Supporting electrolyte	Specific capacitance ($F\ g^{-1}$)	Current density ($A\ g^{-1}$)	Ref.
1.	rGO/carbon black	PVA/H_2SO_4	79	1.0	[47]
2.	CNT-graphene/ Fe_3O_4NPs	Filter paper/ Na_2SO_4	200.4	–	[48]
3.	CNT-Mn_3O_4/ graphene	PAAK-KCL	72.6	0.50	[49]
4.	rGO/ cMWCNT − CFP/ PPy	PAAK-KCL	82.4	0.50	[50]
5.	3D graphene/MnO_2	0.5M Na_2SO_4	130	–	[51]
6.	PANI/GO	1M H_2SO_4	423 ± 7	0.80	[52]
7.	NiO/UDG/NF	KOH	300	10.0	[53]
8.	GH-PANI/GP	PVA/H_2SO_4	864	1.0	[54]
9.	Ni_3S_2/3D rGO	6M KOH	1886	1.0	[55]
10.	rGO/NCO nanostructures	2.0M KOH	282.94	1.0	[56]
11.	RuO_2/g-C_3N_4@GO	6M KOH	704.3	0.50	[57]
12.	GPA	1M H_2SO_4	253	0.50	[58]
13.	PPy@BC/GO composites	1M H_2SO_4	492	0.40	[59]
14.	CC/ZnO@C@NiO NAs	PVA/KOH	677	1.43	[60]
15.	CNG@NCH// r-GOF	1M LiOH	175	0.50	[61]
16.	$NiMn_2O_4$@CoS	1M KOH	1751	1.0	[62]

rGO = Reduced graphene oxide; CB = Carbon black; PVA = Polyvinyl alcohol; CNT = Carbon nanotube; CV = Cyclic voltammetry; potassium polyacrylate = PAAK, MWCNT = Multiwalled carbon nanotube; CFP = Carbon fibre paper; 3D = Three dimensional graphene; PANI = Polyaniline; UDG = Ultrathin derived graphene; NF = Nickel foam; GP = graphene paper; GH = graphene hydrogel; NCO = nickel cobaltite; GPA = Graphene/polypyrrole aerogel; BC = Bacterial cellulose; PPy = Polypyrrole; NAs = Nanorods arrays; CC = Carbon cloth; CNG = Cobalt oxide nanobrush-graphene; NCH = Nickel cobalt hydroxide [$Ni_xCo_{2x}(OH)_{6x}$]; reduced-graphene oxide-film

electrodes. The advanced materials like carbon nanomaterials have to be preferred as supercapacitor because of high specific surface to volume ratio, superior conductivity, high abundancy, considerable electrical/chemical stability, and large pore sizes [46]. Some of the free standing graphene based materials used as supercapacitors are summarized in Table 2 [47–62].

3 Role of Free Standing Graphene as Supercapacitor

In this chapter, we have tried to discuss about the free standing or self-supported graphene based electrode materials and their roles in and as supercapacitors. Research is going on for the designing of supercapacitor devices with lighter weight, high flexibility and miniaturization. Free standing graphene has become a benchmark for the designing of long lifecycle and easily rechargeable supercapacitor, owing to its exceptional surface area, light weight, porous structure, high mechanical strength, and good conductivity [63]. The pores of free standing graphene material act as ion transfer path to increase the specific capacitance. In comparison to the free standing or self-assisted graphene, the template assisted graphene or substrate bound graphene occupies large space. The substrate used to withstand graphene does not contribute towards capacitance rendering decrease in the power density and energy density. In the coming sections, we intend to discuss the role of free-standing graphene in the fabrication of supercapacitors. Sometime, graphene based free standing materials (graphene film, graphene sponge, graphene foam, and graphene paper) are made hybrid with other active electrode materials like metal or metal oxide to form integrated graphene electrode [64]. These materials are also discussed in the coming sections.

3.1 Graphene Foam

Graphene foam is the macroscopic graphene material with a very porous light weight structure well suited for energy storage application. It can be popularly synthesized (as discussed earlier) through CVD using 3D metal foam as the template. In graphene foam, two dimensional graphene sheets are connected to each other to form highly conductive network of 3D foams which provide the graphene material with lots of unique properties like high pore volume, large surface area, good mechanical strength, considerable thermal stability, lofty electronic conductivity, high ion transfer rate, and self-support [65]. The 3D graphene foams are binder free porous continuous structure, where large volume change can take place due to charge discharge phenomenon. Increasing demand of populace cannot be met by pristine graphene based electrode in supercapacitor with limited energy density. Supercapacitor with high charge storage property greatly depends upon the pore size of the material that is only macroporous but mesoporous material can have good charge storage capability [66]. Microporous materials cannot form electrical double layer capacitance due to inability to pass solvated ions. Some of the literature reports, showing role of free standing graphene foam as superior supercapacitor electrode material are discussed below. For example, Manjakkal et al. have synthesized composite of 3D graphene foam with silver conductive epoxy and graphene sheet and used it as electrode [67]. Here, graphene sheet acted as a current collector, while, silver conductive epoxy acted as connector between graphene sheet and graphene foam. The fabricated layered electrode showed specific capacitance of 38 mF cm^{-2} with a current density of about 0.67 mA cm^{-2}, with power density and energy density of 0.27 mW cm^{-2} and 3.4 μW h cm^{-2}, respectively.

Ping et al. have developed free standing 3D graphene foam using nickel foam as the template. After the synthesis, template was removed by etching with solution of $FeCl_3/HCl$. The prepared foam has specific surface area of 412.78 m^2/g and specific capacitance of 128 F/g in 6M KOH electrolyte [68]. Down et al. have synthesized self-supported free standing macroporous graphene foam using CVD devoid of any mechanical stand. The prepared material was used as electrode for supercapacitor designing, with variable electrolytes like [C_4MIM][BF_4], [C_4MIM][PF_6] and [C4MIM][NTf_2]. Using, [C_4MIM][BF_4] as electrolyte, the supercapacitor showed power density value of 29.33 kW kg^{-1} and 40.94 W h kg^{-1} of specific energy density with capacitance of 636 μF [69]. Grakani and his group have synthesized mesoporous heterogeneous arrays of $NiCo_2O_4$–MnO_2 grown on self-standing graphene foam and used for fabrication of energy storage devices like supercapacitor [70]. The prepared electrode from the free standing material possesses exceptional life cycles i.e. 5000 cycles with 94.3% capacitance retention. The prepared supercapacitor exhibited areal capacitance of 5.15 F cm^{-2} at 2 mA cm^{-2} current density and gravimetric specific capacitance of 2577 F g^{-1} at 1 A g^{-1} current density.

Qian et al. have synthesized composites of graphene foam with the use of $Li_4Ti_5O_{12}$ through hydrothermal synthesis procedure and applied it for the fabrication of lithium ion hybrid supercapacitor. The prepared electrode showed specific capacitances of 186 mA h g^{-1} at 0.2 C, 179 mA h g^{-1} at 0.5 C and 175 mA h g^{-1} at 1 C. This electrode also showed energy density of 26 W h kg^{-1} with power density of 2500 W kg^{-1}, in addition of showing good cyclic stability retaining up to 83% capacitance, after 4000 cycle [71]. Ghosh et al. have synthesized three dimensional sea urchin shaped composite of MnO_2@polyaniline (PANI, a very popular conducting polymer) as the positive electrode and 3D free standing graphene foam as the negative electrode in supercapacitor which is illustrated in Fig. 4. The graphene foam and the composite of MnO_2@PANI were prepared by hydrothermal method. The electrode exhibited the specific capacitance of 129.2 F g^{-1} at 0.5 A/g current density with an energy density of 37 W h kg^{-1}. Only 11% of capacitance loss was observed after 5000 cycles [72].

3.2 *Graphene Films*

Graphene based materials have excellent electrical conductivity of 108 Sm^{-1} and high theoretical specific capacitance value of 550 F g^{-1}, hence are used as electrode in supercapacitor. But due to agglomeration and restacking of layers of graphene, the electrolyte ion transfer pathway get clogged, and thus the performance of supercapacitor suffered. This is the major reason, the modification and optimization

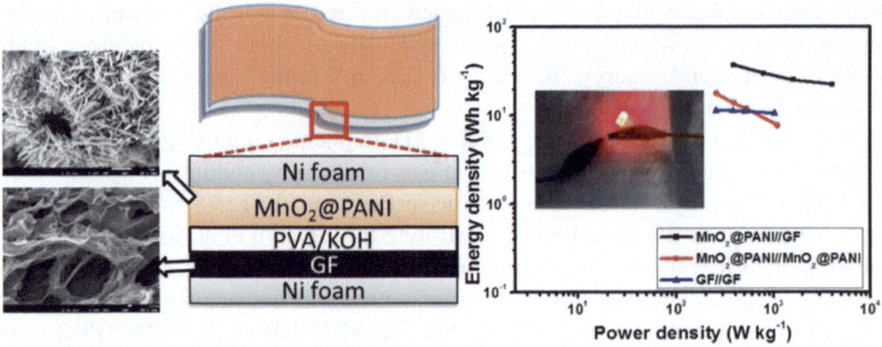

Fig. 4 Illustration showing solid-state flexible asymmetric supercapacitor of 3 dimensional urchin-shaped MnO_2@PANI nanocomposite as anode and 3D graphene foam as cathode materials (Taken permission from Ref. [72])

of graphene to form densely packed graphene films, where all the edges are exposed to form electrical double layers are gaining interest of researchers working in the field of supercapacitors [73]. Graphene thin films have attracted great consideration owing to their captivating optical, mechanical and electrical properties, high thermal conductivity, super flexibility, low cost and have feasible application in energy storage devices. In the form of thin filter paper graphene films can be used in chemical filtration, fuel cells, and/or batteries etc.

Several methods were experimented for the synthesis of graphene thin films but finding the thickness of a film in the nanometer range was the major dilemma among the nanotechnologists. Various top down approaches like direct evaporation, vacuum filtration with various coating method like spin coating, and spray coating have been developed for the synthesis of graphene film. In the bottom up approach, CVD has been the most reasonable technique for the synthesis of thin films, because it can provide thin films with high electrical conductivity and high specific capacitance. For the synthesis of graphene films, at first graphene oxide films are produced and then reduced to form r-GO followed by delaminating of the film from the surface of the substrate. For example,

Zhao et al. synthesized graphene oxide/NaCl/urea films for showing their vivid application as a supercapacitor with specific capacitance value 425 F g^{-1} [74]. The FESEM images of the films derived from several mixtures and the synthetic procedures are illustrated in Fig. 5. Duo et al. have synthesized free standing flexible graphene thin films using the method of electrochemical reduction and electrophoretic deposition for use as electrode for supercapacitor. The prepared electrode showed specific capacitance value of 254 F/g at a scan rate of 10 mV/s. The electrode restored capacitance value to 97%, after 1000 CV cycles [75].

Chao et al. have synthesized densely heaving porous graphene film and applied it for volumetric capacitance supercapacitor. The graphene film was synthesized by template assisted synthesis and several times etched by $Zn(CH_3COO)_2$. The prepared electrode showed 318.8 F cm^{-3} volumetric capacitance in 6.0M KOH at a

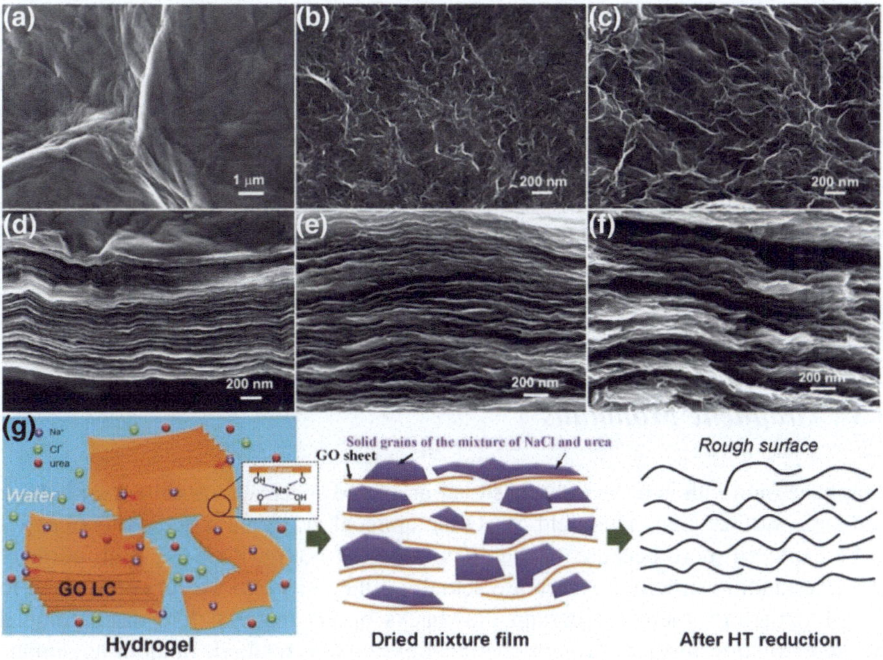

Fig. 5 FESEM images of r-GO films derived from GO; NaCl and GO mixture as (**a**) and (**b**) respectively; **c** FESEM images of mixture of NaCl, GO and urea. **d** Cross-sectional FESEM images of films derived from simply GO. **e** Cross-sectional FESEM images of r-GO films derived from NaCl, GO mixture. **f** Cross-sectional FESEM images r-GO films derived from NaCl, GO and urea mixture. **g** Diagram showing the development procedure for the GO-NaCl-urea mixture from hydrogel to form r-GO film. In inset it shows interlinking between two consecutive GO sheets by means of coattraction through Na^+ is shown in the (Obtained through permission from the Ref. [74])

current density value of 1 A g^{-1}. Additionally, the electrode showed 1000 charge/discharge lifecycle with about 88% retention of capacitance [76]. Jia et al. have synthesized composite films of multiwalled carbon nanotube and graphene through self-assembly method, which exhibited good electrical conductivity and exceptional mechanical flexibility. The prepared hybrid film showed specific capacitance of 249.3 F g^{-1} at a current density of 200 mA g^{-1} and superb electrochemical performance with 91.3% retention of specific capacitance after 8000 cycles [77]. Chen et al. have synthesized paper like ultrathin self-supported graphene films with 40 mm diameter and 40 nm thickness, through top down vacuum filtration method. The ultra-thin film electrode showed specific capacitance of 192 F g^{-1} and capacitance decay of 9%, after 1000 cycles [78]. Xiong et al. have designed free standing flexible Co_3O_4/graphene films for supercapacitor application. Graphene film was fabricated by spin coating technique and then Co_3O_4 nanoarrays were attached to the graphene film substrate to make a hybrid free standing electrode.

The electrode showed very good capacitance value of 652 F g^{-1} after doing 3000 cycles at 2 A g^{-1} current density [79].

Zhu and his coworkers have developed a method for the synthesis of graphene films with the help of foam film method. Foam film is a very easy method for the synthesis of free standing graphene film with variable thickness, just by change in the surfactant concentration. The ultra-thin film electrode of solid state supercapacitor showed capacitance of 56 mF cm^{-2} at 0.2 mA cm^{-2} current density with polyvinyl-gel KOH as electrolyte. The film showed volumetric capacitance value of 2810 mF cm^{-3} at 12 mA cm^{-3} current density in addition to retaining about 90.4% of its initial capacitance value, after 10000 cycles [80].

3.3 Graphene Monoliths

Graphene monolith is a three dimensional graphene sheet with porous morphology and large surface area. This field of 3D hierarchical graphene based material is yet to be explored more because of the coexistence of both criteria i.e. high porosity and freestanding structure, which are actually difficult to achieve. In comparison to the 3D structure, there are certain drawbacks of 2D material like discontinuous stacked structure formed through chemical exfoliation method, random agglomeration, overlaying of sheets, which might be overcome by the use of 3d monoliths [81]. As a result research is going on three dimensional graphene monoliths, which have continuous structure with improved ion transfer capability, negligible agglomeration and better connectivity in between layered structures. Because of the crosslinked structure in 3D graphene monolith, amid the layers of graphene they have high electrical conductivity, better rigidity, and improved [82].

There are different methods reported for the synthesis of graphene monolith like: (1) chemical vapor deposition (CVD), (2) self-assembly of graphene oxide, (3) hydrothermal synthesis, and (4) freeze casting method. Cost effective template free synthesis is the major challenge to the researcher for the synthesis of high mechanical strength three dimensional free standing graphene monoliths. Graphene prepared through Hummers method with the help of certain reducing agent get assembled at different pH and temperature values to form 3D graphene monolith.

On the similar pattern, Zou et al. have synthesized free standing binder free nitrogen-doped reduced graphene oxide (N-RGO) monolith for supercapacitor application. The N-doped graphene was fabricated by mixing graphene oxide with urea using flame bath method. The graphene monolith based electrode showed specific capacitance value of 323 F g^{-1} at a current density of 1 A g^{-1}. The resulted electrodes showed long life and have ~9.2% decrease in specific capacitance after 50,000 cycles [83]. Wang et al. have also fabricated hierarchical N-doped graphene monolith of porous nature through microwave assisted synthesis procedure

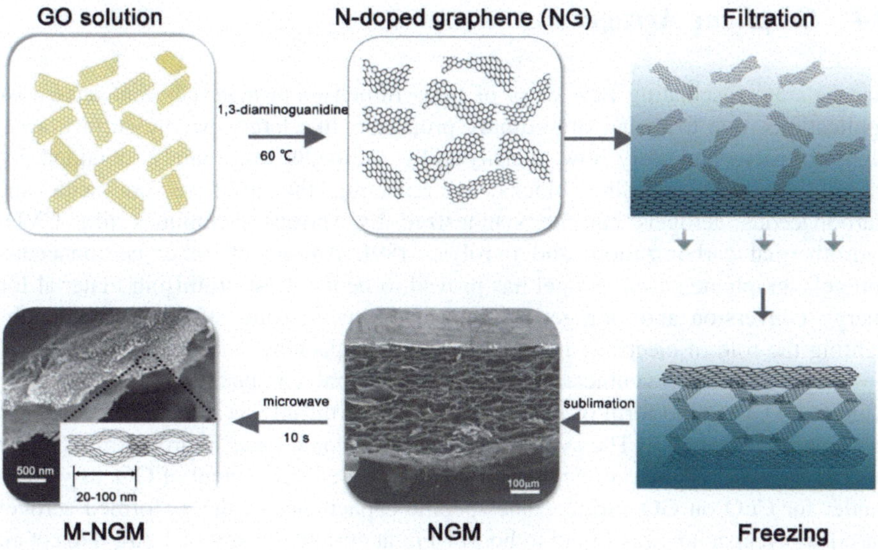

Fig. 6 Schematic representation showing synthesis procedure for N-doped graphene monoliths (Obtained through permission from Ref. [84])

(illustrated in Fig. 6) for preparation of highly efficient flexible supercapacitor. The prepared electrode showed specific capacitance value of 362 F g^{-1} and the areal capacitance of 894 mF cm^{-2}. It illustrated superior cyclic stability with 100% retention of specific capacitance [84].

Deng et al. have synthesized free standing monolith composite of graphene-bismuth vanadate (GR/$BiVO_4$) through hydrothermal technique [85]. Acting as a binder free electrode in supercapacitor, it showed specific capacitance value of 479 F g^{-1} with energy density value of 45.69 W h kg^{-1} and power density of 800 W kg^{-1}. Wen et al. have synthesized nitrogen and phosphorus dual element doped graphene monolith for energy storage application. Melamine phosphate is the only precursor used for dual doping purpose. The fabricated electrode showed high specific capacitance value of 183 F g^{-1} at a current density of 0.05 A g^{-1} with a power density of 11.33 W h kg^{-1} at 1.6 V [86]. Yao and his group have synthesized composite monolith of Ni-pyridine-2, 3-dicarboxylic acid (pda)@3D graphene oxide and showed its performance as supercapacitor electrode. This group has synthesized coordination supramolecular network by means of the ligand pyridine-2, 3-dicarboxylic acid (H_2-pda) and Ni^{2+} as the metal using solvothermal method and combined it with GO in H_2O-DMF solution. The electrode delivered specific capacitance of 952.85 F g^{-1} with energy density of 17.70 W h kg^{-1} and power density value of 1500 W kg^{-1} [87].

3.4 *Graphene Aerogel*

Aerogels are fascinating new class of three dimensional materials that have vast applications, due to their extraordinary properties like large specific surface area, high porosity, and very low density [88]. Aerogels are mainly obtained by assembling small building blocks by replacing the used solvent with air. Carbonaceous aerogels can be synthesized by various techniques like CVD, hydrothermal carbonization, and pyrolysis [89]. Among all other carbonaceous aerogels, graphene based aerogel has proved to be the most promising material for energy conversion and storage. Below are discussed some of the articles highlighting the role of electrode materials for supercapacitor.

Song et al. have synthesized three dimensional composite of copper oxide (CuO) and GO in the form of aerogel and used it fruitfully as an electrode material for supercapacitor [90]. The synthesis procedure adopted was hydrothermal, where self-assembly of GO was done using citric acid as reducing agent of GO, as well as binder for CuO on GO surface. The specific capacitance of the as formed aerogel based supercapacitor was found to be 211 F/g at current density of 1 A/g. Hoa et al. have synthesized 3D polyaniline (PANI) grafted RGO based aerogel for fabrication of high performance supercapacitor. The synthesis procedure was carried out in two different steps. At first, graphene oxide-PANI composite was prepared, after that, RGO-g-PANI aerogels were prepared by hydrothermal technique. The aerogel based electrode used in supercapacitor showed specific capacitance of 1600 F g^{-1} at a current density of 12 A g^{-1}. The material retained specific capacitance of about 91.3%, after 3000 cycle [91]. Xing and coworkers have reported solvothermal synthesis of 3D, free standing melamine functionalized nitrogen doped graphene aerogels with their ability to withstand to different loads.. The material demonstrates good specific capacitance of 170.5 F g^{-1} at a current density of 0.2 A g^{-1} [92]. Similarly, a hybrid, 3D, free standing, high performance electrode material for supercapacitor was synthesized by Choi and coworkers using catechol with graphene oxide. The electrode showed large specific capacitance of 188 F g^{-1} at 1 A

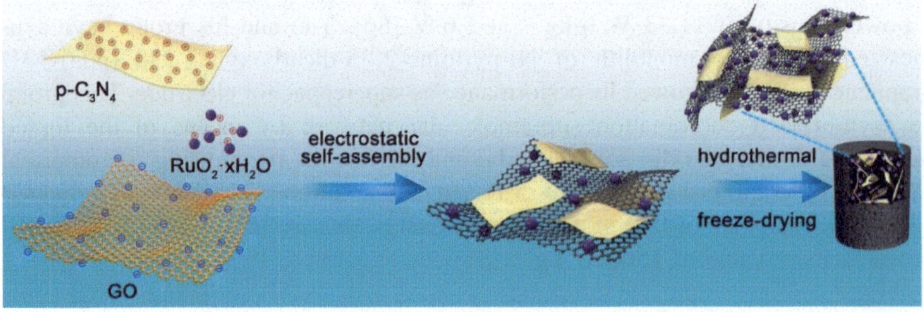

Fig. 7 Schematic diagram showing the synthesis of 3D RuO_2/g-C_3N_4@graphene oxide (Reproduced after taking permission from the Ref. [57])

$g^{-}1$ current density, energy density equivalent to 25 W h kg^{-1} and power density approx. to 18,000 W kg^{-1} [93]. Zhang et al. synthesized 3D graphene aerogel and made its composite with RuO_2/g-C_3N_4 to realize the final product RuO_2/g-C_3N_4@graphene oxide electrode for supercapacitior. The synthesis procedure is shown in Fig. 7. The specific capacitance value of the material was 704.3 F g^{-1} at a current density of 0.5 A g^{-1} [57].

4 Conclusion and Future Aspects

Immense advancements have been accomplished in the field of free standing graphene during the past few years, in which major focus has been on the study of their different expeditious synthesis techniques. The use of various kinds of synthesis approaches and new researches have resulted in free standing graphene with architecture variation, alteration and explanation of physical and chemical properties, exploration of their vast arena of application like energy storage, catalysis, environmental remediation, etc.

In this chapter, we have focused our aim towards role of free standing graphene in supercapacitor only. There are two main things, which we have to consider while using free standing graphene as the high performance electrode for supercapacitor:

1. Synthesis of free standing graphene via template assisted synthesis or template free synthesis,
2. Making composite of free standing graphene with other active materials to enhance their electrical conductivity, flexibility, ions transport ability, and supercapacitor device assemblage.

Because, template assisted synthesis methods require high temperature and hectic etching of template, after synthesis of free standing graphene, however, template free synthesis procedure works under mild condition, comparatively. In templated synthesis process, new templates with small pore sizes, high surface area, small thickness are required. Template free synthesis also has some limitations as they render the graphene with random and inhomogeneous pores of different sizes. So, in near future, we have to work harder to search new controllable synthesis procedures, which can provide uniform pore size and uniform arrangement of layers of free standing graphene. Similarly, improvement in other existing popular synthesis methods is also required. For example, CVD technique provides high electrically conductive free standing graphene. But, now it's high time that material scientist must focus on improving the CVD technique by coupling it with other devices like microwave assisted chemical vapour deposition method (MCVD). CVD is good for graphene synthesis, but, it involves high cost and instrumentation; therefore, hydrothermal synthesis is more popular among the researchers. The only drawback, which has to be eliminated from hydrothermal synthesis technique, is self-assembly of resulting products, in order to avoid poor reproducibility in the structure of the free standing graphene.

Other than focusing on the synthesis of free standing graphene, researchers have to focus on their hybrid preparation also. For this, best active material has to be selected to make hybrid free standing graphene, which can improve the electrode performance by increasing the overall capacitance value, power density, energy density, cycle life and stability.

Inspite of large progress in the area of free standing electrode for supercapacitor there are certain points which needs to be look after like non uniform pore size of the electrode, better surface functionalization for more enhanced supercapacitor property, stability of the electrode in different electrolytic solutions, large scale synthesis approach for better commercialization. Due to lack of multicomponent set up and different optimization technique for the synthesis of free standing electrode there is still long way to go for commercialization of this free standing electrode. To make this field more productive and gleaming in the future different scientific strategies with more intensive research need to be incorporated for ruling out the small shortcommings.

Author Declaration Ms. Karfa has given the major contribution in writing this book chapter along with drawing the Figures and Tables, taking the copyright permission etc.

References

1. Geim, A.K., Novoselov, K.S.: The rise of graphene. Nat. Mater. **6**, 183–191 (2007)
2. Geim, A.K.: Graphene: status and prospects. Science **324**, 1530–1534 (2009)
3. Zhu, Y., Murali, S., Cai, W., Li, X., Suk, J.W., Potts, J.R., Ruoff, R.S.: Graphene and graphene oxide: synthesis, properties, and applications. Adv. Mater. **22**, 3906–3924 (2010)
4. Chen, D., Feng, H., Li, J.: Graphene oxide: preparation, functionalization, and electrochemical applications. Chem. Rev. **112**, 6027–6053 (2012)
5. Huang, X., Yin, Z., Wu, S., Qi, X., He, Q., Zhang, Q., Yan, Q., Boey, F., Zhang, H.: Graphene-based materials: synthesis, characterization, properties, and applications. Small **7**, 1876–1902 (2011)
6. Georgakilas, V., Otyepka, M., Bourlinos, A.B., Chandra, V., Kim, N., Kemp, K.C., Hobza, P., Zboril, R., Kim, K.S.: Functionalization of graphene: covalent and non-covalent approaches, derivatives and applications. Chem. Rev. **112**, 6156–6214 (2012)
7. Bianco, A., Cheng, H.-M., Enoki, T., Gogotsi, Y., Hurt, R.H., Koratkar, N., Kyotani, T.: All in the graphene family–a recommended nomenclature for two-dimensional carbon materials. Carbon **65**, 1–6 (2013)
8. Meyer, J.C., Geim, A.K., Katsnelson, M.I., Novoselov, K.S., Booth, T.J., Roth, S.: The structure of suspended graphene sheets. Nature **446**, 60 (2007)
9. Chen, C., Yang, Q.-H., Yang, Y., Lv, W., Wen, Y., Hou, P.-X., Wang, M., Cheng, H.-M.: Self-assembled free-standing graphite oxide membrane. Adv. Mater. **21**, 3007–3011 (2009)
10. Marcano, D.C., Kosynkin, D.V., Berlin, J.M., Sinitskii, A., Sun, Z., Slesarev, A., Alemany, L. B., Lu, W., Tour, J.M.: Improved synthesis of graphene oxide. ACS Nano **4**, 4806–4814 (2010)
11. Choi, W., Lahiri, I., Seelaboyina, R., Kang, Y.S.: Synthesis of graphene and its applications: a review. Crit. Rev. Solid State **35**, 52–71 (2010)

12. Wei, D., Liu, Y.: Controllable synthesis of graphene and its applications. Adv. Mater. **22**, 3225–3241 (2010)
13. Zhang, Y.I., Zhang, L., Zhou, C.: Review of chemical vapor deposition of graphene and related applications. Acc. Chem. Res. **46**, 2329–2339 (2013)
14. Tang, L., Li, X., Ji, R., Teng, K.S., Tai, G., Ye, J., Wei, C., Lau, S.P.: Bottom-up synthesis of large-scale graphene oxide nanosheets. J. Mater. Chem. **22**, 5676–5683 (2012)
15. Reina, A., Jia, X., Ho, J., Nezich, D., Son, H., Bulovic, V., Dresselhaus, M.S., Kong, J.: Large area, few-layer graphene films on arbitrary substrates by chemical vapor deposition. Nano Lett. **9**, 30–35 (2008)
16. Gass, M.H., Bangert, U., Bleloch, A.L., Wang, P., Nair, R.R., Geim, A.K.: Free-standing graphene at atomic resolution. Nat. Nanotechnol. **3**, 676 (2008)
17. Teng, M., Liu, Z., Wen, J., Gao, Y., Ren, X., Chen, H., Jin, C.: Tailoring the thermal and electrical transport properties of graphene films by grain size engineering. Nat. Commun. **8**, 14486 (2017)
18. Hu, D., Gong, W., Di, J., Li, D., Li, R., Lu, W., Gu, B., Sun, B., Li, Q.: Strong graphene-interlayered carbon nanotube films with high thermal conductivity. Carbon **118**, 659–665 (2017)
19. Chen, K., Chen, L., Chen, Y., Bai, H., Li, L.: Three-dimensional porous graphene-based composite materials: electrochemical synthesis and application. J. Mater. Chem. **22**, 20968–20976 (2012)
20. Yin, S., Wu, Y.-L., Hu, B., Wang, Y., Cai, P., Tan, C.K., Qi, D.: Three-dimensional graphene composite macroscopic structures for capture of cancer cells. Adv. Mater. Interfaces **1**, 300043 (2014)
21. Zhang, S., Li, Y., Pan, N.: Graphene based supercapacitor fabricated by vacuum filtration deposition. J. Power Sources **206**, 476–482 (2012)
22. Dikin, D.A., Stankovich, S., Zimney, E.J., Piner, R.D., Dommett, G.H.B., Evmenenko, G., Nguyen, S.T., Ruoff, R.S.: Preparation and characterization of graphene oxide paper. Nature **448**, 457 (2007)
23. Li, D., Müller, M.B., Gilje, S., Kaner, R.B., Wallace, G.G.: Processable aqueous dispersions of graphene nanosheets. Nat. Nanotechnol. **3**, 101 (2008)
24. Liu, F., Chung, S., Oh, G., Seo, T.S.: Three-dimensional graphene oxide nanostructure for fast and efficient water-soluble dye removal. ACS Appl. Mater. Interfaces **4**, 922–927 (2012)
25. Zhang, W., Zhu, J., Ang, H., Zeng, Y., Xiao, N., Gao, Y., Liu, W., Hng, H.H., Yan, Q.: Binder-free graphene foams for O_2 electrodes of Li-O_2 batteries. Nanoscale **5**, 9651–9658 (2013)
26. Zhao, X., Jia, Y., Liu, Z.-H.: GO-graphene ink-derived hierarchical 3D-graphene architecture supported Fe_3O_4 nanodots as high-performance electrodes for lithium/sodium storage and supercapacitors. J. Colloid Interface Sci. **536**, 463–473 (2019)
27. Wang, Y., Jin, Y., Zhao, C., Pan, E., Jia, M.: 3D graphene aerogel wrapped 3D flower-like Fe3O4 as a long stable and high rate anode material for lithium ion batteriesJ. Electroanal. Chem. **830**, 106–115 (2018)
28. Pan, E., Jin, Y., Wang, Y., Zhao, C., Bo, X., Jia, M.: Facile synthesis of mesoporous 3D CoO/nitrogen-doped graphene aerogel as high-performance anode materials for lithium storage. Microporous and Mesoporous Mater. **267**, 93–99 (2018)
29. Jiang, L., Ren, Z., Chen, S., Zhang, Q., Lu, X., Zhang, H., Wan, G.: Bio-derived three-dimensional hierarchical carbon-graphene-TiO_2 as electrode for supercapacitors. Sci. Rep. **8**, 4412 (2018)
30. Wu, P., Wang, D., Ning, J., Zhang, J., Feng, X., Dong, J., Hao, Y.: Novel 3D porous graphene/Ni_3S_2 nanostructures for high-performance supercapacitor electrodes. J. Alloy. Compd. **731**, 1063–1068 (2018)
31. Foroutan, M., Naji, L.: Systematic evaluation of factors influencing electrochemical and morphological characteristics of free-standing 3D graphene hydrogels as electrode material for supercapacitors. Electrochim. Acta **301**, 421–435 (2019)

32. Sun, S., Wang, P., Wang, S., Wu, Q., Fang, S.: Fabrication of MnO_2/nanoporous 3D graphene for supercapacitor electrodes. Mater. Lett. **145**, 141–144 (2015)
33. Zhang, H., Deng, X., Huang, H., Li, G., Liang, X., Zhou, W., Guo, J., Wei, W., Tang, S.: Hetero-structure arrays of $NiCoO_2$ nanoflakes@nanowires on 3D graphene/nickel foam for high-performance supercapacitors. Electrochim. Acta **289**, 193–203 (2018)
34. Zhang, L., Zhao, Z., Xia, T., Zhang, S., Li, X., Zhang, A.: Anchoring Bi_2WO_6 nanoparticles on 3D graphene frameworks for enhanced lithium storage. Mater. Lett. **210**, 345–349 (2018)
35. Bandyopadhyay, P., Li, X., Kim, N.H., Lee, J.H.: Graphitic carbon nitride modified graphene/NiAl layered double hydroxide and 3D functionalized graphene for solid-state asymmetric supercapacitors. Chem. Eng. J. **353**, 824–838 (2018)
36. Wang, G., Yang, J., Park, J., Gou, X., Wang, B., Liu, H., Yao, J.: Facile synthesis and characterization of graphene nanosheetsJ. J. Phys. Chem. A **112**, 8192–8195 (2008)
37. Liu, C., Li, F., Ma, L.-P., Cheng, H.-M.: Advanced materials for energy storage. Adv. Mater. **22**, E28–E62 (2010)
38. Nikolaidis, P., Poullikkas, A.: A comparative review of electrical energy storage systems for better sustainability. J. Power Technol. **97**, 220–245 (2017)
39. Wang, G., Zhang, L., Zhang, J.: A review of electrode materials for electrochemical supercapacitors. Chem. Soc. Rev. **41**, 797–828 (2012)
40. Zhong, C., Deng, Y., Hu, W., Qiao, J., Zhang, L., Zhang, J.: A review of electrolyte materials and compositions for electrochemical supercapacitors. Chem. Soc. Rev. **44**, 7484–7539 (2015)
41. Kasana, V.K., Kumar, Y., Singh, P., Dixit, S.: Experimental studies on poly (3-hexylthiophene) electrode based supercapacitors: a comparison of electrolytic species (2016)
42. Becker, H.I.: Low voltage electrolytic capacitor. U.S. Patent 2,800,616, issued July 23 (1957)
43. Lee, H.Y., Goodenough, J.B.: Supercapacitor behavior with KCl electrolyte. J. Solid State Chem. **144**, 220–223 (1999)
44. Endo, M., Takeda, T., Kim, Y.J., Koshiba, K., Ishii, K.: High power electric double layer capacitor (EDLC's); from operating principle to pore size control in advanced activated carbons. Carbon Lett. **1**, 117–128 (2001)
45. Lim, Y.S., Tan, Y.P., Lim, H.N., Huang, N.M., Tan, W.T., Yarmo, M.A., Yin, C.-Y.: Potentiostatically deposited polypyrrole/graphene decorated nano-manganese oxide ternary film for supercapacitors. Ceram. Int. **40**, 3855–3864 (2014)
46. Zhang, L.L., Zhao, X.S.: Carbon-based materials as supercapacitor electrodes. Chem. Soc. Rev. **38**, 2520–2531 (2009)
47. Wang, Y., Chen, J., Cao, J., Liu, Y., Zhou, Y., Ouyang, J.-H., Jia, D.: Graphene/carbon black hybrid film for flexible and high rate performance supercapacitor. J. Power Sources **271**, 269–277 (2014)
48. Cheng, H., Dong, Z., Hu, C., Zhao, Y., Hu, Y., Qu, L., Chen, N., Dai, L.: Textile electrodes woven by carbon nanotube–graphene hybrid fibers for flexible electrochemical capacitors. Nanoscale **5**, 3428–3434 (2013)
49. Gao, H., Xiao, F., Ching, C.B., Duan, H.: Flexible all-solid-state asymmetric supercapacitors based on free-standing carbon nanotube/graphene and Mn_3O_4 nanoparticle/graphene paper electrodes. ACS Appl. Mater. Interfaces **4**, 7020–7026 (2012)
50. Yang, C., Shen, J., Wang, C., Fei, H., Bao, H., Wang, G.: All-solid-state asymmetric supercapacitor based on reduced graphene oxide/carbon nanotube and carbon fiber paper/polypyrrole electrodes. J. Mater. Chem. A **2**, 1458–1464 (2014)
51. He, Y., Chen, W., Li, X., Zhang, Z., Fu, J., Zhao, C., Xie, E.: Freestanding three-dimensional graphene/MnO_2 composite networks as ultralight and flexible supercapacitor electrodes. ACS Nano **7**, 174–182 (2012)
52. Wang, Z., Zhang, Q.E., Long, S., Luo, Y., Yu, P., Tan, Z., Bai, J.: Three-dimensional printing of polyaniline/reduced graphene oxide composite for high-performance planar supercapacitor. ACS Appl. Mater. Interfaces **10**, 10437–10444 (2018)

53. Wu, C.H., Deng, S.X., Wang, H., Sun, Y.X., Liu, J.B., Yan, H.: Preparation of novel three-dimensional NiO/ultrathin derived graphene hybrid for supercapacitor applications. ACS Appl. Mater. Interfaces **6**, 1106–1112 (2014)
54. Chi, K., Zhang, Z., Xi, J.B., Huang, Y., Xiao, F., Wang, S., Liu, Y.: Freestanding graphene paper supported three-dimensional porous graphene–polyaniline nanocomposite synthesized by inkjet printing and in flexible all-solid-state supercapacitor. ACS Appl. Mater. Interfaces **6**, 16312–16319 (2014)
55. Zhang, C., Huang, Y., Tang, S., Deng, M., Du, Y.: High-energy all-solid-state symmetric supercapacitor based on Ni_3S_2 mesoporous nanosheet-decorated three-dimensional reduced graphene oxide. ACS Energy Lett. **2**, 759–768 (2017)
56. Foo, C.Y., Lim, H.N., Mahdi, M.A.B., Chong, K.F., Huang, N.M.: High-performance supercapacitor based on three-dimensional hierarchical r-go/nickel cobaltite nanostructures as electrode materials. J. Phys. Chem. A **120**, 21202–21210 (2016)
57. Zhang, J., Ding, J., Li, C., Li, B., Li, D., Liu, Z., Cai, Q., Zhang, J., Liu, Y.: Fabrication of novel ternary three-dimensional RuO_2/graphitic-C_3N_4@ reduced graphene oxide aerogel composites for supercapacitors. ACS Sustain. Chem. Eng. **5**, 4982–4991 (2017)
58. Ye, S., Feng, J.: Self-assembled three-dimensional hierarchical graphene/polypyrrole nanotube hybrid aerogel and its application for supercapacitors. ACS Appl. Mater. Interfaces **6**, 9671–9679 (2014)
59. Liu, Y., Zhou, J., Tang, J., Tang, W.: Three-dimensional, chemically bonded polypyrrole/bacterial cellulose/graphene composites for high-performance supercapacitors. Chem. Mater. **27**, 7034–7041 (2015)
60. Ouyang, Y., Xia, X., Ye, H., Wang, L., Jiao, X., Lei, W., Hao, Q.: Three-dimensional hierarchical structure ZnO@C@NiO on carbon cloth for asymmetric supercapacitor with enhanced cycle stability. ACS Appl. Mater. Interfaces **104**, 3549–3561 (2018)
61. Qu, L., Zhao, Y., Khan, A.M., Han, C., Hercule, K.M., Yan, M., Liu, X.: Interwoven three-dimensional architecture of cobalt oxide nanobrush-Graphene@$Ni_xCo_{2x}(OH)_{6x}$ for high-performance supercapacitors. Nano Lett. **15**, 2037–2044 (2015)
62. Hu, N., Huang, L., Gong, W., Shen, P.K.: High-Performance Asymmetric supercapacitor based on hierarchical $NiMn_2O_4$@ CoS core–shell microspheres and stereotaxically constricted graphene. ACS Sustain. Chem. Eng. **6**, 16933–16940 (2018)
63. Cao, X., Yin, Z., Zhang, H.: Three-dimensional graphene materials: preparation, structures and application in supercapacitors. Energy Environ. Sci. **7**, 1850–1865 (2014)
64. Zhu, Y., Murali, S., Stoller, M.D., Ganesh, K.J., Cai, W., Ferreira, P.J., Pirkle, A.: Carbon-based supercapacitors produced by activation of graphene. Science **332**, 1537–1541 (2011)
65. Xia, X.H., Chao, D.L., Zhang, Y.Q., Shen, Z.X., Fan, H.J.: Three-dimensional graphene and their integrated electrodes. Nano Today **9**, 785–807 (2014)
66. Yao, X., Zhao, Y.: Three-dimensional porous graphene networks and hybrids for lithium-ion batteries and supercapacitors. Chem. Rev. **2**, 171–200 (2017)
67. Manjakkal, L., Núñez, C.G., Dang, W., Dahiya, R.: Flexible self-charging supercapacitor based on graphene-Ag-3D graphene foam electrodes. Nano Energy **51**, 604–612 (2018)
68. Ping, Y., Gong, Y., Fu, Q., Pan, C.: Preparation of three-dimensional graphene foam for high performance supercapacitors. Progress Natl. Sci. Mater. Int. **27**, 177–181 (2017)
69. Down, M.P., Banks, C.E.: Freestanding three-dimensional graphene macroporous supercapacitor. ACS Appl. Energy Mater. **1**, 891–899 (2018)
70. Garakani, M.A., Abouali, S., Xu, Z.-L., Huang, J., Huang, J.-Q., Kim, J.-K.: Heterogeneous, mesoporous $NiCo_2O_4$-MnO_2/graphene foam for asymmetric supercapacitors with ultrahigh specific energies. J. Mater. Chem. A **5**, 3547–3557 (2017)
71. Qian, Y., Cai, X., Zhang, C., Jiang, H., Zhou, L., Li, B., Lai, L.: A free-standing $Li_4Ti_5O_{12}$/graphene foam composite as anode material for Li-ion hybrid supercapacitor. Electrochim. Acta **258**, 1311–1319 (2017)
72. Ghosh, K., Yue, C.Y., Sk, M.M., Jena, R.K.: Development of 3D urchin-shaped coaxial manganese dioxide@polyaniline (MnO_2@PANI) composite and self-assembled 3D pillared

graphene foam for asymmetric all-solid-state flexible supercapacitor application. ACS Appl. Mater. Interfaces **9**, 15350–15363 (2017)
73. Latil, S., Henrard, L.: Charge carriers in few-layer graphene films. Phys. Rev. Lett. **97**, 036803 (2006)
74. Zhao, Y., Liu, J., Wang, B., Sha, J., Li, Y., Zheng, D., Amjadipour, M., MacLeod, J., Motta, N.: Supercapacitor electrodes with remarkable specific capacitance converted from hybrid graphene oxide/NaCl/Urea films. ACS Appl. Mater. Interfaces **9**, 22588–22596 (2017)
75. Dou, Y., Min, L.U.O., Liang, S., Zhang, X., Ding, X., Liang, B.: Flexible free-standing graphene-like film electrode for supercapacitors by electrophoretic deposition and electrochemical reduction. Trans. Nonferr. Metal Soc. China **24**, 1425–1433 (2014)
76. Chao, Y., Chen, S., Chen, H., Hu, X., Ma, Y., Gao, W., Bai, Y.: Densely packed porous graphene film for high volumetric performance supercapacitor. Electrochim. Acta **276**, 118–124 (2018)
77. Jia, Y., Zhou, L., Shao, J.: Direct synthesis of graphene-based hybrid films as flexible supercapacitor electrodes. Synth. Met. **244**, 99–105 (2018)
78. Chen, J., Guo, Y., Huang, L., Xue, Y., Geng, D., Liu, H., Wu, B.: Controllable fabrication of ultrathin free-standing graphene films. Phil. Trans. R. Soc. A **372** (2014). https://doi.org/10.1098/rsta.2013.0017
79. Xiong, D., Li, X., Bai, Z., Li, J., Shan, H., Fan, L., Long, C., Li, D., Lu, X.: Rational design of hybrid Co_3O_4/graphene films: free-standing flexible electrodes for high performance supercapacitors. Electrochim. Acta **259**, 338–347 (2018)
80. Zhu, Y., Ye, X., Tang, Z., Wan, Z., Jia, C.: Free-standing graphene films prepared via foam film method for great capacitive flexible supercapacitors. App. Surf. Sci. **422**, 975–984 (2017)
81. Worsley, M.A., Kucheyev, S.O., Mason, H.E., Merrill, M.D., Mayer, B.P., Lewicki, J., Valdez, C.A.: Mechanically robust 3D graphene macro-assembly with high surface area. Chem. Comm **48**, 8428–8430 (2012)
82. Li, Y., Chen, J., Huang, L., Li, C., Hong, J.-D., Shi, G.: Highly compressible macroporous graphene monoliths via an improved hydrothermal process. Adv. Mater. **26**, 4789–4793 (2014)
83. Zou, X., Zhou, Y., Wang, Z., Chen, S., Li, W., Xiang, B., Xu, L., Zhu, S., Hou, J.: Free-standing, layered graphene monoliths for long-life supercapacitor. Chem. Eng. J. **350**, 386–394 (2018)
84. Wang, X., Ding, Y., Chen, F., Lu, H., Zhan, N., Ma, M.: Hierarchical porous n-doped graphene monoliths for flexible solid-state supercapacitors with excellent cycle stability. ACS Appl. Energy Mater. **1**, 5024–5032 (2018)
85. Deng, L., Liu, J., Ma, Z., Fan, G., Liu, Z.-H.: Free-standing graphene/bismuth vanadate monolith composite as a binder-free electrode for symmetrical supercapacitors. RSC Adv. **8**, 24796–24804 (2018)
86. Wen, Y., Rufford, T.E., Hulicova-Jurcakova, D., Wang, L.: Nitrogen and phosphorous co-doped graphene monolith for supercapacitors. Chem. Sus. Chem. **9**, 513–520 (2016)
87. Yao, H., Zhang, G., Zhang, F., Li, W., Yang, Y., Chen, L.: A novel Ni coordination supramolecular network hybrid monolith of 3D graphene as electrode materials for supercapacitors. Mater. Today Energy **6**, 164–172 (2017)
88. Hu, H., Zhao, Z., Wan, W., Gogotsi, Y., Qiu, J.: Ultralight and highly compressible graphene aerogels. Adv. Mater. **25**, 2219–2223 (2013)
89. Worsley, M.A., Pauzauskie, P.J., Olson, T.Y., Biener, J., Satcher Jr., J.H., Baumann, T.F.: Synthesis of graphene aerogel with high electrical conductivity. J. Am. Chem. Soc. **132**, 14067–14069 (2010)
90. Song, Z., Liu, W., Sun, N., Wei, W., Zhang, Z., Liu, H., Liu, G., Zhao, Z.: One-step self-assembly fabrication of three-dimensional copper oxide/graphene oxide aerogel composite material for supercapacitors. Solid State Commun. **287**, 27–30 (2019)
91. VanHoa, N., Quyen, T.T.H., Hieu, N.V., Ngoc, T.Q., Thinh, P.V., Dat, P.A., Nguyen, H.T.T.: Three-dimensional reduced graphene oxide-grafted polyaniline aerogel as an active material for high performance supercapacitors. Synth. Met. **223**, 192–198 (2017)

92. Xing, L.-B., Hou, S.-F., Zhou, J., Zhang, J.-L., Si, W., Dong, Y., Zhuo, S.: Three dimensional nitrogen-doped graphene aerogels functionalized with melamine for multifunctional applications in supercapacitors and adsorption. J. Solid State Chem. **230**, 224–232 (2015)
93. Choi, J., Yang, M.H., Kim, S.-K.: Pseudocapacitive organic catechol derivative-functionalized three-dimensional graphene aerogel hybrid electrodes for high-performance supercapacitors. Appl. Surf. Sci. **422**, 316–320 (2017)

Organic Electrode Material for Sodium-Ion Batteries

Aneela Sabir, Tahmina Zia, Muhammad Usman, Muhammad Shafiq, Rafi Ullah Khan and Karl I. Jacob

Abstract Exploiting from tremendously abundant and inexpensive sodium reservoirs, sodium ion batteries (NaIBs) are estimated as reassuring candidate for electrochemical energy conservation and storage on large scale. Owing to larger radius and atomic mass of Na^+ than conventionally used materials, NaIBs having inorganic electrode encounter with little capacity and inadequate cycle life. Development of environment friendly, renewable, abundant raw material based batteries are gaining much attention. Organic electrode based sodium ion batteries are one of them. Presently, a lot of work is done on functionalizing organic electrodes, incorporation of nanostructured materials to tune their electrochemical properties. In collation, organic electrode exhibit merits like high capacity, structural design ability and lesser cationic radius limitations. Organic electrodes plagued with solubility issues in electrolytes and lesser conductivity. Here in organic electrodes based on their reactions are divided into three classes; C=O based than C–N=O based and then doping reactions are systematically viewed. In this chapter we summarize the research work to put forward organic electrode material for NaIBs. The conductivity issue can be resolved through increasing conjugated structures. Theoretical capacity can be elevated by expanding active groups. Working voltage can be regulated by tuning grafting overseeing lowest unoccupied molecular orbital (LUMO). Future of organic electrode relies mainly on aprotic electrolyte based full NaIBs with long cycle life.

Keywords Organic electrodes · Sodium ion batteries · Electrochemical energy storage

---

A. Sabir (✉) · T. Zia · M. Usman · M. Shafiq · R. U. Khan
Department of Polymer Engineering and Technology, University of the Punjab, Lahore 54590, Pakistan
e-mail: aneela.pet.ceet@pu.edu.pk

K. I. Jacob
School of Materials Science and Engineering, Georgia Institute of Technology, Atlanta, GA, USA

Inamuddin et al. (eds.), *Self-standing Substrates*,
Engineering Materials, https://doi.org/10.1007/978-3-030-29522-6_12

1 Introduction

Sodium-ion batteries (NaIBs) comes with an edge for extensive energy storage, low cost and highly abundant mineral and similarities with conventionally used batteries on electrochemistry. Nowadays, NaIBs are in full swing. Typically used anode materials are inorganic based giving limited recycling strategies. Comparable with inorganic electrodes, organic electrodes are conferred with various superiority. Including, organic materials comprised of naturally abundant (C, H and O) having flexible molecular structures giving room for Na ions and facilitating facile Na^+ insertion/extraction mechanism. Giving structural diversity with tunable functional groups resulting in tailored electrochemical properties. To upgrade the performance of rechargeable Na-ion batteries anode material with greater energy storage capacity and improved cycle life are considerable. Among numerous electrochemical energy storage (EES) technologies, lithium ion batteries (LiIBs) are contender and drastically increased interests in research field and portable electronics. Although, geographically uneven distribution of lithium reservoir and its cost have created major concerns Indeed, NaIBs go hand in hand with LiIBs since 1980s but left aside due to commercial success of LiIBs.

Presently anode materials used for NaIBs are carbon nano sheets, graphene materials, porous carbon nano sheets, biomass-derived carbon nano-spherules, intercalation-based compounds (e.g. $Na_2Ti_3O_7$), alloying materials (e.g. NaSn, Na_3P, Na_xSb). Including organic materials and polymers (e.g. pteridine, quinone, sodium terephthalates, polyimides). Organic materials act as electrodes are attracting much concern in EES system owing to their design ability, renewability, diversity and environmental benignity. Although low capacity, inferior rate capability and limited cycling stability obstruct their application. Na ion batteries consists of negatively charged anode and positively charged cathode where Na^+ transports through electrolyte and electrons travel through external circuit as shown in Fig. 1. During charging Na^+ moves from cathode to anode and vice versa.

Organic anodes confront three major challenges,

1. Reduced reaction kinetics because of low conductivity of organic materials.
2. Na ion insertion extraction motivates capacity decay on account of particle pulverization instigated by huge volume alterations.
3. Organic Electrode material deprivation upon cycling in organic electrolytes.

Strategies to overcome above challenges are,

1. Low conductivity is overlooked by adding 20–30% conductive carbon and reducing the particle size of organic compound in nanometer for enhancing active surface area, further enhancing reaction kinetics.
2. Nanosized organic material truncate particle pulverization, thus improving cycling stability.
3. Increasing the polarity through salt formation reduces solubility of organic material hindering loss of active material.

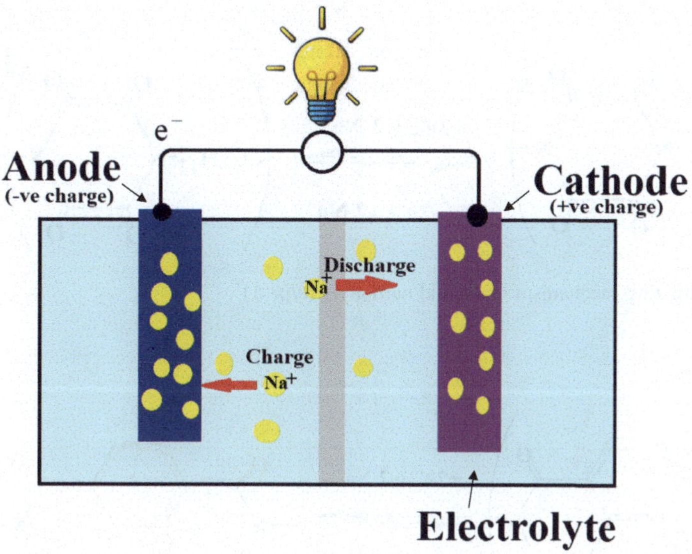

Fig. 1 Schematic representation of sodium ion battery

2 Molecular Design of Electrodes for Organic Sodium Ion Batteries

2.1 Organic Electrodes Constituting of C=O Based Reaction

2.1.1 Carbonyl Compounds

Carbonyl group is frequently used as organic functionality due to their efficient kinetics, structural diversity, high capacity and high reactivity. Virtually all carbonyl compound are n-type organics with reversible single electron reduction per carbonyl giving Na based monovalent anion stabilized by other structural moieties. Depending upon stabilizing mechanism, carbonyl based organic molecules are divided in 3 types.

Type 1: Vicinal carbonyls employing stable enolates shown in Fig. 2.

Type 2: Carbonyl groups directly attached to aromatic ring giving delocalization stabilized well dispersed negative charge as shown in Fig. 3. Aromatic carboxylates, imides, anhydrides fall in this category.

Type 3: Quinones and ketones belongs to this category. They show combined characteristics with Type 1 and 2 compounds as shown in Fig. 4. The main stabilizing force comes from reduction induced structural moieties. Inductive and resonance effect can be used to tune redox behavior of carbonyls. Major pitfall is their solubility in organic electrolytes.

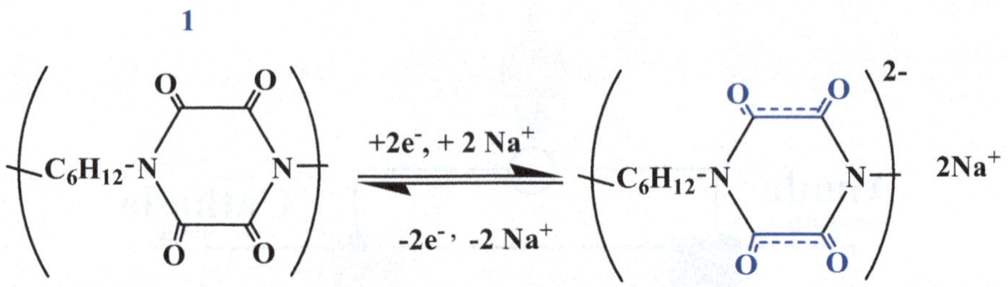

Fig. 2 Stabilizing mechanism of vicinal carbonyls (type 1)

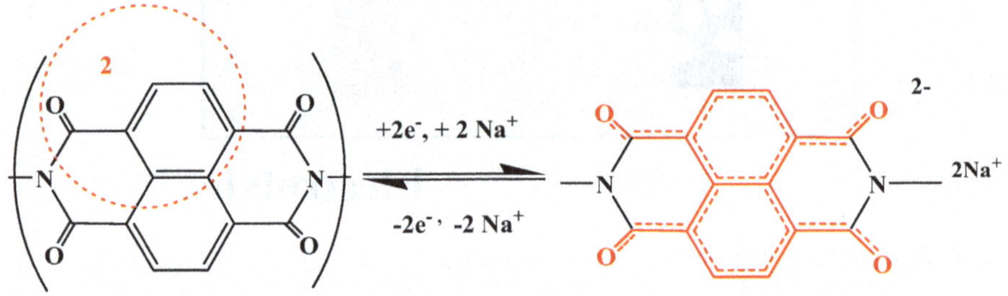

Fig. 3 Stabilizing mechanism of carbonyls in conjugation with aromatic ring (type 2)

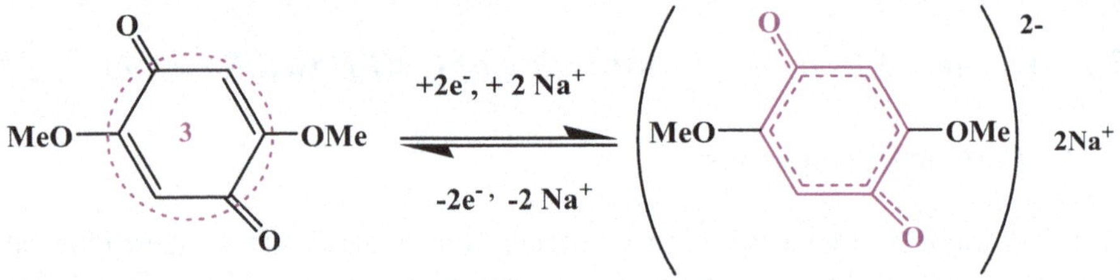

Fig. 4 Stabilizing mechanism of hybrid of type 1 and 2 carbonyls (type 3)

2.1.2 Polyimides

In organic polyimides carbonyl group (C=O) is directly bonded with nitrogen and aromatic rings. Enolization reaction followed by association and disassociation of Na^+ to oxygen atom describes the redox mechanism of imides. Polyimides and their derivatives can be used as electrode material for NaIBs. Due to their inert stable structures and insolubility they exhibit long cycle life and highly constant electrochemical performance. Predominantly used are dianhydride derived polyimides with 1,4,5,8-naphthalenetetracarboxylic dianhydride (NTCDA), 1,2,4,5-benzenetetracarboxylic dianhydride (PMDA) and 3,4,9,10-perylenetetracarboxylic dianhydride (PTCDA) as monomer units as represented in Fig. 5.

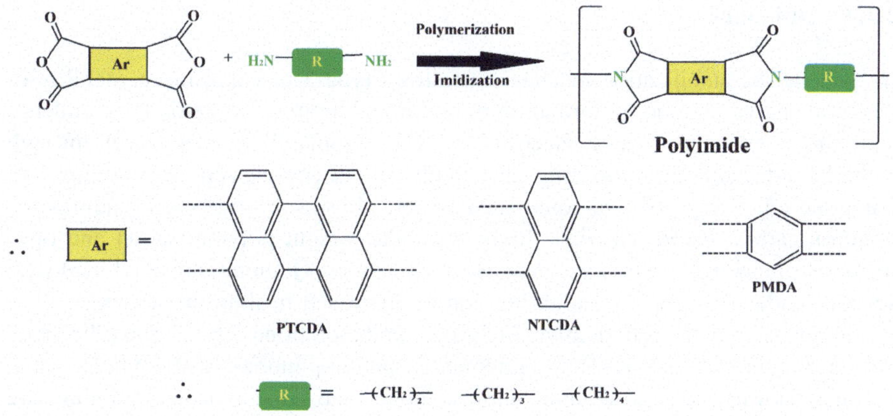

Fig. 5 Imidization of polyimides

NTCDA derived polyimide PNTCDA is used as electrode material for NaIBs. It gives 2 electrons transfer per molecule with discharge capacity of about 140 mAh g^{-1}, discharge voltage of 1.7–2.4 V with 60% capacity deliverance.

To overcome unwanted dissolution for small imides, polyimides with electro active center and inactive linker was used giving additional thermal stability and mechanical properties. By fluctuating the electro active core i.e., increasing aromaticity there was an increase in average discharge voltage.

Redox-inactive groups can be replaced by redox-active groups to avoid capacity loss. Redox active groups like anthraquinone was used with PMDA/NTCDA to form anthraquinonyl imide. It showed a capacity of ~200 mA h g^{-1} at 50 mA g^{-1}. Both polyimides undergo reversible, 4 electron transfer in 1.2–3.2 voltage range as shown in Fig. 6.

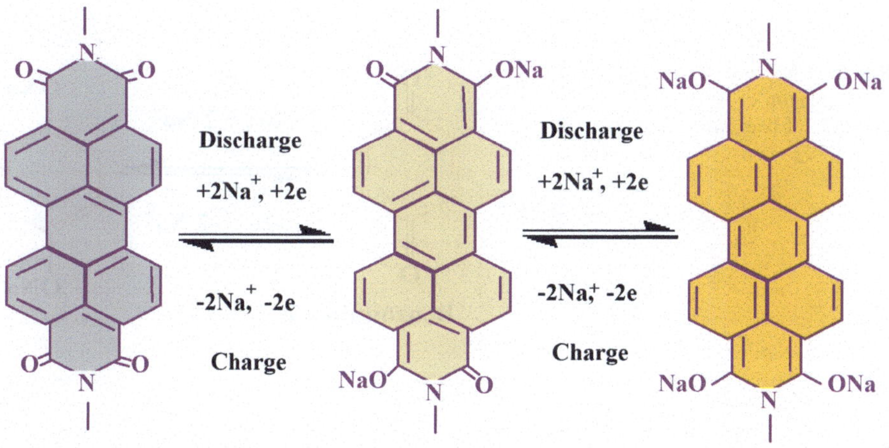

Fig. 6 Schematic representation of working principles of PTCDA based polyimide

2.1.3 Quinones

In quinone the stabilization mechanism follows type 3 mechanism i.e. stability of negative charge is done by formation of additional aromatic system. They undergo solubility issues in organic electrolytes. The dissolved species travel through cathode and anode area resulting into shuttle effect and capacity loss giving low columbic efficiency. To overcome such issues, physiosorption based loadings are done on carbon based materials, high molecular weight polymerization and optimization of electrolytes by increasing their viscosities. Quinones are combined with several binders to give less solubility and to attain self-healing properties.

Quinones and their derivative including anthraquinone-2,6-dicarboxylic acid, 1,4-benzoquinone, 9,10-1nthraquinone, anthraquinone-2-carboxylic acid, 1,4-naphthoquinone, 2,6-diaminoanthraquinone and 2-aminoanthraquinone are studied by researchers with emphasizes on their redox properties, solvation energy, electron affinity, energy density and charge storage capacity. Schematic representation of working principle of benzoquinone is represented in Fig. 7.

Organic electrodes operational voltage can be modulate by electron-withdrawing functional groups. Halogens are strong electron-withdrawing groups with ability to lessen the energy of LUMO. Typically lowest unoccupied molecular orbital [LUMO] energy has direct effect on reduction voltage. Lesser LUMO energy will create more reduction voltage. Theory (DFT) shows redox voltage in following trend.

$$C_6F_4O_2 > C_6Cl_4O_2 > C_6Br_4O_2 > C_6H_4O_2$$

$C_6Cl_4O_2$ provides lower reduction potential than $C_6F_4O_2$ but later is irreversible during charging process, forming NaF.

Salt formation to enhance polarity of inorganic organic hybrid material can overcome organic electrode dissolution in aprotic electrolyte.

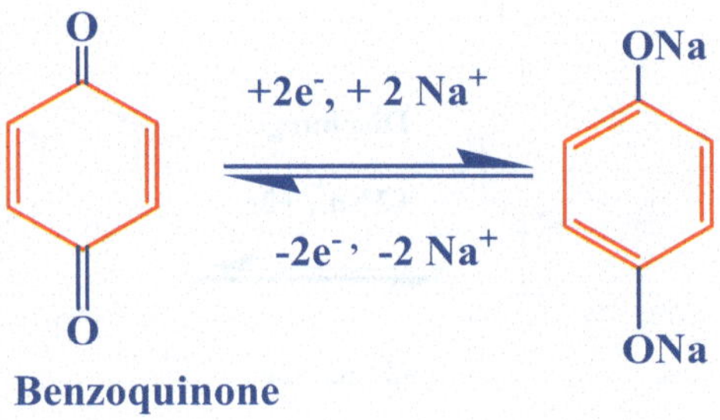

Fig. 7 Schematic representation of working principle of benzoquinone

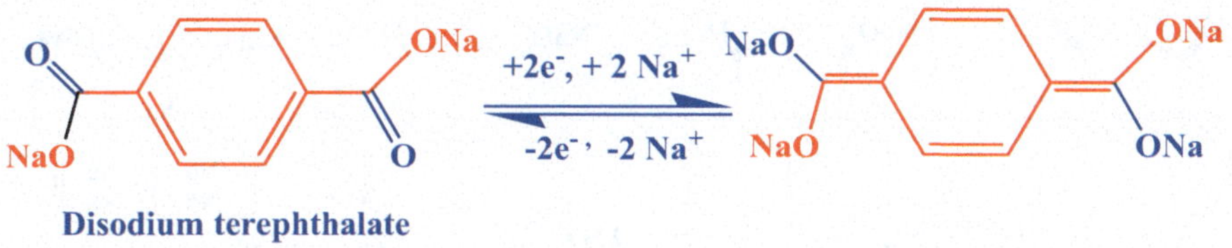

Fig. 8 Schematic representation of working principle of disodium terephthalate

2.1.4 Carboxylates

Sodium carboxylates are widely used electrode in NaIBs. Due to electron donating group (O–Na) directly bonded with carbonyl Na^+ insertion occurs at voltage below 1 V. Similarly, 4,4-biphenyldicarboxylate Na-salt ($Na_2C_{14}H_8O_4$) and disodium terephthalate salt ($Na_2C_8H_4O_4$) deliver 0.5 V discharge voltage with reversible capacity of 220 mA h g^{-1} and Na insertion extraction mechanism was studied, respectively. Na^+ insertion extraction mechanism of disodium terephthalate are shown in Fig. 8.

Solubility issue is controlled by exchange of metallic cation in terephthalate salts. Calcium, zinc, potassium are also incorporated in terephthalate salt. Sodium storage in disodium terephthalate was reported by Mahasin Alam Sk et al. with capacities of $\sim$225 mA h g^{-1}. At lower Na concentration carbonyl site take part in insertion while at high sodium concentration hexagonal ring part dominates.

2.1.5 Anhydrides

Extensively conjugated structures with tunable Na-ion insertion capacity are used as anhydride organic electrodes. PTCDA accommodates 2 Na-ions in between 1.0 and 3.0 V with 150 mA h g^{-1}. By lower discharge voltage more sodium ions can be inserted. At 0.01 V PTCDA results in 15 Na-ion insertion as shown in Fig. 9 with 1017 mA g^{-1} capacity.

Extra capacity is due to generation of interface (solid electrolyte) and inserted Na-ion into aromatic rings. NTCDA also gives extra capacity at 0.01 V discharge voltage with Na^+ insertion extraction mechanism as represented in Fig. 10.

2.2 Organic Electrodes Based on Doping Reaction

2.2.1 Organic Radical Polymers

The main problem using organic radical polymer is self-discharging, low conductivity and highly soluble in organic electrolytes. Kim et al. worked on improving

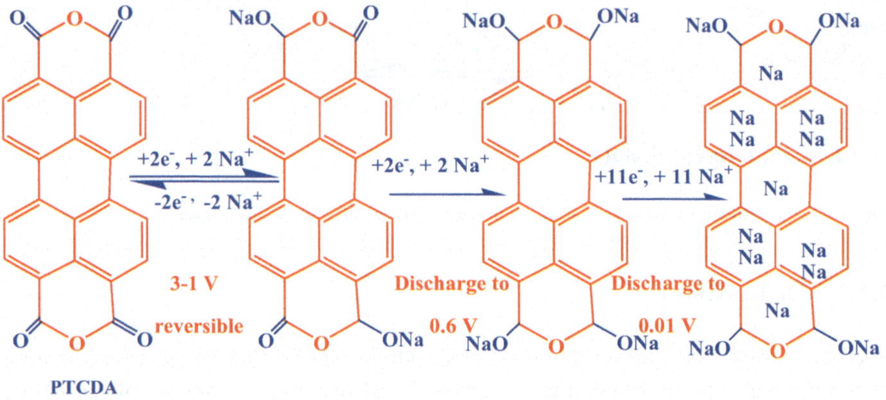

Fig. 9 Schematic representation of working principles of PTCDA

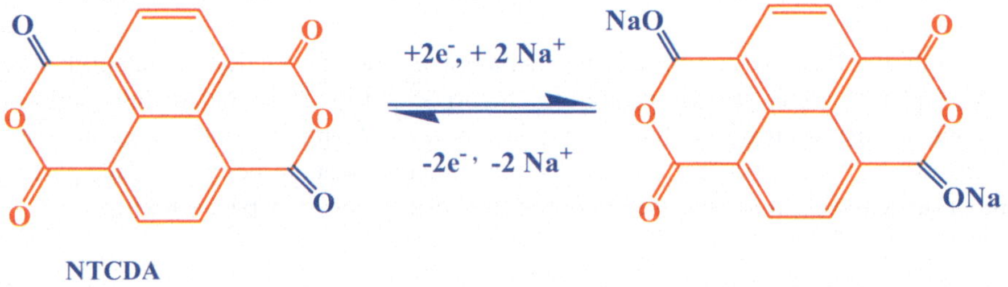

Fig. 10 Schematic representation of working principles of NTCDA

electrochemical performance of PTMA. PTMA encapsulated in carbon nanotubes with 100–300 nm diameter overcome self-discharging issue.

Initially only p-doping reaction occurs with 130 mA h g^{-1} discharge voltage. During discharge both n and p-doping is observed with 220 mA h g^{-1} as shown in Fig. 11. The fast kinetics for radical compounds can be endowed by oxidation induced structural changes with lesser electron rearrangement.

2.2.2 Conductive Polymers

Conductive polymers both n and p doping reactions can be observed at different voltages. During early studies polyacetylene and polyphenylene was studied as conductive electrodes. Polybithiophene, n-type polymer retain 400 mA h g^{-1} at 1.0 V average discharge voltage. Oligopyrene, p-type polymer retain 121 mA h g^{-1} initial reversible capacity at 3.5 V discharge voltage. Polypyrrole are the commonly used conductive polymers but delivers less capacity. Their Na-ion insertion exertion mechanism is shown in Fig. 12. Copolymerization, doping anion and introducing

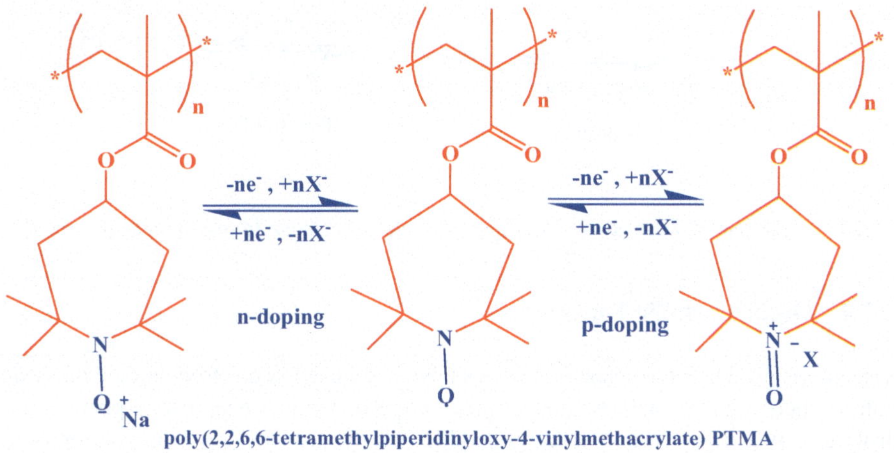

Fig. 11 Schematic representation of working principles of PTMA

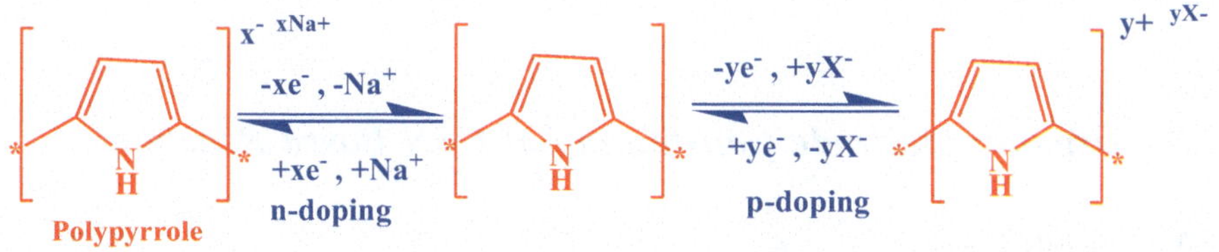

Fig. 12 Schematic representation of working principles of polypyrrole

nanostructured materials are effective strategies to overcome such issues. Furthermore, polypyrrole codoped with sodium p-toluenesulfonate (TsONa) and L-lactic acid (LA) through electropolymerization on Fe foil show high cathode performance.

Additionally polyaniline with nanostructured graphene/polyaniline/polystyrene electrode and copolymer of o-nitro aniline/polyaniline showed enhanced capacity.

2.2.3 Conjugated Micro Porous Polymers

The microporous conjugated organic polymers endowed several benefits, making them suitable for NaIBs. Na-ion transport is facilitated due to high surface area and microporous structure. Dissolution problem is also overlooked as well as more redox active sites gives high theoretical capacity. Triazine ring and benzene based porous, bipolar electrode material gives uniform 1.4 nm micropores. Presently low conductivity is the main issue for microporous electrode batteries (Fig. 13).

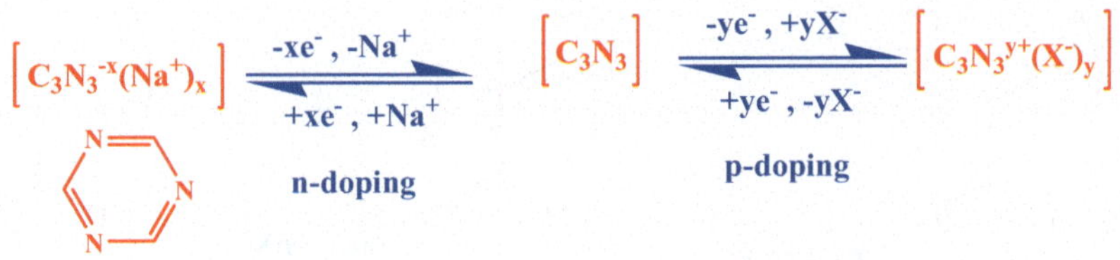

Fig. 13 Schematic representation of working principles of conjugated microporous polymers

2.2.4 Organometallic Polymers

Organometallic like ferrocene electrodes have been used in electrochemical analysis due to their high air stability and ultrafast electrochemical response. Aprotic electrolytes solubilize simple organometallic electrodes so they are commonly employed in redox batteries. High molecular weight organometallic are commonly used cathodes for NaIBs. Poly(ferrocenyl-methylsilane and its derivatives are also studied. In ferrocene Fe exists in Fe^{+2} form. After p-doping reaction valency of Fe is in Fe^{+3} as shown in Fig. 14.

2.3 Organic Electrode Constituting of C=N Based Reaction

2.3.1 Schiff Bases

General repeating unit of Schiff base is (–N=C–Ar–C=N–), where Ar is aromatic rings having 10 πe^- satisfying Huckel rule. This repeating unit with conjugated double bonds is considered as active group while nonconjugated groups are termed as inactive centers. The aromatic inkers are more stable than aliphatic linkers. Aliphatic linker with 1 methyl group gives 0.9 Na-ion insertion while with aromatic linker 1.4 Na-ion insertion/extraction is observed as shown in Fig. 15. This phenomenon is due to planarity. By increasing methyl group as linker (up to 4) results in loss of planarity as well as drop capacity and shift reduction voltage downwards due to inductively electron donating effect of methyl groups.

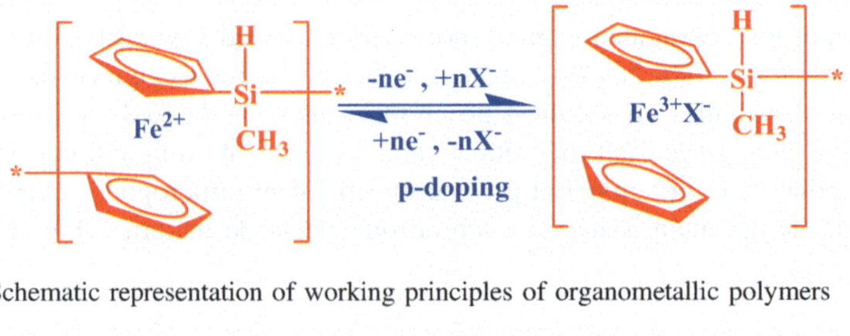

Fig. 14 Schematic representation of working principles of organometallic polymers

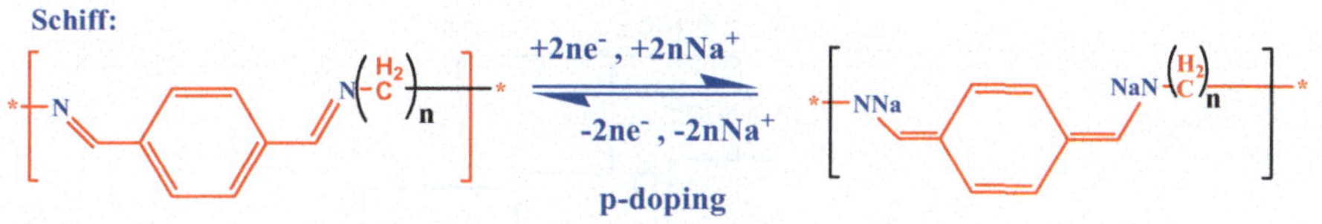

Fig. 15 Schematic representation of working principles of Schiff bases

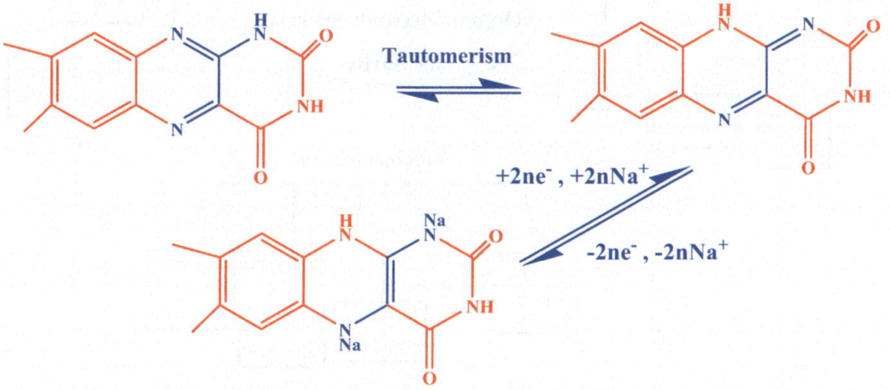

Fig. 16 Schematic representation of working principles of Pteridine derivatives

2.3.2 Pteridine Derivatives

Reversible tautomerism occur in pteridine derivatives as shown in Fig. 16. N is most commonly observed reactive site while O also assists the reaction. In addition to C=N, tetracyanoquinodimethane and tetracyanoethylene having C≡N was also used.

3 Electrode Design for Sodium Ion Batteries

Major issue dealing with sodium ion batteries are

(a) The inherently low conductivity of organic electrodes lower reaction kinetics,
(b) Large volume change during phase transition results in pulverization causing capacity decay.
(c) Organic electrode dissolution in organic electrolytes causes loss of active material.

Designing electrodes draw more attraction as compared to search new organic electrode materials. Tunable statics adopted to modify organic electrode in NaIBs are:

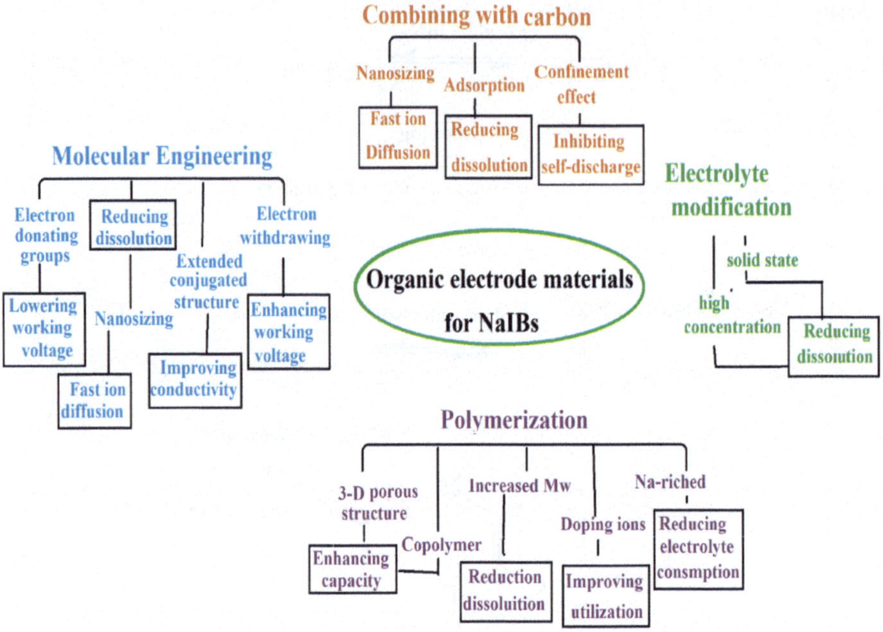

Fig. 17 Electrode designs for Sodium ion batteries

(a) Molecular engineering.
(b) Polymerization.
(c) Carbon hybrid.
(d) Modifying electrolyte.

With the help of these capacity, diffusion rate, conductivity, working voltage and solubility can be reasonably tuned resulting in enhanced cycle life, high rate performance and elevated capacity of NaIBs as schematically represented in Fig. 17.

3.1 Molecular Engineering

In molecular engineering following strategies can be applied

1. Rate capacity of carbonyls can be improved by increasing the π-π aromatic conjugation of core.
2. Rapid ion diffusion can be attained by organic nanostructured salt electrodes.
3. Dissolution of organic electrode can be overcome by salt formation.

4. LUMO energies can be increased or decreased by addition of electron donating and electron withdrawing groups resulting in reduced and enhanced working voltage, respectively.

3.2 Polymerization

Polymerization exhibit following merits:

1. Functionalizing conductive polymers with anions and cations can enhance their utilization and during charging process lower electrolyte consumption of P-Type polymer is observed for Na based polymers.
2. High molecular weight polymer gives lesser dissolution.
3. Theoretical capacity can be enhanced by 3D porous polymers and copolymers.

3.3 Combining with Carbon (Carbon Hybrid)

Combining with carbon material is endowed with following merit:

1. Incorporating carbon materials in organic materials can reduce dissolution with confinement and adsorption effects. Resultantly overcome with self-discharging issue of organic radical polymers.

The prepared material increases the active area of electrode and electrolyte giving high capacity and high rate capability.

3.4 Electrolyte Modification

Concentrated electrolyte or solid electrolyte suppresses dissolution of soft active material, enhancing cycle performance.

4 Future Challenges

With advances in portable electronics rising need is smaller, flexible, wearable and thinner batteries. In comparison with inorganic materials, organic materials endow better film forming ability, sublimability and solubility. Therefore, facile methods are used in designing organic sodium ion batteries, flexible current collectors by printing, casting, vapor deposition, in situ polymerization and vacuum filtration.

Moreover organic materials also endorse contributions in current collectors, binders and separators in flexible batteries.

Organic electrodes in aqueous electrolytes have been gaining researcher's attention. Aqueous electrolytes come with several merits, economical, high ionic conductivity, nonflammable and facile battery assembling. Appropriate organic materials are advised for use in aqueous electrolytes with redox potential between H_2 and O_2. This limits the choice of polymeric electrode materials for sodium ion batteries. To overcome organic electrode based issues composite materials will be used to get better results like quinone with high theoretical capacity can be reinforced with polyimide having high cycle stability. Coating highly conductive polymers on lower conductive electrodes can be employed for enhanced electrochemical properties. Biomass derived organic electrodes will be economical and of high demand.

References

1. Armand, M., Grugeon, S., Vezin, H., Laruelle, S., Ribière, P., Poizot, P., Tarascon, J.M.: Conjugated dicarboxylate anodes for Li-ion batteries. Nat. Mater. **8**, 120 (2009)
2. Chen, J., Liu, Y., Li, W., Wu, C., Xu, L., Yang, H.: Nanostructured polystyrene/polyaniline/graphene hybrid materials for electrochemical supercapacitor and Na-ion battery applications **50** (2015). https://doi.org/10.1007/s10853-015-9092-z
3. Chen, L., Liu, S., Wang, Y., Zhao, L., Zhao, Y.: 2, 3-Dicyano-5, 6-dichloro-1, 4-benzoquinone as a novel organic anode for sodium-ion batteries. J. Electroanal. Chem. **837**, 226–229 (2019)
4. Chen, L., Li, W., Wang, Y., Wang, C., Xia, Y.: Polyimide as anode electrode material for rechargeable sodium batteries. RSC Adv. **4**(48), 25369–25373 (2014)
5. Chen, Y., Manzhos, S.: A comparative computational study of lithium and sodium insertion into Van Der Waals and covalent tetracyanoethylene (TCNE)-based crystals as promising materials for organic lithium and sodium ion batteries. Phys. Chem. Chem. Phys. **18**(13), 8874–8880 (2016)
6. Choi, A., Kim, Y.K., Kim, T.K., Kwon, M.-S., Lee, K.T., Moon, H.R.: 4,4′-Biphenyldicarboxylate sodium coordination compounds as anodes for Na-ion batteries. J. Mater. Chem. A **2**(36), 14986–14993 (2014)
7. Deng, Q., Fan, C., Wang, L., Cao, B., Jin, Y., Che, C.-M., Li, J.: Organic potassium terephthalate ($K_2C_8H_4O_4$) with stable lattice structure exhibits excellent cyclic and rate capability in Li-ion batteries. Electrochim. Acta **222**, 1086–1093 (2016)
8. Dou, X., Hasa, I., Saurel, D., Vaalma, C., Wu, L., Buchholz, D., Bresser, D., Komaba, S., Passerini, S.: Hard carbons for sodium-ion batteries: structure, analysis, sustainability, and electrochemistry. Mater. Today **23**, 87–104 (2019)
9. Gao, H., Xin, S., Xue, L., Goodenough, J.B.: Stabilizing a high-energy-density rechargeable sodium battery with a solid electrolyte. Chem **4**(4), 833–844 (2018)
10. Gao, X., Zhu, G., Zhang, X., Hu, T.: Porous carbon materials derived from in situ construction of metal-organic frameworks for high-performance sodium ions batteries. Microporous Mesoporous Mater. **273**, 156–162 (2019)
11. Geng, J., Bonnet, J.-P., Renault, S., Dolhem, F., Poizot, P.: Evaluation of polyketones with N-cyclic structure as electrode material for electrochemical energy storage: case of tetraketopiperazine unit. Energy Environ. Sci. **3**(12), 1929–1933 (2010)

12. Huang, J., Wei, Z., Liao, J., Ni, W., Wang, C., Ma, J.: Molybdenum and tungsten chalcogenides for lithium/sodium-ion batteries: beyond Mos2. J. Energy Chem. **33**, 100–124 (2019)
13. Ingersoll, N., Karimi, Z., Patel, D., Underwood, R., Warren, R.: Metal organic framework-derived carbon structures for sodium-ion battery anodes. Electrochim. Acta **297**, 129–136 (2019)
14. Jeong, S., Kim, B.H., Park, Y.D., Lee, C.Y., Mun, J., Tron, A.: Artificially coated $NaFePO_4$ for aqueous rechargeable sodium-ion batteries. J. Alloys Compd. **784**, 720–726 (2019)
15. Kim, H., Hong, J., Park, K.-Y., Kim, H., Kim, S.-W., Kang, K.: Aqueous rechargeable Li and Na ion batteries. Chem. Rev. **114**(23), 11788–11827 (2014)
16. Kim, J.-K., Kim, Y., Park, S., Ko, H., Kim, Y.: Encapsulation of organic active materials in carbon nanotubes for application to high-electrochemical-performance sodium batteries. Energy Environ. Sci. **9**(4), 1264–1269 (2016)
17. Kim, J.-K., Scheers, J., Ahn, J.-H., Johansson, P., Matic, A., Jacobsson, P.: Nano-fibrous polymer films for organic rechargeable batteries. J. Mater. Chem. A **1**(7), 2426–2430 (2013)
18. Kim, J.-H., Jung, M.-J., Kim, M.-J., Lee, Y.-S.: Electrochemical performances of lithium and sodium ion batteries based on carbon materials. J. Ind. Eng. Chem. **61**, 368–380 (2018)
19. Kim, K.C., Liu, T., Jung, K.H., Lee, S.W., Jang, S.S.: Unveiled correlations between electron affinity and solvation in redox potential of quinone-based sodium-ion batteries. Energy Storage Mater. (2019)
20. Li, D., Chen, L., Chen, L., Sun, Q., Zhu, M., Zhang, Y., Liu, Y., et al.: Potassium gluconate-derived N/S Co-doped carbon nanosheets as superior electrode materials for supercapacitors and sodium-ion batteries. J. Power Sources **414**,308–316 (2019)
21. Li, G., Xue, R., Chen, L.: The influence of polytetrafluorethylene reduction on the capacity loss of the carbon anode for lithium ion batteries. Solid State Ion. **90**(1), 221–225 (1996)
22. Liang, Y., Zhang, P., Chen, J.: Function-oriented design of conjugated carbonyl compound electrodes for high energy lithium batteries. Chem. Sci. **4**(3), 1330–1337 (2013)
23. Liao, Q., Hou, H., Liu, X., Yao, Y., Dai, Z., Yu, C., Li, D.: L-lactic acid and sodium P-toluenesulfonate co-doped polypyrrole for high performance cathode in sodium ion battery. J. Phys. Chem. Solids **115**, 233–237 (2018)
24. Liu, T., Cheng, X., Yu, H., Zhu, H., Peng, N., Zheng, R., Zhang, J., et al.: An overview and future perspectives of aqueous rechargeable polyvalent ion batteries. Energy Storage Mater. **18**, 68–91 (2019)
25. Luo, C., Fan, X., Ma, Z., Gao, T., Wang, C.: Self-healing chemistry between organic material and binder for stable sodium-ion batteries. Chem **3**(6), 1050–1062 (2017)
26. Luo, C., Huang, R., Kevorkyants, R., Pavanello, M., He, H., Wang, C.: Self-assembled organic nanowires for high power density lithium ion batteries. Nano Lett. **14**(3), 1596–1602 (2014)
27. Luo, C., Wang, J., Fan, X., Zhu, Y., Han, F., Suo, L., Wang, C.: roll-to-roll fabrication of organic nanorod electrodes for sodium ion batteries. Nano Energy **13**, 537–545 (2015)
28. Oyaizu, K., Ando, Y., Konishi, H., Nishide, H.: Nernstian adsorbate-like bulk layer of organic radical polymers for high-density charge storage purposes. J. Am. Chem. Soc. **130**(44), 14459–14461 (2008)
29. Precht, R., Stolz, S., Mankel, E., Mayer, T., Jaegermann, W., Hausbrand, R.: Investigation of sodium insertion into tetracyanoquinodimethane (TCNQ): results for a TCNQ thin film obtained by a surface science approach. Phys. Chem. Chem. Phys. **18**(4), 3056–3064 (2016)
30. Price, J.T., Paquette, J.A., Harrison, C.S., Bauld, R., Fanchini, G., Gilroy, J.B.: 6-Oxoverdazyl radical polymers with tunable electrochemical properties. Polym. Chem. **5** (18), 5223–5226 (2014)
31. Shirakawa, H., Louis, E.J., MacDiarmid, A.G., Chiang, C.K., Heeger, A.J.: Synthesis of electrically conducting organic polymers: halogen derivatives of polyacetylene, (CH). J. Chem. Soc. Chem. Commun. **16**, 578–580 (1977)

32. Sk, M.A., Manzhos, S.: Exploring the sodium storage mechanism in disodium terephthalate as anode for organic battery using density-functional theory calculations. J. Power Sources **324**, 572–581 (2016)
33. Su, D., Zhang, J., Dou, S., Wang, G.: Polypyrrole hollow nanospheres: stable cathode materials for sodium-ion batteries. Chem. Commun. **51**(89), 16092–16095 (2015)
34. Wang, H., Shuang, Y., Ma, D., Huang, X., Meng, F., Zhang, X.: Tailored aromatic carbonyl derivative polyimides for high-power and long-cycle sodium-organic batteries **4** (2014). https://doi.org/10.1002/aenm.201301651
35. Wang, H.-g., Yuan, S., Si, Z., Zhang, X.-b.: Multi-ring aromatic carbonyl compounds enabling high capacity and stable performance of sodium-organic batteries. Energy Environ. Sci. **8**(11), 3160–3165 (2015)
36. Wang, J., Lv, C., Zhang, Y., Deng, L., Peng, Z.: Polyphenylene wrapped sulfur/multi-walled carbon nano-tubes via spontaneous grafting of diazonium salt for improved electrochemical performance of lithium-sulfur battery. Electrochim. Acta **165**, 136–141 (2015)
37. Wang, L., Zou, J., Chen, S., Yang, J., Qing, F., Gao, P., Li, J.: Zinc terephthalates $ZnC_8H_4O_4$ as anodes for lithium ion batteries. Electrochim. Acta **235**, 304–310 (2017)
38. Wang, S., Wang, L., Zhang, K., Zhu, Z., Tao, Z., Chen, J.: Organic $Li_4C_8H_2O_6$ nanosheets for lithium-ion batteries. Nano Lett. **13**(9), 4404–4409 (2013)
39. Wang, X., Shang, Z., Yang, A., Zhang, Q., Cheng, F., Jia, D., Chen, J.: Combining quinone cathode and ionic liquid electrolyte for organic sodium-ion batteries. Chem **5**(2), 364–375 (2019)
40. Wang, Z., Li, A., Gou, L., Ren, J., Zhai, G.: Computational electrochemistry study of derivatives of anthraquinone and phenanthraquinone analogues: the substitution effect. RSC Adv. **6**(92), 89827–89835 (2016)
41. Wu, D., Luo, K., Du, S., Hu, X.: A low-cost non-conjugated dicarboxylate coupled with reduced graphene oxide for stable sodium-organic batteries. J. Power Sources **398**, 99–105 (2018)
42. Wu, F., Zhao, C., Chen, S., Lu, Y., Hou, Y., Hu, Y.-S., Maier, J., Yu, Y.: Multi-electron reaction materials for sodium-based batteries. Mater. Today **21**(9), 960–973 (2018)
43. Xie, J., Gu, P., Zhang, Q.: Nanostructured conjugated polymers: toward high-performance organic electrodes for rechargeable batteries. ACS Energy Lett. **2**(9), 1985–1996 (2017)
44. Xu, F., Xia, J., Shi, W.: Anthraquinone-based polyimide cathodes for sodium secondary batteries. Electrochem. Commun. **60**, 117–20 (2015)
45. Zhang, C., Lu, C., Zhang, F., Qiu, F., Zhuang, X., Feng, X.: Two-dimensional organic cathode materials for alkali-metal-ion batteries. J. Energy Chem. **27**(1), 86–98 (2018)
46. Zhang, S., Huang, W., Hu, P., Huang, C., Shang, C., Zhang, C., Yang, R., Cui, G.: Conjugated microporous polymers with excellent electrochemical performance for lithium and sodium storage **3** (2014). https://doi.org/10.1039/c4ta06058j
47. Zhang, Y., Gao, Z.: High performance anode material for sodium-ion batteries derived from covalent-organic frameworks. Electrochim. Acta **301**, 23–28 (2019)
48. Zhao, Q., Hu, X., Zhang, K., Zhang, N., Hu, Y., Chen, J.: Sulfur nanodots electrodeposited on Ni foam as high-performance cathode for Li–S batteries. Nano Lett. **15**(1), 721–726 (2015)
49. Zhao, Q., Whittaker, A., Zhao, X.S.: Polymer electrode materials for sodium-ion batteries **11** (2018). https://doi.org/10.3390/ma11122567
50. Zhao, R., Zhu, L., Cao, Y., Ai, X., Yang, H.X.: An aniline-nitroaniline copolymer as a high capacity cathode for Na-ion batteries **21** (2012). https://doi.org/10.1016/j.elecom.2012.05.015
51. Zhong, H., Wang, G., Song, Z., Li, X., Tang, H., Zhou, Y., Zhan, H.: Organometallic polymer material for energy storage. Chem. Commun. **50**(51), 6768–6770 (2014)
52. Zhu, H., Yin, J., Zhao, X., Wang, C., Yang, X.: Humic acid as promising organic anodes for lithium/sodium ion batteries. Chem. Commun. **51**(79), 14708–14711 (2015)

The manufacturer's authorised representative in the EU is Springer Nature Customer Service Centre GmbH, Europaplatz 3, 69115 Heidelberg, Germany. If you have any concerns regarding our products, please contact ProductSafety@springernature.com

Printed and bound by CPI Group (UK) Ltd, Croydon, CR0 4YY
05/10/2026
02228130-0011